THE ROUTLEDGE INTERNATIONAL HANDBOOK OF ISLAND STUDIES

From tourist paradises to immigrant detention camps, from offshore finance centres to strategic military bases, islands offer distinct identities and spaces in an increasingly homogenous and placeless world. The study of islands is important, for its own sake and on its own terms. But so is the notion that the island is a laboratory, a place for developing and testing ideas, and from which lessons can be learned and applied elsewhere.

The Routledge International Handbook of Island Studies is a global, research-based and pluri-disciplinary overview of the study of islands. Its chapters deal with the contribution of islands to literature, social science and natural science, as well as other applied areas of inquiry. The collated expertise of interdisciplinary and international scholars offers unique insights: individual chapters dwell on geomorphology, zoology and evolutionary biology; the history, sociology, economics and politics of island communities; tourism, wellbeing and migration; as well as island branding, resilience and 'commoning'. The text also offers pioneering forays into the study of islands that are cities, along rivers or artificial constructions.

This insightful *Handbook* will appeal to geographers, environmentalists, sociologists, political scientists and, one hopes, some of the 600 million or so people who live on islands or are interested in the rich dynamics of islands and island life.

Godfrey Baldacchino is Professor of Sociology at the University of Malta; UNESCO Co-Chair (Island Studies and Sustainability) at the University of Prince Edward Island, Canada; President of the International Small Islands Studies Association (ISISA); and Founding Executive Editor of *Island Studies Journal* (2006–2016).

THE ROUTLEDGE INTERNATIONAL HANDBOOK OF ISLAND STUDIES

A World of Islands

Edited by Godfrey Baldacchino

LONDON AND NEW YORK

First published 2018 by Routledge

2 Park Square, Milton Park, Abingdon, Oxfordshire OX14 4RN
52 Vanderbilt Avenue, New York, NY 10017

Routledge is an imprint of the Taylor & Francis Group, an informa business

First issued in paperback 2020

British Library Cataloguing-in-Publication Data
A catalogue record for this book is available from the British Library

Library of Congress Cataloging-in-Publication Data
A catalog record has been requested for this book

ISBN: 978-1-4724-8338-6 (hbk)
ISBN: 978-0-367-65989-9 (pbk)

Typeset in Bembo
by Apex CoVantage, LLC

This book is dedicated to John Barry Bartmann (1941–2015).
Dr Bartmann (Barry) was for many years a professor of political science at the
University of Prince Edward Island (UPEI), Canada. He was the academic chair
and visionary (with Harry Baglole) of the conference titled An Island Living,
held on Prince Edward Island in September 1992. He helped leverage some
Can$1,500,000 to finance this event and the ensuing North Atlantic Islands
Programme. He was instrumental in setting up the educational programme in
island studies at UPEI, including its island studies minor (1999), its Master
of Arts in Island Studies (2003) and the Canada Research Chair in Island
Studies (2003–2013).

Dr Bartmann provided specialist advice to the Constitutional Committees
and Governments of Åland, Faroe Islands and Iceland. He was a recognised
authority on sub-national island jurisdictions and the international relations of
small (or micro) states.

He also wrote the 'war and security' chapter in A World of Islands *(2007).*

Last known photo of Barry Bartmann, taken at a family gathering by Donna Lindo in Montpellier VT, USA, on 1 August 2015. With permission.

CONTENTS

Contents

CONTRIBUTORS

Robert Aldrich is a writer and Professor of European History at the University of Sydney, Australia. His research includes the history of empire, gender and sexuality.
robert.aldrich@sydney.edu.au

Bénédicte André teaches French and Francophone Studies at the Department of International Studies, Macquarie University, Sydney, Australia. Her research interests include literary and cultural studies, island francophonies and spatial theory.
benedicte.andre@mq.edu.au

Godfrey Baldacchino is Pro Rector and Professor of Sociology at the University of Malta, Malta and UNESCO Co-Chair (Island Studies and Sustainability) at the University of Prince Edward Island, Canada.
godfrey.baldacchino@um.edu.mt

Mitul Baruah is Assistant Professor of Sociology and Anthropology, Ashoka University, Haryana, India. He hails from Majuli river island in the Brahmaputra river, and his research interests include political ecology, water, river islands, state theories, rural livelihoods and social movements.
mitul.baruah@ashoka.edu.in

Andrew J. Berry is at the Department of Organismic and Evolutionary Biology, Harvard University, Cambridge MA, USA.
berry@oeb.harvard.edu

R. J. 'Sam' Berry is Emeritus Professor of Genetics at University College London, UK and is past president of the Linnean Society, the British Ecological Society and the European Ecological Federation.
rjberry@ucl.ac.uk

Geoff Bertram is a Senior Associate at the Institute of Policy Studies, Victoria University of Wellington, New Zealand.
Geoff.Bertram@vuw.ac.nz

Stephen Blackmore is a Professor of Botany. He served as Keeper of Botany at the Natural History Museum in London and the 15th Regius Keeper of the Royal Botanic Garden Edinburgh, UK. He is chair of the Botanic Gardens Conservation International (BGCI).

Laurie Brinklow is at the Institute of Island Studies, University of Prince Edward Island, Charlottetown, Canada, where she teaches in the Master of Arts programme in island studies. She is also a poet and a publisher. brinklow@upei.ca

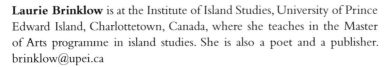

Eric Clark is Professor of Human Geography at Lund University, Lund, Sweden, and affiliated with the Lund University Centre for Sustainability Studies. He is Managing Editor of *Journal of Urban Affairs* and Chair of the International Geographical Union Commission on Islands. eric.clark@keg.lu.se

Tara Coleman is a professional teaching fellow and researcher, School of Environment, University of Auckland, New Zealand. t.coleman@auckland.ac.nz

John Connell is a Professor of Geography at the School of Geosciences, University of Sydney, Australia. His research interests include rural development, migration and inequality, as well as decolonisation and nationalism. john.connell@sydney.edu.au

Quentin C. B. Cronk is a botanist at the Department of Biology, University of British Columbia, Canada.
quentin.cronk@ubc.ca

Arthur Dahl is President of the International Environment Forum, and a retired Deputy Assistant Executive Director of the United Nations Environment Programme (UNEP).
dahla@bluewin.ch

Christian Depraetere is a hydrologist and geomorphologist at the Institut de Recherche pour le Développement (IRD), Montpellier, France.
christian.depraetere@ird.fr

Veronica della Dora is Professor of Human Geography at Royal Holloway, University of London, UK.
Veronica.DellaDora@rhul.ac.uk

Klaus Dodds is Professor of Geopolitics at Royal Holloway, University of London, UK. He holds a Leverhulme Trust Major Research Fellowship (2017–2020).
K.Dodds@rhul.ac.uk

Rosemary G. Gillespie is at the Department of Environmental Science, Policy and Management, University of California at Berkeley, USA.
gillespie@berkeley.edu

Sonya Graci is at the Ted Rogers School of Hospitality and Tourism Management, Ryerson University, Toronto ON, Canada.
sgraci@ryerson.ca

Adam Grydehøj is Director of Island Dynamics, Copenhagen, Denmark and Executive Editor of *Island Studies Journal* since 2016.
agrydehoj@islanddynamics.org

Miranda Johnson is a historian of indigenous peoples and settler colonialism in the Anglophone post/colonial world, at the University of Sydney, Australia.
miranda.johnson@sydney.edu.au

Robin Kearns is Professor of Geography in the School of Environment at the University of Auckland, New Zealand. His teaching and research focuses on links between people, place and wellbeing. He is an editor of the journal *Health and Place*.
r.kearns@auckland.ac.nz

Ilan Kelman is at University College London, UK and the University of Agder, Norway. His research interests include disaster diplomacy and sustainability for island communities.
islandvulnerability@yahoo.com

Susie Khamis teaches public communication at the University of Technology Sydney, Australia. Her research interests include branding, cultural diversity and consumer culture.
Susie.Khamis@uts.edu.au

Siri M. Kjellberg is a doctoral candidate at the Department of Human Geography, Human Ecology Division, and at the Lund University Centre of Excellence for Integration of Social and Natural Dimensions of Sustainability (LUCID) in Lund, Sweden.
siri_m.kjellberg@hek.lu.se

Roselyn Kumar is a researcher at the Sustainability Research Centre, University of the Sunshine Coast, Queensland, Australia.
Roselyn802@yahoo.com

Adrian M. Lister is research leader in fossil mammals at the Department of Earth Sciences, Natural History Museum, London, UK.
a.lister@nhm.ac.uk

Patrick T. Maher is Associate Professor of Community Studies at, Cape Breton University, Sydney NS, Canada. He is chair of the International Polar Tourism Research Network and the lead for the University of the Arctic's Thematic Network on Northern Tourism.
Pat_Maher@cbu.ca

Elizabeth McMahon teaches cultural and literary studies at the School of the Arts and Media, University of New South Wales, Sydney, Australia. Her research interests are in Australian literature, island studies and gender studies.
e.mcmahon@unsw.edu.au

David Milne is Professor Emeritus of Political Science at the University of Prince Edward Island, Canada, and has served as a visiting professor at the University of Malta, Malta. He has a sustained interest in small, often island, jurisdictions.
dmilne70@gmail.com

Jenia Mukherjee teaches humanities and social science at the Indian Institute of Technology, Kharagpur, India. Her research interests include environmental history, political ecology and urban studies.
jenia@hss.iitkgp.ernet.in

Patrick D. Nunn is Professor of Geography at the University of the Sunshine Coast, Queensland, Australia.
pnunn@usc.edu.au

Diana M. Percy is at the Department of Botany, University of British Columbia and the Beaty Biodiversity Museum, Canada, as well as at the Department of Life Sciences, Natural History Museum, London, UK.
diana.percy@botany.ubc.ca

Bernard Poirine is Professor Emeritus of Economics at the University of French Polynesia, Tahiti.
bernard.poirine@upf.pf

James A. Randall is Professor of Island Studies and UNESCO Co-Chair (Island Studies and Sustainability) at the University of Prince Edward Island, Charlottetown, Canada.
jarandall@upei.ca

Graeme Robertson is founder and Executive Director of Global Islands Network (GIN) and a research associate of the Institute of Island Studies, University of Prince Edward Island, Canada.
graeme@globalislands.net

Stephen A. Royle is Professor Emeritus of Island Geography at Queen's University Belfast, Northern Ireland, UK.
s.royle@qub.ac.uk

Ramanathan Swaminathan is visiting research fellow at Uppsala University, Gotland Campus, Sweden, Advisor for independent research collective Partners for Urban Knowledge, Action and Research (PUKAR) and editor of *Urban Island Studies*.
swaminathan.ramanathan@etnologi.uu.se

Wouter Veenendaal teaches at the Institute of Political Science, Leiden University, The Netherlands, as well as a research fellow at the Royal Netherlands Institute of Southeast Asian and Caribbean Studies (KITLV).
veenendaal@kitlv.nl

Edward Warrington is at the Department of Public Policy, University of Malta, Malta. He has a special interest in the governance, institutional design and administrative history of small states and islands.
edward.warrington@um.edu.mt

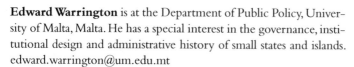

PREFACE

Introduction: a lunch invitation

Lunch was served at the Teasdale residence at Pelican Lagoon, Kangaroo Island (KI), Australia, on Saturday 8 July 2017. At table were: an Australian academic couple (a sociologist and education specialist respectively), a retired geography professor from Northern Ireland, UK; a Maltese–Canadian couple (a social scientist and an early childhood educator), the Executive Secretary of the Wadden Academy, which supports research in the Wadden Region (a UNESCO World Heritage site) in the Netherlands, and the Director of Film in Friesland, a Dutch organisation.

The members of this unlikely international and multidisciplinary gathering were united in their passion for the study of islands, the promotion of island studies, and the desire to transit any such knowledge and insights to local and international publics.

The event was the conclusion of the 15th 'Islands of the World' (IoW) Conference, a flagship event organised by the International Small Islands Studies Association (ISISA), this time in collaboration with the Kangaroo Island Council. The Teasdales had been heavily involved in the conference organisation from the KI end; Klaas and Jacqueline had just been handed the responsibility to help plan and organise the 16th IoW conference in Leeuwarden and the island of Terschelling, the Netherlands, in June 2018. Godfrey and Anna are the President and Newsletter Editor of ISISA respectively; while Stephen, now an Emeritus Professor, has been the world's only academic with a chair in Island Geography, at Queen's University Belfast.

Like the gathering around this table, the IoW ISISA Conference in Kangaroo Island was a richly diverse affair: around 100 attendees, from Kangaroo Island itself, the rest of Australia and some 20 other countries, with the broadest imaginable disciplinary reach, converged on Kingscote for the event. The elderly white faces around the dinner table represented one segment of the conference audience. Both the KI Council and ISISA offered scholarships, which brought in (more youthful) student scholars from places like Canada, India and Fiji.

Institution building

ISISA remains the only global, not-for-profit association that brings together all those interested in island studies in its broadest sense. Its 100 or so members hail from all over the world, and a

Figure 0.1   Ready for lunch at Pelican Lagoon on 8 July 2017. From left: Bob Teasdale, Stephen A. Royle, Jennie Teasdale, Godfrey Baldacchino, Anna Baldacchino and Klaas Deen. Missing from photo (and taking the picture): Jacqueline Schrijver. Photo used with permission.

rotation of conference venues ensures that new members join as others move on. The organisation provides a platform for a healthy exchange of ideas, questions and information about anything dealing with islands; the IoW conference, typically held once every two years, serves as the culmination and showcase of its efforts at bringing island people and island scholarship together in island contexts. The lunch party plans to meet again at the next IoW.

One indicator of the health of an area of scholarly inquiry is the number of associations and organisations that operate explicitly within it, and the many others that may connect with it directly or obliquely, offering bridges to, and conversations with, other disciplines. Within academic island studies, ISISA (2018) is joined by the Small Island Cultures Research Initiative (SICRI 2018) and RETI, the Network of Island Universities (RETI 2018), an initiative of the University of Corsica, France. ISISA (along with RETI) also supports *Island Studies Journal* (ISJ, published since 2006), a peer-reviewed bi-annual journal and which has been joined by *Shima: International Journal for Research in Island Cultures* (since 2007), the *Journal of Marine and Island Cultures* (since 2012) and *Urban Island Studies* (since 2016). ISJ (2018) is listed in the Web of Science© and the Social Science Citation Index© (Thomson Reuters 2018).

Then, there are the courses and the books to accompany these. At the Islands and Small States Institute, University of Malta, in Malta; at the Institute of Island Studies, University of Prince Edward Island, in Canada; at the postgraduate programme of the University of Tasmania in Australia; at the Research Centre for Pacific Islands, Kagoshima University, Japan; at the Universities of Greenland, the Faroe Islands and Okinawa (Japan), amongst others, students are invited to pursue courses or undertake research in island studies. In many other centres of learning and higher education, from Wellington, New Zealand to the multi-island campus of the University of the Aegean, Greece, and from the multi-national campus of the University of the West Indies to that of the University of the South Pacific, half a world away, many more students are exposed to the 'island imagination' in particular courses of study. For those looking at potential textbooks featuring an island focus, there is Royle (2001), the more affordable and student-friendly Royle (2014) and the more environmental science-focused Ratter (2018), along with this book, of course. The ecological and evolutionary aspects of island populations are treated by Williamson (1981); King and Connell (1999) review islands and migration; and Connell (2013) looks at islands in the context of contemporary environmental threats. Gillespie and Clague (2009) have delivered an encyclopaedia of islands. And, for the proverbial cherry on the island studies cake – perhaps it should be a Smith Island cake, from the island (and cake) so named in Chesapeake Bay, Maryland, USA? – consider the four-volume island studies anthology curated by Kelman and Baldacchino (2016). Pinter/Cassell, in the UK, had started an island studies book series in the late 1990s (series editor: Lino Briguglio); while, more recently, US-based publisher Rowman and Littlefield International (RLI) has ushered in a 'Rethinking the Island' series (co-editors: Elaine Stratford, Elizabeth McMahon, Godfrey Baldacchino) (RLI 2018).

These are the academic underpinnings of an interest in island studies that received a boost with the small island country equivalent to the UN's 1992 Rio Summit on Sustainable Development: the 1994 Barbados Conference that provided a common voice to the world's Small Island Developing States (SIDS), a recognised UN category. The lobby – acting notably through AOSIS, the Alliance of Small Island States, with 44 members and various associated members – has been remarkably vocal at the meetings of the Conference of the Parties (COP), the supreme decision making body of the UN's Framework Convention on Climate Change (UNFCCC). COP 23 in Bonn, Germany, in November 2017 was chaired by Fiji, an AOSIS member. While at COP 21 in Paris, France, in December 2015, came the first recognition by the international community that committing to an increase in global mean temperatures of up to 1.5°C since the dawn of the industrial revolution would be a more successful, and still attainable, scenario than the more lax commitment to 2°C (which would leave many SIDS in peril). With almost a quarter of UN member states being made up exclusively of island or archipelagic jurisdictions, this focus is inevitable: 2014 was declared *International Year of Small Island Developing States*, the first Year ever dedicated to a group of countries. With dozens of small island states around, islandness (along with small state size) has become a historically and geographically specific form of political organisation, political mobilisation and political thought (Dalby and Tuathail 1998).

But, already well before that, islands had been front and centre in human imagination. The smallness, boundedness, isolation and strategic location of most islands – across both known and the imagined worlds – have rendered them the subject of myth and lore, both sacred and profane, places of both respite and danger throughout long periods of history (Trompf 1993, Gillis 2004, Patke 2018). The sanctuary afforded to Leto, pregnant with wanton Zeus' twins, on the island of Delos; the island landings of wanderers Ulysses of Ithaca and Aeneas of Troy (and later, founder of Rome) in their multi-year journeys across the Mediterranean and Aegean seas after the Siege of Troy; the journeys of St Brendan and other Irish Monks to the Hebrides, Orkneys, Faroes, possibly to Iceland and as far as Newfoundland, Canada; the exploits of giant Ymir in the 'land of fire and ice' that is Iceland as narrated in the Norse myths of the *Poetic Eddas*; the island-hopping adventures of Portuguese sailors down the coast of West Africa, and eventually past the Cape of Good Hope and onward to India (and the land of spices, and still more 'Indian islands', hence *Indo-nesia*); Thomas More's *Utopia* (1516) and Daniel Defoe's *Robinson Crusoe* (1719), two semi-fictional accounts that both gave their names to literary and film genres; down to the *Jurassic park* franchise, first depicted in the novel of the same name (Crichton 1990) and the blockbuster TV series *Lost* (2004–2010) with most of the action unfolding on an island that held a secret deep in its core. 'The island' has become one of the multi-layered and timeless metaphors for human existence (Le Juez and Springer 2013, p. 1), catapulting island spaces as premier alluring tourism destinations.

That islands have secrets waiting to be discovered was not lost on the natural and human sciences. Charles Darwin was pleasantly surprised by the discernments that islands and archipelagos could provide to zoogeography. Half a world away, Alfred Wallace reached a similar conclusion: together they proposed the basic principles of the theory of evolution (Darwin 1845, Wallace 1880). Margaret Mead travelled through island Melanesia to find out about 'primitive' practices of adolescence and family life that could shed insights on the malaise of Western civilisation (Mead 1928). Experiments involving mangrove 'islands' in Florida led Daniel Simberloff and Edward Wilson to validate features of the 'theory of island biogeography' (Simberloff and Wilson 1969, MacArthur and Wilson 1967). The Icelandic horse (*Equus ferus caballus*), the extinct pygmy elephant from Sicily and Malta (*Palaeoloxodon falconeri*), and *Homo floresiensis* (the 'hobbit' from Flores) provide evidence of island species dwarfism; while the Komodo dragon (*Varanus komodoensis*) and the giant Fijian long-horned beetle (*Xixuthrus heros*) are exemplars of island species gigantism (or giantism).

Definitional conundrums

The planet currently has some 80,000 permanently inhabited islands, on which live some 600 million humans: some two-thirds of these on Java, Honshu, Great Britain, Luzon and Sumatra alone. This collective necklace of 'our world of islands' comprises enormous differences in land area, geology, climatic conditions, biota, histories and cultures. There are also seasonally inhabited islands; temporary (or part-time) islands, given tidal effects (Mont St Michel, Lindisfarne); overwhelmingly urban islands and island cities (Singapore, Hong Kong, Manhattan, old town Stockholm); still others constructed or 'augmented' by humans from sand or concrete (Okinotorishima, Fiery Cross Reef) or, *Utopia*-like, resulting from collapsed land bridges or the carving out of isthmuses/narrow necks of land (Euboea). Some other islands are 'drowning' while others are 'appearing': Kutubdia is one of various islands being devoured by the sea in the Bay of Bengal, Bangladesh: the island has roughly halved in size over the last 20 years as the rising waves overwhelm houses and fertile farmland. Meanwhile, Shelly Island has formed near Cape Point, North Carolina USA, in 2017, with a high tide or storm-driven water elevation that piled up sediment up to near the Atlantic Ocean surface (Owens 2017, Coffey 2017). Many more islands and islets are uninhabited, surviving as relatively undisturbed ecosystems; in sharp contrast, Ap Lei Chou (Hong Kong), Manhattan (New York State, USA), Malé (Maldives) and Malta count among the world's most densely populated islands (see Figure 0.2).

Other islands have been destroyed in the course of natural action: think of the volcanic explosions that obliterated Thera (now called Santorini) some 4,000 years ago and Krakatoa in 1883 . . . while others have been born *de novo* out of similar natural action: at least nine new islands have thus formed in the last 20 years (Stevens 2016). Humans have been just as busy meddling with island formations: damaging and destroying islands with nuclear tests – Bikini Atoll, Marshall Islands, remains uninhabitable 60 years after 23 nuclear detonations (Macdonald 2016; see Figure 0.3) – while building others for the affluent who crave exceptional and secluded spaces, such as the Palm Islands in Dubai (Jackson and Della Dora 2009). Finally, islands can also be bought or sold: ask Farhad Vladi for a quote (Vladi 2017).

One might be excused for thinking that defining an island is a self-evident, even a sterile and philosophical task; but one that is actually notoriously difficult; yet the question 'what is an

Figure 0.2 The island of Malé, capital of the Republic of Maldives, home to some 130,000 residents on a land area of 5.8 km² (2.2 square miles).

Source: Wikipedia; photo by Shahee Ilyas. Available from: https://commons.wikimedia.org/w/index.php?curid= 621195.

Figure 0.3 25 July 1946: a mushroom cloud and water column appear after the underwater *Baker* nuclear explosion in the Bikini atoll, Marshall Islands. Photo taken from Bikini island, 5.6 km (3.5 miles) away.

Source: Wikimedia Commons. Available from: https://commons.wikimedia.org/wiki/File:Operation_Crossroads_Baker_Edit.jpg.

island?' should continue to be asked (Baldacchino 2012a, p. 60). The geographical definition of islands – as being pieces of land surrounded by water (even at high tide) that are smaller than a continent – is extended by biologists who consider mountaintops, oases, lakes, caves and the like as additional insular habitats (Carlquist 1974, pp. 4–5); and, in some cases, even more enisled than the water surrounded kind. Literary scholars remind us of the imagined island spaces of *Atlantis*, *Lilliput*, *Lincoln* and *Never never land* (Plato 360 BC?, Swift 1726, Verne 1875, Barrie 1915); as well as *Villings*, where the narrator is, like the island, also a probable invention *within* the story (Bioy Casares 1940). There is then the much larger trove of fictional genres, where islands feature prominently in popular, gripping tales of crime fiction, thrillers, romance and fantasy (Crane and Fletcher 2017). Others will reminisce over the part-island myth, part-island reality that seep through such riveting, timeless novels as *Journey to the Centre of the Earth*, *Anne of Green Gables*, *The Godfather* and *Death of a River Guide* (Verne 1871, Montgomery 1908, Puzo 1969, Flanagan 1994). Add metaphor to the mix, and other 'islands' can be teased out in kitchen designs, dividing roads and in the pancreas. Even for those islands that pertain to the strict material-geographical register, their boundaries will shift and change under the influence of tides, natural erosion or accretion, or human action; their economies and cultures can be impacted by such fixed links as bridges, tunnels and causeways to mainlands (or other islands); and, over longer spans of time, they disappear, reappear, adjoin other lands or separate from them. The two land masses of EurAsia-Africa and America behave as super islands when suggesting environmental differences which influence contrasting evolutionary paths (Diamond 1997). It has also been argued that the world's only islands in geological time are its tectonic plates, "in constant motion across molten seas" (Okihiro 2010, p. 755).

Islands also suffer from a gross essentialisation of their alleged attributes. In academic circles, islands have suffered from a widespread and stubborn incredulity: why should one study islands at all? Meanwhile, for others, there is a facile and passionate conviction that islands (and their inhabitants) are naturally (and even totally) different from mainlands (and their people) (Churchill Semple 1911). The former stance – that islands are not a valid analytic category – is a favourite argument of some scholars steeped in traditional disciplines; for them, the 'island studies' foray into what may come across as a materialist geography is a distraction from more significant, post-modernist and post-structuralist epistemologies. While the latter position – that islandness sets basic terms of reference, and is an "ontologically secure marker of selfhood" (Whittaker 2016) – is the clarion call and predilection of various political actors who may wish to fan an island-based nationalism, a sense of 'ethnie' (Smith 1986) that builds upon what is usually a subdued, but nevertheless often keenly felt, 'us-them', island-mainland tension:

> Our geography has shaped our psychology; we have the character of an island nation: independent, forthright, passionate in defence of our sovereignty. We can no more change this British sensibility than we can drain the English Channel.
>
> *(David Cameron, Prime Minister of the UK, 23 January 2013)*

Here, readers may have noticed the usage of the word *islandness*: a term that I, along with an expanding list of others, opt to use to represent all that constitutes an island; a distilled and judgement-free sense of island living. This is preferred to the more commonly used term *insularity*, since this particular word comes along with a considerable adverse baggage in the English language, which is then unfairly foisted on island spaces and their peoples: backwardness, parochialism, small-mindedness. Islandness is the more neutral choice, and so does not linguistically pre-condemn islands and islanders to a negative assessment.

Unpacking islandness

A review of the literature on the patterns and challenges of spatial development in small islands suggests that islandness, the emergent distinctiveness or essence of islands, can be translated into distinguishing patterns of spatial development. Unpacking islandness in this sense gives us the quartet of boundedness, smallness, isolation and fragmentation, along with their "amplification by compression" (Percy et al. 2007, p. 193; also Fernandes and Pinho 2017). It is these five space-related variables, and their various and inter-related combinations and permutations, which provide us with the materialist rendition of what islands are, and what they can be. These notions are reviewed in turn below.

Boundedness

The boundedness of islands comes across as their most significant defining feature. The encircle-ment of islands by the aquatic medium sets the stage for a 'land–sea' dialectic: a "terraqueous-ness" (DeLoughrey 2007) that defines an island's biome, its history, its economic development, and – in spite of the hubris of modern information and communication technology – may well still predict its future.

The naturally circumscribed, sea-girt mass suggests easier control and management. The ter-ritorial specificity of an island is obvious: does that not explain why there are only ten populated islands in the world that are shared, some with sufferance, between more than one country (Baldacchino 2013)? The condition, in combination with small size (see more below), begets and nurtures notions of hegemony, despotism and enclave meta-geography (Sidaway 2007). In many historical accounts, islands are prizes, there to be taken; although island civilisations, from Minoan Crete and Britain to Venice and Japan, have also nurtured empires. Larger-than-life island politicians have obtained and held on to power by controlling different aspects of social, economic and political life (Singham 1968, Baldacchino 2012b): they blur the lines between personal and political interests in ways that are, alas, all too common in small communities; to the chagrin of continentals who may insist on clear division of powers and role conflict avoidance:

> Put a man on an island, give him power over other men, and it won't be long before
> he realises that the limits of that power are nowhere to be seen.
> *(Captain Flint, Black Sails 2017)*

Meanwhile, and before the coming into force of the United Nations Conference on the Law of the Sea (UNCLOS), outsiders have coveted islands not as much for their intrinsic resources or exclusive economic zones but for their strategic location, serving as footholds of expansion to the contiguous interior, and securing vital defensive or forward positions close to perceived 'hotspots' (Anderson 2000). The sea makes islands more difficult to invade; though, once inva-sions take place, they tend to be more difficult to resist: Britain has thus thwarted invasion (and thus resisted conquest) since AD 1066 (Royle 2001, p. 152); likewise, Cuba under Castro has resisted the US; Taiwan/Republic of China has rebuffed mainland/People's Republic of China. The sea also makes islands more difficult to escape from: and so many islands have served natu-rally as parts of an "enforcement archipelago" (Mountz 2011): whether as prisons (Alcatraz, Gor-gona, Robben, Ognenny Ostrov) and more recently as migrant detention centres (Christmas Island, Lampedusa, Lesvos, Nauru). When mythical Daedalus needed to abscond from the island of Crete, he had to take flight with waxed wings.

The water boundary of islands has made them premier sites for experiments, both natural and political. Species that may find difficulty in crossing aquatic surfaces – ants, seeds, snails – may experience only chance and occasional landings on islands; but, having done so, may thrive by occupying empty ecological niches. While, even for species where water is not a barrier – some mammals, most birds – the colonisation of hitherto empty islands (and archipelagos) becomes a story of adaptive radiation and evolving endemism: a celebration of genetic diversity over time, as well as a slate of narratives of extinction.

Nature's boundaries are porous and will remain so; in contrast, human borders suggest zero porosity – intimating anything less would diminish the power of the modern state – and yet, they will begrudgingly condone leakages. Overlaying the natural land–sea boundary of island spaces with political identity and jurisdictional powers equips islands, and their governments, with the legitimacy of protecting and disciplining their land space and its contiguous maritime zone (Doty 1996). Sub-national island jurisdictions (Bermuda, Jeju, Niue) will seek to guard their large oceanic exclusive economic zones (with the support of metropolitan patron states of which they form part) or lease their use – say, for fishing – to foreign countries. Sovereign island states will maintain a self-evident presence of the state at their air and sea ports, regulating inflows of persons and cargo, inspecting bills of lading, checking passports and stamping visas.

Unless connected by 'fixed links' to mainlands, island borders unfold at the beach/shore; and, for islands not that close to other lands, only at the seaport and airport: the latter become the only points of access to or from the island, and suffer from the challenges of transportation. The seaport (including cruise ship terminal) and airport are the chokepoints for the flow of the many thousands of tourists who may descend on island spaces, overwhelming local residents and fragile local habitats through sheer weight of numbers (Boissevain 1996, Baldacchino 2014).

Islands do well to remain sensitive to 'border crossings': it is because of a ship appearing on their horizon – or of a plane landing in their airport – that their history then changes forever. Think of how Rapa Nui's long isolation was ended on Easter Sunday, 1722, when Dutch explorer Jacob Roggeveen 'discovered' the island. Epidemics, explorers, raiders, colonisers, missionaries, invaders, international investors, consultants, tourists . . . their movements to and from islands follow a well-worn path.

Smallness

Smallness is a silent feature of much of what passes as island studies. At one end of the size continuum, in places like New Guinea, Greenland or Great Britain, it may be hard for one to feel on an island, and much less to act like an islander. And yet, the 2016 decision by the UK to trigger 'Brexit' – withdrawing from membership of the European Union (EU) – may be seen as part of an 'island–mainland' nervousness (Delanty 2017).

Scientific dividers, floors or thresholds to establish the definitive size of a small island do not exist, although they have been proposed (there are similar issues and challenges with the definition of a small state). Land area, resident population and economic heft are three such common candidates; but, again, none of these is unambiguous. Is Greenland, the world's largest island at 2 million km^2 but with a population of less than 60,000, small or large? How to come to terms with places like Kiribati (pronounced *Ki-ri-bas*), a Pacific archipelago of hardly 100,000 residents but steward over an oceanic space larger than the land area of Europe? It may be a 'small island state' in the eyes of the UN, but is it not also a 'large ocean state' (Cook 2016)? What is the economic value of having thousands of registered companies managing trillions of US dollars, as may happen in island financial centres like the Cayman Islands? (e.g. Roberts 1995). And do we not sympathise with Fog Olwig (2007) when she realises that it is impossible to study the communities of the US Virgin Islands, without referring to the larger diaspora of islanders living in New York? Moreover, smallness ushers in dizzy oscillations in economic and demographic trends: net out-migration in a small island society can be quickly turned around into net immigration, or vice versa, with low absolute numbers; one decent investment can transform an island's economic fortunes (or a company closure could spirit them away just as fast). The people of Nauru may hold the world's records for having been both the richest and the poorest per capita, and being so within a relatively short time span (Connell 2006).

At the other end of the scale, what differentiates an island from an islet, a reef or a rock? According to UNCLOS, an island must be "a naturally formed area of land, surrounded by water, which is above water at high tide" and must "sustain human habitation or economic life on its own" (Part 8, Article 121). If not, then such protuberances from the sea are 'rocks' which shall have no claims to an exclusive economic zone or continental shelf (UNCLOS 2017). At stake are huge swathes of oceanic waters and their seabed resources.

Consider Taiping Island (太平島, also known as Itu Aba) (see Figure 0.4). It is the largest of the naturally occurring Spratly Islands in the South China Sea. The island has been administered by the Republic of China (ROC/Taiwan) since 1956; but it is also claimed by the People's Republic of China, the Philippines and Vietnam. The 46-hectare island (0.46 km²) is low and flat, and has an elliptical shape, being some 1,290 m long and 366 m wide. The island hosts a shelter for fishers, a clinic, a post office, a Guanyin temple, a weather station, satellite communication facilities and radar surveillance equipment; but a 1,120-m long runway is its most distinctive feature. Since 2011, Taiping Island has become a visiting site for Taiwanese college students, teachers and researchers who participate in a study camp organised by Taiwan's Ministry of National Defence. At any time, the island's population is a few hundred people, mainly military and coast guard personnel (Song 2015).

However, in 2016, as part of a landmark ruling about China's claims to the South China Sea, the Permanent Court of Arbitration in The Hague declared that Taiping Island/Itu Aba was, in spite of its name, not an 'island' but a 'rock' since it does not meet the definition of an island

Figure 0.4   Taiping Island (or Atu Iba).

Source: Tomo News, 23 September 2016. Available at: www.youtube.com/watch?v=pNaSKAFPZCc.

under UNCLOS. This decision is rejected by the ROC (and by the PRC). At stake are huge swathes of oceanic waters and their seabed resources.

As islands tend towards smallness – and towards flatness, since high altitude islands have a larger land area than flat ones – their land area tends towards being exclusively coastal. There may, literally, be no hinterland to speak of; thus obliging islanders to interact deeply and intimately with any encroaching parties (apart from interacting intimately amongst themselves) out of sheer necessity. This predicament explains the long, deep and thorough exposure to colonialism and its post-independence lingering (inclusive of colonial language, religion and pedagogies) in many (now) small island states (Baldacchino and Baldacchino 2017); or, its endurance so as to forestall support for island independence and full sovereignty in other, small island territories (Baldacchino 2004, Prinsen and Blaise 2017). This also explains why the only way out of the tense and toxic mix of political totality, economic monopoly and social intimacy on a small island is simply exile (Baldacchino 1997); or, better 'ex-isle' (Bongie 1998); relocating to the next town simply will not work.

Isolation

According to Eurostat, the statistical agency of the European Union, an island is not an island if it is less than one kilometre from the mainland (and, if it has fewer than 50 permanent residents, is attached to the mainland by a rigid structure, or is home to the capital of an EU state) (Eur-Lex 2012). The proximity variable is meant as a proxy for isolation and therefore the extent to which island life can actually unfold (as against it being a replica of that unfolding on the mainland close by).

Most islands are indeed not too far from the mainland; and a socio-economic study of such 'near islands' is in progress (Starc 2019). The majority are termed continental islands: they currently hug the shore line, and they would have been part of the same shore many years ago, before the waters rose and/or their previous land connection eroded or collapsed. Others are located in lakes, in rivers or their estuaries; so, again, these islands are very susceptible to and dependent on their immediate mainland surroundings. It is oceanic islands, born out of dramatic volcanic activity or slower coralline accretion, that tend to crop up much further away from landmasses, appearing at the intersection of tectonic plates and/or where coral reefs can grow. Here, isolation is palpable: first, in terms of flora and fauna, and how such species evolve in ways that progressively distinguish them from any mainland cousins; and, second, in terms of human settlement, and how communities learn to survive, cope with and seize any opportunities presented by their geographical predicament (Baldacchino and Bertram 2009). Communication – especially where there is no airport and sheltered harbour – can be rare and intermittent; a passing ship perhaps. In other situations, the exoticism heaped on such islands and their biotic exuberance renders them beguiling tourism destinations: many visitors can be drawn in on long haul flights, thanks to that tantalising sea, sand, sun and palm tree.

Effective connectivity is critical to isolated islands and their populations. If they are expected to compete economically, then they must get their goods and services to markets which lie elsewhere, given the absence of plausible markets at home. And yet, such islands must grapple with relatively high transport costs, irregularity of flight or ferry services and/or single monopolistic service provisions (e.g. Roberts et al. 2016). In such situations, private enterprise can be constrained to operate only as petty self-employment, artisanal craft and trade, and any seasonal opportunity that may fleetingly present itself. But other, similarly isolated islands – such as Iceland or Mauritius – have done well: brandishing financial services, bunkering, niche manufacturing and/or other 'strategic rents' (apart from tourism).

Fragmentation

We may say 'islands', but we often really mean 'archipelagos', a set of islands that represents a fragmentation of the domestic space. Indeed, a careful scrutiny of the world's islands reveals that most are actually members of archipelagos, assemblages that range from a minimum size of two (say, St Kitts and Nevis), to over 50,000 in the archipelago sea off South-West Finland, with the Åland mainland being its largest component (Depraetere and Dahl 2007, p. 77). This archipelagic condition and its effects on, say, adaptive radiation, has long been recognised in the natural sciences (e.g. Seehausen 2004); but it has been an epistemological blind spot in the social sciences, until recently (Pugh 2013, Stratford et al. 2011), and is now leading to a flurry of research output that is exploiting the fresh and powerful perspective of the 'archipelagic turn', with its appreciation of inter-island rivalry, centrifugal tendencies and the jealous safeguarding of the territorial or 'inner sea' (Baldacchino 2016, Roberts and Stephens 2017, Thompson 2010).

Indeed, contemporary scholarship tends to settle upon two rather overworked topological relations of islands. The first presents a clear focus on an island's singularity, its unique history and culture, crafted and inscribed by the border between *land* and *sea*. The second distinguishes an island from a *mainland/continent*, and dwells on its differences from, and dependencies on, the larger player. The concept of the archipelago evokes a third topological relation that is much less commonly deployed than the previous two. It foregrounds interactions between and among islands themselves. The relation of island to island is characterised by repetition and assorted multiplicity, which intensify, amplify and disrupt relations of land and water, as well as island and continent/mainland.

The notion of the archipelago is now a core concern of the international law of the sea, where the island members of an archipelago are located in such proximity to each other that they can naturally be seen as a unit, assigning them specific rights and obligations, such as designating sea lanes (Sand 2012). In political geography, the archipelagic lens offers fresh insights on the jurisdiction of Taiwan (Baldacchino and Tsai 2014), the UK (Pocock 2005, p. 29), Australia's historically shifting identity/ies as island, continent, nation and archipelago (Perera 2009); and on the consciousness of the Japanese state as a *shimaguni* (island nation) and its possible bearing on how it tackles tensions in the China Sea (Suwa 2012).

Every archipelago has a centrifugal death wish: "in an archipelago, the temptation is always great, at worst to secede and at best to disregard the political jurisdiction of the centre" (LaFlamme 1983, p. 361). This intra-archipelago dynamic is at the root of uneven island–island politics: how do the (often democratically elected) politicians of multi-island jurisdictions balance the wishes of their various island publics and constituencies with the rationale of hub-and-spoke transport logistics (versus costly repetitive infrastructure), tourism differentiation (versus repetition), complimentary (rather than similar) and cooperative (rather than competitive) economic development trajectories? (Baldacchino and Ferreira 2013). All islanders know that they experience tense relations with their island neighbours. Often made fun of in popular idiom, these tensions and rivalries may find expression in discriminatory practices of various kinds, official or otherwise; and these become more likely if specific islands claim, or are represented as claiming, a linguistic, ethnic, religious, historical, occupational and/or economic status that is distinct from that pertaining to other islands. Island nationalism knows no limits, as various attempts at secession demonstrate: Mayotte (resident population: about 212,000) did not join the other Comoros islands to independence and is now an integral part of France, although the Republic of the Comoros continues to claim the island (Muller 2012); Tuvalu (population: 10,000) successfully engineered its separation from Kiribati (population: 100,000) before becoming an independent state in 1978 (McIntyre 2012); Nevis (population: 12,000), failed by a whisker to pass a referendum in 1998 that would have seen it secede from St Kitts (population: 32,000) (Premdas 2000).

Amplification by compression

It is easy to adopt the notion of the island as microcosm; "a little world within itself" (Darwin 1845, p. 454), a laboratory for similar processes which unfold elsewhere in such profusion and complexity that they could never be as well analysed; a mini-version of the world out there, a synecdoche where the island is a fragment which however stands for the whole. Islands easily, *too* easily, lend themselves to serve as references to other, larger places: they are the bellwethers; the miner's canary; the conveniently located laboratory. Easter island is 'Earth island' (Flenley and Bahn 1992).

Fair enough: but boundedness, smallness, isolation and fragmentation also come together in complex ways. There is a tendency for the local island elite to straddle political and economic sectors; for the operation of a tightly webbed social fabric where residents could be just 'one degree of separation' away from each other at worst; for politicians to be neighbours, school colleagues and (for those of us who are academics) past students; for a 'soft state' where decisions can be easily traced directly to individuals, who are then held to account, for better or for worse; for personality politics to outperform media campaigns; for familiarity with 'gatekeepers' to offer alternative routes to individual satisfaction (apart from the implicitly politically correct legal-rational ones); and where it is advisable not to make enemies, for you may have to live with them for the rest of your life.

Hence the cautionary note: islands are not merely scaled down versions of larger, continental places. They have an 'ecology' of their own; which means that islands comprise a target that is suggestive of deserving particular strategies and epistemologies. It does *not* mean that all islands are the same. Indeed, and echoing Hay (2013, p. 212), every island comes along with an "irreducible uniqueness". It would also be a far poorer world if islands merely reflected continental goings on at a convenient and manageable scale. And those who approach islands in such a rash, raw and naïve manner may be led to believe, like Lemuel Gulliver, that they have 'figured' islands out, and in a short time (Swift 1726). They may never know or find out that they are mistaken (Baldacchino 2008, p. 42).

Caveats

In spite of the widespread fascination with facts, definitions and cause–effect relationships, one must be wary of peddling reductively deterministic renditions of island-related parameters. Space and place are experienced, constructed and interpreted; and island spaces and places no less so. Islandness is far more nuanced than its strictly materialist renditions may suggest. Exceptions are likely to turn up for every attempt at a water-tight designation of any such 'island ecology'. And internal divisions should not be glossed over or laid aside for the sake of island unitarism.

Subjective experience is critical. Connell (2013, p. 262) concludes his book with a poignant story drawn from his experience in 1991 on an island in Woleai, a Micronesian atoll in Yap state, some 300 metres long, where he spent some months. Towards the end of his sojourn, he was invited to meet a chief at the other end of the island from where he was staying. He surprised himself by thinking that this was 'far away' and he resented having to walk so far.

In any case, and with these caveats, I can vouch from experience for the power of an island epistemology: a framework through which to interpret the world, foregrounding the geographical in the argument, reminding ourselves that specific events may be unfolding on *island* spaces (and not just anywhere), and that islandness may be somehow impacting and nuancing such events. We would neglect the island *focus* for a mere island *locus* (Ronström 2013) at our peril. And a comparative, 'island–island' analysis may present interesting insights: it is harder to entertain than a simpler, single, case study; and indeed it is a less common enterprise; but: the results may be worth the effort. I embarked on this road when, as a PhD applicant, I was forced to come up quickly with a viable doctoral thesis topic at the University of Warwick, UK, in 1990: in my *Labouring in Lilliput*, I looked at labour relations in small island states (subsequently published as Baldacchino 1997). I have not looked back since.

The lure of the island

The lure of the island as a privileged inner space is a deduction that kicks in even in the very act of drawing an island (Baldacchino 2005). What starts off as a *line* heading in no particular direction and then starts turning on itself is transformed into a *border*, strictly defining an inner and outer world. The 'logic of inversion' that is involved in this (seemingly innocuous) act of drawing (Ingold 1993) turns "the pathways along which life is lived into boundaries within which life is contained" (Ingold 2008, p. 1796). Separation for transformation is foundational to the contemplation of islands:

> To dream of islands – whether with joy or in fear – is to dream of pulling away, of being already separate, far from any continent, of being lost and alone – or to dream of starting from scratch, recreating, beginning anew.
>
> *(Deleuze 2004, p. 9)*

Physical detachment can provide the environment that confuses and unsettles; yet, it invites a deep questioning of paradigmatic knowledge.

When I introduce the study of islands in my lectures, I continue to be amazed by the readiness with which, when asked to draw 'the island', students of any cultural, ethnic, educational or linguistic background habitually frame and contain whole islands neatly on a sheet of paper. They ascribe a roughly circular outline to 'the island', sometimes punctured by the contours of an inlet or a fjord, or a break in the coastline or coral reef, to permit a safe harbour to seagoing vessels. And they add smaller islands or parts of larger mainlands to serve as context. This profiling suggests that an island is really a piece of land of the right size, fitting the conceptual and epistemological project – one of refuge, escape, transformation? – to which it is meant to be deployed (Baldacchino 2005). Thus, the island is itself the quintessence of design; the prototype workscape for artists of all stripes; it is an edginess that charts a special space which can then be mentally embraced and accommodated.

No wonder that islands also abound as sacred spaces, sites of monasteries and destinations of pilgrimages; a sense of other-worldly exclusivity now being replicated in very worldly upscale hotels and hideouts for the rich and famous (Jackson and Della Dora 2009). But: the same ease of accommodation of the island trope into one's mental psyche also morphs humans into daring engineers who, finding themselves on islands, ditch their moral and ethical compass: devising wild genetic experiments; taking on god-like qualities that are doomed to fail; descending into bestiality. Think of *The Island of Dr Moreau* (Wells 1896) and *Lord of the Flies* (Golding 1954). The most consummate landscape artist and civil engineer of all however may be Robinson Crusoe who subjects and bends his desert island to his will, his epic island endeavour serving as a paean to the virtues of unbound industrial and bourgeois capitalism (Hymer 1971). The Robinsonade is the name assigned to a host of derivative stories based on similar victories of man (yes, usually male) over nature (McMahon 2016). A similar contemporary experiment is being run by the US National Aeronautics and Space Administration (NASA) which has islanded sets of trainees on the slopes of Mauna Loa, on the Big island of Hawai'i – six in a habitat dome, completely isolated, for stretches of eight months at a time – simulating the conditions of a future colony on the red planet Mars, itself an island in space (Chason 2017, Sammler 2018).

Island studies has come a long way. Frank Zappa had said, half-jokingly, that any 'real and self-respecting' country must have its own airline and its own beer at the very least, and perhaps some kind of football team or some nuclear weapons (Brainy Quotes 2017). Perhaps a 'real and self-respecting' discipline must have its own journal, professors and courses, and perhaps an anthology and a handbook. Island studies now has all of these.

Godfrey Baldacchino

References

Anderson, E.W. (2000) *Global geopolitical flashpoints: an atlas of conflict*. London, The Stationery Office.

Baldacchino, G. (1997) *Global tourism and informal labour relations: the small-scale syndrome at work*. London, Mansell.

Baldacchino, G. (2004) Autonomous but not sovereign? A review of island sub-nationalism. *Canadian Review of Studies in Nationalism*, 31(1–2), 77–90.

Baldacchino, G. (2005) Islands: objects of representation. *Geografiska Annaler: Series B, Human Geography*, 87(4), 247–251.

Baldacchino, G. (2008) Studying islands: on whose terms? Some epistemological and methodological challenges to the pursuit of island studies. *Island Studies Journal*, 3(1), 37–56.

Baldacchino, G. (2012a) The lure of the island: a spatial analysis of power relations. *Journal of Marine and Island Cultures*, 1(2), 55–62.

Baldacchino, G. (2012b) Islands and despots. *Commonwealth & Comparative Politics*, 50(1), 103–120.

Baldacchino, G. (ed.) (2013) *The political economy of divided islands: unified geographies, multiple polities*. New York, Palgrave Macmillan.

Baldacchino, G. (2014) Capital and port cities on small islands sallying forth beyond their walls: a Mediterranean exercise. *Journal of Mediterranean Studies*, 23(2), 137–151.

Baldacchino, G. (ed.) (2016) *Archipelago tourism: policies and practices*. London, Routledge.

Baldacchino, A., and Baldacchino, G. (2017) Conceptualising early childhood education in small states: focus on Malta and Barbados. In T.D. Jules and P. Ressler (eds.), *Re-reading education policy and practice in small states: issues of size and scale*, Frankfurt, Germany, Peter Lang, pp. 97–109.

Baldacchino, G., and Bertram, G. (2009) The beak of the finch: insights into the economic development of small economies. *The Round Table: Commonwealth Journal of International Affairs*, 98(401), 141–160.

Baldacchino, G., and Ferreira, E.C.D. (2013) Competing notions of diversity in archipelago tourism: transport logistics, official rhetoric and inter-island rivalry in the Azores. *Island Studies Journal*, 8(1), 84–104.

Baldacchino, G., and Tsai, H.-M. (2014) Contested enclave metageographies: the offshore islands of Taiwan. *Political Geography*, 40(1), 13–24.

Barrie, J.M. (1915) *Peter and Wendy*. London, Scribner.

Bioy Casares, R. (1940/2003) *The invention of Morel*. Translated by Ruth L.C. Simms. New York, New York Review of Books.

Black Sails (2017) TV Series, 2014–2017. Details available from www.imdb.com/title/tt2375692/ Quoted text available from: www.tveskimo.com/2015/02/12/black-sails-xi-s2-e3-reconciliation/.

Boissevain, J. (Ed.). (1996) *Coping with tourists: European reactions to mass tourism*. New York, Berghahn.

Bongie, C. (1998) *Islands and exiles: the creole identities of post/colonial literature*. Stanford, CA, Stanford University Press.

Brainy Quotes (2017) Frank Zappa Quotes. Available from: www.brainyquote.com/quotes/quotes/f/frankzappa134155.html.

Cameron, David (2013) Speech, London. Available from: www.gov.uk/government/speeches/eu-speech-at-bloomberg.

Carlquist, S. (1974) *Island biology*. New York, Columbia University Press.

Chason, R. (2017, September 27) This is the true story of six strangers picked to live in a NASA dome. *Washington Post*. Available from: www.washingtonpost.com/news/speaking-of-science/wp/2017/09/27/this-is-the-true-story-of-six-strangers-picked-to-live-in-a-nasa-dome/?utm_term=.923168e95117.

Churchill Semple, E. (1911) *Influences of the geographic environment, on the basis of Ratzel's system of anthropogeography*. New York, H. Holt.

Coffey, H. (2017, 30 June) Shelly Island appears off the tip of Cape Pont. *The Independent UK*. Available from: www.independent.co.uk/travel/news-and-advice/shelly-island-new-cape-point-bermuda-triangle-dangerous-buxton-north-carolina-a7816501.html.

Connell, J. (2006) Nauru: the first failed Pacific state? *The Round Table: Commonwealth Journal of International Affairs*, 95(383), 47–63.

Connell, J. (2013) *Islands at risk? Environments, economies and contemporary change*. Cheltenham, Edward Elgar.

Cook, C.M. (2016) *Marketing in large ocean states: theory, practice, problems and prospects*. Austin, TX, Sentia Publishing.

Crane, R., and Fletcher, L. (2017) *Island genres, genre islands: conceptualisation and representation in popular fiction*. Lanham, MD, Rowman and Littlefield International.

Crichton, M. (1990) *Jurassic Park*. New York, Alfred A. Knopf.

Dalby, S., and Tuathail, G.Ó. (eds.) (1998) *Rethinking geopolitics*. New York, Psychology Press.

Darwin, C. (1845) *Journal of researches into the natural history and geology of the countries visited during the voyage of H.M.S. Beagle round the world, under the command of Capt. Fitz Roy, R.A.* London, John Murray. Available from: http://darwin-online.org.uk/content/frameset?pageseq=1&itemID=F14&viewtype=text.

Defoe, D. (1719) *The life and strange surprising adventures of Robinson Crusoe of York*. London, W. Taylor.

Delanty, G. (2017) *A divided nation in a divided Europe: emerging cleavages and the crisis of European integration*. London, Anthem Press.

Deleuze, G. (2004) *Desert islands and other texts, 1953–1974*. Cambridge, MA, MIT Press/Semiotexte.

DeLoughrey, E.M. (2007) *Routes and roots: navigating Caribbean and Pacific island literatures*. Honolulu, HI, University of Hawai'i Press.

Depraetere, C., and Dahl, A.L. (2007) Island locations and classifications. In G. Baldacchino (ed.), *A world of islands: an island studies reader*. Charlottetown, Canada and Luqa, Malta, Institute of Island Studies, University of Prince Edward Island and Agenda Academic, pp. 57–105.

Diamond, J. (1997) *Guns, germs and steel: the fates of human societies*. New York, W.W. Norton.

Doty, R.L. (1996) Sovereignty and the nation: constructing the boundaries of national identity. *Cambridge Studies in International Relations*, 46(1), 121–147.

Eur-Lex (2012, June 21) Opinion of the European Economic and Social Committee on 'specific problems facing islands' (own-initiative opinion), 2012/C 181/03. *Official Journal of the European Union*. Available from: http://eur-lex.europa.eu/legal-content/EN/TXT/?uri=CELEX%3A52012IE0813.

Fernandes, R., and Pinho, P. (2017) The distinctive nature of spatial development on small islands. *Progress in Planning*, 112(1), 1–18.

Flanagan, R. (1994) *Death of a river guide*. Carlton, Australia, McPhee Gribble.

Flenley, J., and Bahn, P. (1992) *Easter island, Earth island*. London, Thames and Hudson.

Fog Olwig, K. (2007) *Caribbean journeys: an ethnography of migration and home in three family networks*. Durham, NC, Duke University Press.

Gillespie, R.G., and Clague, D.A. (eds.) (2009) *Encyclopedia of islands*. Berkeley, CA, University of California Press.

Gillis, J.R. (2004) *Islands of the mind: how the human imagination created the Atlantic world*. New York, Palgrave Macmillan.

Golding, W. (1954) *Lord of the flies*. London, Faber and Faber.

Hay, P. (2013) What the sea portends: a reconsideration of contested island tropes. *Island Studies Journal*, 8(2), 209–232.

Hymer, S. (1971) Robinson Crusoe and the secret of primitive accumulation. *Monthly Review*, 23(4), 11–36.

Ingold, T. (1993) The art of translation in a continuous world. In G. Palsson (ed.), *Beyond boundaries: understanding translation and anthropological discourse*, Oxford, Berg, pp. 210–230.

Ingold, T. (2008) Bindings against boundaries: entanglements of life in an open world. *Environment and Planning D: Society and Space*, 40(8), 1796–1810.

ISISA (2108) *International small islands studies association*. Available from: www.isisa.org.

ISJ (2018) *Island Studies Journal*. Available from: www.islandstudies.ca/journal.

Jackson, M., and della Dora, V. (2009) 'Dreams so big only the sea can hold them': man–made islands as anxious spaces, cultural icons, and travelling visions. *Environment and Planning A*, 41(9), 2086–2104.

Kelman, I., and Baldacchino, G. (eds.) (2016) *An anthology of island studies*, 4 volumes. London, Routledge.

King, R., and Connell, J. (eds.) (1999) *Small worlds, global lives: islands and migration*. London, Pinter.

LaFlamme, A.G. (1983) The archipelago state as a societal subtype. *Current Anthropology*, 24(3), 361–362.

Le Juez, B., and Springer, O. (eds.) (2015) *Shipwreck and island motifs in literature and the arts*. Amsterdam, The Netherlands, Brill/Rodopi.

MacArthur, R.H., and Wilson, E.O. (1967/2015) *Theory of island biogeography*. Princeton, NJ, Princeton University Press.

Macdonald, C. (2016, June 8) Bikini Atoll is still uninhabitable: radiation on island exceeds safety standards nearly 60 years after nuclear tests. *Daily Mail (UK)*. Available from: www.dailymail.co.uk/sciencetech/article-3630359/Bikini-Atoll-uninhabitable-Radiation-island-exceeds-safety-standards-nearly-60-years-nuclear-tests.html.

McIntyre, W.D. (2012) The partition of the Gilbert and Ellice Islands. *Island Studies Journal*, 7(1), 135–146.

McMahon, E. (2016) *Islands, identity and the literary imagination*. New York, Anthem Press.

Mead, M. (1928) *Coming of age in Samoa*. New York, William Morrow and Company.

Montgomery, L.M. (1908) *Anne of Green Gables*. Boston, MA, L.C. Page & Company.

More, T. (1516) *Utopia*. Self-published. Original in Latin. English translation by G. Burnett. Available from: www.gutenberg.org/ebooks/2130.

Mountz, A. (2011) The enforcement archipelago: detention, haunting and asylum on islands. *Political Geography*, 30(3), 118–128.

Muller, K. (2012) Mayotte: between Europe and Africa. In R. Adler-Nissen and U. Pram Gad (eds.), *European integration and postcolonial sovereignty games*. London, Routledge, pp. 187–202.

Okihiro, G.Y. (2010) Unsettling the imperial sciences. *Environment and Planning D: Society and Space*, 28(5), 745–758.

Owens, J. (2017, 19 April) Bangladesh's disappearing islands fuel climate change fears. *VOA News*. Available from: www.voanews.com/a/bangladesh-disappearing-islands-fuel-climate-change-fears/3816377.html.

Patke, R. S. (2018) *Poetry and islands: materiality and the creative imagination*. Lanham, MD: Rowman and Littlefield International.

Percy, D.M., Blackmore, S., and Cronk, Q. (2007) Island Flora. In G. Baldacchino (ed.), *A world of islands: an island studies reader*. Luqa, Malta and Charlottetown, Canada, Institute of Island Studies, University of Prince Edward Island and Agenda Academic, pp. 175–198.

Perera, S. (2009) *Australia and the insular imagination: beaches, borders, boats and bodies*, New York, Palgrave Macmillan.

Plato (360 BC?) *Dialogues: Timaeus and Critias*. B. Jowett, translator. Available from: www.activemind.com/Mysterious/topics/atlantis/timaeus_and_critias.html.

Pocock, J.G.A. (2005) *The discovery of islands*, Cambridge, Cambridge University Press.

Premdas, R. (2000) Self-determination and secession in the Caribbean: the case of Nevis. In R. Premdas (ed.), *Identity, ethnicity and culture in the Caribbean*. St Augustine, Trinidad and Tobago, School of Continuing Studies, University of the West Indies, pp. 447–484.

Prinsen, G., and Blaise, S. (2017) An emerging 'islandian' sovereignty of non-self-governing islands. *International Journal*, 72(1), 56–78.

Pugh, J. (2013) Island movements: thinking with the archipelago. *Island Studies Journal*, 8(1), 9–24.

Puzo, M. (1969) *The godfather,* New York, G. P. Putnam & Sons.

Ratter, B.M.W. (2018) *Geography of small islands: outposts of globalisation*, Heidelberg, Germany, Springer International.

RETI (2018) Reseau d'excellence des territoires insulaires [Network of island universities]. Available from: http://reti.univ-corse.fr/.

RLI (2018) Rethinking the island. Available from: www.rowmaninternational.com/our-publishing/series/rethinking-the-island/.

Roberts, B.R., and Stephens, M.A. (eds.) (2017) *Archipelagic American studies*. Durham, NC, Duke University Press.

Roberts, S., Telesford, J.N., and Barrow, J.V. (2016) Navigating the Caribbean archipelago: an examination of regional transportation issues. In G. Baldacchino (ed.), *Archipelago tourism: policies and practices*, Farnham, Ashgate, pp. 147–163.

Roberts, S.M. (1995) Small place, big money: the Cayman Islands and the international financial system. *Economic Geography*, 71(3), 237–256.

Ronström, O. (2013) Finding their place: islands as locus and focus. *Cultural Geographies*, 20(2), 153–165.

Royle, S.A. (2001) *A geography of islands: small island insularity*. London, Routledge.

Royle, S.A. (2014) *Islands: nature and culture*. London, Reaktion Books.

Sammler, K. (2018) How Mars beckons: science, colonialism and island and outer space imaginaries. Paper presented at 16th 'Islands of the World' conference in Terschelling, The Netherlands, 10–14 June.

Sand, P.H. (2012) Fortress conservation trumps human rights? The 'marine protected area' in the Chagos archipelago. *Journal of Environment and Development*, 21(1), 36–39.

Seehausen, O. (2004) Hybridisation and adaptive radiation. *Trends in Ecology & Evolution*, 19(4), 198–207.

SICRI (2018) *SICRI network: small island cultures research initiative*. Available from: http://sicri-network.org/.

Sidaway, J.D. (2007) Enclave space: a new metageography of development? *Area*, 39(3), 331–339.

Simberloff, D.S., and Wilson, E.O. (1969) Experimental zoogeography of islands: the colonisation of empty islands. *Ecology*, 50(2), 278–296.

Singham, A.W. (1968) *The hero and the crowd in a colonial polity*. New Haven, CT, Yale University Press.

Smith, A.D. (1986) *The ethnic origins of nations*. Oxford, Blackwell.

Song, Yann-Huei (2015, May 7) 'Taiping island: an island or a rock under UNCLOS?,' *Asia maritime transparency initiative*. Available from: https://amti.csis.org/taiping-island-an-island-or-a-rock-under-unclos/.

Starc, N., and collaborators (2019) *Near islands: a study from Croatia*. Lanham, MD, Rowman and Littlefield International, forthcoming.

Stevens, S. (2016, July 5) 10 new islands formed in the last 20 years. *Mother Nature Network*. Available from: www.mnn.com/earth-matters/wilderness-resources/stories/10-new-islands-formed-last-20-years.

Stratford, E., Baldacchino, G., McMahon, E., Farbotko, C., and Harwood, A. (2011) Envisioning the archipelago. *Island Studies Journal*, 6(2), 113–130.

Suwa, J.C. (2012) Shima and aquapelagic assemblages. *Shima: The International Journal of Research Into Island Cultures*, 6(1), 12–16.

Swift, J. (1726) *Gulliver's travels*. London, B. Motte.

Thompson, L. (2010) *Imperial archipelago: representation and rule in the insular territories under US dominion after 1898*. Honolulu, HI, University of Hawai'i Press.

Thomson Reuters (2018) Social sciences citation index – journal list. Available from: http://science.thomsonreuters.com/cgi-bin/jrnlst/jlresults.cgi.

Trompf, G. (1993) *Islands and enclaves*. Delhi, India, Stirling Publishers.

UNCLOS (2017) *Part 8: regime of islands*. Available from: www.un.org/depts/los/convention_agreements/texts/unclos/part8.htm.

Verne, J. (1871) *A journey to the centre of the earth*. London, Griffith and Farran.

Verne, J. (1875) *The mysterious island*. London, Sampson Low, Marston, Low & Searsle.

Vladi (2017) *Vladi private islands*. Available from: www.vladi-private-islands.de/.

Wallace, A.R. (1880/1975) *Island life, or, the phenomena and causes of insular faunas and floras: including a revision and attempted solution of the problem of geological climates*. New York, Palgrave Macmillan.

Wells, H.G. (1896) *The island of Dr Moreau*. London, Heinemann, Stone & Kimball.

Whittaker, N.J. (2016) *The island race: geopolitics and identity in British foreign policy discourse since 1949*. Unpublished PhD thesis. Sussex, University of Sussex.

Williamson, M. (1981) *Island populations*. Oxford, Oxford University Press.

ACKNOWLEDGEMENTS

The first edition of this book was conceived in 2007 (see Figure 0.5). The idea was to provide a truly global introduction to the study of islands, from the perspective of the various established disciplines, in both the natural and social sciences. Not all fields were covered. Being comprehensive was always going to be a difficult task, and not just because of the sheer extent of material: I also had to turn down a literature chapter draft because it was not global enough in its scope, for example. In any case, the hefty tome that resulted, with its 578 pages plus detailed index, has served its purpose well. It has been out of print and out of stock, with only pdf versions of the chapters available for purchase, for several years. Ashgate was the first publisher to accept taking over the task of publishing a revised edition. Then in 2014, Routledge, having taken over Ashgate, consented to add this (completely rewritten and much more comprehensive) text as an 'island studies' volume to its range of international handbooks.

The thrust of this current volume remains the same: a disciplinary introduction to the study of islands, from archaeology to zoology, permitting readers to navigate chapters with which they will have different degrees of familiarity. The offerings of this new volume are more expansive: there are now 21 chapter contributions, instead of the original 16, and including river islands, wellbeing, literature, branding, sociology and politics for the first time.

Nevertheless, there will be readers who may still feel that some important area of inquiry has been neglected. I would be the first to admit the sheer impossibility of a comprehensive text: that admission, in itself, is already evidence of the way in which island studies as a discipline and area of research focus has expanded of late. I would also advise any such disappointed readers to take up their pens, or keyboards, and help address gaps in island studies scholarship, doing so particularly by submitting manuscripts to the suite of island studies journals that are now available.

I was also criticised for having too many 'grey-haired, bearded, white and Anglo men' as authors to the 2007 volume. Some of these have persevered, 11 years on: they are hard to dislodge from my list of contributors when they are so good in their respective fields and have graciously accepted to contribute to this book. In such cases, they are a decade older and greyer (this editor included). However, a conscious attempt has also been made to have as many of the chapters as possible written by two authors; and, in so doing, younger and female scholars have joined the contributing team. Equally capable scholars from so-called 'developing countries'

Figure 0.5   Cover of the book *A world of islands: an islands studies reader*, precursor to this volume, co-published in Canada and Malta by the Institute of Island Studies, University of Prince Edward Island and Agenda Academic. Cover image is from the sculpture *Embarkation for Cythera*, by Canadian artist Richard E. Prince, selected from a series of works titled *A taxonomy of islands*. Visit: https://richardeprince.com/2017/02/12/2000s/#jp-carousel-300 Reproduced with permission.

have also been added to the list of contributors. Together, these have enriched the contributions to this text.

My thanks to Faye Leerink, Commissioning Editor, and Ruth Anderson, Editorial Assistant, both at Routledge/Taylor & Francis: Faye and Ruth have been my guides and reference beacons throughout the compilation of this work. I also thank Kate Fornadel for ably shepherding the content through the proofs stage. And a special thanks to the 40 contributing authors, along with the various copyright holders of accompanying images and photographs, who have made this compendium possible.

Sadly, one of our esteemed contributors, R.J. 'Sam' Berry (co-author of Chapter 6, and father of Andrew Berry, co-author of Chapter 4) passed away in March 2018. It was such a pleasure to work with Sam, and I am confident that he would have been delighted to see this book come to light. Our condolences to his family and many friends.

Finally, a disclaimer: every effort has been made to trace copyright holders and to obtain their permission for the use of copyright material. The publisher apologises for any errors or omissions in this endeavour and would be grateful if notified of any corrections or additions that should be incorporated in future reprints or editions of this book.

Godfrey Baldacchino
Marsaskala, Malta
January 2018

NOTES ON THE COVER

The study of islands includes the examination of the relationships between land and sea, between human and non-human life, between local and other forms of governance. Such scrutiny is significantly impacted by the vantage point and positioning of the scholar and researcher.

These multiple dynamics are captured in this spectacular photo, taken on 28 January 2018 by Stephen A. Royle. The town nestled in the valley with towering cliffs on both its west and east is Jamestown (population: 630), capital of the British Overseas Territory of Saint Helena, Ascension and Tristan da Cunha, located on the island of Saint Helena in the South Atlantic Ocean. Located 1,900 km (1,200 miles) from Africa and 2,900 km (1,800 miles) from South America, Saint Helena – a rocky inter-plate outcrop on the mid-Atlantic ridge, with a land area of 123 square kilometres (47 square miles) – is one of the world's most isolated islands.

Humans arrived on the island some 500 years ago. The port at Jamestown has since been the only navigational link with the outside world for the island community of 'Saints': until an airport (code: HLE) was built and (after a few hiccups) opened in 2017 to a restricted range of aircraft.

History has it that Saint Helena was discovered by the Portuguese in 1502, and then entered the British realm, being colonised by the East India Company in 1659. After the decisive Battle of Waterloo in 1815, Napoleon was exiled on St Helena until his death in 1821. Its quite unique evolutionary conditions left a lasting impression on the young Charles Darwin on his visit aboard HMS *Beagle* in 1836. A 2014 zoological survey established that, on the cliffs around Jamestown and beyond, the island had 502 endemic species, not found anywhere else on the planet. These include *Pseudolaureola atlantica*, a spiky and bright yellow woodlouse with a current estimated population of 90: one of the rarest animals in the world, and in danger of imminent extinction.

Professor Royle was in Jamestown participating in an international conference discussing the changing nature of St Helena with its newly built airport now open for commercial flights. He proudly reported to me that he climbed the 699 high steps of Jacobs Ladder from Jamestown to get this view of the capital from the top of Ladder Hill.

He describes St Helena as 'the ultimate island'.

PART I

Foundations of islands and island life

1

DEFINITIONS AND TYPOLOGIES

Stephen A. Royle and Laurie Brinklow

Introduction: islands as geographical spaces

What is an island? This book is written in English and we should thus interrogate this language, and the society from which it emanated, to identify the origin and meaning of this simple English word. The part referring to insularity, which itself is based on the Latin word *insula*, meaning 'island', is the first syllable, and a clue to its origins can be found through an inspection of a map of the archipelago off northwest Europe where English developed (see Figure 1.1). This exercise reveals that many of the small islands have '-ay' or '-ey' endings to their names.

This is a souvenir of the voyages of the Vikings or Norse around these coasts in the Dark Ages, in the 8th to 11th centuries, a period of Viking expansion from their base in Scandinavia, a time when the sea was a highway rather than a barrier and when small islands were ports of call for refreshment and refuge. The Vikings left their mark, including their place names, as the Old Norse word for island is *ey* (modern Danish has *ø*), with roots in *ea*, a proto-Indo-European word for river, thus representing water. This word has not entered the English language by itself: although there is an island off Dublin in Ireland, close to Lambay, another island with a Norse name, called Ireland's Eye. It is part of the word 'eyot' which is a river island, as in Chiswick Eyot in the Thames in west London. More significantly, it has entered the modern language, via the Old Norse (and Old Frisian) *ey* and the Old English *ie*, *i* or *ieg*, in combination with 'land', a solid part of the earth, to form 'island'. In this word 'is' is pronounced as 'eye'. Matters are clearer if it is realised that an older spelling was 'iland': the *Oxford English Dictionary* traces that back to AD 888. The added 's' results from a 15th-century association of the word with the French-derived *île*, sometimes *isle*. 'Island' means the same as the older word, *iland*, which it paralleled and, before 1700, replaced. So John Donne in 1624 wrote 'no man is an Iland'. The derivation thus associates land with water. Those islands with the *ey-* or *ø*-derived endings do not normally take 'island' as part of their names. To do so would be tautological: it is Jersey, not Jersey Island.

Definition and typology

An island is 'a piece of land surrounded by water'. The source here is the *Compact Oxford English Dictionary* and, typical of this popular dictionary, the definition is short, practical and seemingly comprehensive. The *Oxford English Dictionary* is only marginally more complex, having the primary meaning as 'a piece of land completely surrounded by water'. However, one word, despite its comprehensive meaning, seems not to have been sufficient for the English, for within the *Compact Oxford* are also found:

- archipelago: an extensive group of islands;
- atoll: a ring shaped reef or chain of islands formed of coral;
- isle: literary: an island, often a small one;
- islet: a small island;
- key: a low-lying island or reef, especially in the Caribbean [in which area it is often spelt 'cay', though the pronunciation does not change]:
- reef: a ridge of jagged rock or coral just above or below the surface of the sea;
- rock: mass of rock projecting above the ground or water.

Moreover, the *Oxford Compact Thesaurus* adds the word 'holm' under its entry for 'island'. Island writer, Bill Holm, an American of Viking, Icelandic ancestry, noted that this is another word from the Vikings (its origin is *holmr*), who, as a seafaring society, needed to subdivide the world of islands into different types. Holm made much of his surname; the first words of his *Eccentric islands* are: "Call me Island. Or call me Holm. Same thing . . . Holm, an Old Norse masculine noun . . . means small island or inshore island" (2000, p. 3). Off the coast of Wales, within the compass of the Dark Age Viking realm, the substantial island of Anglesey takes the *ey*-derived ending as do Bardsey, Ramsey and Caldey. But others, true to the origin of the word 'holm', take that form: Grassholme, Skokholm, Gateholm, Steep Holm and Flat Holm are all "small . . . or inshore" islands just off the coast (see Figure 1.1). The Swedish capital, Stockholm, is built on a series of such islands.

It is not unusual in the development of society and language for even a straightforward construct to require a variety of terms to capture its complexity: the best-known example must be the categorisation of 'snow' into a multiplicity of different types, each with its own word, in Inuktitut and other Arctic languages. English needed several distinct words, not just synonyms, for varieties of pieces of land surrounded by water, marking the importance of the construct to a language which developed in an island setting amongst a trading and colonising people:

> Great Britain came to think of itself in insular rather than continental terms, creating for itself an archipelagic empire beginning with nearby Ireland and eventually extending throughout the Atlantic.
>
> *(Gillis 2003, p. 29)*

In contrast to English, in languages emanating from societies in which islands have not been important, there are few words denoting insularity. Slovakia in eastern Europe is landlocked, and inspection of a standard dictionary reveals just one Slovak word for 'island', *ostrov*, with developments therefrom, such as *ostrovčan* for 'islander'; also *súostrovie*, for 'archipelago'. To the south, Slovenia has only a short coastline on the Adriatic, without offshore islands, and this country has but one island and that only periodically, for it is in a lake which, due to the peculiarities of the local limestone topography, emerges seasonally and surrounds a hillock within its basin. The occasional island and its village are both called *Otok*; that is also the Slovene word for 'island' and it does duty for most insular matters, although *atol* and *arhipelag* have been introduced from foreign languages.

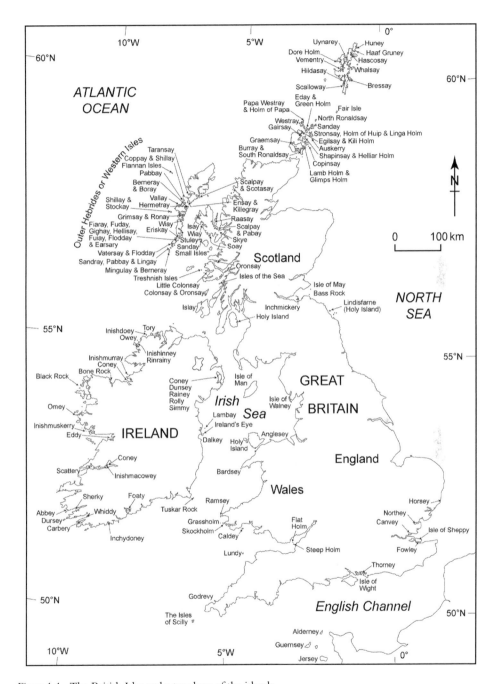

Figure 1.1 The British Isles and a typology of the islands.

Source: © Stephen A. Royle. (Drawn by Maura Pringle, Queen's University Belfast, Northern Ireland, UK.)

In English, which of the range of words is used for a particular piece of land surrounded by water depends, perhaps, on its formation (see Nunn and Kumar, this volume) as with 'reef' or 'atoll', or its size. Isles are usually small islands and in the region where English developed we find the Isles of Scilly and the Isle of Wight off England; the Isle of Man in the Irish Sea; also the Western Isles and other smaller groups off Scotland (see Figure 1.1).

Some geographical spaces are troubling regarding the second syllable of 'island'. Van Duzer (2004) has written on floating islands, buoyant vegetation and organic matter that can be substantial enough to support trees, even livestock. This would see them accepted under the Scottish census definition (see below), but not being land, can they ever be islands? To further complicate matters, there are places not always completely surrounded by water: naturally occurring periodic, usually tidal islands. As stated, Otok in Slovenia is a seasonal island; ensiling is normally caused by tidal movement as with Lindisfarne (Holy Island) off northeast Britain, where islanding happens on a twice-daily basis, while Modo in Korea is only connected to other land twice a year (see Figure 1.2). Are these places true islands? The answer must be a non-periodic, permanent yes. A trip to Lindisfarne certainly presents visitors with the feeling of being on a small island, for access depends on the times when tides are low; little different than waiting for the ferry on non-periodic islands.

To the Vikings, a piece of land surrounded by water was not regarded as an *ey* or island unless it was sufficiently distant and distinct for the sound separating it from the mainland to be navigable by a ship with its rudder in place. In the Scottish census of 1861 a piece of land surrounded by water graduated to being worthy of the name 'island' only if it had sufficient pasturage to support at least one sheep (King 1993). Such functional thresholds for the recognition of an island are not just of historic interest, for the United Nations Convention on the Law of the Sea (Part VIII, Article 121) does not allow exclusive economic zones or continental shelves to be awarded to rocks incapable of sustaining human habitation or economic life on its own. A case in point here is the granite oceanic crag of Rockall, 461 km (286 miles) west of Scotland, which is claimed by Denmark (for the Faroes), Iceland, Ireland and the UK (see Figure 1.3). At 74 m^2 (88.5 square yards), Rockall is small enough to have been mistaken for an enemy ship in the First World War, and it came close to being rammed in what would have been a vain attempt to sink it, after it failed to respond to signals (Gordon 1961). British sovereignty was declared in 1972,

Figure 1.2 Crossing on foot to Modo Island (모도), Korea, at low tide is a twice-yearly ritual.

Source: Caithlyn Debeer, https://caitlyndebeer.wordpress.com/tag/korea-festival/ reproduced with permission.

Figure 1.3   Rockall: the numerous birds shown – puffins, kittiwakes, gannets and guillemots – provide a helpful sense of scale.

Source: Wikimedia Commons, https://commons.wikimedia.org/wiki/File:Rockall_-_the_most_difficult_island_in_ the_world_to_sleep_on_-_geograph.org.uk_-_1048830.jpg.

with a 1974 claim to exclusive exploitation of a 134,680 km^2 (52,000 square miles) zone around Rockall. In 1975, a former British Special Forces soldier spent 40 days on this 'island' – for so it was declared in 1972 UK legislation – to demonstrate its habitability and to support the British claim. In 1997, the UK's claim that Rockall was an island was dropped, but the rock is still regarded as part of Scotland (Royle 2014a).

In contrast, writers seeking to diminish an island's significance might resort to a lesser term, as with Philip Carteret's description of the first European sighting of the unpromising Pitcairn Island in the Pacific in 1767 (tellingly named after the young man at the masthead, the son of the Captain of Marines, who first spotted it, rather than the ship's captain or a member of the establishment): "a small high uninhabited Island . . . scarce better than a large rock in the ocean" (Carteret 1965, p. 150). Ironically, it was that very dismissal of Pitcairn Island that made it attractive to the *Bounty* mutineers who settled there in 1790, seeking seclusion. Their leader, Fletcher Christian, knew of the passage, Carteret's book being carried on board (Smith 2003). Another case would be that of the Irish nationalist Alice Stopford Green's characterisation of the 122 km^2 (47 square mile) St Helena in the South Atlantic as a 'slag-heap' in her attempt to magnify for political reasons the deprivations faced by Boer prisoners of war incarcerated thereupon by the British from 1900 to 1902 during the South African War (Stopford Green 1900, Royle 1998).

Taking 'island' in its conventional use in English does not enable us to calculate for certain even how many islands there are. It would depend on which of the 'isles' down to 'rocks' categories were included. And even if the count were restricted to pieces of land customarily regarded as 'islands' rather than the lesser categories, discrepancies would emerge. The Vikings, as described, distinguished between *ey* and *holm* but, to stay with the Welsh examples used above, note that Ramsey and Bardsey are little bigger than Skokholm. So: why would these places, which today are all uninhabited and with the steering of a Viking longboat a forgotten art, not all belong to the same category of island landmass? Convention only helps at the upper end of the scale, for Australia at 7,686,843 km^2 (2,967,907 square miles) is conventionally regarded as continental, which makes Greenland at 2,175,600 km^2 (840,004 square miles) the world's largest island. But then should large islands be regarded as being insular at all? Shakespeare's John of Gaunt made much of the advantages for England (part of the island of Great Britain) of it being cut off by the 'silver sea', which is both a 'wall' and, a line later, a 'moat'; comparisons stressing the role of the water as a defensive barrier around the island. That it is a permeable barrier was shown by the many successful invasions up to and including that of the Normans from France in 1066. But the sea still defended, and during the Second World War, having to move his armies across this barrier was one of the factors that dissuaded Adolph Hitler from invading. At this time, there was certainly considerable public consciousness of Great Britain being an island, for it was seen to gain significant strategic advantage therefrom. However, to generations born since the Second World War, the functionality of islandness would seem to have declined: there is now even a railway tunnel connecting England and France.

Islandness and identity

Great Britain, like New Guinea, Honshu, Luzon or Java, is of a scale that it being an island does not usually matter to its people's day-to-day lives. For smaller islands, however, scale can matter. In her study of islands off northwest France, Françoise Péron (2004, pp. 328, 330) puts it thus:

> the notion of an island shall be discussed, deliberately restricting ourselves to small, inhabited islands: those specks of land large enough to support permanent residents, but small enough to render to their inhabitants the permanent consciousness of being on an island . . . The omnipresence of the sea intensifies the feeling of being cut off from the rest of the world. The maritime barrier is always there, solid, totalising and domineering, tightening the bonds between the island folk, who thus experience a stronger sense of closeness and solidarity.

Péron begins to deal here with island identity, reflecting, perhaps, ethnographic traditions in island studies. For instance, Dodds and Royle (2003) mention Anthony Cohen's 1987 classic investigation of Whalsay, one of the Shetland Islands off Scotland, and John Messenger's study of Inishere in the Irish Aran Islands (imperfectly disguised as Inis Beag, Irish for 'small island'). They then go on to note that such island "ethnographic research is used to illustrate how 'belonging' is rooted in a powerful sense of community and kinship which serve to consolidate a common heritage, shared places and social knowledge" (2003, p. 488). It is apposite to note that the traditional inhabitants of the Turks and Caicos Islands and of the British Virgin Islands are called 'Belongers'. In a similar vein, *shima* (島), the Japanese word for island, can also mean community: there are a number of places on the large islands that have 'shima' in their names, such as Kagoshima, which speaks to this association. Islands of a size large enough for their inhabitants not to have Péron's 'permanent consciousness' of insularity, her logic has it, should, perhaps, not be regarded as true islands. However the identity associated with British insularity certainly remained a psychological issue as was amply demonstrated in the June 2016 referendum when a majority of the British electorate who voted, voted for 'Brexit': for the UK to leave the European Union.

Boundedness

Being encircled by water can result in a geographical, psychological, and societal boundedness that contributes to a strong attachment to one's island. This literal/littoral boundary is "nature's emphatic and unambiguous way of telling Islanders that they are a separate and unique people . . . a geographic situation [that] dictate[s] both a sense of unity and separateness, of inclusion and exclusion" (Baglole and Weale 1973, pp. 105–106). The resulting boundedness creates "the island effect" (Baldacchino 2007), a term borrowed from the field of biology, where life on an island is intense, distilled, exaggerated. Baldacchino (2013) has also called it "the ABCs of island living: amplification (or articulation) by compression", also borrowed from biology (see Percy, Cronk and Blackmore, this volume). The defined edge of the island provides a natural limit, concentrating that sense of belonging to what is knowable, intimate. "Being geographically defined and [archetypically] circular, an island is easier to hold, to own or manipulate as much as to embrace and to caress" (Baldacchino 2005, p. 247). Islands are a mirror with an emphatic frame, with boundedness accentuating the experience and intensifying the perceived unity of those thus bounded and hemmed in with respect to outsiders. Geography is "a mirror for man [sic]" (Tuan 1971, p. 181), the island physicality "reflecting and revealing human nature and seeking order and meaning in the experiences that we have of the world" (Relph 1976, p. 4).

Indeed, a further effect of the water boundary is to bind an island's inhabitants together, giving them a shared sense of identity (a form of solidarity at its most pronounced). Akin to Lamarck's theory of genetic inheritance influenced by the environment, this is what Sugars (2010, p. 8) referred to as "a scientifically outmoded concept which nevertheless finds expression in discussions of cultural origins and transmission". Sugars also observes that, "despite the fact that Charles Darwin refuted the notion that 'knowledge' and 'memory' were inheritable traits, this belief has informed much modern thinking about group identities, and, indeed, notions of collective destiny" (Sugars 2010, p. 9). These longstanding connections cross geographical and temporal boundaries and, through a binding together, islanders become bound *to* their island, experiencing a "subterranean connection" to place (MacLeod 1999, p. 163) and to the past that is, for some, unbreakable. These subterranean connections are often passed down through stories that incorporate a deep and intimate knowledge of bounded places that exemplify strong emotional connections. Islanders draw identity from a deep intimacy with the island that has been cultivated through stories (including the Icelandic *sagas*) that bind people to place, and are keenly felt in a tight network of community and kinship, itself strengthened in the retelling of these tales.

Connectedness

But just as islands are bounded and separate, they are also connected. Despite John Donne's devotion, "no man is an iland, intire of it selfe", no island can be entirely independent from the rest of the world. From dugout canoes and galleons to today's jetliners and the Internet, the geographical particularity of islands is in flux. In the last 50 years, changes in transportation structures have changed life irrevocably on islands, and definitions created from intra-island separation and remoteness are constantly undergoing revision. It is true that, on some of the larger islands, becoming aware of one's island status happens only when one has to get off one. To cross the body of water – necessarily by boat or airplane – becomes a conscious decision, often involving serious logistical planning and a substantial financial outlay. At the heart of one's awareness is the realisation that one must cross water in order to leave, which takes a more concerted effort than just taking the metro, train, bus or car and heading to wherever one needs to go. In populated areas, rural areas become subsumed by urban sprawl; suburbs blend into cities as boundaries spread. But on islands – sea-level rise notwithstanding – this sprawl can only go so far: the boundary is indisputable, immutable.

Islands have often been regarded as places of dwelling caught in a tense land–sea opposition. In that tense relationship, islands are portrayed as vulnerable to outside pressure: ranging from colonisation to natural disasters and sea-level rise. In most instances, the ocean is seen as the delimiting factor, the cause of the vulnerability and a hindrance to land-based activities, particularly when transportation on and off the island is involved. Beer (1990, p. 272) called for an "equal foregrounding of land and sea" when she discussed the impact of British imperialism and considered how "the sea offers a vast extension of the island, allowing the psychic size of the body politic to expand, without bumping into others' territory".

Tongan scholar and writer Epeli Hau'ofa (1999, p. 31) continued the paradigm shift when he wrote about the Pacific Ocean traditionally serving as a road, binding Pacific peoples together instead of isolating them. In so doing, Hau'ofa breathed new energy and hope into Polynesian and Micronesian academic pursuits, radically redefining how they are regarded by Europeans, as well as the islands' own "indigenous elites" (Edmond and Smith 2003, p. 9). Hau'ofa (1999, p. 31) writes:

> There is a world of difference between viewing the Pacific as "islands in a far sea" and as "a sea of islands". The first emphasizes dry surfaces in a vast ocean far from the centres of power. Focusing in this way stresses the smallness and remoteness of the islands. The second is a more holistic perspective in which things are seen in the totality of their relationships.

Hau'ofa (1999, p. 37) ends his essay with the clarion call:

> We are the sea, we are the ocean. We must wake up to this ancient truth and together use it to overturn all hegemonic views that aim ultimately to confine us again, physically and psychologically, in the tiny spaces that we have resisted accepting as our sole appointed places and from which we have recently liberated ourselves. We must not allow anyone to belittle us again and take away our freedom.

A call was recently made to expand or reframe the study of islands to include and acknowledge archipelagos, unsettling the "fundamental disjuncture in spatiality; the island split between two basic forces": land and water (Stratford et al. 2011, p. 115). The authors note that the word stems from the Greek *arche*, meaning 'original, principal' (as in *arch*bishop), and 'pelago', meaning 'deep, abyss, sea' (ibid., p. 120). Hayward (2012, p. 5) went on to coin the phrase 'aquapelago', to refer to an "assemblage of the marine and land spaces of a group of islands and their adjacent waters". Hayward deploys this term to foreground the water's 'spatial depths', such that "the aquatic spaces between and around a group of islands are utilised and navigated in a manner that is fundamentally interconnected with and essential to the social group's habitation of land and their senses of identity and belonging". The term might become part of the lexicon when discussing marine topics such as fishing zones, shipping routes and seabed resource extraction from an aqueo-centric lens (Hayward 2012).

Because islands appear and disappear with sea-level rise, or other natural processes such as volcanism, Indigenous Australians have long regarded the ocean and land as one, with invisible threads of connection. Likewise, shore dwellers and fishers often do not distinguish between land and water, "actually mentally seeing shoals and eddies and sunken ships and rocks that are exposed only at low tide: not [as] barriers but features" (Theroux 2000, p. 148). "In the ancient world", writes Gillis (2014, p. 156): "land and water were never as differentiated as they are today"; the world was then viewed, he notes, as 'terraqueous'. It is the industrial revolution and its continental paradigm that has shifted Western consciousness decidedly inland (Vannini et al. 2009).

What is islandness?

'Islandness' is a term that is notoriously difficult to define or to pin down. There have been numerous attempts, such as Conkling's (2007 p. 191) "metaphysical sensation that is so hard to put into words". Islandness is meant to embody the essence of island living, the attributes that make an island what it fundamentally is, and which it has by necessity, without which it loses its identity. One clear advantage of the term is value neutral: unlike 'insularity' which bears negative connotations of parochialism and small-mindedness.

So for many, the essence of being surrounded by water is a blessed state: people find comfort and safety within a defined edge. A correspondent on Barra in the Scottish Outer Hebrides reported to the authors on her "sense of community . . . knowing (all) one's neighbours is such a rare thing in the modern, mobile world [and] it gives one the sense of belonging and home". Islanders enjoy the relative ease of access to nature on islands and the possibility of living in tune with the rhythms of the ocean and land. They may see islands as paradise, safe, cocooned within the confines of the edge. But just as often the opposite is true: the island is a place from which one must escape. Many people who experience this feeling are the ones who leave: for education, for employment, or just to get away from meddling family and stifling community (see Connell, Baldacchino and Veenendaal, this volume). In seeming opposition to each other, these 'tropes of islandness' include tradition/modernity, dependency/autonomy, roots/routes, globalisation/particularity, vulnerability/resilience. Yet, rather than binary, these traits are just as often experienced on a continuum, and, similar to one's identity, they may change throughout one's life, dependent on life-cycle stage.

Islanders identify who they are in relation to the rest of the world; their island is a platform on which they firmly plant their feet. Sense of place and belonging are profound human needs, as Edward Relph (1976, p. 41) suggests when he observes that a "deep relationship with place is as necessary as close relationships with people; without such relationships human existence is bereft of much of its significance". Yi-Fu Tuan (1974, pp. 114–5) also writes of places where people experience strong attachment – forests, mountains and deserts – and then posits that the "ideal places" that have "persistent appeal" in human imagination are "the seashore, the valley, and the island". The island is often imbued with deep spiritual meanings, particularly in early civilisations where sometimes it is the seat of the Gods. Many islands today persevere with strong associations to faith, mystery, the sacred and the sublime.

The role of the sea

The role of the ocean in creating 'islandness' – the mystique, the allure, that is so hard to define – has to do with water, the most basic of elements from which all life has sprung. It has to do with the separation from the mainland: the 'island-self' and 'mainland-other' (Stratford 2008, p. 161). It is also the dynamism that comes from seeming binaries: the physical space of an island that is set apart, isolated and bounded by a medium that, by its make-up, also connects (Hay 2013). Meeting places are dynamic spaces, and islands have several. Of course, the meeting place islanders know best is the shore, physically a rich ecotone: from the Latin *eco*, derived from the Greek *oikos* or house; and *tone*, or tension, in Greek (Gillis 2014, p. 159). This is where two ecosystems meet and intertwine, and where land and sea life are often interdependent. In Newfoundland, this is called the 'landwash'. As a liminal space (from the word limen, meaning threshold), the shore is the ocean's doorstep, and vice versa. The edge is a threshold, the doorway to an untamed, unruly ocean, one of the planet's least explored geographies. The edge throws up an unobstructed view of the horizon: unattainable yet always beckoning. And to get to an island, to succumb to its allure, a crossing must be endured. Even the shortest distance over water – the crossing – can be an adventure as the body breaches what *should* be impassable.

Societal islandness

Living within a set parameter, defined by water, frames island life emphatically. While some feel safe and comforted, manifested in a strong sense of community, tribalism and strong kinship webs, others chafe at the edges, which, when tightened, can press inward until emotions push to the top, resulting in passionate extremes such as anger, hatred or love. People who have survived – and thrived, often in the most appalling conditions – take pride in knowing that they have made it through yet another weather-related disaster with no one coming to the rescue; it is just another test of one's mettle to be able to survive against the odds. Islanders the world over have demonstrated time and time again an innate strength, independence and resourcefulness in response to externally caused crises: thinking of islands as insular is fast being replaced by seeing them as adaptable and resilient: mostly out of necessity, but also out of fleeting opportunity (Baldacchino 2015). Conkling (2007, p. 198) calls it "'lifeboat ethics'; where the sense of islanders' individual fates is intimately and inextricably tied up with those with whom they are cast and with whom they have (almost) no choice to accept, since all succeed or fail together".

Among the tropes of islandness is the insider/outsider dynamic. Often those who have chosen to live on an island have to learn and 'earn' their islandness (Conkling 2007, p. 198), and some of those, "like all converts, burn with a harder flame for island institutions and values than does the natal experience" (Putz 1984, p. 26). However, everyone on an island today would have an ancestor who has moved to that island. As Dening (1980, p. 31) has written about oceanic islands: "Everything on an island has been a traveller. Every species of tree, plant and animal on an island has crossed the beach." When you consider time spent on an island as a determinant of who should belong, then, it is slightly ironic that the insider/outsider discourse should be a constant in determining characteristics of islandness. Although it is often done with a sense of humour, "suspicion of outsiders" (Péron 2004, p. 330) is an underlying thread of island life. Perhaps it is in reaction to being subjected to mainlander attitudes that range from outright colonisation to being the butt of jokes. Perhaps it is a way of protecting an island's finite amount of land and resources. Perhaps it is an assertion of individuality that comes from being (or believing yourself to be) a little more special because one lives on an island. Whatever the reason, islands continue to attract outsiders, which results in either an enriching or negative experience, for both sides.

Just as the 'island effect' has an impact on evolutionary processes, so, too, has island isolation contributed to some aspects of culture being maintained on islands longer than on some mainlands. For example, Old English is the basis for the Newfoundland English and accent that prevails on this east coast Canadian island, and is preserved most intensely on Newfoundland's small isolated offshore islands and coastal communities called outports. Indeed, scholars come to Newfoundland and Labrador specifically to study Old English. On Canada's Cape Breton Island, the Scottish fiddling tradition has been described as "more Scottish than the Scots". When Scots left the highlands and islands during the 18th and 19th century 'clearances' and emigrated to the new world – Nova Scotia (New Scotland), in particular – they brought with them the traditions from their homeland, out of which comes the commonly held belief that the isolation of Cape Breton led to the Scots tradition being better preserved than in the homeland.

A living, breathing culture is not meant to be contained or held; it must adapt and evolve in order to survive. In the face of creeping globalisation and its threat of all-enveloping homogenisation "islandness is a stable geophysical and cultural variable that is an anchoring comfort in the current turbulent context of shifting boundaries and politico-economic fusion and fission" (Baldacchino and Greenwood 1998, p. 10). This is particularly so amongst islanders who define themselves against a presumed 'other': whether it is the inhabitants of a neighbouring mainland or another island with which they relate jurisdictionally, economically, linguistically or culturally; often it is only in standing up to outsiders that culture can remain vibrantly distinctive. Ireland and its troubled history with Great Britain would be a case in point. After all, in a postcolonial world, islands can utilise their 'otherness' in ways that lead to self-determined, culturally aware societies; the more we study them, the more we see that islands prove to be complex dynamic entities that have much to offer the rest of the world.

Fixed links

Then there are the islands no longer isolated because a bridge, causeway or tunnel has been built out to/from them, so functionally, at least, they have lost their individual status. All the major islands of Japan – including its 'mainland' islands of Hokkaido, Honshu, Shikoku and Kyushu – are now permanently linked. Links have transformed Bermuda into a single island, from the archipelago which is its actual geographic nature.

Links do not necessarily destroy island identity. Cape Breton Island in eastern Canada was ceded to Britain by France under the Treaty of Paris in 1763 and until 1785 was ruled as part of the colony of Nova Scotia. It then became a separate colony before being joined once more to Nova Scotia in 1820. It was joined physically in 1955 when the Canso Causeway was opened across the short but deep Canso Strait (see Figure 1.4). Thus legally and politically for many generations, and operationally for several decades, people from Cape Breton have been just another group of Nova Scotians. However, many Cape Bretoners have a strong identity with the island, within a regional identity with the Maritimes (the three provinces of Nova Scotia, New Brunswick and Prince Edward Island) and a national identity with Canada. "'No finger of land' was going to change their independent attitude", Lotz and Lotz reported, continuing with the story of an old lady saying in her prayers the day the causeway reached the island, "thank God for having at last made Canada a part of Cape Breton" (1974, p. 14). Few islanders would label themselves as Nova Scotian; Cape Breton has had for many decades a troubled economy and its inhabitants have resented the Nova Scotia government's role in dealing with the island's problems, emanating from the different world of the city of Halifax. Frank was rather mild when he concluded: "Cape Bretoners have continued to offer some elements of resistance to the powerful centralising forces of Canadian society" (1985, p. 218).

Identity is thus another factor to consider when deciding what is or remains an island. Let us consider another Canadian case. Prince Edward Island, Canada's smallest and only fully enisled province, entered Canadian confederation in 1873 with an assurance that the federal authorities would be responsible for an all-year passenger and mail service to the island, across the Northumberland Strait (see Figure 1.5).

The Strait is typically frozen from January to April: at confederation, only iceboats made the winter journey, sailing over open stretches and being dragged on runners over ice (Macdonald 1997). The federal government brought boats with ice-breaking capacity, but there were schemes to provide a fixed link from soon after confederation. All went nowhere, although a

Figure 1.4 Canso Causeway on its 50th anniversary in 2005, with bridge in background, linking Cape Breton Island to Mainland Nova Scotia, Eastern Canada.

Source: The Chronicle Herald, http://thechronicleherald.ca/novascotia/1304758-canso-causeway-at-60-how-it-got-made.

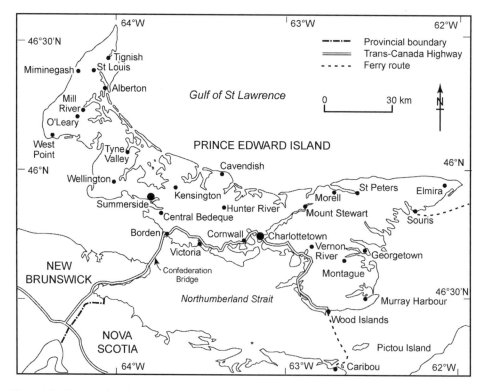

Figure 1.5 Prince Edward Island and the Confederation Bridge.

Source: © Stephen A. Royle. (Drawn by Maura Pringle, Queen's University Belfast, Northern Ireland, UK.)

causeway scheme was actually started in 1963 before it was cancelled in 1968. Interest revived in the 1980s when Public Works Canada asked for tenders for a bridge or tunnel; a causeway, it was realised, would have a deleterious impact on the ice conditions in the strait. A plebiscite on PEI in 1988 saw 59.5 per cent vote in favour of a fixed link. Some claimed the question was worded to gather in voters favouring a bridge or a tunnel, although a tunnel was never seriously considered. This is not the place to review the controversy about this project, mostly involving environmental concerns versus economic benefits (Federal Environmental Review Office 1993); suffice it to say that a bridge proposal was accepted, work commenced in October 1993 and the Can$840 million structure was completed on schedule in May 1997 since when Prince Edward Island has been connected to New Brunswick by the Confederation Bridge. Is Prince Edward Island still an island and its people still islanders? David Weale, who in 1988 stood for election to publicise the need for debate, argued that the identity of the province was intimately tied up with its island geography. He stated that "we are islanders . . . living in a place set apart by time and nature . . . it is our islandness that is the centre of the debate" (Macdonald 1997, p. 34). Now with the bridge an established structure, have matters changed? Baum (1997) considered the "fact of difference" an important part of the appeal and identity of islands and those with fixed links lose this, once a journey to them becomes a banal road trip. However, the size and scale of the 12.9 km Confederation Bridge mitigate against this, and a drive across the Northumberland Strait could be far from banal. The Prince Edward Island annual statistical review for 1999 (Government of Prince Edward Island 1999) wrote of the "large increase in the number of tourist parties visiting the Island since the opening of the Confederation Bridge". The annual reviews show return crossings to have settled to around 1.5m each year. That for 2015 noted 1,008,000 same-day visits by car for 2014 and 552,000 overnight visits. Tellingly, these much exceed numbers visiting by any other means, which were 22,000 same-day and 122,000 overnight that year (Government of Prince Edward Island 2016, p. 85), so the bridge is now central to the island's access for tourism. Since the bridge opened, there have been few scholarly or technical studies done on its impact on the Island's economy, society, environment, politics or sense of identity (Baldacchino and Spears 2007, p. 48). However, Prince Edward Island residents have stuck to describing themselves as 'Islanders' – always with a capital letter – a people who had and retain a 'permanent consciousness of being on an island'. And the hefty toll for the bridge, at Can$47 (2018), collected upon departure from 'the Island', contributes to that permanent consciousness.

Islands as laboratories

Baldacchino takes Donne's metaphor forward when he writes: "islands are not islands" in the sense that, although bounded spaces, islands are not "closed unto themselves" (2004, p. 272). The fact that islands are not closed systems does not invalidate the 'island as laboratory' concept, used by field scientists, including Charles Darwin and Alfred Russel Wallace (Grove 1995). Nunn, himself an island scholar, said of Darwin: "We can but marvel at his attempts to explain the wonders with which he was confronted. When he reached the Galápagos Islands, he came to realise that the insular realm was truly a new world" (Nunn 2004, p. 312). Accepting that islands are not closed, Wallace wrote in his *Island life*:

> As compared with continents . . . [islands] have a restricted area and definite boundaries, and in most cases their biological and geographical boundaries coincide . . . their relations with other lands are often direct and simple and even when they are more complex are far easier to comprehend than those of continents.
>
> *(1892, pp. 241–242)*

Modern island science might cavil at island external relations being direct and simple. Nevertheless, island studies made a telling contribution to biological sciences, both as field examples or for the specifics of insularity, including work on island biogeography theory (MacArthur and Wilson 1967), now used in both island and continental contexts (Harris 1984, Quammen 1996).

Because of their isolation – and potential for discretion and even secrecy – islands were sites of scientific experimentation as early as the 1500s when astronomer Tycho Brahe (1546–1601) set up his observatory on the now Swedish island of Ven. From field testing of anthrax on Scotland's Gruinard Island to the detonation of nuclear weapons in the Pacific Ocean – including the Bikini and Enewetak Atolls – islands have been looked upon as ideal laboratories – with little regard for long-term consequences of contamination and forced resettlement of their populations (Royle 2014b). In recent years, islands have become the canaries in the coal mine when it comes to climate change. Resettlement of 'climate change refugees' has become necessary as rising sea levels claim low-lying islands such as the Marshall Islands, Kiribati, Tuvalu, Tonga, the Federated States of Micronesia and the Cook Islands (in the Pacific Ocean); Antigua and Nevis (in the Caribbean Sea); and the Maldives (in the Indian Ocean) (IPCC 2001). Extreme weather events and sea-level rise that contribute to coastal erosion on islands threaten coastal property and infrastructure; indeed, Prince Edward Island lost 20.67 km^2 between 1968 and 2010 (UPEI 2015). Since 1900 the capital city of Charlottetown has witnessed a 30 cm rise in sea level and scenarios suggest up to 100 cm can be anticipated by 2100 (Atlantic Climate Adaptation Solutions Association n.d.).

In the human sciences, a parallel story of islands as 'living labs' unfolds. For Darwin and Wallace, one might offer Bronislaw Malinowski and Margaret Mead, anthropologists who studied Pacific island societies in the 1920s (Malinowski 1922, Mead 1928). Baldacchino (2004, p. 276) argues that Mead was keen not just to learn *about* islands and islanders, but *from* them. "Mead didn't go to Samoa just to study Samoa. Rather she wanted to understand the whole human race" (Pipher 2001, p. xviii).

Paradises, prisons and metaphors

Islands have engendered a romantic tradition associated with tags such as Paradise Island and Treasure Island emanating from popular culture (Royle 2014b). This is significant in the modern world where a positive image is welcome for tourism. By contrast, another metaphor has the island as prison, again relating to isolation and boundedness. This metaphor is strengthened by the fact that islands have often been used as, or as a location for, prisons and prisoners, including, of course, the 'island continent' – Australia – and some of its offshore islands such as Norfolk Island and Tasmania. Other examples are Alcatraz, Devil's Island and St Helena . . . and not just for Napoleon (Royle 1998). There is also a centuries-old tradition of using islands as quarantine spaces as well as locales for corralling those with communicable diseases: note, for example, the Lazaretto in the Venetian Lagoon (which started operating in AD 1423), Kamau Taurua in Aotearoa/New Zealand and the Mokola'i leper colony in Hawai'i. The contemporary attraction of and fascination with islands as desirable and alluring spaces has reversed the age-old tradition of using islands as depositories for the rejected, unwanted or suspected of disease.

The word 'island' is not just used in a geographical context, but also in metaphor – which, from the Latin *metaphora*, means 'to carry across'. The *Compact Oxford English Dictionary* provides a secondary definition for 'island': "a thing that is isolated, detached or surrounded". There are many geographical examples of this usage, a number of peninsulas and other isolated spaces have 'island' in their names: Islandmagee and Inishowen are two Irish examples (*inish* means island), whilst most of the state of Rhode Island is firmly attached to the North American continent.

Moreover, the concept of being surrounded has led to terrestrial islands being widely described in the field sciences: islands of vegetation types, or 'sky islands': isolated mountain tops whose ecosystems differ from those of neighbouring lowlands. Kitchen islands, traffic islands and the planet as 'Earth Island' embody the concepts of island without being surrounded by water: all are presumed safe havens set within a larger space.

Literary islands

The isolation and scale of an island can also be used in literature – and the films that are often subsequently adapted from them – as a convenient location to constrain and contain the cast of characters. An island's isolation can be a device in which societal norms can be or need to be shed, as in William Golding and his *Lord of the flies* (1954), a dystopian counter-narrative to the romantic *Coral Island* (Ballantyne 1858). The isolated island notion has also been used to set out utopias, most notably in the original *Utopia* by Sir Thomas More (1516). Shakespeare had this device in *The tempest* (1611) – a play thought to be based on the events surrounding the shipwreck of Sir George Somers on Bermuda in 1609 – where an island foregrounded the tension between the possibilities "for erecting an alternative Utopian society on the one hand and for starkly encountering the difficulties of sheer animal survival on the other" (Grove 1995, p. 34).

Additionally, the island and interactions with it can represent a metaphor for personal and/ or spiritual development as with Daniel Defoe's *Robinson Crusoe* (1719). Modern literature continues this tradition, with one example being the imaginatively illustrated *Hippolyte's island* by Barbara Hodgson (2001) in which the eponymous Hippolyte Webb sets off to rediscover the Aurora Islands, once thought to exist between the Falklands and South Georgia and searched for before in fiction as well as fact: see Edgar Allan Poe's *The narrative of Arthur Gordon Pym of Nantucket* (1838). Potential readers are assured on the back cover that Hippolyte's "unforgettable voyage takes him not only through unfamiliar seas but through the unchartered territory of his own mind and heart".

In today's popular fiction set on islands, four genres in particular stand out: crime, thrillers, romance and fantasy (Crane and Fletcher 2017). Agatha Christie's *Evil under the sun* (1941) is set on 'Smugglers' Island', although she also employed trains, boats and, of course, isolated country houses to achieve the same objects. Thrillers include Dennis Lehane's psychological thriller *Shutter island* (2003), home of the Ashecliffe Hospital for the Criminally Insane; on this isolated island, from which escape is impossible, just how did a convicted murderess get away? Romance on an island embodies the trope of island as a place for the four 's': sun, sand, sea, and sex, and what better way to escape the humdrum of daily life than with a good romance novel set on an island: particularly if one is reading it on a beach? Elin Hilderbrand's appropriately titled *The island* (2010), set on the rustic Tuckernuck Island, off the coast of Nantucket Island, Massachusetts, USA, fits the bill perfectly. And, finally, islands and archipelagos are the ideal setting for fantasy and speculative fiction, embodying tropes of utopia and dystopia mentioned above; Ursula Le Guin's *A wizard of Earthsea* (1968) that launched the Earthsea series, and Diana Wynne Jones and Ursula Jones's *The islands of chaldea* (2014), are novels where impossible magic can only be possible on an island set out of space, out of time.

Conclusion

The word 'island' comes along with a perfectly acceptable geographical definition. However, the usage of the word, at least in the language in which we are communicating, does not accord completely with its definition. Whilst an island is 'a piece of land surrounded by water', some accepted islands are not completely surrounded; not all pieces of land surrounded by water are called, or are regarded as, islands. Not all of those who live on islands see themselves as islanders; while those on mountain tops, desert oases or urban ghettos may feel quite enisled (Tuan 1974). There is no accepted wisdom as to at what size islands have to give way to a lesser category; and there is similar ambiguity as to whether large islands should be seen as true islands when their inhabitants fail to manifest a sense of island identity, with hardly any trace of Péron's 'permanent consciousness' of insularity. Finally, the self-referred identity of people living on islands that now bear 'fixed links' must also be taken into account: some think that, once an island becomes attached to a mainland, it loses its sense of islandness and forever; others would hotly dispute this. Nevertheless, islands and islandness are among the master metaphors of Western civilisation (Connell 2003, Hay 2006, Baldacchino 2012). Gillis (2004, p. 1) writes: "Western culture not only thinks about islands, but thinks *with* them."

References

Atlantic Climate Adaptation Solutions Association (n.d.) *Sea level rise and storm surge hazard mapping in Prince Edward island.* Available from: https://atlanticadaptation.ca/en/islandora/object/acasa%3A635.

Baglole, H., and Weale, D. (1973) *The island and confederation: the end of an era.* Charlottetown, Canada, Williams and Crue.

Baldacchino, G. (2004) The coming of age of island studies. *Tijdschrift voor Economische en Sociale Geografie,* 95(3), 272–283.

Baldacchino, G. (2005) Editorial: islands, objects of representation. *Geografiska Annaler,* 87B(4), 247–251.

Baldacchino, G. (2006) 'Editorial: Islands, island studies, island studies journal.' *Island Studies Journal,* 1 (1), 3–18.

Baldacchino, G. (2007) *A world of islands: an island studies reader.* Charlottetown, Canada and Luqa, Malta, Institute of Island Studies and Agenda Academic.

Baldacchino, G. (2012) The lure of the island: a spatial analysis of power relations. *Journal of Marine and Island Cultures,* 1(1), 55–62.

Baldacchino, G. (2013) Capital and port cities on small islands sallying forth beyond their walls: a Mediterranean exercise. *Journal of Mediterranean Studies,* 23(2), 137–151.

Baldacchino, G. (2015) Small island states and territories: vulnerable, resilient, but also doggedy perseverant and cleverly opportunistic. In G. Baldacchino (ed.), *Entrepreneurship in small island states and territories,* London, Routledge, pp. 1–30.

Baldacchino, G., and Greenwood, R. (1998) Introduction. In G. Baldacchino and R. Greenwood (eds.), *Competing strategies of socio-economic development for small islands,* Charlottetown, PE, Institute of Island Studies, pp. 9–28.

Baldacchino, G., and Spears, A. (2007) The bridge effect: a tentative score card for Prince Edward island. In G. Baldacchino (ed.), *Bridging islands: the impact of fixed links,* Charlottetown, Canada, Acorn Press, pp. 47–66.

Ballantyne, R.M. (1858) *Coral Island* (1995 edition). Ware, Wordsworth Books.

Baum, T. (1997) The fascination of islands: a tourist perspective. In: D.G. Lockhart and D. Drakakis-Smith (eds.), *Island tourism: trends and prospects,* London, Pinter, pp. 21–36.

Beer, G. (1990) The island and the aeroplane: the case of Virginia Woolf. In H. Bhabha (ed.), *Nation and narration,* London, Routledge, pp. 265–290.

Carteret, P. (1965) *Carteret's voyage round the world 1766–1769,* Vol 1. Cambridge, Hakluyt.

Christie, A. (1941) *Evil under the sun.* London, Collins.

Cohen, A.P. (1987) *Whalsay: symbol, segment and boundary in a Shetland community.* Manchester, Manchester University Press.

Conkling, P. (2007) On islanders and islandness. *Geographical Review,* 97(2), 191–201.

Connell, J. (2003) Island dreaming: the contemplation of Polynesian paradise. *Journal of Historical Geography,* 29(4), 554–581.

Crane, R., and Fletcher, L. (2017) *Island genres, genre islands: conceptualisation and representation in popular fiction.* New York, Rowman and Littlefield International.

Defoe, D. (1719) *The life and surprising adventures of Robinson Crusoe of York, mariner*. London, W. Taylor. (Modern edition, 1983) Oxford, Oxford University Press.

Dening, G. (1980) *Islands and beaches: discourse on a silent land. Marquesas 1774–1880*. Carlton, VIC, Melbourne University Press.

Dodds, K., and Royle, S.A. (2003) Introduction: rethinking islands. *Journal of Historical Geography*, 29(4), 487–498.

Donne, J. (1624) *Devotions upon emergent occasions*. London, Thomas Iones.

Edmond, R., and Smith, V. (2003) Editors' introduction. In R. Edmond and V. Smith (eds.), *Islands in history and representation*, London, Routledge, pp. 1–18.

Federal Environmental Review Office (1993) *The Northumberland strait crossing project: report of the environmental assessment panel*. Toronto, ON, Ministry of Supply and Services.

Frank, D. (1985) Tradition and culture in the Cape Breton mining industry in the early twentieth century. In K. Donovan (ed.), *Cape Breton at 200: historical essays in honour of the island's Bicentennial 1785–1985*, Sydney, NS, University College of Cape Breton Press, pp. 203–218.

Gillis, J.R. (2003) Taking history offshore: Atlantic islands in European minds, 1400–1800. In R. Edmond and V. Smith (eds.), *Islands in history and representation*, London, Routledge, pp. 19–31.

Gillis, J.R. (2004) *Islands of the mind: how the human imagination created the Atlantic world*. New York, Palgrave Macmillan.

Gillis, J.R. (2014) Not continents in miniature: islands as ecotones. *Island Studies Journal*, 9(1), 155–166.

Golding, W. (1954) *Lord of the flies*. London, Faber.

Gordon, S. (1961) Rockall. In C.H.C. Williamson (ed.), *Great true stories of the islands*, London, Arco, pp. 104–106.

Government of Prince Edward Island (1999) *25th Annual statistical review*. Charlottetown, Canada, Department of the Provincial Treasury.

Government of Prince Edward Island (2016) *42nd Annual statistical review 2015*. Charlottetown, PEI Statistics Bureau, Department of Finance. Available from: www.princeedwardisland.ca/sites/default/files/publications/stats2015.pdf.

Grove, R.H. (1995) *Green imperialism: colonial expansion, tropical island Edens and the origins of imperialism, 1600–1860*. Cambridge, Cambridge University Press.

Harris, L.D. (1984) *The fragmented forest: island biogeography theory and the preservation of biotic diversity*. Chicago, IL, University of Chicago Press.

Hau'ofa, E. (1999) Our sea of islands. In V. Hereniko and R. Wilson (eds.), *Inside out: literature, cultural politics, and identity in the new Pacific*, Lanham, MD, Rowan and Littlefield, pp. 27–38.

Hay, P. (2006) A phenomenology of islands. *Island Studies Journal*, 8(2), 209–232.

Hay, P. (2013) What the sea portends: a reconsideration of contested island tropes. *Island Studies Journal*, 8(2), 209–232.

Hayward, P. (2012) Aquapelagos and aquapelagic assemblages. *Shima: The International Journal of Research into Island Cultures*, 6(1), 1–11.

Hilderbrand, E. (2010) *The island*. New York, Reagan Arthur Books/Little, Brown and Company.

Hodgson, B. (2001) *Hippolyte's island*. Vancouver, BC, Raincoast Books.

Holm, B. (2000) *Eccentric islands: travels real and imaginary*. Minneapolis, MN, Milkweed Editions.

Intergovernmental Panel on Climate Change (2001) Threatened small island states. In *Climate change 2001: impacts, adaptation and vulnerability*. Cambridge, Cambridge University Press. Available from: www.ipcc.ch/ipccreports/tar/wg2/index.php?idp=671.

King, R. (1993) The geographical fascination of islands. In: D.G. Lockhart, D. Drakakis-Smith and J. Schembri (eds.), *The development process in small island states*. London, Routledge, pp. 13–37.

Le Guin, U.K. (1968) *A wizard of Earthsea*. Berkeley, CA, Parnassus Press.

Lehane, D. (2003) *Shutter island*. New York, HarperCollins.

Lotz, P., and Lotz, J. (1974) *Cape Breton Island*. Vancouver, BC, Douglas, David and Charles.

MacArthur, R.H., and Wilson, E.O. (1967) *The theory of island biogeography*. Princeton, NJ, Princeton University Press.

Macdonald, C. (1997) *Bridging the strait: the story of the Confederation Bridge project*. Toronto ON, Dundurn Press.

MacLeod, A. (1999) *No great mischief*. Toronto, ON, McClelland and Stewart.

Malinowski, B. (1922) *Argonauts of the Western Pacific: an account of native enterprise and adventure in the archipelagoes of Melanesian New Guinea*. London, Routledge.

Mead, M. (1928) *Coming of age in Samoa: a psychological study of primitive youth for Western civilisation*. New York, HarperCollins.

Messenger, J.C. (1969) *Inis Beag: isle of Ireland*. New York, Holt Rinehart and Winston.

More, T. (1516) *Utopia*. Louvin, T. Martin, in Latin, first English edition 1551; modern edition (1996) London, Phoenix.

Nunn, P.D. (2004) Through a mist on the ocean: human understanding of island environments. *Tijdschrift voor Economische en Sociale Geografie*, 95(3), 311–325.

Péron, F. (2004) The contemporary lure of the island. *Tijdschrift voor Economische en Sociale Geografie*, 95(3), 326–339.

Pipher, M. (2001) Introduction. In M. Mead (ed.), *Coming of age in Samoa: a psychological study of primitive youth for Western civilisation*, New York, HarperCollins, pp. xv–xix.

Poe, E.A. (1838) *The narrative of Arthur Gordon Pym of Nantucket*. New York, Harper and Bros.

Putz, G. (1984) On islanders. In *Holding ground: the best of the island journal, 1984–2004*. Rockland, ME, Island Institute, pp. 28–31.

Quammen, D. (1996) *The Song of the dodo: island biogeography in an age of extinctions*. London, Pimlico.

Relph, E. (1976) *Place and placelessness*. London, Pion.

Royle, S.A. (1998) St Helena as a Boer prisoner of war camp, 1900–2: information from the Alice Stopford Green papers. *Journal of Historical Geography*, 24(1), 53–68.

Royle, S.A. (2014a) 'This mere speck on the surface of the waters': Rockall aka Waveland. *Shima: International Journal of Research on Island Cultures*, 8(1), 71–82.

Royle, S.A. (2014b) *Islands: nature and culture*. London, Reaktion Books.

Shakespeare, W. (1611) *The tempest*. Edited by A. Gurr (1984), Cambridge, Cambridge University Press.

Smith, V. (2003) Pitcairn's guilty stock: the island as breeding ground. In R. Edmond and V. Smith (eds.), *Islands in history and representation*. London, Routledge, pp. 116–132.

Stopford Green, A. (1900) A visit to the Boer prisoners at St Helena. *The nineteenth century*, XLVII, 972–983.

Stratford, E. (2008) Islandness and struggles over development: a Tasmanian case study. *Political Geography* 27(2), 160–175.

Stratford, E., Baldacchino, G., McMahon, E., Farbotko, C., and Harwood, A. (2011) Envisioning the archipelago. *Island Studies Journal*, 6(2), 113–130.

Sugars, C. (2010) Genetic phantoms: geography, history, and ancestral inheritance in Kenneth Harvey's *The town that forgot how to breathe* and Michael Crummey's *Galore*. *Newfoundland and Labrador Studies*, 25(1), 7–36.

Theroux, P. (2000) *Fresh air fiend: travel writings*. New York, Houghton Mifflin Harcourt.

Tuan, Y.-F. (1971) Geography, phenomenology, and the study of human nature. *The Canadian Geographer*, 15(3), 181–192.

Tuan, Y.-F. (1974) *Topophilia: a study of environmental perception, attitude and values: study of environmental perception, attitudes, and values*. Englewood Cliffs, NJ, Prentice-Hall.

University of Prince Edward Island (UPEI) (2015) UPEI Climate Research Lab reports PEI coastal erosion for 2014 greater than anticipated. Available from: www.upei.ca/communications/news/2015/11/upei-climate-research-lab-reports-pei-coastal-erosion-2014-greater-anticipated.

Van Duzer, C. (2004) *Floating islands: a global bibliography*. Los Altos Hills, CA, Cantor Press.

Vannini, P., Baldacchino, G., Guay, L., Royle, S.A., and Steinberg, P.E. (2009) Recontinentalizing Canada: Arctic ice's liquid modernity and the imagining of a Canadian archipelago. *Island Studies Journal*, 4(2), 121–138.

Wallace, A.R. (1892) *Island life: or the phenomena and causes of insular floras and faunas, including a revision and attempted solution of the problem of geological climates*. London, Palgrave Macmillan (first published 1880).

Wynne Jones, D., and Jones, U. (2014) *The islands of Chaldea*. New York, HarperCollins.

2

LOCATIONS AND CLASSIFICATIONS

Christian Depraetere and Arthur Dahl

Introduction: definitions

There is a continuum of geographic entities that can fit the definition of 'a piece of land surrounded by water': from the continent of Eurasia to the rock on a beach lapped by the waves that becomes a child's imaginary island. Nevertheless, drawing two lines between something that is too large to be an island and something too small to be an island is, ultimately, an arbitrary decision. In this chapter, we explore the geographic realities behind our concept of islands. We develop a terminology for islands that corresponds to scientific observations gleaned from geological, biological and human perspectives.

The tools of remote sensing satellite imagery (e.g. NASA 2017a) now provide data sets with globally consistent measurements of the 'world archipelago' made up of all those pieces of land surrounded by water. At the same time, our understanding of geotectonic movements, sea level changes and island-building processes allows us to see the islands of today not as eternal entities but as one static image captured from ongoing processes of islands growing and shrinking, joining and being separated, appearing and disappearing through time. In looking across the spectrum of continents, islands and other bodies, we can start with a purely geographic definition of an island as any piece of land surrounded by water, whatever its size or its distance to the closest mainland. Using criteria from physical geography, the number of oceanic islands in any particular size range can be counted, and details concerning their distribution explored.

From this platform, one can explore more functional definitions of 'islandness', or various scales and forms of isolation, as expressed for instance in the processes underlying island biogeography and the amazing diversity of species for which many islands are famous (MacArthur and Wilson 1967). Similar processes lie behind the human populations of islands, the history of their settlement and the unique island cultures they have generated. Today, islandness also finds economic, social and political expressions that are significant in addressing the challenges of sustainable development.

While the concept of an island may seem simple in theory, the many variations on the theme of islandness raise puzzling questions. What happens to an island if it is not surrounded by water all the time, or the channel that separates it fills in, or a causeway is constructed? Does a piece of land become an island if it is artificially cut off from surrounding land? The study of islands on their own terms has also acquired its own name: nissology (Depraetere 1991a, 1991b; McCall

1994, 1996). Understanding the deeper significance of islands means considering such issues as isolation and the resulting creation of diversity, the effects of fragmentation and the influence of marginality. Each of these would, in turn, affect various geographic, evolutionary and cultural processes in different ways.

The world archipelago

When looking at a globe or world map, the emerged lands of our blue planet constitute a sort of 'world archipelago' (see Map 2.1). The general description of this archipelago as it is today supposes no *a priori* definition of pieces and bits of land: continents, islands, islets, atolls, motus, reefs, keys and rocks are inherited from a rather imprecise usage in the past (see Royle and Brinklow, this volume). In any case, they present an obvious hierarchy according to their size (see Table 2.1). There are mainlands surrounded by smaller units, whatever the scale considered.

The two mainlands of the world archipelago are the Old World (including Eurasia with its southern peninsula of Africa) and the New World of North America (with its southern peninsula). Both Antarctica and Greenland are mostly ice caps with a large proportion of their bedrock below sea level. They are thus not, strictly speaking, emergent land and will not be considered as mainlands, even though they form an emergent part of the Earth. The next ranked mainland is Australia, sometimes referred to as a 'continental island'. This small continent represents only 43 per cent of the surface of South America but is ten times larger than the next ranked land, New Guinea. This appears to be the major size gap between two land areas according to their surface proportion (see Figure 2.1). This basic and robust relative surface proportion method provides an objective definition of what are the mainlands of the world archipelago compared to other land that usage tends to consider as islands *per se*. This method is also useful in describing smaller archipelagos (like Japan, British Isles and the Caribbean).

Are there any other size gaps among lands below 1 billion km²? The surface proportion of tenth-ranking Honshu (the main island of Japan, R_{10}=227,899km²) relative to ninth-ranking Sumatra (Indonesia, R_9=430,802km²) is 53 per cent. This threshold occurs just after a class of five islands: New Guinea (as part of the Sahul plate), Borneo and Sumatra (as parts of the Sunda Shelf), Baffin within the northern Canadian Shield, and Madagascar, off East Africa, as a micro-plate on its own. These reflect various tectonic and geomorphologic contexts (see Nunn and Kumar, this volume). Another size gap occurs between 23rd-ranked Mindanao (Philippines, R_{23}=94,550km²) and 24th-ranked Ireland (R_{24}=83,577km²), though the proportion of 88 per cent is not as sharp.

What is the significance of this pattern of land areas? Various processes with specific time and geographical scales are responsible for the structure of the world archipelago as it is today. The combination of processes involved in the formation of an island depends on its size: the smaller the size, the better the chance there is only one dominant process. Major mainlands and large islands are directly derived from the break-up of the ancient continents of Gondwana and Laurasia. Smaller islands may also be fragments of tectonic plates more or less submerged at the present sea level such as the Seychelles. They also come from processes related to volcanism (Galápagos, Iceland, Mauritius), coral reef formation (Bahamas, Micronesia, Tuamotous) or the carving of coastlines by glaciers (Norway, Chile). All these endogenous, exogenous and biological processes interact with worldwide sea level changes to define the land pattern in space and time. Each process tends to produce islands of different size ranges, which may explain some of the size gaps and peaks observable in the distribution of lands by area at global, regional and local scales. Timing may also be important. Since small islands erode faster than large ones because of their higher coastline-to-area ratio, the many small islands created by the retreat of glaciers

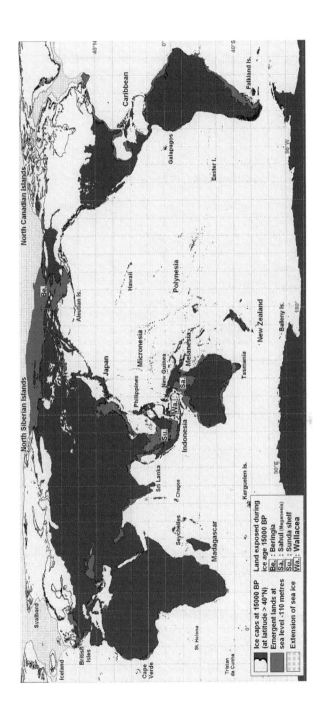

Map 2.1 The evolution of the 'world archipelago' between 15,000 BP [Before Present] and today.

Source: © Christian Depraetere.

Table 2.1 Size ranking of continents and islands of the 'world archipelago'.

Continents and islands	Size S_{km^2}	% Earth %	Rank R_i	Rank proportion $100 \star S(R_i)/S(R_{i-1})$	Comments or context
Ocean	454,204,533	76.1			*excluding Caspian and Aral seas*
Old World	77,355,469	13.0			
Eurasia	48,032,081	8.0	1		
Africa	29,323,387	4.9	2	61.0	
New World	37,255,401	6.2			
North America	19,574,227	3.3	3	66.8	
South America	17,681,174	3.0	4	90.3	
Antarctica	10,540,223	1.77			*mostly ice caps*
Australia	7,605,661	1.27	5	43.0	*the 'island continent'*
Greenland	2,104,005	0.35			*mostly ice caps*
Islands>=0.1km²	7,733,461	1.30			*computed from GSHHS*
New Guinea	783,408		6	10.3	*part of the Sahul plate*
Borneo	735,853		7	93.9	*part of the Sunda Shelf*
Madagascar	592,495		8	80.5	*micro-plate*
Baffin	477,549		9	80.6	*northern Canadian Shield*
Sumatra	430,802		10	90.2	*part of the Sunda Shelf*
Honshu	227,899		11	52.9	*part of the Pacific rim of fire*
Victoria	219,135		12	96.2	*northern Canadian Shield*
Great Britain	218,571		13	99.7	*part of Western European shelf*
Ellesmere	199,289		14	91.2	*northern Canadian Shield*
Sulawesi	170,493		15	85.6	*part of Wallacea*
South Island (NZ)	149,955		16	88.0	*micro-plate*
Java	127,207		17	84.8	*part of the Sunda Shelf*
North Island (NZ)	113,886		18	89.5	*cut off from mainland by glacier*
Newfoundland	109,315		19	96.0	*cut off from mainland by glacier*
Cuba	105,797		20	96.8	*Micro-plate*
Luzon	105,548		21	99.8	*part of Wallacea*
Iceland	101,794		22	96.4	*volcanic mid-Atlantic ridge*
Mindanao	94,550		23	92.9	*part of Wallacea*
Ireland	83,577		24	88.4	*part of Western European shelf*
Hokkaïdo	77,661		25	92.9	*part of the Pacific rim of fire*
etc.	. . .	. . .	. . .		
Islands<0.1km²	28,570	0.005			*estimated from power law of fractal surface*

at the end of the last ice age have not yet been impacted significantly by a reduction in their number via erosion.

Order does not come without underlying patterning: a theoretical approach can help to describe the complex origin of the world archipelago. Several authors, including Mandelbrot (1982), suggest that the world looks like a fractal surface with inherently similar properties that are shared at various scales. Thus, changing magnification will still result in typically large but few chunks of mainland plus smaller but more numerous islands, both set in a largely aquatic frame (see Map 2.2). This can help answer a key question: *how many islands are there?*

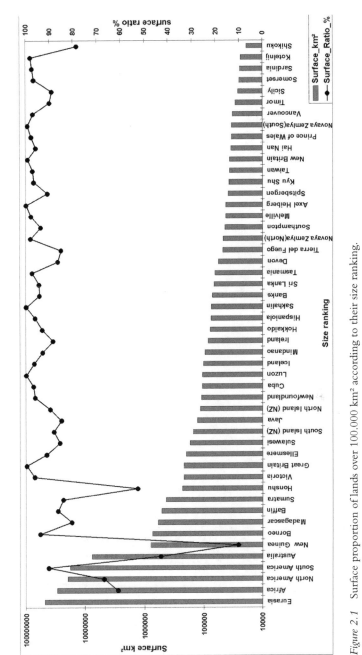

Figure 2.1 Surface proportion of lands over 100.000 km² according to their size ranking.

Source: © Christian Depraetere.

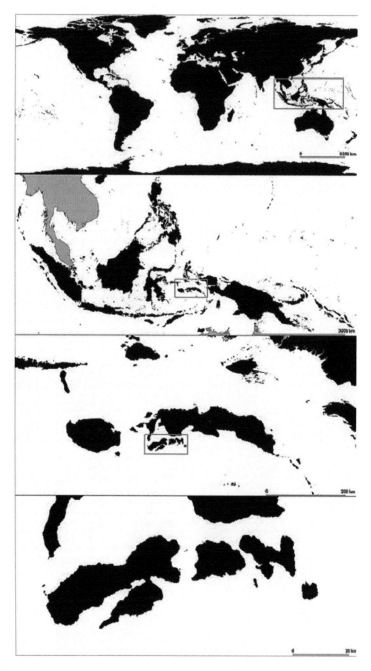

Map 2.2   Islands as a fractal property of the land surface of the Earth.

Source: © Christian Depraetere.

How many islands?

Looking at maps from global to local scales suggests some sort of constancy in the structure of emerged lands. The search for islands is an endless quest, down to very small patches of land only visible at a large scale (see Map 2.2). This observation fits with the underlying concept of self-similarity implied by fractal theory.

The fractal dimension D of the world archipelago can be estimated from the relationship between island size and frequency, given a cumulative size-frequency distribution (Burrough 1986). This has been calibrated on the Global, self-consistent, hierarchical, high-resolution shoreline database (GSHHS) (Wessel and Smith 1996) which is homogenous and includes all pieces of land greater than 0.1 km², as shown on a Log(Frequency)/Log(Area) graph (see Figure 2.2). Thus, we can expect only one island with a land area of around 10,000,000km², but close to 100 islands with a land area of 9,000 km². The frequency of lands with area **a** above the value **A** can be estimated from equation (1):

$$(1) \quad F(a > A) = \alpha.A^{-\beta}$$

Where: $\alpha = 26702$; $\beta = 0.6295$

(Calibration on land between 0.1km² and 100,000km² from GSHHS data)

From equation (1), the fractal dimension D is 1.26 (D=2β) and is consistent with the previous estimation of 1.3 (Mandelbrot 1975). The extrapolation of this relationship to islands smaller than 0.1 km² is questionable, since some authors suggest that "the land surface is not unifractal" (Evans 1995). Assuming that it is unifractal, the number of islands according to their size can be estimated (see Table 2.2).

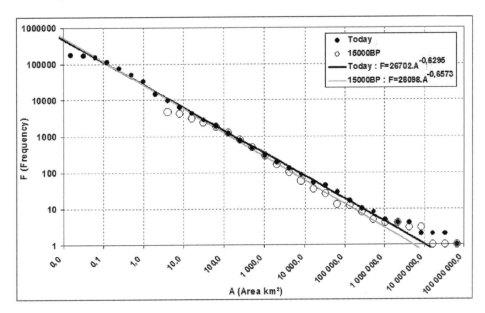

Figure 2.2   The distribution of land areas today and 15000 BP follows a power law as one might expect of the contours of a fractal surface (Departures from regression lines for small areas are due to under-sampling.)

Source: © Christian Depraetere.

Table 2.2 Classification of lands according to size magnitude.

Magnitude 10^m	Class	of	Area	Number (from GSHHS)	Fractal (from $F=\alpha.A^\beta$)	Surface km²	Example	Prefix	+	Term
7		A>=	10^7 km²	3	1	125,151,092	America	*standard*		continent
6	10^6 km²	<A<	10^7 km²	2	3	9,709,666	Australia	micro	–	continent
5	10^5 km²	<A<	10^6 km²	17	15	4,868,996	Madagascar	giga	–	island
4	10^4 km²	<A<	10^5 km²	53	62	1,653,299	Iceland	mega	–	island
3	10^3 km²	<A<	10^4 km²	219	264	674,559	Mauritius	*standard*		island
2	10^2 km²	<A<	10^3 km²	1,135	1,126	337,947	Barbados	micro	–	island
1	10 km²	<A<	10^2 km²	4,251	4,796	129,128	Nauru	nano	–	island
0	1 km²	<A<	10 km²	16,359	2,435	49,959	Pitcairn	giga	–	islet
–1	0.1 km²	<A<	1 km²	63,324	8,072	19,573	Heligoland	mega	–	islet
–2	1 hectare	<A<	0.1 km²	90,446	371,005	16,060	Sala-y-Gómez	*standard*		islet
–3	1,000 m²	<A<	1 hectare		1,580,809	7,100		micro	–	islet
–4	100 m²	<A<	1,000 m²		6,735,650	3,139	Rockall	nano	–	islet
–5	10 m²	<A<	100 m²		28,699,843	1,388		mega	–	rock
–6	1 m²	<A<	10 m²		122,286,783	613		*standard*		rock
–7	0.1 m²	<A<	1 m²		521,050,143	271		micro	–	rock

We may thus expect about 370,000 'islets' ranging in size from 1 to 10 hectares (that is, from 10,000m² to 100,000m²). Some of these are well known, despite their small size: take Liberty Island (4.9 hectares) at the entrance to New York Harbour, or the two islets of Sala-y-Gómez (4 and 11 hectares respectively, see Figure 2.3), the only land between the remote Chilean territory of Rapa Nui (Easter Island) and the islands closer to continental Chile of San Ambrosio, Robinson Crusoe and Alejandro Selkirk. A few of the nearly 7 billion 'nano-islets' are even sources of international disputes: Ireland, Iceland and Denmark contest the sovereignty of the UK over the bare, windy and misty bird-nesting refuge of Rockall (0.08 hectare) in the northeast Atlantic.

Answering the question 'how many islands?' requires setting a minimum size. Common sense suggests that there is a physical limit: a half-submerged grain of sand cannot be an island. In our counting, we stop arbitrarily at 0.1 m² (1 square foot): just enough for a bird or a child to have a rest on one of the expected hundreds of billions of 'micro-rocks'. We can safely assume that there are some 680 billion such 'islands'. This number originates either from observations for lands greater than 0.1 km², or from extrapolation from the smaller ones (see Table 2.2). This allows us to compute some statistics across a variety of sizes:

- three continents (land area greater than 1,000,000 km²): Old World, New World and Australia. (Greenland and Antarctica are considered as ice caps and so not counted as continents)
- 5,675 islands (with an area from 10 km² to 1,000,000 km²): for a total land area of 7,700,000 km²
- 8,800,000 islets (with an area from 10^{-4} km² to 10km²): for a total land area of 95,000 km²
- 672,000,000 rocks (with an area from 10^{-7} km² [one square foot] to 10^{-4} km²): for a total land area of 2,300 km².

Figure 2.3 The double islet of Sala-y-Gómez, Chile.

Source: Photo by Enrdes. Wikipedia Commons, https://en.wikipedia.org/wiki/File:Salaygomez.jpg.

This formal statistical exercise of island counting and defining shows that it is scale depend-ant and presupposes empirically defined limits. For instance, the official number of islands in Indonesia is sharply defined as 18,108, but without any reference to a minimum size (DK Atlas 2004). This limit is about 2.5 hectares computed from a size-frequency distribution of Indone-sian islands.

This pattern of land and sea has changed over time, since it is known that, during the last glacial maximum of 15,000 BP, the sea level was 110 metres lower than today and ice shelves covered large parts of Europe and Northern America (see Map 2.1). The NOAA/NGDC ETOPO2 bathymetry data set used for that purpose (NOAA 2017) – with digital databases of seafloor and land elevations on a two-minute latitude/longitude grid – is not as accurate as the GSHHS used to define the seashore as it is today. Therefore, it is only relevant for paleo-emerged lands with an area greater than 50 km². This lowering of sea level has had dramatic consequences on the mainlands of the world archipelago (see Map 2.1):

- Connection of Eurasia and America via the Bering Strait ('Beringia').
- Emergence of the Sunda Shelf including Borneo, Sumatra, Java and Bali.
- Merging of New Guinea, Australia and Tasmania ('Sahul' paleo-continent).
- Disappearance of the English Channel and North Sea, linking the British Isles to Eurasia.

One major change in the oceanic domain was the emergence of large land areas where today only tiny sparse islands can be found. This phenomenon is exemplified in the Indian Ocean by the Seychelles, Maldives and Chagos archipelagos where several islands of more than 10,000 km² existed at that time. Another stunning case is the Grand Banks southeast of Newfoundland, now part of Canada, which formed a large island of 150,000 km² now completely submerged.

At that time, there was a less overwhelming predominance of the largest islands. This is due to the merging of many of them (Borneo, Sumatra, Great Britain, North Canada, North Siberia) into the main continent formed by Eurasia, Africa and America; new mainlands such as Sahul (New Guinea, Australia and Tasmania); and the merging of archipelagos into unique land masses and clusters (Indonesia, Japan, New Zealand, Philippines).

Islands in space and time

The statistical distribution of land *masses* within the world archipelago is consistent with the fractal hypothesis. This not true for island *locations* however, which are not evenly distributed around the Earth and do not follow a simple spatial distribution law. Islands tend to be aggre-gated in specific regions or to form large archipelagos where most of them are concentrated.

The distribution of islands according to latitude shows that most of them are located in the northern hemisphere, despite the fact that this has a lower proportion of ocean (Figure 2.4). The most striking result is the high occurrence of islands between latitude 50°N and 80°N with a sharp peak between 58°N and 66°N, latitudes also with the lowest ratio of ocean to land. This suggests that *pericontinental* islands (which tend to hug coastlines) are much more numerous than *oceanic* islands (that lie in mid-ocean).

The density of islands of about 90 per 10,000 km² between 58°N and 66°N is much higher than anywhere else. Most of these islands are coastal, creating a patchy landscape made of tiny islands separated by narrow channels. They are typical of the fjordlands of Norway, Greenland, Baffin, Labrador and western Canada where they have been carved from the continent during the last glaciation on the rims of the ice cap (see Nunn and Kumar, this volume). The same type of island structure occurs in the southern hemisphere in Chile, the Kerguelen and Falkland

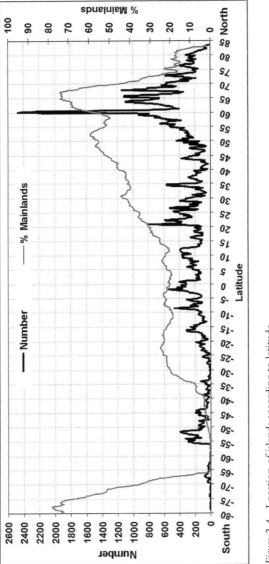

Figure 2.4 Location of islands according to latitude.

Source: © Christian Depraetere.

Islands, and on the west coast of the Antarctic Peninsula. Another consequence of reduced glaciation is the steady uplift of the Scandinavian shield at 1 cm/year in the northern part of the Baltic Sea. One result is the progressive rise of numerous low-lying islands along the coast, as beautifully exemplified by the Åland Archipelago: 3,000 patches of land over a surface of 15,000 km² create the world's highest density of islands (see Map 2.5).

The frequency of islands by latitude in the tropical zone is quite erratic, depending mostly on the location of the main clusters of islands, such as the Caribbean (10°N to 25°N), Insulindia, from Indonesia to the Solomon Islands (15°N to 10°S) and the numerous archipelagos of the South Pacific (5°S to 25°S). The reef-building activity of living coral from northernmost Bermuda (32°N) to southernmost Lord Howe (32°S) also has a noticeable effect by constructing or preserving many islands with a volcanic basement. The absence of coral reefs may partly explain the minimum frequencies of islands observed at subtropical latitudes in both hemispheres.

The proximity of islands to continental landmasses is another significant geographic detail. For that purpose, a useful criterion is the coastal maritime zone, legally defined as the territorial sea (12 nautical miles, or 22.2 km), which also corresponds to the distance from which the coast is visible at sea level. Since an island within this zone also has a territorial sea, such islands may extend this continental coastal zone beyond 12 miles, and so on until there are no more islands within the zone. This criterion defines two classes: the 'pericontinental' islands located within this continental coastal zone and subject to strong continental influences, and 'open ocean' islands distant from immediate continental areas.

Most islands are located near continents, while 'open ocean' islands are less numerous but include a much larger total island area (see Table 2.3). This contrast between the two main island classes can be summarised by calculating a coastal/archipelagic island density (ID) using equation (2):

$$(2)\ \mathbf{ID\ =\ 10,000\ \times\ NI\,/\,(CZ-IA)}$$

Where:
NI: number of islands in a given area
CZ: area in km² of the 12 nautical mile coastal zone including the area of the islands themselves
IA: area in km² of the islands

The reference surface is 10,000 km², which approximates a square degree at the equator.

The density along continental coasts (ID=85 islands/10,000km²) is more than twice that of archipelagos (ID=36).

These island-forming processes lead to sharp contrasts in the distribution and density of islands. For example, the US west coast from San Francisco, California, along Oregon and up to Anchorage, Alaska, has hardly any islets before Cape Flattery (at the northwest tip of the State of Washington, USA); while, to the north, and already within the Juan de Fuca Strait, there is a labyrinth of innumerable islands, islets and rocks. The same is true along most of the world's coasts, except for Africa which has fewer than 2,000 pericontinental islands with a total land area of less than 10,000 km² (see Map 2.3). (Antarctica may be affected by under-sampling. Moreover, 30 per cent of its coast lies under ice caps or ice shelves, as is the case for Ross Island.) African coastlines with significant islands are limited to the Red Sea, the Mediterranean coast near Suez, and portions of East Africa. The small continent of Australia has more coastal islands than Africa in both numbers and area.

Much of the open sea is empty of land above water. Some parts of the southern Pacific are 3,000 km from the closest land. There are isolated islands such as uninhabited Bouvet, a Norwegian

Table 2.3 Islands according to their distance to/from the nearest mainland.

Type of islands according to distance to mainland	Number of islands (NI)	12 nautical mile Coastal Zone (CZ km²)	Island Area (IA km²)	(NI %)	(CZ %)	(IA %)	Island Density ID=10000 × NI/ (CZ-IA)
World archipelago	86,732	22,415,363	7,738,683	100.0	100.0	100.0	59
'Pericontinental' islands	58,913	7,580,845	647,009	67.9	33.8	8.4	85
Europe	15,422	1,000,058	102,633	17.8	4.5	1.3	172
Greenland	4,392	383,602	51,061	5.1	1.7	0.7	132
North America	16,872	1,612,737	204,968	19.5	7.2	2.6	120
South America	6,902	898,321	155,525	8.0	4.0	2.0	93
Asia	10,247	1,661,328	72,179	11.8	7.4	0.9	64
Australia	2,556	603,872	28,187	2.9	2.7	0.4	44
Africa	1,940	747,398	9,488	2.2	3.3	0.1	26
Antarctica	544	631,733	22,518	0.6	2.8	0.3	9
'Open ocean' islands	27,819	14,834,518	7,091,675	32.1	66.2	91.6	36
North Canada	4,942	2,069,373	1,346,028	5.7	9.2	17.4	68
Caribbean	1,877	516,595	213,763	2.2	2.3	2.8	62
Indonesia, Philippines, PNG, Solomons	7,387	5,134,339	2,908,367	8.5	22.9	37.6	33

possession, 1,600 km from Antarctica and 2,500 km from southern Africa; or Tristan da Cunha, the world's most remote inhabited archipelago, 2,000 km from the nearest inhabited land, Saint Helena, and 2,400 km from the nearest continental land, South Africa. Open ocean islands only occupy a small proportion of the oceanic expanse: even with their 15 billion km² of coastal zones, they barely represent 3.3 per cent of the ocean surface (1.7 per cent for pericontinental zones).

Islands tend to amass in clusters or lines, giving birth to the fuzzy term 'archipelago'. On occasion, it reflects the tangible reality of a group of islands isolated from others, as in the case of Cape Verde or Hawai'i. There are other examples where this term is used more loosely, such as the Dodecanese archipelago in Greece.

There are three main 'mega-archipelagos' in the world: the group bounded by Sumatra, Timor, the Solomons and Philippines; the archipelagos of northern Canada; and the Caribbean. Together, they contain more islands, and cover a larger area, than all other open ocean archipelagos put together, and include most of the world's largest islands. Their island densities range from 33 to 68 islands/10,000km², but with major differences among the component archipelagos. In the Caribbean (see Map 2.4), Cuba and part of the patchy coral keys of the Bahamas reach a density of 100 islands/10,000km² compared to 15 for the massive Hispaniola and Jamaica groups, while the spotty line of mostly volcanic islands in the Lesser Antilles has a density of 28/10,000km². Puerto Rico and its tiny eastern neighbours in the Virgin Islands lie in between these, at 56 islands/10,000km².

What has caused such contrasts in island locations and density? While plate tectonics provides a global explanation, the location, density and spatial structure of islands depend on climatic, geological, oceanic, hydrological and biological processes.

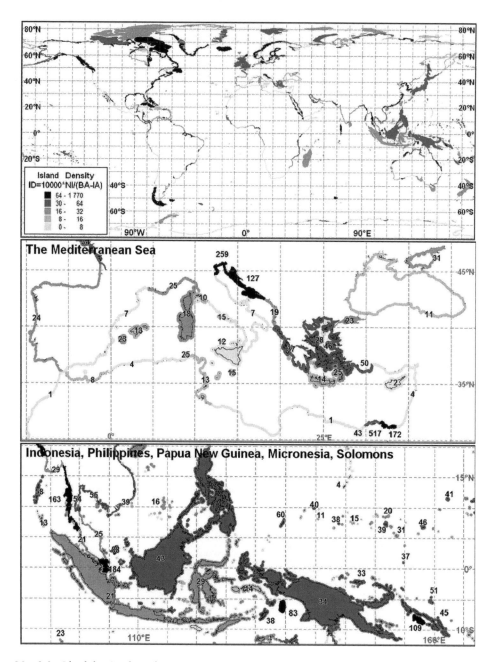

Map 2.3   Island density along the continental coast and within archipelagos.

Source: © Christian Depraetere.

Consider the case of archipelagos or sets of islands well defined by origin (see Table 2.4). The bulky ice caps and the post glacial uplift resulting from their melting created most of the numerous pericontinental islands at high latitudes, with densities of 300 islands/10,000km² or even more (case of Åland, see Map 2.5). Coral reef formation only in tropical waters of over 20°C

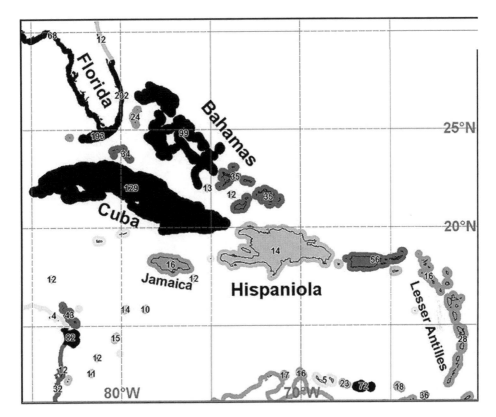

Map 2.4 Island density in the Caribbean.

Source: © Christian Depraetere.

Table 2.4 Examples of islands according to their dominant geological origin.

Type of islands according to processes	Example	NI	CZ km²	IA km²	Largest km²	Island density	Mean ID
Glacial	Åland (Finland)	2,874	20,412	4,112	879	1,763	
processes	Norway	4,236	124,400	22,229	2,239	415	300
including post glacial	Chile	4,511	289,916	127,630	47,107	278	
isostatic uplift	West Canada	3,107	222,158	97,943	31,947	250	
Coral	Bahamas	1,099	157,711	14,390	3,618	77	
processes	Maldives	489	66,045	215	6	74	70
including uplifted coral	Tuamotou (French Polynesia)	566	99,008	562	56	57	
Volcanic	Lesser Antilles	129	61,597	6,049	1,469	23	
processes	Galápagos (Ecuador)	41	44,295	8,087	4,739	11	15
	Cape Verde	17	33,872	4,139	1,006	6	
Alluvial processes	Mississippi (USA)	505	20,382	796	41	258	
at the mouth	Lena (Russia)	476	21,377	1,394	71	238	250
of main rivers	Orinoco (Venezuela)	65	3,034	542	81	261	

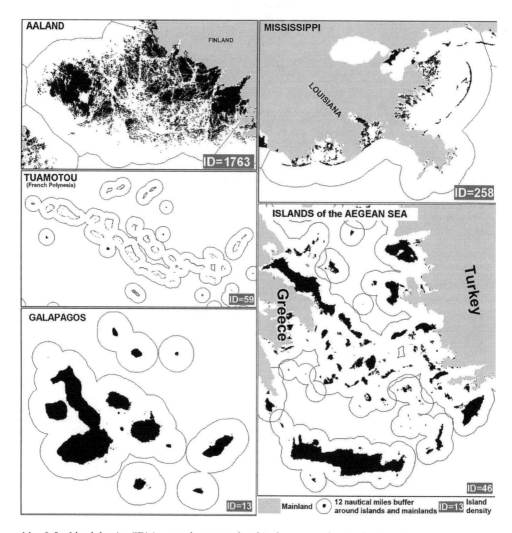

Map 2.5 Island density (ID) in coastal zones and archipelagos according to dominant processes.
Source: © Christian Depraetere.

generates archipelagos with an average density of 70 islands/10,000km² in at least three cases: generally small atolls as in the Maldives; uplifted coral islands reaching up to 50 km² as in the Tuamotous; or even islands of more than 1,000 km² in the Bahamas. Compared to these two types, volcanic islands can be found at any latitude. They are mostly located far from the mainland in isolated archipelagos or islands in areas of subduction at the borders of continental plates ('the rim of fire', in the Pacific), along oceanic ridges, or at hot spots within oceanic plates. They give birth to islands of a few thousand square kilometres, like Big Island in Hawai'i, or more if they merge as Isabella, the largest of the Galápagos. For that reason, the density of their archipelagos is only 15 islands/10,000 km². The alluvial islands formed at the front shore of main continental river deltas, while they may result from a terrestrial hydrological system, are a major local element of the pericontinental seascape, with a high density of 250 islands/10,000 km² in the case of the Mississippi River on the Gulf of Mexico, the Orinoco in Venezuela and the Lena in Northern Siberia.

Another useful measure is the coastal index (Ic) proposed by Doumenge (1984, 1989), based on the relationship between the perimeter (length of coastline) (P) and the surface area of an island (S) (equation 3):

$$(3)\ \mathbf{Ic\ =\ S\ /\ P}$$

This coastal index has the advantage of being simple in formulation, while also easily interpreted in geographical terms. "The importance of the line of coastal contact in relation to the surface area of an island is an expression of the degree of direct influence that maritime affairs exercise on the island" (Doumenge 1989, p. 41). The coastal index gives the length of the coast for 1 km^2 of land. As a consequence, the smaller the island, the larger is its coastal index. However, this index cannot be used unless the coast has been measured at the same scale of maps or resolution of images (Depraetere 1991a), because the length (P) of the coastline, as a fractal object, depends on the scale (Mandelbrot 1967). Fortunately, uniform global data sets derived from satellite imagery offer a solution to this problem.

Island life

The analysis of the geographic features of islands, their size, form and composition, their proximity to or isolation from other islands or continents, and their situation in terms of winds, currents, climate and migratory pathways, all help to explain their biological populations. Again, the dynamic processes involved have to be understood in the context not only of islands as they are at present, but also as they have changed over time. Islands that were once part of a continent until they were cut off by rising sea levels will be much richer biologically for having started with a continental fauna and flora. Those that are actually ancient continental fragments that have drifted away from adjacent land masses may even preserve primitive life forms that became extinct or were replaced as evolution proceeded elsewhere. Volcanic islands and atolls, in contrast, start with nothing and accumulate biota with the passage of time. Moreover, some islands may be repeatedly submerged and exposed as sea levels have risen and fallen during the ice ages, and with each exposure the process of terrestrial colonisation must start over again. With all these processes operating, it is difficult to find any two islands that are completely identical. Each one represents variations on a theme or a series of themes, and the result is an endless diversity of island life forms and ecosystems with many surprises.

The principles of island biogeography also apply somewhat to the marine biology of islands. Many coastal species stick to shallow water or require a hard substrate to attach themselves; others depend on coastal habitats for food or shelter. For such species, islands may be just as isolated as for terrestrial species, with immigration perhaps dependent on how long their larvae can drift in ocean currents before they either settle or die. One variance is that there may be seamounts or reefs that do not qualify as islands because they fail to emerge above the ocean surface – examples of these lie in the northwest Pacific – but which can still serve as stepping stones for marine migrants. So the geographical relationships at a regional scale may be distinct from those of terrestrial forms.

The same principles operate at all the geographic scales of the world archipelago, whether for the flora of islets and rocks or the distribution of mammalian species among the continents and largest islands. The map of mammalian similarities across the world archipelago includes five groups of islands that illustrate their relationships with neighbouring continents (see Map 2.6).

For a start, the Philippines and southeast Asian islands are similar in mammalian fauna with a score of 100 per cent. This is consistent in that they were all part of the former Wallacea

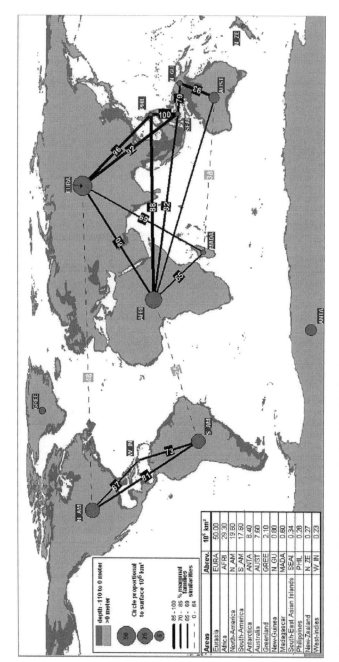

Map 2.6 Mammalian similarities across the 'world archipelago'.

Source: © Christian Depraetere.

archipelago bounded by two major biogeographic lines: the Wallace line to the west and the Weber line to the south and east. They also show strong similarities with Eurasia and the islands of Sunda Shelf to the west (92 per cent or more) and weaker relationships of 50 per cent to 79 per cent with their southeastern neighbours of New Guinea and Australia, both split apart from Sahul. This suggests that the Weber line was more of an obstacle to mammalian dispersal than its western counterpart, and is coherent with the average width of straits up to 15,000BP, which were significantly narrower between Wallacea and the Sunda Shelf (approx. 20km) than between Wallacea and Sahul (approx. 100km). The two parts of Sahul, New Guinea and Australia, have similar mammalian faunas with a score of 93 per cent, despite different climates. This reflects their recent separation only 8,000 years ago. The two other insular biogeographic units were more isolated from their nearest continents both today and during the last ice age: the West Indies are 70km from South America and 120km from North America; Madagascar is 350km from Africa. The mammalian similarities between those insular units and their continents are respectively 73 per cent, 67 per cent and 65 per cent. These figures demonstrate the relationship between species similarities and such simple geographical features as the width of straits during the late Quaternary era. Other factors should also be taken into account, in particular the paleo-geography of plate tectonic movements during the Tertiary era when mammals were evolving. If geographical constraints are obvious for a class of animals at the global scale, they are even more significant at other scales for genera, species or varieties. The specificity of islands compared to the vast terrestrial continuum of continents is largely the outcome of their relative isolation by water gaps that constitute drastic dispersal barriers for most species.

A geographic analysis of the different facets of 'islandness' can have considerable predictive and explanatory value in understanding island biogeography. Islands in turn have been impor-tant laboratories of evolution whose study has done much to advance the biological sciences (see Berry and Gillespie; Percy, Cronk and Blackmore; and Berry and Lister; this volume).

Stages of human discovery and settlement

The same tools of geographic analysis that shed light on island biogeography can also help to explain the interaction of geographic, technological and cultural factors behind the migrations of people into island regions, migrations that have been the subject of many analyses not only by historians but by anthropologists, archaeologists, linguists and geneticists, especially in the Pacific (Gibbons 1994, Connell, this volume). For people to settle on an island, they have to get there. Beyond the distance that people can swim, this requires some technology in raft, canoe or boat-building, propulsion by paddle or sail, and skills in navigation or orientation. Even with the technology, there is the psychological and cultural capacity to overcome the fear of the unknown and to want to explore new opportunities, and particularly to venture out of sight of land. There are enormous gaps in time between the migrations that were possible overland, those that could take place by navigating within sight of land, and those courageous adventurers who were ready to set out over the horizon in the hope of finding a new home.

With three-dimensional, geographic data sets of islands – such as the Enhanced Land Eleva-tion Data from the Space Shuttle Radar Topography Mission, available for latitudes 61°N to 60°S (NASA 2017b) – and information on the sea level at the time of human migrations that has been documented with archaeological evidence, it is possible to calculate where there were land bridges between areas that today are islands, and where someone on the highest point of an island or coast could see the next piece of land across the water, and thus know that, in set-ting out from the shore in a particular direction, a landfall was certain, a process that we can call *island hopping*.

According to paleo-anthropology, it seems that the pre-*homo sapiens* was trapped on the mainland of the world archipelago, and it took quite some time before mastering the ability to get out of the African peninsula cradle via the isthmus of Sinaï or the Bab el Mandeb narrow strait.

Before considering the stages of human saga over the marginal isolated lands, as we may called islands, one needs to know the equation (4) which gives the 'visibility range' (Lewis 1994) on our spherical planet:

$$(4) \ V_{km} = 3.57 \times Z_m^{0.5}$$

Where:

V_{km}:   distance of visibility in km
$Z_m^{0.5}$:   elevation in metres.

We are thus able to roughly determine theoretically which islands are visible from the highest or closest point on an adjacent island or continental coast. Two other factors need to be added to the calculation. The first is the notion of intervisibility: a point where two other specific points are within sight. It is relevant to the island hopping process since two islands may only be within eyesight simultaneously when a navigator is half-way in between. The second factor relates to the timing of the observation by island hoppers, with reference to the elevation to be considered when sea level has changed. In that case, the elevation as given today must take into account the relative variation of sea level in the past. To estimate the actual elevation during the last glacial maximum at 15,000 BP, the difference of elevation due to see level change (SLC) must be subtracted from the elevation as calculated today:

$$Z \ (15,000BP) = Z(Present) - SLC \ (15,000BP) \text{ with } SLC \ (15,000BP) = -110 \text{ metres.}$$

These notions can be applied to specific regions such as southeast Asia and Sunda Shelf, including the southern line of islands constituted by Sumatra, Java, Bali and Borneo (see Map 2.7). The migration of pre-*homo sapiens* was blocked there by critical water gaps along what is called now the Wallace line: the straits between Bali and Lombok (see Map 2.8) and between Borneo and the Philippines. Once the *homo sapiens* Asian peoples had developed the technology to cross the water, island hopping began about 50,000 BP and continued through 15,000 BP, covering all of the south and east lands of Wallacea, Sahul and as far as the Solomon Islands, involving the ancestors of the dark-skinned Aborigines, Papuans and Melanesians. This migration reached its limit beyond the Solomon Islands where there was a major 'island gap' before the Santa Cruz Archipelago. Island hopping stopped there because there was no more land visible on the horizon to encourage another step. The statement that only *homo sapiens* was able to 'island hop' has however been recently shaken. *Homo floresiensis*, a dwarf form of *homo erectus* dated from 12,000 BP, was discovered on the island of Flores (Brown et al. 2004) beyond the Bali-Lombok water gap. Scientists are thus considering various hypotheses that would explain how the Flores exemplar was possibly a side effect of *homo sapiens* migrations.

The next stage in the island quest of humankind remains undoubtedly the giant advance that led to the Polynesian migration across the Pacific. This came about because these people had the technology to master long sea voyages with everything necessary to colonise a new island (Dening 2004). What remains outstanding is their courage and cultural framework that enabled these peoples to hope that there would be another island beyond the one they were

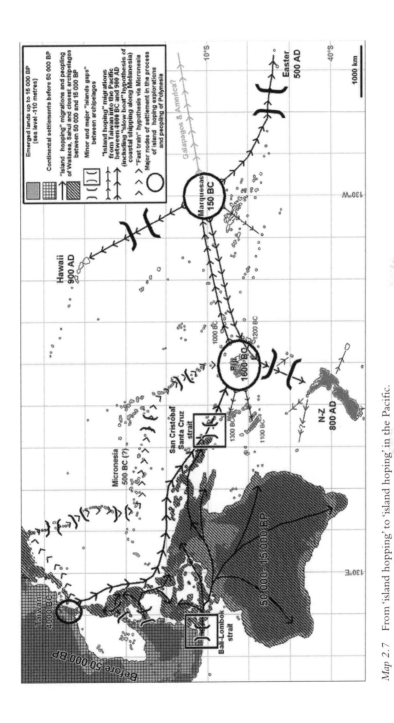

Map 2.7 From 'island hopping' to 'island hoping' in the Pacific.

Source: © Christian Depraetere.

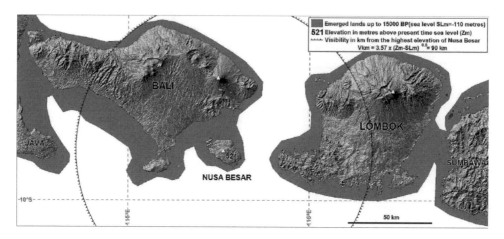

Map 2.8 The strait between Bali and Lombok islands, up to 15,000 BP.

Source: © Christian Depraetere.

leaving behind, and possibly also the confidence that they could sail back if they did not find land in time. This great migration from 4000 BC to AD 1200, can be understood as being fuelled by the hope of finding an island, or *island hoping*. These Polynesians reached all over, except the easternmost Pacific islands close to South and Central America. Two migration routes have been proposed for these people who probably originated in Taiwan: an 'express' route down through Micronesia to Polynesia from near Japan, thus bypassing Melanesia; and a 'slow' route along the fringes of Melanesia, with some interchange along the way with Melanesian peoples (see Map 2.7) (Gibbons 1994).

One geographic factor that may have encouraged this island hoping is the linear alignment of most island groups, such that the probability of finding another island was considerably greater along the axis of known islands. This of course requires a strong cultural sense of geography as well as orientation and navigation, which the Polynesian and Micronesian people acquired and maintained (Lewis 1994). The stick-charts of Micronesia illustrate the ability of their navigators to synthesise their knowledge into a reticular structure made of points and lines (see Figure 2.5).

The knowledge of seas and their currents, winds, stars and other signs acquired during their oceanic migrations, whether voluntary or not, gave them an empirical knowledge as useful as that of modern island cartography (see Map 2.9).

In general, Melanesian peoples were culturally more land based, developing many different local cultural and linguistic forms corresponding to the fragmented terrain of their high islands. Their island hoping, when they finally reached Fiji, Vanuatu and New Caledonia (1600 BC–1100 BC), was contemporaneous with the eastward Polynesian migrations and may well have been inspired and enabled by transfers of Polynesian technologies and cultural elements. Significantly, the Melanesian legends and myths of those archipelagos retain no trace of their island migrations (Bonnemaison 1986, Nunn 2003). The Melanesians became totally rooted in their new lands without looking back and with no expectation of discovering new lands beyond the horizon. New Zealand was only about 1,400 km from New Caledonia with Norfolk Island half way in between; but it was only discovered by the Maori around AD 800.

Beyond the Pacific, lie other examples of both island hopping and island hoping. One of the most dramatic examples is the settlement of Madagascar by people coming from Borneo after a

Figure 2.5 Stick–chart from the Marshall Islands made of shells representing islands and sticks indicating currents and lines of swell.

Source: Photo by Daderot, Wikimedia Commons. Exhibit in the Ethnological Museum, Berlin, Germany. Photo taken in the museum without restriction. https://commons.wikimedia.org/wiki/File:Stick_chart_used_for_naviga tion,_Marshall_Islands,_undated_-_Ethnological_Museum,_Berlin_-_DSC01161.JPG.

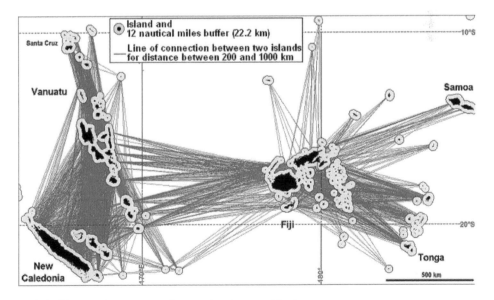

Map 2.9 The reticular structure of connections between islands.

Source: © Christian Depraetere.

long sea migration along the southern coast of Asia and the east coast of Africa. They probably crossed the Mozambique Strait via the Comoros Archipelago by island hoping before landing on Madagascar some 1,800 BP. Today, the Malagasy language is part of the Nusantarian (or Austronesian) family that includes Malay and Polynesian, but with significant lexical influences from African and even Indian origins.

The Mediterranean islands were only settled rather recently, despite most of them being visible from adjacent mainlands. Cyprus was colonised early: about 6000 BC, probably by Neolithic groups coming from Anatolia. Since it cannot be seen by coastal shipping from the continent, the Balearic archipelago was the last to be inhabited, around 3000 BC.

By heading westward from their native peninsula of Scandinavia, the Vikings initiated island hoping in the North Atlantic; their descendants are proud to recall these voyages in their famous sagas: Faroes in AD 800, Iceland in AD 860 and Greenland in AD 982. Their ultimate landing at the L'Anse aux Meadows site in Newfoundland around AD 1000 after sailing along the wooded coast of Labrador proved that they were able to reach the New World, without realising that it was much more than a large island.

With the Europeans, the process of exploring and peopling the uninhabited island margins of the known world changes in nature, with more mercantile, exploitative and political objectives rather than (or apart from) settlement (see Warrington and Milne, this volume). As an example, and with such protagonists as Henry the Navigator (AD 1394–1460), Portugal had a deliberate policy priority to explore both the African coast and the country's oceanic neighbourhood. While the discovery of Madeira and the Azores during the 14th century could be classified as an island-hoping process, the pace and nature of exploration start to change during the 15th century with an engagement with the Cape Verde Archipelago and the line of volcanic islands including Fernando Poo, Príncipe, São Tomé and Annobon in the Gulf of Guinea. It was no longer island hoping *per se*, but a systematic exploration for strategic and logistic reasons: an '*island claiming and naming*' exercise that lasted until the 19th century (see Map 2.10).

The apex of this new process between AD 1488 and 1522 saw outstanding navigators like Bartolomeu Dias, Christoforo Colombo (Columbus), Vasco de Gama and Fernão de Magalhães (Magellan) who pioneered the exploration of oceanic space on behalf of the Iberian powers. This led steadily to the discovery and exploitation of the last islands that remained out of human reach.

The combination in the long term of all these processes of island hopping, hoping, claiming and naming is responsible for the present intricate geopolitical situation that many islands find themselves in today.

The current island situation

The same geographic processes that have operated in the past on the evolution of island biodiversity and the mosaic of island cultures are now influencing island societies in new ways as they integrate into a globalising world. Air travel, radio, telephone, satellite connections, the internet and other new technologies of transportation and communications have reduced island isolation in some cases and increased it in others.

For example, considerable work has been done on island economies which, because of their small size, openness, limited diversity and lack of economies of scale, have been considered to be vulnerable to outside perturbations, and arguably cannot be competitive in world markets, except in specialised niches like and within tourism (Briguglio 1995). Others have considered

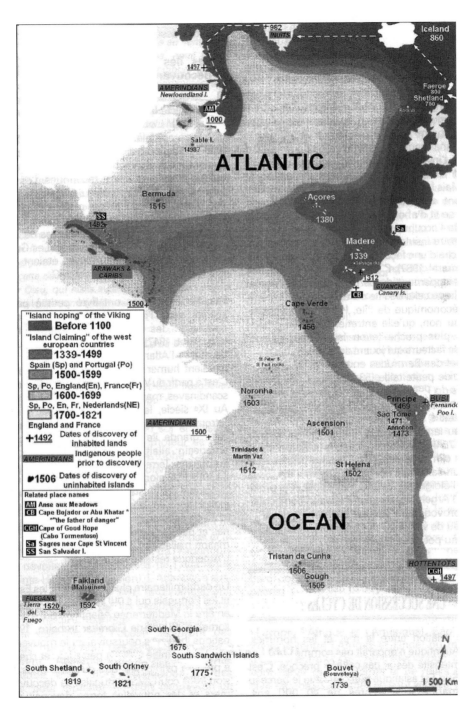

Map 2.10 European explorations and discoveries in the Atlantic Ocean.

Source: Based on Marrou (1998).

the inevitable openness of small island societies to external trade as a spur for strategic flexibility, economies of scope and the venturing into novel forms of entrepreneurship: financial services; citizenship by investment schemes; sale of country level internet domain names (Baldacchino 2015).

Culturally, the picture is more mixed. On the one hand, the wave of global culture in popular music, dress, films and television, and now the internet has swept over islands as it has the rest of the inhabited world. Yet, islands are as subject to the digital gap as other remote areas, and the cost of these services where there are no economies of scale is very high. Emigration to find work or wider opportunities has led to many island countries having larger populations overseas than at home (see Connell, this volume). Nevertheless, the special conditions that spawned distinct island cultures remain, and many island cultural traits have demonstrated surprising resilience, and even a renaissance. Island peoples continue to make a significant contribution towards global cultural diversity.

Politically, how do you weigh the influence of a nation of ten thousand or a hundred thousand people against one of a hundred million or a billion? Yet, population is not everything, as will be evident in the analysis below. Even politically, small island developing states (SIDS) have learned that there is power in numbers, and in fora where each nation has one vote, SIDS can represent a significant voting block of some 40 sovereign states. Starting from regional groupings of island states and territories in the Pacific and Caribbean, SIDS have built a range of global political processes to define their particular situation and identify policy responses. Islands were included as a program area in Agenda 21, the action plan for environment and development adopted at the Rio Earth Summit in 1992 (UN 1992). This led to the Conference on Sustainable Development of Small Island Developing States in Barbados in 1994, which adopted the Barbados Programme of Action (UN 1994), assured the regular inclusion of SIDS issues in subsequent international negotiations, and ushered a follow-up international meeting in Mauritius ten years later, in January 2005 (UN 2005), and a third in Samoa in 2014 (UN 2014).

Moreover, clearly aware of their limitations on the world stage, small island states joined forces in 1991 to set up the Alliance of Small Island States (AOSIS), now with 39 member countries (see Table 2.5) and five subnational island jurisdictions (SNIJs) as observers, giving them a stronger and unified international voice. The Alliance includes four low-lying coastal states which are mostly continental (Belize, Guinea Bissau, Guyana and Surinam). AOSIS adopts common positions on key international issues such as climate change and is able to defend them with its votes at the United Nations and the meetings of the Conferences of the Parties (COP) as part of the United Nations Framework Convention on Climate Change (UNFCCC). In these and similar ways, AOSIS demonstrates strength in numbers that can help to counterbalance the power of major countries at both global and regional scales.

While independent island states have found a political voice at an international level, others have persevered as SNIJs, remaining part of larger states or distant fragments of former empires (Baldacchino and Milne 2009). Thus, while transportation and communications links have reduced the isolation of many island communities, their marginality is expressed today in more economic and political terms.

The present geopolitical situation of islands can be mapped (for SNIJs globally in Map 2.11; and, for the Caribbean only, in Map 2.12). A distinction should be made between island states and continental states, as they do not have the same views on oceanic issues, both in terms of scale and necessity. The UK is an exception, because it is an island state yet often acts like a continental power. The fact that many islands are the dependent possessions in one way or another of continental countries is a major reality of ocean geopolitics. In some cases, this is a simple extension

Table 2.5 The UN's Small Island Developing States (SIDS).

AIMS (Atlantic, Indian Ocean and South China Sea)	Pacific region	Caribbean region
1. Cape Verde	9. Cook Islands	24. Antigua and Barbuda
2. Comoros	10. Fiji	25. Bahamas
3. Guinea-Bissau	11. Kiribati	26. Barbados
4. Maldives	12. Marshall Islands	27. Belize
5. Mauritius	13. Micronesia, Federated States of	28. Cuba
6. São Tomé and Principe	14. Nauru	29. Dominica
7. Seychelles	15. Niue	30. Dominican Republic
8. Singapore	16. Palau	31. Grenada
	17. Papua New Guinea	32. Guyana
	18. Samoa	33. Haiti
	19. Solomon Islands	34. Jamaica
	20. Timor-Leste	35. Saint Kitts and Nevis
	21. Tonga	36. Saint Lucia
	22. Tuvalu	37. Saint Vincent and the Grenadines
	23. Vanuatu	38. Suriname
		39. Trinidad and Tobago

of their coastal waters to nearby islands, as in Australia, India, Chile and Portugal. In other cases, these are remote island territories far away and without geographic continuity, often the heritage of the colonial past of island claiming. Three countries control overseas island territories in many oceans: France, the UK and USA. Some islands in the Caribbean are still attached to the Netherlands, the Faroes and Greenland are still linked to Denmark while Norway has Bouvet and Peter Islands near Antarctica. One consequence of this is that the truly island states have neighbouring SNIJs that are, to some degree, controlled by global or local continental countries. This mix of small island developing states and world powers (via SNIJs) makes regional policy-making difficult, as exemplified in the Caribbean, Pacific and Indian Oceans. All the near-Antarctic islands below 35°S are under the control of European countries or former dominions of the UK (Australia, New Zealand and South Africa). This overwhelming extension of continental national sovereignty over islands at the world scale has direct consequences for access to oceanic economic resources as they are defined by the 200 nautical mile (370 km) exclusive economic zone.

The political interest of states in island issues is related to the importance of islands in their national territory. For most countries, their islands are relatively insignificant in terms of area, population or economic activities, and are usually marginalised in political processes. However they take on added significance when they extend national sovereignty over coastal resources or undersea oil and gas reserves, or serve as strategic outposts. Consider the tension over a bunch of rocks in the East China Sea claimed by both China (and Taiwan) and Japan (Baldacchino 2017); and the even more complex dispute over the South China Sea and its various shoals, islands and reefs between all the regional players (Rolf and Agnew 2016).

While islands may be at the small end of those geographic entities on our planet of human significance, they do help us to understand our problems and challenges at many other scales. Islands do symbolise that balance of isolated independence and integration into larger systems that are essential characteristics of all physical existence, whether geographically or metaphorically. Our planet too is an island in space, and we may have yet to learn to live within its limits.

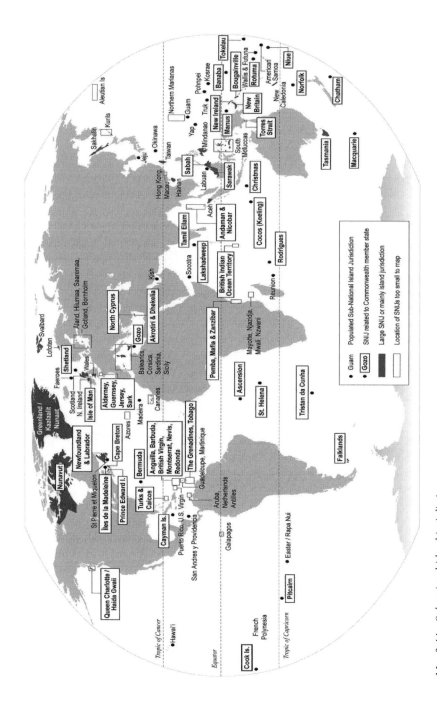

Map 2.11 Subnational island jurisdictions.

Source: © Godfrey Baldacchino. Reproduced with permission.

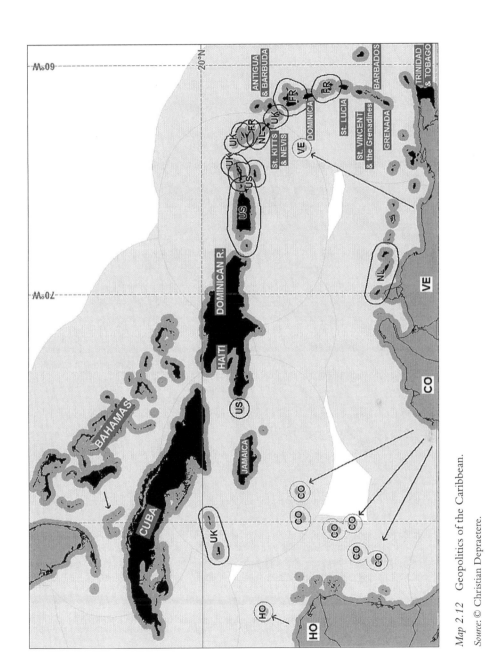

Map 2.12 Geopolitics of the Caribbean.

Source: © Christian Depraetere.

References

Baldacchino, G. (2015) *Entrepreneurship in small island states and territories.* London, Routledge.

Baldacchino, G. (2017) *Solution protocols to festering island disputes: 'win-win' solutions for the Diaoyu/Senkaku islands.* London, Routledge.

Baldacchino, G., and Milne, D. (eds.) (2009) *The case for non-sovereignty: lessons from sub-national island jurisdictions.* London, Routledge.

Bonnemaison, J. (1986) *Les fondements d'une identité, territoire, histoire et société dans l'archipel de Vanuatu (Mélanésie)*, Essai de géographie culturelle, Livre 1: *L'arbre et la Pirogue*, Paris, Editions de l'ORSTOM, Collection Travaux et Documents, No. 201.

Briguglio, L. (1995) Small island developing states and their economic vulnerabilities. *World Development*, 23(9), 1615–1632.

Brown, P., Sutikna, T., Morwood, M.J., and Soejono, R.P. (2004) A new small-bodied hominid from the Late Pleistocene of Flores, Indonesia. *Nature*, 431, 1055–1061.

Burrough, P.A. (1986) *Principles of geographical information systems for land resources assessment.* Oxford, Clarendon Press.

Dening, G. (2004) *Beach crossings: voyaging across times, cultures and self.* Melbourne, Australia, Miegunyah Press.

Depraetere, C. (1991a) Le phénomène insulaire à l'échelle du globe: tailles, hiérarchies et formes des îles océanes. *L'Espace Géographique*, 2, 126–134.

Depraetere, C. (1991b) NISSOLOG: Base de Données des Îles de plus de 100 km². Presented at XVII Pacific Science Congress (Pacific Science Association). Honolulu, Hawai'i. MSDOS Computer programme and unpublished manuscript, Centre de Montpellier, France, Editions de l'ORSTOM.

DK Atlas (2004) *The ultimate pocket book of the world atlas and fact-file.* London, Dorling Kindersley, Cambridge International Reference on Current Affairs (CIRCA).

Doumenge, F. (1984) Unity and diversity of natural features of tropical islands. In *Nature and people in tropical islands: reflections and examples.* [In French]. Bordeaux, France, CRET Bordeaux III and CEGIT, pp. 9–24.

Doumenge, F. (1989) Basic criteria for estimating the viability of small island states. In J. Kaminarides, L. Briguglio and J.N. Hoogendonk (eds.), *The economic development of small countries: problems, strategies, policies*, Delft, The Netherlands, Eburon, pp. 39–56.

Evans, I., and McClean, C. (1995) The land surface is not unifractal: variograms, cirque scale and allometry. *Zeitschrift für Geomorphologie*, Supplement, 101, 127–147.

Gibbons, A. (1994) Genes point to a new identity for Pacific pioneers. *Science*, 263(5143), 32–33.

Lewis, D. (1994) *We, the navigators: the ancient art of land finding in the Pacific*, 2nd edn. Honolulu, HI, University of Hawai'i Press.

MacArthur, R.H., and Wilson, E.O. (1967) *The theory of island biogeography.* Princeton, NJ, Princeton University Press.

Mandelbrot, B. (1967) How long is the coast of Britain? Statistical self-similarity and fractional dimension. *Science*, 155(3775), 636–638.

Mandelbrot, B. (1975) Stochastic models for the Earth's relief, the shape and the fractal dimension of the coastlines, and the 'number–area' rule for islands. *Proceedings of the National Academy of Sciences of the USA*, 72, 3825–3828.

Mandelbrot, B. (1982) *The fractal geometry of nature.* New York, W.H. Freeman & Co.

Marrou, L. (1998) Les Îles Atlantiques Océaniques. In A. Miossec and L. Marrou (eds.), *L'Atlantique: un regard géographique*, Paris, UGI-CNFG, pp. 281–294.

McCall, G. (1994) Nissology: a proposal for consideration. *Journal of the Pacific Society*, 17(1), 1–14.

McCall, G. (1996) Clearing confusion in a disembedded world: the case for nissology. *Geographische Zeitschrift*, 84(2), 74–85.

NASA (2017a) WORLD WIND for global LANDSAT imagery. Available from: http://worldwind.arc.nasa.gov/.

NASA (2017b) Enhanced Shuttle Land Elevation Data available from 61°N to 60°S. Available from: www2.jpl.nasa.gov/srtm/.

NOAA (2017) NOAA/NGDC ETOPO2 bathymetry data. National Centers for Environmental Information. Available from: www.ngdc.noaa.gov/mgg/image/2minrelief.html.

Nunn, P.D. (2003) Fished up or thrown down: the geography of Pacific Island origin myths. *Annals of the Association of American Geographers*, 93(2), 350–364.

Rolf, S., and Agnew, J. (2016) Sovereignty regimes in the South China Sea: assessing contemporary Sino-US relations. *Eurasian Geography and Economics*, 57(2), 249–273.

UN (1992) *Agenda 21*. New York, United Nations. Available from: www.un.org/esa/sustdev/documents/agenda21/.

UN (1994) *Programme of action for sustainable development of small island developing states*. New York, United Nations. Available from: www.un.org/documents/ga/conf167/aconf167-9.htm.

UN (2005) *Report of International meeting to review implementation of programme of action for sustainable development of small island developing states*. Port Louis, Mauritius, January. New York, United Nations. Available from: http://daccessdds.un.org/doc/UNDOC/GEN/N05/237/16/PDF/N0523716.pdf?OpenElement.

UN (2014) *SIDS Accelerated Modalities of Action (SAMOA) Pathway A/RES/69/15*. New York, United Nations. Available from: www.un.org/ga/search/view_doc.asp?symbol=A/RES/69/15&Lang=E.

Wessel, P., and Smith, W.H.F. (1996) A global self-consistent, hierarchical, high-resolution shoreline database. *Journal of Geophysical Research*, 101(B4), 8741–8743. Available from: www.soest.hawaii.edu/wessel/gshhg/Wessel+Smith_1996_JGR.pdf.

3

ORIGINS AND ENVIRONMENTS

Patrick D. Nunn and Roselyn Kumar

Introduction: island origins

Islands in different parts of the world often show similarities in origin although conversely, islands that appear superficially similar may sometimes have quite different origins. Commonalities of origin and physical development occur mostly among the community of 'oceanic islands', those islands that originated within the ocean basins (Nunn 1994). Older 'continental islands' are parts of the continents that have become islands through the uneven submergence of continental margins.

Most oceanic islands develop either along convergent plate boundaries or in intraplate (mid-plate) locations. Convergent plate boundaries are places where one slab (or 'plate') of oceanic crust (or 'lithosphere') is thrust beneath another. The downgoing plate is eventually pushed so far into the Earth that it melts, producing magma that sometimes finds its way back to the ocean floor where it erupts and may produce a volcanic island. Intraplate locations ('hotspots') are sites where the Earth's crust is uncommonly thin and where liquid rock from the underlying asthenosphere may force its way to the surface to form a volcanic island.

Origins cannot always be readily determined from examination of modern, above-sea islands. Often, the key to an island's origin lies buried deep beneath a thick cover of younger rocks, sometimes far below sea level, so the use of models of island genesis is common. For example, many atoll islands (islands formed on ring or atoll reefs) are made solely from superficial material that accumulated within the past few thousand years yet rest on ancient coral reef that, in turn, rises upwards from the underwater flanks of long-submerged volcanic islands; it is from the geochemical character of these volcanic rocks that we can learn about the ancient origins of particular atolls.

This section discusses the origins of islands by appearance and composition, beginning with nascent ocean-floor islands – from which all oceanic islands develop – through mature above-sea oceanic islands and older sunken islands. These are not primarily age distinctions but developmental stages that may not be attained by every oceanic island.

Young under-sea oceanic islands

All oceanic islands form as a result of volcanic activity on the deep ocean floor. Much ocean-floor volcanism is non-threatening and undramatic, often associated with upwelling of magma (liquid rock) along a fissure that has extended downwards to tap an underground magma source. As the early eruptions continue and the amount of lava produced increases, so parts of the fissure become blocked, and eruptions become concentrated at particular points. It is these point eruptions that then allow the growth of small volcanic edifices which may one day form giant oceanic islands.

The weight of the ocean water overlying eruptions on the ocean floor renders these as non-explosive events. The principal material produced is pillow lava, so called because the magma is forced out of the volcano in discrete blobs, the outside of which immediately solidifies upon coming into contact with cold ocean water. The inside remains liquid for much longer while the blob rolls down the flank of the volcano, coming to rest at its foot and forming one of many 'pillows'.

Yet, as an undersea volcano grows upwards, its crestal vent may eventually reach a point about 600 m beneath the ocean surface where there is no longer sufficient overlying water to suppress explosive eruptions. As the volcano grows above this level and into shallower water, eruptions become explosive – and spectacularly visible above the ocean surface (see Figure 3.1). One reason why many shallow-water eruptions are explosive has to do with the reaction between liquid rock (heated perhaps to 1,200°C), and cold ocean water, which leads to the production of fragmental volcanic material. These 'volcaniclastics' commonly drape the core (made from pillow lavas and intrusive igneous rocks) of undersea volcanic islands, but may also float to the ocean surface to form floating pumice mats, even an island.

Figure 3.1   January 2005: An eruption of the Kavachi Volcano, located just below the ocean surface in the
Solomon Islands.

Source: © Simon Albert. Reproduced with permission.

Islands made from newly-erupted volcaniclastic material may disappear through wave erosion once the eruption ends. Such 'jack-in-the-box' islands appear and disappear regularly in parts of the Southwest Pacific, such as Solomon Islands and Tonga, as well as in the Mediterranean Sea where Graham Island (Ferdinandea) appeared above the ocean surface southwest of Sicily, Italy, for six months in 1831–1832. For such an island to persist above the ocean surface, as has Surtsey off the south coast of Iceland in the North Atlantic, it is necessary for the eruptive vent(s) to be cut off from ocean water so that lavas will be produced instead of fragmental material (see Figure 3.2).

Mature above-sea oceanic islands

A newly-emerged volcanic island will commonly betray its origin, although various processes soon conspire to begin to disguise this. Denudation – the physical wearing away of the land – can erase the distinctive form of a volcanic island, even to the extent of reversing the original topography. Successive drowning and emergence associated with sea-level changes or long-term tectonic movements can also help disguise the origin of an oceanic island.

In the coral seas, generally those where ocean-surface waters remain between 20–27°C all year, the growth of coral reef around, sometimes even over, a subsiding volcanic island can transform it into an atoll island, one where the presence of a largely-submerged island edifice is manifested only by a ring of coral reef. Should an atoll island emerge as a result of experiencing uplift, it will then appear as a high limestone island (see Figure 3.3).

Such transformations are relatively common in the most tectonically-active areas of the ocean basins, particularly near convergent plate boundaries. The island of Jamaica in the Caribbean is an oceanic island, despite being unusually large (11,500 km²). Slow steady sinking of Jamaica beginning around 55 million years ago led to deposition of the 'Yellow Limestone' around its ancient volcanic core. The Yellow Limestone was overlain by the coral-reef dominated 'White Limestone' as subsidence continued, the entire island being submerged 40–25 million years ago. As a result of uplift associated with nearby plate convergence, the island subsequently re-emerged (Robinson 1994).

Most above-sea oceanic islands will begin to subside once volcanic activity ceases. Subsidence may occur because the island is being moved on an ocean plate from an area of shallower ocean (in which volcanic activity is taking place) to an area of deeper ocean (where it is not). Thus islands that move away from a hotspot generally become both smaller in area and lower in altitude, as their bases are carried into deeper water. The northwestern islands in the Hawai'i group, for example, are mostly atolls or low volcanic islands that were perhaps once as large as the islands of Hawai'i (the 'Big Island') and O'ahu in the southeast of the group where the hotspot is located (see Figure 3.4).

Ancient continental islands

Islands of continental origin are generally larger and almost always older than oceanic islands. They may rise from continental shelves, as do the islands of Sardinia (Italy) and Tasmania (Australia), or from isolated pieces of continental lithosphere that have become detached from the parent ones, such as Madagascar and New Caledonia. The origins of these islands are generally as diverse and as complex as the continents themselves. Many well-studied structural alignments on continents are also exposed on offshore islands. For example, ancient Mediterranean islands (such as Minorca, Spain) appeared above the ocean surface at the same time and for the same

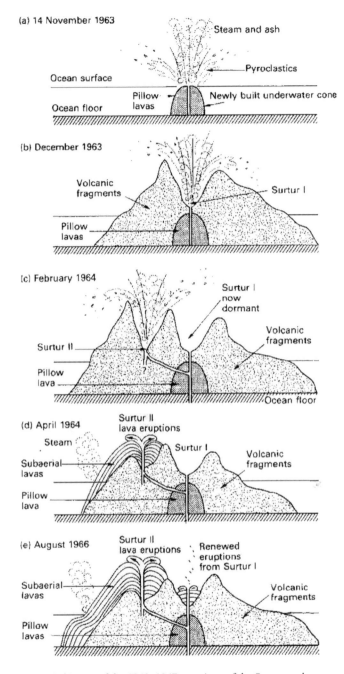

(a) 14 November 1963

Steam and ash

Pyroclastics

Ocean surface

Pillow lavas

Ocean floor

Newly built underwater cone

(b) December 1963

Volcanic fragments

Surtur I

Pillow lavas

(c) February 1964

Surtur I now dormant

Volcanic fragments

Surtur II

Pillow lava

Ocean floor

(d) April 1964

Surtur II lava eruptions

Steam

Surtur I

Volcanic fragments

Subaerial lavas

Pillow lava

(e) August 1966

Surtur II lava eruptions

Renewed eruptions from Surtur I

Subaerial lavas

Volcanic fragments

Pillow lavas

Figure 3.2 Diagrammatic history of the 1963–1967 eruptions of the Surtsey volcanoes.

Source: Nunn (1994), used with permission. © Patrick Nunn.

A – The earliest eruptions visible above the ocean surface were explosive owing to mixing of ocean water and magma.

B – Pyroclastic and ash eruptions from the main eruptive centre of Surtur I built a cone of unconsolidated volcanic fragments.

C – Eruptions from Surtur I ceased and a new eruptive centre (Surtur II) came into being. Since the ocean still had access to this centre, mixing with the upwelling magma occurred and eruptions continued to be of an explosive character, resulting in the accumulation of unconsolidated fragmental material easily eroded by the ocean.

D – Finally, the Surtur II centre became isolated from the sea and lava eruptions replaced those of pyroclastic materials. Lavas armoured the surface of the existing cone, rendering it significantly less vulnerable to marine erosion.

E – Eruptions renewed in the Surtur I area and lavas began to cover most parts of Surtsey. Lavas extruded on land that entered the sea gave rise to steam clouds. No explosive activity occurred because the surface of the lavas was already cooled.

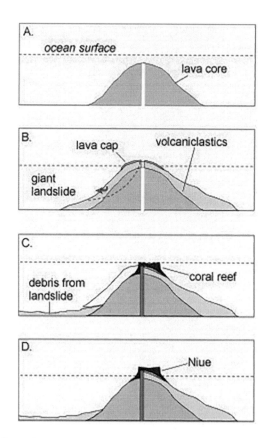

Figure 3.3   Stages in the evolution of Niue Island.

Source: © Patrick Nunn.

A – The earliest Niue Volcano was one which rose from the deep-ocean floor and from which lava was erupted.

B – When the summit of the Niue Volcano entered the hydro-explosive zone, lava eruptions were replaced by the eruption of volcaniclastics until the summit of the volcano grew above the ocean surface when lava eruptions resumed. The giant landslide in the south and west occurred around this time.

C – The debris from the landslide extended southwest. As the volcano subsided below sea level, an atoll reef became established at the ocean surface. This condition was maintained for around three million years.

D – About 2.3 million years ago, the island began to be uplifted as it started to ascend a crustal flexure to the west. Niue became a high limestone island now reaching 70 m above sea level.

reasons as the Alps, while the island of Trinidad exhibits the same structures as the adjacent parts of northern South America.

Some continental islands that subsequently found themselves close to unusually active convergent plate boundaries have become draped with upthrust pieces of the ocean floor called ophiolites. Well-studied island ophiolites are the Troodos Massif of Cyprus (Robertson 2002) and much of the main island (La Grande Terre) of New Caledonia (Gautier et al. 2016).

Some continental islands show signs of having been islands – and therefore subject to the same processes as older oceanic islands – for a considerable time. Only when Australia collided with the Banda Arc about three million years ago did the island of Timor begin emerging, a process that can be calibrated by the staircases of fossil coral reefs found along its coasts (Chappell and Veeh 1978).

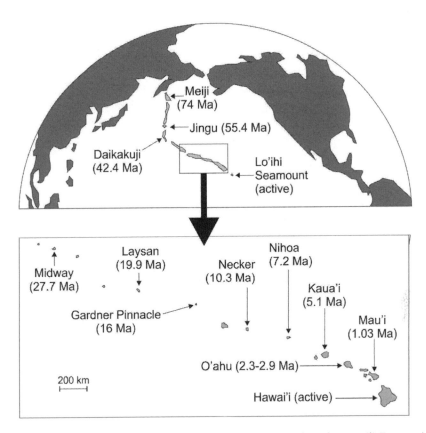

Figure 3.4   Ages of the most recent dated eruptions of shield volcanoes along the Hawai'i-Emperor island-guyot chain (dates in millions of years ago: Ma). The upper map shows the location of the Hawaiian Ridge within the North Pacific; the lower map shows the geography of the younger, largely emerged part of the island chain.

Source: © Patrick Nunn.

Island landscapes

Like other landscapes, island landscapes vary in character primarily because of geology and climate. Amongst the principal geological controls are age, rock type (lithology), structure and tectonic history. Traditionally the most important climate controls on landscape were regarded as temperature and precipitation but, as geomorphic studies of islands have increasingly shown, climate history, particularly climate extremes, has also had profound influences on the character of modern island landscapes (Huggett 1991). Half a century ago, most scientists took a simple and uniform view of climate and discounted its past variations when trying to explain the evolution of particular island landscapes; whereas today, the evidence of past climates is widespread, ranging from former glaciation (such as u-shaped valleys) on islands like Arran in Scotland (MacDonald and Herriot 1983) to glacial-period wetness on tropical Hawai'i Island (Sheldon 2006). Yet, it is the variations attributable to lithology and evolutionary complexity that are most visible in many modern island landscapes, for which reason they form the basis of the discussion in this section.

Volcanic island landscapes

The nature of landscapes of active volcanoes varies, depending on that of the eruptive materials. Island volcanoes built solely from viscous lava often form dome-shaped edifices (cumulo-domes) while those built from both lavas and fragmental material (stratovolcanoes) are usually higher and have steeper, characteristically concave, flanks. The volcanoes on the Caribbean island of Montserrat, that include the Soufrière Hills Volcano which erupted spectacularly in 1997, are stratovolcanoes, as are most of the active volcanoes in the Philippines including the highly-active Mayon on Luzon Island. Mt Egmont on the North Island of New Zealand is a mixture, its base being domed and representing an earlier phase of eruption to that involving mostly fragmental material which built the stratovolcano above. The largest volcanoes on Earth commonly form oceanic islands, such as those of the Hawai'i island chain, and are known as shield volcanoes, from the shape of the edifice they form, built largely from basalt lava.

Near the end of the active life of an island volcano, a caldera may form, either from explosions tearing out the heart of the old volcano or a collapse of the summit into an emptied magma chamber below. The landscape of the island of Nisyros (Aegean Sea) is dominated by a 15 km^2 caldera. On Lihir (Niolam) Island in Papua New Guinea, the formation of the Luise Caldera exposed a huge gold deposit (Corbett et al. 2001).

Once a volcano ceases activity, its form changes. Around symmetrical volcanoes, a radial drainage pattern will normally develop and, through time (as some streams are 'captured' by their neighbours), relict pieces of the flanks of the original volcano (planezes) will be isolated from erosion and may persist for much longer than the rest; planezes on the island of St Helena (South Atlantic), that last erupted some seven million years ago, are used as pastures on this otherwise deeply-eroded volcanic island. Most oceanic volcanoes subside when activity wanes or ceases, resulting in drowning of island coasts and the formation of bays where there were once valleys; stellate (star-shaped) islands like Ono (southern Fiji) may form (Nunn 1994).

Although an island comprising one or more extinct volcanoes may no longer pose a threat to its inhabitants from eruption, the steep-sided character of some oceanic volcanoes means that they are also notoriously unstable. The Hawai'i island chain, for example, is one of the steepest-sided structures on the Earth's surface and has experienced periodic catastrophic flank collapses. One of these, perhaps around 105,000 years ago, may have created a wave more than 300 m high that washed back over the Hawaiian islands of Lana'i and Moloka'i, leaving behind gravels containing innumerable coral fragments (Moore et al. 1994). The same wave may have crossed the Pacific, driving up on every coast it encountered (Young and Bryant 1992).

Collapse of volcanic island flanks can occur with both active and extinct volcanoes. For example, the high volcanic islands of the Canary Islands have experienced enormous flank collapses many times in the past, some of which may have been caused only by gravity, others of which may have been triggered by the intrusion of magma deep within the volcanic edifice causing its flank to bulge and eventually collapse (Carracedo 1994). Flank collapses of island edifices in the Pacific have caused abrupt disappearances of entire islands (Nunn 2009).

Limestone island landscapes

Limestone is permeable, meaning that the surface landscapes of limestone islands are comparatively slow to evolve and may continue to represent the form of the island when it emerged above the ocean surface long after this event occurred. An example is the island of Niue (South Pacific), a coral atoll before it began emerging about 600,000 years ago. Niue now lies 70 m above the ocean surface; yet, the former lagoon and the former ring reef (named the Mutalau

Reef) are clearly visible in the modern landscape (Nunn and Britton 2004). The fringes of emerging limestone islands in warmer ocean–surface waters are often marked by staircases of fossil coral reefs; those on the island of Choiseul (Solomon Islands) extend 800 m above sea level (Stoddart 1969), while those on Halmahera (Indonesia) extend to 1,000 m (Hall et al. 1988).

The surfaces of many limestone islands are low relief; areas of lower ground (sinkholes or dolines) mark the places into which surface water is concentrated before moving down below the surface. In certain limestones, funnel-like sinkholes may develop close together and extend downwards several metres, giving rise to particular types of karst landscape, such as the cockpit country of Jamaica (Fleurant et al. 2008). Long before humans arrived, the remote limestone island of Nauru (central Pacific) was home to millions of seabirds whose phosphate-rich excrement filled the sinkholes. Mining of Nauru for phosphate during the 20th century has re-exposed the original karst (pit-and-pinnacle) landscape (Weeramantry 1992; see Figure 3.5).

The subterranean parts of limestone islands may change more rapidly than their surfaces. Rain water percolates down from the ground surface and forms a freshwater lens that rests on top of limestone saturated with seawater. The unsaturated (vadose) zone above the freshwater lens is commonly riddled with cracks along which water trickles downwards. Sometimes, the cracks grow into narrow elongate and steeply-plunging caves: the 'blue holes' of the Bahamas and Caicos islands are vadose-zone caves formed during the last Ice Age and drowned by subsequent sea-level rise (Mylroie et al. 1995).

Within limestone islands, the surface of the freshwater (Ghyben–Herzberg) lens – the water table – is commonly dome-shaped, meaning that water which reaches it through the vadose zone usually then trickles through the limestone at an angle. The concentration of water flow

Figure 3.5 Contemporary pit-and-pinnacle landscape on Nauru island, west Pacific, being created by phosphate extraction.

Source: Photo posted by CdaMVvWgS, Wikipedia Commons, https://commons.wikimedia.org/wiki/Nauru#/media/File:Nauru-phosphatefields.jpg.

down the dome of limestone-island water tables means that these are places where erosion (through solution and roof collapse) of the limestone is concentrated. Large water-table (or epi-phreatic) caves often form in such places, leading downslope from the centre of an island to its coast, typically with a river, representing water flowing at the surface of the water table, running through it. Tatuba Cave in the interior of Viti Levu Island (Fiji) comprises a lower active part containing a river and several dry caves above, a result of the area's uplift (Nunn 1998). Simi-lar emerged caves are found along the coast of Isla de Mona (Puerto Rico) where most have developed along the contact between the principal reef-limestone/dolomite contact. This is an example of a geological structure influencing processes of cave formation (Frank et al. 1998).

Sometimes, miniature archipelagoes are created when karst landscapes are drowned, as in the case of southern Vava'u Island (Tonga) and Phang Nga Bay on the Krabi coast of Thailand (Nunn 1998, Harper 1999).

About 66 per cent of the continental Caribbean island of Cuba (111,000 km²) is covered by limestone and its landscape reflects this lithology, its great age, and its tropical climate (Itturralde-Vinent 1997). Most tobacco fields in Pinar del Río province in Cuba are in areas dominated by collapsed limestone separated by isolated remnants of the original surface, termed *mogotes* (residual limestone hills).

Composite island landscapes

Many islands cannot be readily classified as either volcanic and limestone. Many such composite islands have distinctive landscapes, at least in places, on account of their varied lithologies.

Larger islands of this kind, sometimes of continental origin, typically have an older core sur-rounded by rocks of younger age, often formed only after the landmass became an island. Typical of these is Jamaica (discussed above), Lesbos in the Aegean, and New Guinea, many fringing parts of which are formed by emerged coral reefs (Löffler 1977). In a reverse of this situation, the sedimentary core of the Shetland Islands (Scotland) that dates back to the Cambrian Era (590–500 million years ago), has become fringed by younger (Tertiary) lavas and intrusive igne-ous rocks (Stoker et al. 1993).

Among the smaller composite islands in ocean basins are the *makatea* islands that comprise a volcanic centre fringed by uplands made of emerged coral reefs; several examples are found in the southern Cook Islands of the Central Pacific (Nunn 1994, Stoddart et al. 1985).

Simple island landscapes

There are many, usually smaller islands that have simple landscapes on account of their compara-tively homogenous composition. Such islands can be divided into young coral islands, which owe their existence largely to the growth of living reef, and others, typically those formed from detrital material introduced to nearshore areas by large rivers. Other such islands include ice islands and islands made from floating vegetation and/or pumice which may have been impor-tant in the dispersal of certain organisms from island to island, as well as lake and river islands discussed below (Stehli and Webb 1985, Van Duzer 2004).

(a) Young coral island landscapes

Islands exist on many broad coral reef flats and are composed primarily of reef detritus driven onto them by large (storm) waves and then concentrated in particular areas by wave action. On Funafuti Atoll (Tuvalu, central Pacific), a storm surge associated with Tropical Cyclone Bebe in

October 1972 led to the creation of a 'rubble bank' along the edge of the reef off the island's east coast. Over the next few years, this rubble bank was moved slowly landward by wave action and became incorporated into the main island (Baines and McLean 1976). Analysis of the geology of other reef islands in Tuvalu show that they formed from the successive accretion and distribution of rubble banks of varying grade and size. On smaller reef platforms, the central lagoon-depression enclosed by these islands can eventually become infilled with reef detritus (McLean and Hosking 1991).

Newly formed coral islands of this kind are known as cays and are transient islands that often migrate across reef flats, and even are sometimes removed from them in their entirety. Sometimes, cays persist long enough for conglomerates (such as beachrock) to form just beneath their surfaces, and these help armour the cays and protect them from erosion; well-armoured cays that have persisted for centuries rather than years are known as *motu*. Most inhabited coral islands in countries like Kiribati and the Marshall Islands in the western Pacific are *motu*, and research has focused increasingly on their structure and how it might be influenced by sea-level rise (Nunn 1994, Dickinson 1999).

(b) Non-coral island landscapes

Denudation of continents and larger islands is manifested by the suspended sediment load of large rivers. When these rivers reach the sea, much of this sediment is deposited on the ocean floor which may then shoal to produce islands in places. At a later stage these islands often become incorporated into the mainland as they are subsumed by the prograding shoreline of the river delta.

In some places, such river (or estuarine) islands may become colonised by dense mangrove forests that stabilise them and help them endure. While not always desirable places for humans to live, mangrove islands are places where other organisms often thrive, and the destruction of such habitats is frequently lamented. Some of the 54 islands in the Sundarbans, the 10,000 km² mangrove forest at the mouth of the Brahmaputra-Meghna-Ganges in India and Bangladesh, are the objects of conflict between local inhabitants and conservationists (Ghosh 2015). Owing to their isolation yet relative proximity to the mainland, such islands may also be important places of refuge or retreat; the 6th-century Irish saint Senan founded a succession of monasteries on river-mouth islands, including one on Scattery Island in the Shannon Estuary.

Controls on long-term environmental evolution of islands

The island environments that we see today manifest the subtle interplay of nature and humans and it is consequently difficult to generalise about their long-term evolution. Yet, understanding environmental evolution is a necessary precondition to suitable and successful environmental management, increasingly a priority on many islands (Nunn 2004). The four critical controls on long-term environmental evolution of most islands are geology, climate (including changes in sea level), and extreme or rapid events. For those islands that have been inhabited by people for a significant length of time, pre-modern human impacts are often also a significant contributor to environmental development. Each of these four distinct controls is discussed separately below.

(a) Geological controls

The key element in understanding the evolution of island environments is time, specifically how much has elapsed since that island appeared. Geological history can provide an answer to that, and also to the various earth-surface processes that have moulded island environments. For example, many island environments have been affected by volcanism, a few have been pushed upwards from beneath the ocean without a hint of volcanic activity, while others have alternated between periods in which each of these processes dominated. For instance, a former volcanic island may sink beneath the ocean surface, becomes covered with reef limestone, and then emerges: the island of Vava'u (Tonga, South Pacific) is composed of emerged reef limestone that covers an ancient caldera (Cunningham and Anscombe 1985).

Many volcanic rocks are easily moulded by natural denudational processes – the moonscape-like surface of Iceland is an example – whereas most limestones are more resistant and, on account of their permeability, comparatively unaffected by processes of surface denudation. Most volcanic islands in Samoa (South Pacific) have deep-cut valleys, a consequence of the low-resistant lavas from which many formed. The massive peridotites (olivine-rich igneous rocks) that cover 30 per cent of the main island (La Grande Terre) in New Caledonia have produced ultrabasic (very low silica) soils containing elements that are toxic to many plants (Gautier et al. 2016).

The geological structure of some islands exercises the dominant control on their landscapes and the ways they have evolved. The island of Iceland (North Atlantic) is one of the very few places on the Earth's surface where a mid-ocean ridge (divergent plate boundary) emerges above the ocean surface, something manifested by a 150-km wide rift valley on the flanks of which there is considerable volcanic activity (Bott 1985). Ancient fold belts around the eastern Mediterranean margins extend onto islands offshore; for example, the islands of Crete and Rhodes are parts of the Hellenic Arc that is exposed conspicuously in mainland Greece (Barka and Reilinger 1997).

Many islands, particularly oceanic islands, lie in tectonically-active locations, and their environments manifest the long-term effects of tectonic processes. Owing to their location in a compressive tectonic zone associated with plate convergence, Miocene coral reefs on the large island of New Guinea emerged and became covered with impermeable volcanic cap rocks, creating hydrocarbon reserves (Hill et al. 1996). On a smaller scale, the island of Moala in southeast Fiji is bisected by a rift valley that has opened progressively as the island has drifted up a flexure in the ocean floor formed as a result of nearby plate convergence (Nunn 1995). The removal of Last-Glacial ice cover from some islands has resulted in their emergence, marked in cooler ocean waters by staircases of 'raised beaches', as are common on islands such as Lasqueti Island off the coast of British Columbia, western Canada (Hutchinson et al. 2004).

(b) Climate and sea-level controls

While geology may account for variations in the character of island environments, it is climate that largely drives the processes by which much of that change is accomplished. Weathering of islands in hot climates involves different processes – at commonly faster rates – than in colder climates. High levels of mean annual precipitation, especially when that is concentrated seasonally and/or in storms, generally result in faster and more profound changes in the environments than occur in all-year drier climates. For example, ground-surface lowering is 3–4 mm/year in the wet New Guinea highlands (Pickup et al. 1980) but only 0.04–0.19 mm/year in the Hawaiian Islands (Li 1988), the latter representative of most islands.

The oscillating temperatures of the last 10 million years, particularly during the Quaternary (last two million years), caused the Earth's climate to swing between cool ice ages (glacials) and warm interglacials, such as that (the Holocene) in which we live. Changing temperatures forced vegetation zones to shift, and with them those organisms that could not adapt to the changed climate. Although the effects of these climate changes on certain islands have been somewhat offset by the dominance of maritime influences in their climates, most high-latitude islands were subject to alternating glaciation and deglaciation during this period. Even high subtropical islands like Hawai'i experienced such climate changes and associated processes which have left their mark on the island's modern environment (Sheldon 2006). In general, the temperatures of lower-latitude islands did not change much during the Quaternary although many experienced aridity, the legacy of which is visible today. The grasslands that exist in many parts of the Pacific Islands were once thought to be wholly anthropogenic – created only after humans arrived and began burning the native forests – although now many such grasslands are suspected to be far older, a relic ecosystem that developed during the arid Last Glacial Maximum about 18,000 years ago (Nunn 1994).

Quaternary temperature fluctuations also drove sea-level changes that fundamentally altered the geography of the ocean basins and their islands. During periods of low sea level (the 'ice ages'), island areas increased and islands became closer together, facilitating biotic dispersal. Island climates changed because of the increased altitude. Conversely, during periods of comparatively high sea level (interglacials), islands were smaller, farther apart, and many islands which were 'high' 18,000 years ago during the Last Glacial Maximum may today only poke a few metres above the ocean surface.

(c) Extreme or rapid events

Viewed in the context of long-term environmental evolution, extreme or rapid events may in fact have lasted decades rather than days but still have left a profound and long-lasting legacy.

A good example of a climatic event that was both extreme and rapid compared to times before and after is the Younger Dryas, an approximately 1,000-year long reversion to almost full glacial climate that began around 11,000 years ago during the period of postglacial warming. Tropical islands were among those affected by rapid temperature fall during the Younger Dryas; sea-surface temperatures 10,200 years ago in the Vanuatu archipelago were around 5°C cooler than today (Beck et al. 1992). Changes in the rates of upwelling around Caribbean islands during the Younger Dryas (Overpeck et al. 1989) would have had significant impacts on their biotas.

A similar, more recent example is the 'AD 1300 Event', a period of rapid cooling and sea-level fall that affected the Pacific Islands (and perhaps elsewhere) for around 100 years beginning about AD 1250. Among the direct environmental effects were the emergence of islands made of surficial materials, the emergence of nearshore coral reefs, and degradation of lagoonal ecosystems that forced a food crisis for coastal-dwelling humans that led to centuries of conflict (Nunn 2007).

More rapid, catastrophic events have also figured in the long-term evolution of island environments, although the identification of such events is often controversial. Giant waves, from storms or tsunami, have sometimes left behind diagnostic deposits; an example comes from the Okupe Lagoon in New Zealand (Goff et al. 2000). Collapse of island flanks, even entire islands, is an important process in their long-term evolution; such events have been isolated by geological survey, as in the case of Johnston Atoll in the Central Pacific (Keating 1987), and by myth, as with the now-vanished islands of Tuanahe (Southern Cook Islands) and Vanua Mamata (Vanuatu) (Nunn 2009).

(d) Pre-modern human impacts

Owing to the vulnerability of most island ecosystems, the effects of human colonisation are often regarded as having been immediate and massive. While there are many case studies suggesting that this is the case – such as the charcoal 'spikes' in swamp sediments on several western Pacific islands (Hope et al. 1999) – there is evidence to suggest that the connection is more ambiguous (Nunn 2001). Much vegetation change that led to landscape change was brought about by human commensals, such as rats, or accidentally-introduced exotic plants. The origin of the singular gullies (named *lavaka*) on Madagascar was debated for a long time, with human impact as a leading explanation; it is now clear that *lavaka* have climatic origins although they may have been enlarged by human activities (Wells and Andriamihaja 1993).

Several authorities have pointed to Easter Island and the 14th-century collapse of its society as an example of its inhabitants committing 'ecocide' by cutting down all the island's trees to support statue construction (Bahn and Flenley 1992, Diamond 2005). Such explanations, while salutary and more readily apprehended on smaller islands than on continents, are still contentious and alternatives have been mooted (Stenseth and Voje 2009). Principal among these is the climatic explanation which sees sea-level fall (and water-table fall) during the AD 1300 Event, reducing the amounts of food available to island peoples. This in turn forced lifestyle changes, sometimes what might be described as societal 'collapse', marked by conflict in many places (Nunn 2007).

The AD 1300 Event led to people throughout the Pacific abandoning their coastal settlements and establishing new ones inland, commonly in fortifiable locations such as hilltops and caves. This led to an abrupt impact on the inland parts of many islands, resulting in their degradation and often an associated response in downstream and coastal areas (Kumar et al. 2006). Deliberate adaptations for the enhancement of agricultural production also remain visible on many islands today. This is true especially of agricultural terracing, introduced to many tropical Pacific Islands during the arid Little Climatic Optimum, warm period (AD 750–1250) that preceded the AD 1300 Event and the ensuing Little Ice Age (AD 1350–1800) (Nunn 2007).

The beginning of plantation agriculture on many islands, often coincident with the start of colonial history, resulted in major changes to their environments. Illustrative of this are the rates of erosion from various environments in the Philippines: areas under primary forest (largely undisturbed) lose around 3 tons/hectare/year; while areas of open grassland (converted from forest) lose 84 tons/hectare/year; and overgrazed areas lose 250 tons/hectare/year (Coxhead 2000).

Influences on contemporary island environments

Many island environments today bear little resemblance to their historical counterparts. Wholly urbanised islands like Manhattan (New York), Malé (Maldives) and Stockholm (Sweden) and some in Hong Kong are extreme examples of this situation. For most others, it makes sense to separate human influences from natural (non-human) ones. Among the latter, climate change and sea-level rise, as well as catastrophic events, are selected for discussion.

(a) Climate change and sea-level rise

Within the past hundred years, most islands have experienced surface temperature warming and sea-level rise. This has caused a range of problems although it is sometimes difficult to separate non-human from human causes. For example, vegetation change during the 20th century on islands is more likely a consequence of direct human impact although some may have been due to warming and, particularly in low-lying coastal areas, to groundwater salinisation resulting from sea-level rise.

Warming has also affected ocean-surface temperatures, and this increase is implicated in recent changes to a number of shallow-water marine ecosystems, notably coral reefs. When ocean-surface temperatures exceed 30°C for prolonged periods, as they have done increasingly over the past few decades, corals will sometimes be stressed to the point that they eject their symbiotic algae, thereby losing their colour (bleaching) and dying. Such 'coral bleaching' is likely to affect island reefs more frequently in future decades if ocean-surface temperatures continue rising, and many coral reefs are likely to become barren and nearby beaches starved of calcareous sediment (Hoegh-Guldberg 2011).

It is also likely that warming over the past few decades has contributed to both the changes in the frequency and intensity of tropical cyclones (hurricanes or typhoons) as these usually develop only in ocean areas where the surface temperature exceeds 27°C. In the tropical Pacific, recent warming has increased the area within this condition is met resulting in more frequent tropical cyclones over the past 50 years that occurred increasingly 'out of season' and affect islands farther east than what was long regarded as the cyclone-prone region. Tropical Cyclone Ofa, which slammed into the islands of Samoa in 1990, was the first such storm to affect them for more than 35 years. Since 2007, in line with global climate models, tropical cyclone frequency in the Pacific has declined, although the intensity of such storms has increased. Tropical Cyclone Winston, that affected Fiji in February 2016, had the strongest winds ever recorded in the southern hemisphere (Anonymous 2016).

The problem with sea-level rise being portrayed as 'the' issue with which islands will have to cope in the 21st century is that it encourages poorly-informed and cash-strapped island governments to overlook more pressing problems, both those associated with climate change (such as changes in precipitation regime) and those not linked to climate change (such as population growth) (Nunn 2004, Baldacchino and Kelman 2014). That said, sea-level rise represents an important challenge for many island communities, particularly those which are ill-equipped to meet it. In many poorer island countries, communities which are only marginally within the cash economy often raise huge sums to build seawalls, believing these will protect against coastal erosion, both today and in the future. Most such seawalls collapse within a few years of and are often subsequently abandoned (see Figure 3.6a); a cheaper and more sustainable solution is to plant (often re-plant) the mangrove forest that once fringed these islands' coasts (see Figure 3.6b). As part of a multi-goal coastal rehabilitation programme on Marinduque Island in the Philippines, mangroves are being allowed to grow out over potentially toxic copper tailings in Calancan Bay

(b) Catastrophic events

Owing to their youth and location, many islands are especially vulnerable to catastrophic events, and many island environments (and peoples) reflect their influence. Many islanders live in the shadow of active volcanoes, sometimes because there is nowhere else to go but often because of the fertile soils and dependable water supply characteristic of many such locations. Examples include the people occupying Tofua Island (Tonga) and those whose food gardens are scattered across the slopes of Merapi Volcano (Java). In the modern age, volcanic catastrophes affecting islands are mostly containable in that they can be predicted and appropriate action taken to avoid unnecessary impacts and loss of life.

Yet, some of the world's largest eruptions have occurred on islands and produced effects that were often felt around the world. The 1883 eruption of Krakatau Island (Indonesia) (see Figure 3.7) produced devastating tsunami and eventually destroyed the island (Simkin and Fiske

Figure 3.6 Adaptation to erosion along an island shoreline: Yadua Village, Viti Levu Island, Fiji.

Source: © Patrick Nunn.

A – The earliest response was to build a seawall, using largely rock from the fringing coral reef. The seawall was undermined by wave erosion, collapsed and was rebuilt repeatedly until it became clear that this was not an effective long-term option. This view shows part of the degraded seawall, and the land behind, into which storm waves penetrate and erode the coastal flat on which the village lies.

B – The villagers are now replanting a mangrove forest along the worst affected part of the shoreline. This option is sustainable, effective and will enhance the nearshore ecology. The only difficulty is that it may take 25 years for the mangrove fringe to reach maturity. Mangroves are grown in a nursery and then planted out at regular intervals when they are about 80 cm tall.

Figure 3.7 The eruption of the Indonesian island volcano of Krakatau (Krakatoa) in August 1883 was one of the largest in recent time. Shock waves circled the planet seven times; the eruption was heard in Perth (Australia), more than 3,000 km away; and in this part of Indonesia the dawn did not appear for three days because of all the ash in the air. More than 36,000 people perished, many drowned by the huge tsunami waves generated by the collapse of the islands into an undersea caldera.

Source: This figure was published in 1888 as Plate 1 in the Royal Society's report of the Krakatoa eruption. Wikimedia Commons, https://en.wikipedia.org/wiki/1883_eruption_of_Krakatoa.

1983). Yet it was considerably smaller than the AD 1452 eruption of Kuwae in Vanuatu – one of the largest in the past 10,000 years – that also destroyed the island, producing a 72-km² submarine caldera (Eissen et al. 1994) and ejecting so much ash into the atmosphere that northern-hemisphere weather was drastically affected for several years (Pang 1993). The largest eruption of the past two million years occurred at Toba (Sumatra) and formed a 3,000-km² caldera; pyroclastic flows covered an area of 20,000 km², and ash covered an area of about 4 million km²

(Lee et al. 2004); so many people were evidently killed by the eruption and the climate changes it produced that it has been cited as the cause of a bottleneck in human evolution (Ambrose 1998).

Large earthquakes may also have catastrophic effects on islands, sometimes causing them to rise, sink or tilt. Many islands close to convergent plate boundaries rise co-seismically (when earthquakes are coincident with uplift). Parts of the coast of the island of Taiwan are stepped, indicating the effects of repeated co-seismic-uplift events (Huang et al. 1997) and terracing the landscape in ways that facilitate plantation agriculture. During the 1964 Prince William Sound Earthquake in the northeast Pacific, parts of Montague Island rose 11.3 m in a few seconds while parts of Kodiak Island sunk abruptly by more than 2 m (Plafker 1972). Co-seismic subsidence was also the cause of the destruction of Port Royal in Jamaica in 1692 (Gragg 2000).

The history of Pukapuka Atoll (Cook Islands), passed down orally through generations, is dominated by one event named in the vernacular *te mate wolo* (the Great Death) that marks the time when a huge wave swept across the atoll, carrying away most of its inhabitants (Beagle-hole and Beaglehole 1938). Such waves can obliterate entire islands, and can strip them to their unweathered sedimentary foundations. Yet, paradoxically it might seem, they can also create and enlarge such islands by driving reef detritus onshore. Much of the variation is attributable to the morphology of the ocean floor over which the wave approaches the island, and the amount of sediment it is carrying when it reaches the island coast.

Island environmental futures

Many of those responsible for managing island environments are grappling with multiple environmental-related problems, almost all accentuated by islandness, with insufficient financial or human resources at their disposal (Nunn 2004). Most professionals who are managing island environments have also been trained in continental environments and assume – often with disastrous consequences – that islands are merely continents in miniature (Doumenge 1987). There is no shortage of bleak prognostications for the future of island environments: from the possibility of climate change causing the Gulf Stream to weaken bringing cold wet summers to the Western Isles of Scotland; the flooding of Caribbean island wetlands with huge ecological consequences (Nicholls et al. 1999); to the likelihood that entire nations of low-lying islands like Tokelau and Tuvalu could disappear (Lewis 1990).

A way forward is to recognise island environments – perhaps with isolated and archipelagic subsets – as distinct and requiring effective solutions to environmental problems devised and implemented by persons with the long-term interests of islands at heart. Many islands are regarded as economically 'under-developed': this explains why sustainable environmental development is invariably considered as secondary – despite many fine speeches – to income-generating activities by those island leaders.

Typically as a result of unsustainable demands being made on them, the environments of most inhabited islands have deteriorated over the past few decades, something that is likely to continue on many as population densities increase and climate change has an increasingly significant effect on livelihoods (Nunn 2013). While this is likely to one day lead to radical responses from island leaders, particularly around resource conservation and stewardship, it will inevitably also lead to island abandonment. Nowhere is this more apparent than on inhabited atoll islands (such as those in the island nations of Kiribati, Marshall Islands, Maldives and Tuvalu) where rising sea level is often causing both freshwater lenses and habitable land to decrease in size (Connell 2016).

References

Ambrose, S.H. (1998) Late Pleistocene human population bottlenecks. *Journal of Human Evolution*, 34(6), 623–651.

Anonymous. (2016). Tropical cyclone Winston causes devastation in Fiji. *Weather*, 71(4), 82–82.

Bahn, P.G., and Flenley, J. (1992) *Easter island, Earth island*. London, Thames and Hudson.

Baines, G.B.K., and McLean, R.F. (1976) Sequential studies of hurricane deposit evolution at Funafuti Atoll. *Marine Geology*, 21, M1–M8.

Baldacchino, G., and Kelman, I. (2014) Critiquing the pursuit of island sustainability: blue and green, with hardly a colour in between. *Shima*, 8(2), 1–21. Available from: http://shimajournal.org/issues/v8n2/c.-Baldacchino-&-Kelman-Shima-v8n2-1-21.pdf.

Barka, A., and Reilinger, R. (1997) Active tectonics of the Eastern Mediterranean Region deduced from GPS, Neotectonic and Seismicity data. *Annali Geofisica*, 40, 587–610.

Beaglehole, E., and Beaglehole, P. (1938) *Ethnology of Pukapuka*. Honolulu, HI, B.P. Bishop Museum, Bulletin No. 150.

Beck, J.W., Edwards, R.L., Ito, E., Taylor, F.W., Recy, J., Rougerie, F., Joannot, P., and Henin, C. (1992) Sea-surface temperature from coral skeletal strontium/calcium ratios. *Science*, 257(5070), 644–647.

Bott, M.H.P. (1985) Plate-tectonic evolution of icelandic transverse ridge and adjacent regions. *Journal of Geophysical Research*, 90(B12), 9953–9960.

Carracedo, J.C. (1994) The Canary Islands: An example of structural control on the growth of large oceanic-island volcanoes. *Journal of Volcanology and Geothermal Research*, 60(3–4), 225–241.

Chappell, J., and Veeh, H.H. (1978) Late quaternary tectonic movements and sea-level changes at Timor and Atauro Island. *Geological Society of America Bulletin*, 89(3), 356–368.

Connell, J. (2016). Last days in the Carteret Islands? Climate change, livelihoods and migration on coral atolls. *Asia Pacific Viewpoint*, 57(1), 3–15.

Corbett, G., Hunt, S., Cook, A., Tamaduk, P., and Leach T. (2001) Geology of the Ladolam gold deposit, Lihir island, from exposures in the Minifie open pit. In G. Hancock (ed.), *Geology, exploration and mining conference: proceedings*. Port Moresby, Papua New Guinea and Parkville, Australasian Institute of Mining & Metallurgy, pp. 69–78.

Coxhead, I. (2000) Consequences of a food security strategy for economic welfare, income distribution and land degradation: The Philippine case. *World Development*, 28(1), 111–128.

Cunningham, J.K., and Anscombe, K.J. (1985) Geology of 'Eua and other Islands, Kingdom of Tonga. In D.W. Scholl and T.L. Vallier (eds.), *Geology and offshore resources of Pacific island arcs: Tonga Region*, Houston, TX, Circum-Pacific Council for Energy and Mineral Resources, pp. 221–257.

Diamond, J. (2005) *Collapse: how Societies choose to fail or succeed*. New York, Viking.

Dickinson, W.R. (1999) Holocene sea-level record on Funafuti and potential impact of global warming on Central Pacific Atolls. *Quaternary Research*, 51(2), 124–132.

Doumenge, F. (1987) Quelques contraintes du milieu insulaire. In *Îles Tropicales: Insularité, 'Insularisme'*. Bordeaux, CRET, Université de Bordeaux III, pp. 9–16.

Eissen, J.P., Monzier, M., and Robin, C. (1994) Kuwae, l'éruption volcanique oubliée. *La Recherche*, 270, 1200–1202.

Fleurant, C., Tucker, G.E., and Viles, H.A. (2008). A model of cockpit karst landscape, Jamaica. *Geomorphologie-Relief Processus Environnement*, 1, 3–14.

Frank, E., Mylroie, J., Troester, J., Alexander, E.C., and Carew, J. (1998) Karst development and speleogenesis, Isla de Mona, Puerto Rico. *Journal of Cave and Karst Studies*, 60(2), 73–83.

Gautier, P., Quesnel, B., Boulvais, P., and Cathelineau, M. (2016) The emplacement of the Peridotite Nappe of New Caledonia and its bearing on the tectonics of obduction. *Tectonics*, 35, 3070–3094.

Ghosh, P. (2015) Conservation and conflicts in the Sundarban Biosphere Reserve, India. *Geographical Review*, 105(4), 429–440.

Goff, J.R., Rouse, H.L., Jones, S.L., Hayward, B.W., Cochran, U., McLea, W., Dickinson, W.W., and Morley, M.S. (2000) Evidence for an earthquake and saltwater tsunami about 3100–3400 years ago, and other catastrophic saltwater inundations recorded in a coastal lagoon. *New Zealand Marine Geology*, 170(1–2), 231–249.

Gragg, L. (2000) The port royal earthquake. *History Today*, 50, 28–34.

Hall, R., Audley-Charles, M.G., Banner, F.T., Hidayat, S., and Tobing, S.L. (1988) Late Paleogene-quaternary geology of Halmahera, Eastern Indonesia: initiation of a Volcanic Island Arc. *Journal of the Geological Society of London*, 145, 577–590.

Harper, S.B. (1999) Morphology of tower karst in Krabi, Southern Thailand. In *Geological society of America Abstracts with Programs*, Denver, Colorado, Annual Meeting, October, pp. A–52.

Hill, K.C., Simpson, R.J., Kendrick, R.D., Crowhurst, P.V., O'Sullivan, P.B., and Saefudin, I. (1996) Hydrocarbons in New Guinea, controlled by basement fabric, mesozoic extension and tertiary convergent margin tectonics. In P.G. Buchanan (ed.), *Petroleum exploration, development and production in Papua New Guinea*, Proceedings of the 3rd PNG Petroleum Convention, Port Moresby, September, pp. 63–76.

Hoegh-Guldberg, O. (2011) Coral reef ecosystems and anthropogenic climate change. *Regional Environmental Change*, 11(1), 215–227.

Hope, G., O'Dea, D., and Southern, W. (1999) Holocene vegetation histories in the Western Pacific: alternative records of human impact. In J.-C. Galipaud and I. Lilley (eds.), *The Pacific from 5000 to 2000 BP: colonisation and transformations*, Paris, Editions de IRD, pp. 387–404.

Huang, C.Y., Wu, W.Y., Chang, C.P., Tsao, S., Yuan, P.B., Lin, C.W., and Kuan-Yuan, X. (1997) Tectonic evolution of accretionary prism in the arc-continent collision terrane of Taiwan. *Tectonophysics*, 281(1–2), 31–51.

Huggett, R.J. (1991) *Climate, Earth processes and Earth history*. Berlin, Germany, Springer-Verlag.

Hutchinson, I., James, T.S., Clague, J.J., Barrie, V., and Conway, K.W. (2004) Reconstruction of late quaternary sea-level change in southwestern British Columbia from sediments in isolation Basins. *Boreas*, 33(3), 183–194.

Itturralde-Vinent, M. (1997) Introducción a la geología de Cuba. In G. Furrazola-Bermúdez and K. Nuñez Cambra (eds.), *Estudios sobre la geología de Cuba*, Havana, Centro Nacional de Información Geológica, pp. 35–68.

Keating, B. (1987) Structural failure and drowning of Johnston Atoll, Central Pacific Basin. In B.H. Keating, P. Fryer, R. Batiza and G.W. Boehlert (eds.), *Seamounts, islands and atolls*, Washington, DC, American Geophysical Union, Monograph No. 45, pp. 49–59.

Kumar, R., Nunn, P.D., Field, J.E., and de Biran, A. (2006) Human responses to climate change around AD 1300: a case study of the Sigatoka Valley, Viti Levu Island, Fiji. *Quaternary International*, 151, 133–143.

Lee, M.-Y., Chen, C.-H., Wei, K.-Y., Iizuka, Y., and Carey, S. (2004) First toba super-eruption revival. *Geology*, 32(1), 61–64.

Li, Y.-H. (1988) Denudation rates of the Hawaiian islands by rivers and groundwater. *Pacific Science*, 42(3–4), 253–266.

Löffler, E. (1977) *Geomorphology of Papua New Guinea*. Canberra, Australia, Australian National University Press.

MacDonald, J.G., and Herriot, A. (1983) *Geology of Arran*. Glasgow, Geological Society of Glasgow.

McLean, R.F., and Hosking, P.L. (1991) Geomorphology of Reef islands and atoll motu in Tuvalu. *South Pacific Journal of Natural Science*, 11(1), 167–189.

Moore, J.G., Bryan, W.B., and Ludwig, K.R. (1994) Chaotic deposition by a giant wave, Moloka'i, Hawai'i. *Geological Society of America Bulletin*, 106, 962–967.

Mylroie, J.E., Carew, J.L., and Moore, A.I. (1995) Blue holes: definition and genesis. *Carbonates and Evaporites*, 10(2), 225–233.

Nicholls, R.J., Hoozemans, F.M.J., and Marchand, M. (1999) Increasing flood risk and wetland losses due to global sea-level rise: regional and global analyses. *Global Environmental Change*, 9, S69–S87.

Nunn, P.D. (1994) *Oceanic islands*. Oxford, Blackwell.

Nunn, P.D. (1995) Lithospheric flexure in Southeast Fiji consistent with the tectonic history of Islands in the Yasayasa Moala. *Australian Journal of Earth Sciences*, 42(4), 377–389.

Nunn, P.D. (1998) *Pacific island landscapes*. Suva, Fiji, Institute of Pacific Studies, The University of the South Pacific.

Nunn, P.D. (2001) Ecological crises or marginal disruptions: the effects of the first humans on Pacific Islands. *New Zealand Geographer*, 57(2), 11–20.

Nunn, P.D. (2004) Through a mist on the ocean: human understanding of island environments. *Tijdschrift voor Economische en Sociale Geografie*, 95(3), 311–325.

Nunn, P.D. (2007) *Climate, environment and society in the Pacific during the last millennium*. Amsterdam, The Netherlands, Elsevier.

Nunn, P.D. (2009) *Vanished islands and hidden continents of the Pacific*. Honolulu, HI, University of Hawai'i Press.

Nunn, P.D. (2013). The end of the Pacific? Effects of sea level rise on Pacific Island livelihoods. *Singapore Journal of Tropical Geography*, 34(2), 143–171.

Nunn, P.D., and Britton, J.M.R. (2004) The long-term evolution of Niue Island. In J. Terry and W. Murray (eds.), *Niue island: geographical perspectives on the rock of Polynesia*. Paris, INSULA, International Scientific Council for Island Development, pp. 31–74.

Overpeck, J.T., Peterson, L.C., Kipp, N., Imbrie, J., and Rind, D. (1989) Climate change in the circum-North Atlantic region during the last deglaciation. *Nature*, 338, 553–557.

Pang, K.D. (1993) Climatic impact of the mid-15th century Kuwae caldera formation as reconstructed from historical and proxy data. *Eos: Transactions of the American Geophysical Union*, 74(43), Supplement F, 106.

Pickup, G., Higgins, R.J., and Warner, R.F. (1980) Erosion and sediment yield in the fly river drainage basins, Papua New Guinea. *Publications of the International Association of Hydrological Sciences*, 132, 438–456.

Plafker, G. (1972) Alaskan earthquake of 1964 and Chilean earthquake of 1960: implications for arc tectonics. *Journal of Geophysical Research*, 77(5), 901–925.

Robertson, A.H.F. (2002) Overview of the genesis and emplacement of Mesozoic ophiolites in the Eastern Mediterranean Tethyan Region. *Lithos*, 65(1–2), 1–67.

Robinson, E. (1994) Jamaica. In S.K. Donovan and T.A. Jackson (eds.), *Caribbean geology: an introduction*, Kingston, Jamaica, University of the West Indies Publishers' Association, pp. 111–127.

Sheldon, N.D. (2006) Quaternary glacial-interglacial climate cycles in Hawai'i. *Journal of Geology*, 114(3), 367–376.

Simkin, T., and Fiske, R.S. (1983) *Krakatau 1883: the volcanic eruption and its effects*. Washington, DC, Smithsonian Institution Press.

Stehli, F.G., and Webb, S.D. (1985) A kaleidoscope of plates, faunal and floral dispersals, and sea-level changes. In F.G. Stehli and S.D. Webb (eds.), *The great American biotic interchange*, New York, Plenum Press, pp. 3–16.

Stenseth, N.C., and Voje, K.L. (2009) Easter Island: climate change might have contributed to past cultural and societal changes. *Climate Research*, 39(2), 111–114.

Stoddart, D.R. (1969) Geomorphology of the Solomon Islands coral reefs. *Philosophical Transactions of the Royal Society of London*, 255 B, 355–382.

Stoddart, D.R., Spencer, T., and Scoffin, T.P. (1985) Reef growth and karst erosion on Mangaia, Cook Islands: a reinterpretation. *Zeitschrift für Geomorphologie*, Supplementband, 57, 121–140.

Stoker, M.S., Hitchen, K., and Graham, C.C. (1993) *The geology of the Hebrides and West Shetland shelves, and adjacent deep-water areas*. British Geological Survey, United Kingdom Offshore Regional Report, No. 2, London, HMSO.

Van Duzer, C. (2004) *Floating islands: a global bibliography*. Los Altos Hills, CA, Cantor Press.

Weeramantry, C.G. (1992) *Nauru: environmental damage under international trusteeship*. Melbourne, Australia, Oxford University Press.

Wells, N.A., and Andriamihaja, B. (1993) The initiation and growth of gullies in Madagascar: are humans to blame? *Geomorphology*, 8(1), 1–46.

Young, R.W., and Bryant, E.A. (1992) Catastrophic wave erosion on the southeastern coast of Australia: impact of the Lanai tsunamis circa 105ka. *Geology*, 20(3), 199–202.

4

EVOLUTION

Andrew J. Berry and Rosemary G. Gillespie

Introduction

In the opening paragraph of the *Origin of Species* (1859), Charles Darwin refers to the role of his voyage on the *Beagle* in inspiring his ideas. This conjures up an image of the young Darwin hard at work on the Galápagos, frenetically hunting up specimens on the islands' parched lava fields, the *Beagle* picturesquely anchored all the while off-shore. In fact, despite the modern 'Eureka!' myth surrounding Darwin's experiences on the Galápagos – the erroneous potted version of the popular story has Darwin making his great breakthrough there and then – Darwin did not come to appreciate the significance of what he had seen and collected in the Galápagos and else-where until after his return to England. Neither the Galápagos nor other islands play especially prominent roles in the key early chapters of *On the Origin of Species*, in which Darwin lays out the basic logic of evolutionary thinking.

This is not to say, however, that the Galápagos, or islands in general, are completely over-looked in Darwin's masterpiece. But it is not until near the end of the book, in the second of two chapters on biogeography (chapter 12), that islands come into their own in a section where every paragraph, in true Darwin style, contains a fresh insight. This is arguably the founding document of modern evolutionary analysis of island biotas:

> Although in oceanic islands the number of kinds of inhabitants is scanty, the propor-tion of endemic species (i.e. those found nowhere else in the world) is often extremely large. If we compare, for instance, the number of the endemic land-shells in Madeira, or of the endemic birds in the Galápagos Archipelago, with the number found on any continent, and then compare the area of the islands with that of the continent, we shall see that this is true. This fact might have been expected on my theory for, as already explained, species occasionally arriving after long intervals in a new and isolated dis-trict, and having to compete with new associates, will be eminently liable to modifica-tion, and will often produce groups of modified descendants.
>
> *(Darwin 1859/1985, p. 379)*

Darwin was not alone among his contemporaries in his understanding that islands can provide special insights into evolution. The naturalist who co-discovered the theory of evolution by natural selection with him, Alfred Russel Wallace, spent eight formative years as a biological collector in southeast Asia where he was impressed repeatedly by the biological differences among islands. For example, in his 1857 paper on the Aru islands off western New Guinea, he emphasised "how totally the productions of New Guinea differ from those of the Western Islands of the Archipelago, say Borneo" despite the similarity of their "climate and physical features". Under the natural theological paradigm of the day, organisms were supposedly designed by the Creator for the specifics of their environment, implying that similar environments should be inhabited by similar organisms. In highlighting differences, Wallace recognised that evolutionary processes – the historical quirks of migration, colonisation, selection, speciation and extinction – were a key determinant of what species occur where.

Wallace would later publish an influential book on the subject, *Island life* (1880). In it, he categorised islands as *continental* or *oceanic*. This distinction applies both to an island's location – continental islands are typically close to the mainland, oceanic ones distant from it – and, often, to its origin. Oceanic islands are typically formed *de novo* by geological processes such as volcanism whereas continental islands are often 'pinched off' pieces of mainland: for example, what was once a headland or peninsula connected to the mainland by a low-lying area has become an island as sea level has increased.

In evolutionary biology, an island can be defined simply as a patch of suitable habitat surrounded by an inhospitable matrix that cannot readily or easily be crossed (Gillespie and Clague 2009). The reason for the broad definition is simply that the same biological processes apply, whatever the matrix of isolation for a given organism. 'Habitat islands' may, for example, include mountaintops, with species adapted to life at high altitude unable to cross the stretches of lowland separating mountains. Regardless, many of the key insights and much of the data on island evolution has come from studies of islands in the vernacular sense: pieces of land surrounded by water.

The familiar terms for islands – 'oceanic', 'continental' and 'habitat' – can be confusing and/or overlapping, so we prefer to categorise islands according to origin and geography (see Table 4.1).

There are several broad attributes of islands that dictate their biological properties:

- *Isolation*. This may be a function of distance or of the specifics of the ecological situation: a population of fish in a lake, for example, is more isolated than a population of amphibians in the same lake, because the terrestrial phase of the amphibians' life cycle may facilitate inter-lake dispersal. The matrix – in this case, dry land – is differentially permeable for different groups of organisms. Isolation may also be affected by *in situ* evolution: sometimes the

Table 4.1 Categorising islands by geological origin and geography.

Origin	Geography	Outcome	Examples
De novo (from scratch)	Isolated	Adaptive Radiation	Hawaiian Islands, African Rift Lakes
	Less Isolated	Extinction/Colonisation Dynamics	Neighbouring Sky Islands (mountain tops colonised *de novo* post-glaciation)
Fragmentation	Isolated	Relicts	New Zealand
	Less Isolated	Extinction/Colonisation Dynamics	Barro Colorado Island, in Lake Gatun, Panama

dispersal abilities of island inhabitants may change, such as in the evolution of flightlessness in island insects. Isolation, because it limits or prohibits gene flow, facilitates evolutionary diversification. Two diverging populations derived recently from a common ancestor will probably not diverge into two distinct species if there is gene flow between them. For this reason, we see a correlation between the number of endemics on an island and its isolation (see Figure 4.1).

- *Age*. Evolution takes time, so older islands tend, for example, to have higher levels of diversity and endemism than younger ones because there has been more opportunity for diversification to occur. In the case of less isolated *de novo* islands, which are colonised on ecological timescales, age is again key because diversity accumulates over time through immigration until a colonisation/extinction equilibrium is reached.
- *Size*. Ecologists have long recognised that the number of species in a given region is strongly predicted by its land area (MacArthur and Wilson 1967). In addition, the size of an island is correlated with the range of ecological opportunities it has to offer (compare a coral atoll to Hawai'i's Big Island) and with the probability of colonists arriving: a small island is a small target.
- *Initial state*. Islands formed *de novo* (from scratch) start as biological blank slates: each one represents a new evolutionary–ecological experiment, as species arrive and evolve. The flora and fauna of fragment islands derived from established ecosystems, on the other hand, tend to be a vestige of what was present when the island was generated.
- *Scale dependence*. What functions as an island for one species may not function as one for another. An isolated tree may be a true island from the point of view of a beetle species that conducts its entire life cycle on the single tree, the nearest other potential host trees being too far away for ready dispersal; but that same tree is merely an outlier from the point of view of a bird species that forages for insects in woodland.

As Darwin and Wallace noted, populations on islands are – not surprisingly – often biologically remarkable. Islands have thus legitimately been described as evolution's 'natural laboratories'.

Natural laboratories

Islands – like any informative laboratory demonstration of a scientific principle – facilitate making the distinction between signal and noise in ecological and evolutionary studies. To simplify the following discussion, we focus here on isolated *de novo* islands.

- *Small and simple*. Islands may, for example, lack habitats that are represented on the nearest mainland: a flat, low island has no high altitude habitats. The number of species on an island is primarily a function of the island's size and its distance from the nearest source of colonising species. A small, remote island, then, is typically poorer in terms of number of species.
- *Closed systems*. Migration constantly muddies the waters in studies of the evolutionary dynamics of populations on mainlands. A given allele (genetic variant) may be favoured by natural selection in a population but it is prevented from going to fixation (that is, a 100 per cent frequency in the population) because alleles from neighbouring populations (in which, courtesy of differing ecological conditions and a correspondingly distinct selection regime, different alleles are favoured) constantly leak in. On islands, the nearest neighbours may lie further or too far away, and so affect these evolutionary processes to a much lesser extent, if at all.

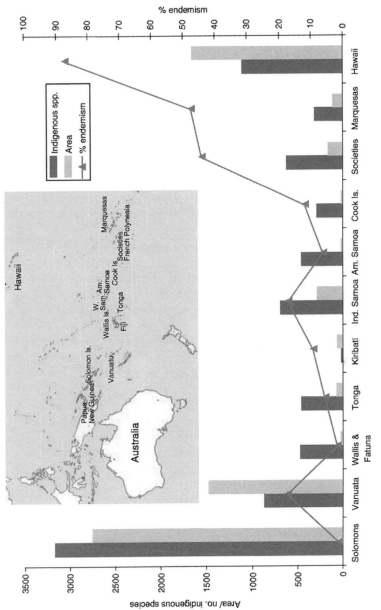

Figure 4.1 Plant transect across selected Pacific islands. Larger islands have disproportionately larger numbers of species. The high levels of endemism on remote islands are evidence of the role of evolution within these islands.

Source: Gillespie (2007); © Rosemary G. Gillespie.

- *Spaces for novelty*. Islands often represent novel and remarkable experiments in evolution because of the peculiarities of their origins and/or colonisation. These natural experiments are especially valuable to the evolutionary biologist who is typically in no position to conduct real-time experimental studies of evolution because the timescales of evolutionary change are so long.

- *Disharmony*. This strange-sounding term refers to the distinct taxonomic composition of an island ecosystem relative to continental source regions; i.e. an overrepresentation of some groups and an underrepresentation of others. The quirks of colonisation may result in an island ecosystem being ecologically incomplete or imbalanced. Imagine a newly formed remote volcanic island: what species will colonise it? Some groups are much better at long-distance dispersal than others. Birds are relatively good dispersers; flightless terrestrial invertebrates (think earthworms), on the other hand, are poor ones. Birds are therefore likely to be better represented among the early colonists than earthworms, even though our new island may harbour many inviting habitats appropriate for the worms. But dispersal ability is not the only significant factor: the other is chance. Of all the land birds that could conceivably colonise our island, only a few will. They, probably, are simply the lucky ones: a few individuals of species A, say, were blown in a storm from the adjacent mainland to the island. Individuals of species B would have been equally capable of taking advantage of the ecological opportunities proffered by the new island; fate alone – B's were not caught in the same storm that drove A's to the island – dictates that it is A, not B, that establishes a bridgehead on the island.

- *Crucibles of evolutionary change*. Over time, geographic isolation permits (and may promote) the genetic divergence of populations. Indeed, such isolation is a key ingredient in the process underlying the evolutionary generation of diversity: speciation. It is only with the formation of a new species that significant divergence and specialisation can occur. Because islands are isolated from each other (such as within an archipelago) and from the mainland, island environments are especially conducive to speciation and therefore to evolutionary innovation.

- *Replication*. Islands are microcosms – pared down, simplified systems – that allow us to study aspects of the evolutionary process. Key to this is replication – the availability of many islands for comparison. Given that each island represents a separate evolutionary experiment, if we are to make generalisations about evolution, the patterns observed should not be specific to a particular island, but, rather, should be features shared by many different islands.

Colonisation and relaxation: island absences and disharmony

MacArthur and Wilson (1967) showed that there is an equilibrium number of species for a given island, with that number being dictated by the size of the island and its distance from a source of colonisers (see Berry and Lister, this volume). Empirical verification of these ideas came from the experimental de-faunation of mangrove islets by Simberloff and Wilson (1969). Over time after de-faunation, each islet was re-colonised, coming eventually to host approximately the same number of species – its equilibrium number – as it had prior to de-faunation. But: what of island colonisation under natural conditions?

In two instances, it has been possible to follow the colonisation of 'empty' islands from the very start. In August 1883, an enormous volcanic explosion literally ripped apart the Indonesian island of Krakatau located between Sumatra and Java. Tens of thousands of people in the region were killed by the resulting tsunami. The remnant of what had been a mountain on Krakatau, called Rakata, was a sterile, ash-buried husk. Life on Rakata had been eliminated (Simkin and Fiske 1983). But, slowly, it came back. Wilson (2010) recounts the story of the biological recovery of Rakata. Visiting within a year of the eruption, the naturalist on a French ship found nothing alive on Rakata except for a tiny spider, which must have 'ballooned' by launching itself into the air attached to buoyant thread of silk from a neighbouring island. Within three years, however, naturalists had recorded 15 species of grasses and shrubs. The colonisation-cum-succession process was under way. Particular species, sometimes called pioneers, are adapted for taking advantage of habitats in which competition is not intense. These species, however, sow the seeds, figuratively, of their own destruction because their presence – the impact they have on the soil, and in providing shade and other aspects of micro-climate for new arrivals – facilitates the establishment of new species, which eventually out-compete and displace the pioneers. By 1919, Rakata was home to forest patches surrounded by nearly continuous grassland. Ten years later, the succession process having progressed, forest dominated, and the last remaining patches of grassland were gradually being taken over by the forest (Thornton et al. 1993).

The birth of Surtsey, named after Surtur, the fire giant of Icelandic mythology, was almost as spectacular as the death of Krakatau. On 14 November 1963, 33 km off the southeast coast of Iceland, a submarine volcano erupted, creating in the process an island that today is 2.8 km² and 147m above sea level at its highest point. A new island had been born.

The biological colonisation of Surtsey has been carefully monitored (Baldursson and Ingadottir 2006). Some 300 invertebrate species have now been recorded on the island, with about half that number successfully establishing permanent colonies. Curiously, it was the arrival of nesting gulls that has proved critical to the development of the fauna on Surtsey. Gull guano is the major source of fertiliser for top soil enhancement, and it is around the gull colonies that life is at its richest and most diverse.

The rate of colonisation of Rakata and Surtsey will eventually taper off and the number will remain more or less constant around the equilibrium level. There will, however, still be turnover in species: some species will go extinct and new ones will arrive, but the net number of species will remain fairly constant.

Regardless of the final number of species, many island biotas, especially on remote, *de novo* islands, are disharmonic: they are imperfect samples of source biotas, and are usually biased towards species with high dispersal abilities. An additional, deterministic factor can contribute to island disharmony: small islands may not harbour the resources to support certain components of ecosystems. Imagine a limited population of predators arriving on a small island. They will rapidly decimate the resident herbivore population and it will be a toss-up whether the herbivores will go extinct (thereby inevitably condemning the predator to extinction as well) or the predators will go extinct first, permitting the herbivore population to bounce back.

In newly formed fragment (that is, continental) islands, we see *relaxation* (Diamond 1972). The species count on the island decreases with time as taxa are squeezed out in response to the smaller area of the island and its resource limits. This process of species loss may be complicated by the life cycles of the species involved. For example, the extinction of pollinators on an island might doom a plant species that depends upon them, but, if the plant is long-lived, individuals will persist for a long time after the species' extinction has become inevitable. Tilman et al. (1994) termed this *extinction debt*: the future cost of species loss incurred by current ecological events.

Barro Colorado Island, a 1,500 hectare lowland rainforest island in Gatun Lake, Panama, illustrates these ideas. Formed in 1914 when the Panama Canal project created Gatun Lake (see Figure 4.2), Barro Colorado was until then part of the region's continuous forest. With its 'islandisation' came the loss of its top mammalian predator, the jaguar, presumably because the

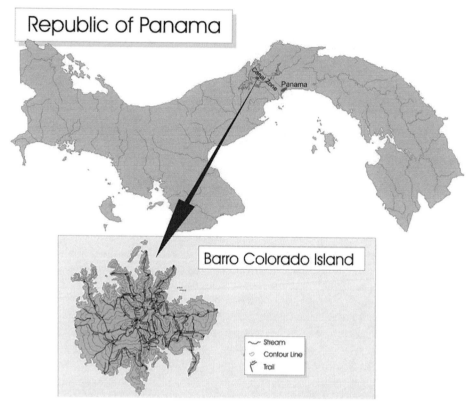

Figure 4.2   Barro Colorado Island, Gatun Lake, Panama.

Source: © Nativa Tours, Panama, www.nativatours.com. Reproduced with permission.

island was not large enough to support prey sufficient for a viable population (Diamond 1984). And extensive changes have been documented to the island's avifauna: Robinson (1999) reports the disappearance of 65 species of birds.

   These effects can result in adaptive anomalies on islands. An animal on Barro Colorado may be adapted, courtesy of having evolved in the region's forests, to avoiding jaguars. Those adaptations, however, are redundant – vestigial – on Barro Colorado. Darwin noticed the same phenomenon in island plants: they are adapted for interaction with species that are absent:

> Many remarkable little facts could be given with respect to the inhabitants of remote islands. For instance, in certain islands not tenanted by mammals, some of the endemic plants have beautifully hooked seeds; yet few relations are more striking than the adaptation of hooked seeds for transportal by the wool and fur of quadrupeds.
>
> *(Darwin 1859/1985, p. 381)*

Hanging on: island relicts

Living on a small island can cut off natural populations from the mainstream of life. This is especially true of ancient, isolated fragment islands – once, but no longer, connected to the mainland – such as Madagascar or New Zealand. Evolution may proceed elsewhere but the

Figure 4.3 Tuatara (*Sphenodon punctatus*).

Source: Photo by Rod Morris, Poor Knights Islands, New Zealand, 1985. By kind permission of the New Zealand Department of Conservation.

inhabitants of the island remain blissfully ignorant of events over the sea (though of course evolution may also be occurring independently on the island). Islands sometimes are, as a result, the home of ancient taxa which were long ago widespread but have been driven to extinction elsewhere by competitively superior invaders or descendants. Islands are thus in effect evolutionary time capsules, some of them preserving a glimpse of the past.

One extraordinary example of an island relict is the tuatara, *Sphenodon* (see Figure 4.3). This is a reptile represented by two species whose populations are scattered over some of New Zealand's small offshore islands. Although it looks superficially lizard-like, the tuatara in fact belongs to an ancient group of reptiles, the Sphenodontia, dating from the time of the dinosaurs; its closest relatives disappeared from the fossil record 60 million years ago. The details of its biology are certainly not lizard-like: for example, the tuatara male lacks a penis such that copulation involves simple cloaca–cloaca contact. Tuataras apparently occurred earlier on the main islands of New Zealand but are now limited to off-lying islets. The process of marginalisation thus continues: where once the Sphenodontia were globally distributed, the entire lineage has been relegated first to New Zealand, and now to New Zealand's own satellite islands (Newman 1987).

Niche expansion: making the most of ecological opportunities

The extent to which ecological communities are structured by competition remains controversial. Islands, however, have provided unequivocal evidence of *character displacement*, a process that is mediated by competition. The various species of Darwin's finch are inconsistently distributed among the Galápagos islands: on some islands, species A may co-occur with species B, while, on others, either A or B is absent. Comparing the beak morphologies of populations of two species,

Geospiza fuliginosa and *G. fortis*, reveals an interesting pattern. When populations of each species are allopatric (geographically separated from each other) – *G. fuliginosa* on Los Hermanos; *G. fortis* on Daphne – their beak morphologies are similar, suggesting that they are each separately foraging on similar resources on each island. When, however, the two species are sympatric (occurring together), as is the case of the two species on Santa Cruz, their bills differ significantly (Schluter et al. 1985). Competition on Santa Cruz has driven the two species apart: natural selection has effectively intervened to minimise competition between them (see Figure 4.4).

Unusual ecological opportunities create disharmony on islands. In an ecologically packed ecosystem on the mainland, by contrast, individuals that deviate from a species' ecological specialty will stray into another species' niche and lose out competitively. Imagine two species of birds, one with a small bill (adapted for eating small seeds) and one with a large one (for large seeds). Both species are present in a mainland ecosystem. Those small-billed individuals whose bills are relatively large may try to exploit large seeds, but they are inevitably out-competed by the large seed specialist. Now, however, imagine that the small-billed species is on an island from which the large-billed one is absent. The previous competitive constraint operating against the small-billed birds that attempt to feed on large seeds no longer applies. The small-billed species is, under the influence of natural selection, free to expand its niche. Natural selection will favour the exploitation of the untapped large seed resource as this extends the overall resource base available to individuals. We therefore expect to see instances of *niche expansion* born of this form of competitive release.

One of the most spectacular sets of examples of niche expansion can be found on the Hawaiian Islands, where, remarkably, there are no native ants (Wilson 1990), no native amphibians, no native terrestrial reptiles, and no native mammals, except a bat and a seal. (Human-facilitated introduction of ants has changed this within historical times. Today, Hawai'i's native invertebrate fauna is severely threatened by ant invaders, particularly *Linepithema humile* and *Pheidole megacephala* (Krushelnycky and Gillespie 2008).) Ants typically occupy an insect scavenger/predator

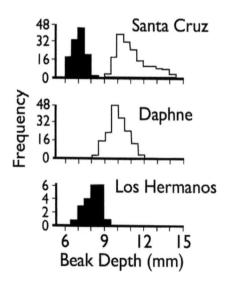

Figure 4.4 Character displacement in two species of Darwin's finches – *Geospiza fortis* (white bars) and *G. fuliginosa* (black bars) – on three Galápagos islands.

Source: Redrawn by Andrew Berry, from Schluter et al. (1985).

niche; Hawai'i, then, was missing an entire piece of the ecological jigsaw puzzle. The result has been evolutionary opportunism: the expansion of the niches of other species.

Eupithecia is a genus of geometrid moth that has forsaken the herbivorous habits of its relatives (indeed, of virtually all Lepidoptera) in favour of predation on Hawai'i. These are inchworm caterpillars whose 'looping' gait has been modified into a wait-and-strike predatory mechanism. This is an extraordinary evolutionary transformation. Evolution has long fashioned caterpillars as specialised plant-eaters and worm-like foliage chompers; but *Eupithecia* has subverted the entire scheme to become an ambush predator (Montgomery 1982).

Megalagrion oahuense is a Hawaiian damselfly. Larvae of the order *Odonata* are aquatic; only the adult lives out of the water. Not so in the case of *M. oahuense*, whose larvae can be found hunting insect prey on the forest floor of the wet mountain forests that are their home (Wilson 2010). Here, we have another remarkable shift: from aquatic to terrestrial.

New Zealand is almost as ant-poor as the Hawaiian Islands and its native mammal fauna consists of just three species of bat, one of which is presumed to be extinct. Thus, New Zealand has no native rodents. (But the story is the same since the arrival of humans, about 1,000 years ago in New Zealand's case. Introduced rodents are one of the many crises confronting New Zealand's native fauna.) Rodents are a major guild of generalised scavenger-omnivore in terrestrial ecosystems. Presumably as a result of the absence of rodents, some of New Zealand's native species have undergone niche expansion, to fill this ecological vacuum created by the absence of rodents and other small mammals. *Mystacina tuberculata*, the lesser short-tailed bat, has adopted a weirdly rodent-like habit. Unlike any other bat, it forages primarily on the ground, shuffling through the leaf litter of the forest floor using its folded front wings as forelimbs. *M. tuberculata* is a bat doing its best to be a mouse (Daniel 1990) (see Figure 4.5).

Figure 4.5 New Zealand lesser short-tailed bat (*Mystacina tuberculata*).

Source: Photo by Brian D. Lloyd, New Zealand, 1997. By kind permission of the photographer and the New Zealand Department of Conservation.

Rodent mimicry on New Zealand has gone beyond mammals. A group of remarkable crickets, the wetas (see Figure 4.6), has diversified in New Zealand into over 100 species in two families: *Rhaphidophoridae* and *Anostostomatidae*. Among them is the world's heaviest insect, the giant weta, *Deinacrida heteracantha*: 71g and up to 15cm long. Forest-dwelling wetas often live in holes, are mainly nocturnal, and forage generally on the forest floor for plant matter and/or dead insects. Wetas have evolved to become surrogate rodents (Gibbs 2001).

Adaptive radiation: evolution in action

Thus the several islands of the Galápagos Archipelago are tenanted, as I have elsewhere shown, in a quite marvellous manner, by very closely related species; so that the inhabitants of each separate island, though mostly distinct, are related in an incomparably closer degree to each other than to the inhabitants of any other part of the world.

(Darwin 1859/1985, p. 387)

The most striking evolutionary phenomenon associated with remote *de novo* islands (but not exclusively so) is adaptive radiation, a burst in evolutionary diversification yielding a range of species derived from a single common ancestor and showing distinct but related adaptations (Schluter 2000). Three ingredients are critical to the process:

- *Ecological opportunity.* As with niche expansion above, natural selection will drive evolutionary changes that take advantage of unexploited ecological resources. Adaptive radiation

Figure 4.6   Pseudo-rodent: New Zealand's giant weta (*Deinacrida heteracantha*, also known as *Wetapunga*) has evolved to take advantage of the ecological opportunities afforded by the absence of native rodents.

Source: Photo by Dinobass, Wikimedia Commons, https://commons.wikimedia.org/wiki/File:Wetapunga.jpg.

cannot therefore typically occur in an ecologically saturated environment: that is, one in which all the available resources are fully exploited by the species already in place. On an island with unfilled ecological space, taxa evolve to fill that space. Disharmony is amplified as ecological space fills up: there are, for example, hundreds of species of the moth in a single genus, *Hyposmocoma*, in the Hawaiian Islands, with different species taking advantage of different ecological opportunities, even preying on snails (Haines et al. 2014). So the ecosystem becomes increasingly disharmonic as diversity increases.

- *Isolation.* The extent of evolutionary diversification that can occur within a species is limited. Think of our own species: different regional groups have indeed undergone divergent evolution – for skin colour, facial type, etc. – but the underlying unity of the species is incontrovertible. The most different humans are infinitely more similar to each other, genetically and otherwise, than either of them is to a chimpanzee. It is *speciation* – the process that results in a single gene pool splitting into two separate ones – that is critical to biological diversification. Simply put, speciation is a by-product of genetic divergence between two populations: once two populations are sufficiently genetically distinct, their members cease to be able to inter-breed, to exchange genes. Once that threshold is passed, we deem speciation to have occurred. Models of *how* speciation occurs remain hotly debated. How often, for example, do new species arise sympatrically, despite the homogenising effect of gene flow between the diverging populations? One aspect of speciation theory, however, that is not controversial is the importance of geographical isolation. Speciation, in fact, is the inevitable outcome of long-term allopatry simply because different mutations will accumulate independently in each population. Islands facilitate this process: island populations are geographically isolated from their sister populations on the mainland, and, in the case of archipelagos, from sister populations on other islands (see Figure 4.7).

- *A new environment.* An island population may be confronted by environmental conditions that differ significantly from those encountered by its sister population on the mainland. Natural selection may then enhance genetic divergence between offspring and parent

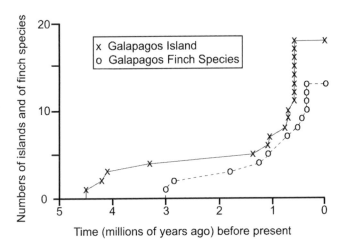

Figure 4.7 The role of allopatry in the diversification of Darwin's finches. The rate of generation of new species of finch is correlated with the availability of islands to colonise – i.e., with the accompanying opportunities for allopatry.

Source: Modified by Andrew Berry from Grant and Grant (1996).

populations. We know from empirical studies that adaptive responses to new island conditions may be remarkably rapid (Millien 2006). Stuart et al. (2014) found that significant ecological character displacement in foraging behaviour could occur in a mere three years in an *Anolis* lizard species responding to the presence of an experimentally introduced congener (a member of the same genus).

A simple model for a classical adaptive radiation is as follows. A small number of individuals arrives on an island, which may be in the early phases of biological colonisation. In practice, successful colonisation events, at least for sexual species, involve several individuals, though of course in principle a single gravid female would suffice. A key determinant of whether or not the incipient population can establish itself is the successional stage of the island. A group of seed-eating birds, for example, will only thrive if there is already an established plant community; should those birds arrive earlier in the succession, before seed-bearing plants have become established, their proto-population is doomed.

Natural selection will favour those individuals capable of exploiting an available resource. Say our species is small-billed. On the mainland, those individuals of the small-billed species with unusually larger bills that tried to compete with members of another species, with a large bill, for large seeds lost out competitively: the large-seed specialists were superior large-seed harvesters. However, on the island, in the absence of the large-billed, large-seed specialist, those same members of the small-billed species, with relatively larger bills, find themselves in seed-eating nirvana: they have hardware – a slightly larger bill – to cope with the large seeds, and there is no competition for this resource. In principle, one of two things can happen at this stage. Evolution could favour a Jack-of-all-trades bill, such that the island population's bill becomes, say, intermediate in size between the original small and large bills. Such an intermediate bill would permit the exploitation of both small and large seeds, though it is likely that the Jack-of-all bill would be competitively inferior on small seeds to a small-seed specialist and on large seeds to a large-seed specialist. The aphorism 'Jack (Jill) of all trades, master or mistress of none' applies as much to evolution as it does to human endeavour. An alternative outcome is the evolution of two 'master' populations: a small-seed specialised population and a large-seed specialised one. Natural selection, in short, would re-create on the island the situation on the mainland.

This latter process of producing two specialised populations is most readily achieved when the two diverging populations are geographically separated from each other. If the populations are sympatric, interbreeding between them constantly undoes the divergence instilled by natural selection. Sympatric speciation can only occur if there is sufficiently strong disruptive selection (i.e. selection in favour of the two extremes of a distribution, and against its middle) to counteract the homogenising influence of gene flow. In sympatry, then, the 'Jack of all' outcome is more likely: no speciation, and a single, generally adapted species. In general, we expect instances of sympatric speciation to be rare, as confirmed by studies of patterns of speciation on islands (Coyne and Price 2000).

Geographic isolation facilitates speciation. This can result in new adaptive traits – the ability to exploit large seeds in our example – becoming biologically incarnated in a new species. In an archipelago, it is easy to see how populations can become isolated from one another. Imagine that our initial immigrant population was established on island P of an archipelago; now imagine that some individuals migrate to the neighbouring island, Q. Colonisation of Q is not the same kind of low probability event as the original arrival-from-the-mainland one, which may have involved the crossing of hundreds or thousands of kilometres of open ocean; getting from P to Q is, relatively speaking, a trivial island hop. Despite the relative proximity of P to Q, however, the two populations may nevertheless be fully isolated from each other, each one therefore at liberty to follow its own evolutionary trajectory free from genetic interference from the other. In principle, then, the populations on P and Q can evolve into species that are distinct from each other and from the parent species on the mainland. It is this mix of adaptive evolution and genetic divergence in allopatry that is critical to adaptive radiation on islands.

Note that our incipient species on islands P and Q may in fact produce rather similar adaptations. They may both, for example, evolve the intermediate-sized bill that serves adequately for processing both large and small seeds. A later phase of evolution may then drive them apart. If the P and Q species come into contact with each other in what is termed secondary contact, natural selection may act to enhance the distinction between them. But if the two populations have not been isolated for long enough – that is, individuals from P are still capable of reproducing with individuals from Q (speciation has not yet occurred) – the two incipient species will merge into one. The two partially diverged gene pools will become one. If, alternatively, the P and Q populations are sufficiently diverged for speciation to have occurred, merging into one is no longer a possibility because they are incapable of exchanging genes. What happens on secondary contact in this case? It depends on each population's ecological predilections. Assume that the P and Q populations have both evolved the intermediate bill type. The two species will be competing with approximately equivalent efficacy for the same set of resources. Basic ecological theory mandates that one of two things will happen under these circumstances: one of the two populations will go extinct, out-competed by the other, or, under the influence of natural selection, one or both of the populations will undergo character displacement in order to minimise inter-specific competition. If the latter, we may come to see the P species specialise on small seeds and the Q one on large.

This process of genetic divergence in allopatry, coupled with adaptive evolution in response, in particular, to the ecological opportunities born of island living, may result in speciation. Classically, we see a suite of closely related species – all of them descended from the single colonising ancestor species – that seem to mimic (because they have adopted equivalent ecological roles) a disparate group of mainland species.

Island biologists take special pleasure in documenting adaptive radiations on their favourite islands and we have many excellent examples. Why, however, stray too far from the classics?

Darwin's finches, Galápagos

For historical reasons, Darwin's finches occupy a privileged position in the pantheon of adaptive radiations. In fact, they were initially merely a source of confusion for Darwin, and he was never keen to make the finches too central a plank of his argument. It was David Lack, the British ornithologist whose monograph (Lack 1947) first made coherent sense of the evolutionary tale embodied by the finches, who popularised the name, Darwin's finches. It was a solution to a geographical problem. The first tendency was to call them Galápagos finches but this would be inaccurate since one of the 14 species, *Pinaroloxias inornata*, is endemic to Cocos Island, several hundred kilometres to the northeast of the Galápagos. Calling them Darwin's finches was therefore a convenient compromise, and it is the name that has given them iconic status in evolutionary biology.

The finches' ancestors arrived in the Galápagos around three million years ago. Genetic estimates suggest a founding population of some 30 birds. The Galápagos are relatively simple in terms of habitat, the entire archipelago supporting only 29 species of land bird; 22 are endemic and 14 of those are finches (see Figure 4.8). As adaptive radiations go, the Darwin's finch radiation is not especially morphologically striking: there is relatively little differentiation between species and it is virtually impossible, even for experienced Galápagos naturalists, to distinguish

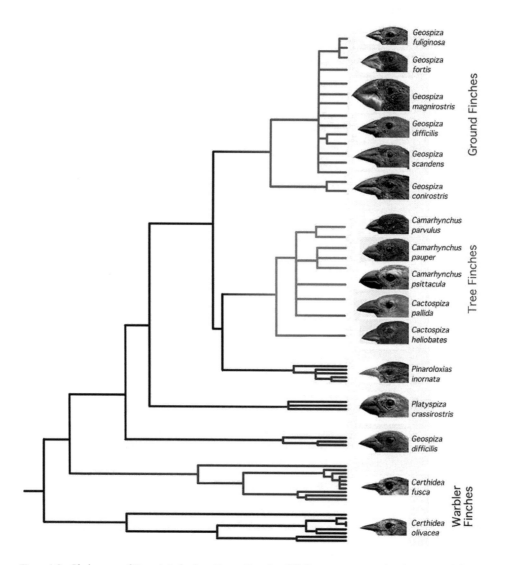

Figure 4.8 Phylogeny of Darwin's finches. Note *Geospiza difficilis* appears anomalously in two different positions in the tree. This reflects ongoing hybridisation among the members of this group of very closely (and recently diverged) species (Lamichhaney et al. 2015).

Source: Modified by Ken Petren, University of Cincinnati, Ohio, USA, from Petren et al. (2005).

among the species (Grant 1999; for a popular account, see Weiner 1995). Most of the adaptive evolution has involved changes in bill morphology which in turn are associated with changes in feeding behaviour and diet. The group as a whole feeds on virtually everything the Galápagos has to offer, from insects and seeds to the pulp of cactus pads and flower pollen. The most remarkable feeding adaptation is seen in the woodpecker finch, *Camarhynchus pallidus*, which uses a strip of wood – often a cactus spine – as a tool in its attempts to winkle insects out of their refuges. Sharp-billed ground finches are known to peck ectoparasites off other components of the local fauna, such as the giant tortoises or iguanas; on two of the northern islands of the archipelago, this behaviour has developed into vampirism with the finches taking a blood meal from the base of the tail feathers of nesting seabirds.

Peter and Rosemary Grant, whose long-term studies of the finches have yielded one of evolutionary biology's finest datasets, have observed that individuals from different species will very occasionally hybridise and successfully rear their hybrid offspring (Grant and Grant 2015). This recurrent bleeding of genetic material from one species into another may account for the lack of marked morphological divergence among the species, and suggests that the adaptive radiation of these species is still a work in progress (Grant et al. 2005). Presumably this is what accounts for the multiple appearances of one species, *Geospiza difficilis*, in different locations on the phylogeny (see Figure 4.8). There is no evidence, in contrast, of hybridisation among comparable Hawaiian honeycreeper species (see below).

For historical reasons, the finches have enjoyed special status as the much trumpeted textbook example of adaptive radiation. Thus, one might be easily forgiven for thinking that theirs is the only evolutionary tale of any significance from the Galápagos. Nothing could be further from the truth: the archipelago is home to many other groups that have undergone similarly dramatic adaptive radiations (Parent et al. 2008).

Honeycreepers, Hawaiian Islands

The Hawaiian Islands' extreme remoteness, coupled with its permissive climate, make it a living textbook of evolutionary biology. We have already noted its lack of ants; its fauna, however, is heavily disharmonic in many ways. There are just a handful of freshwater fish, no amphibians and reptiles, no land mammals; and ants are by no means the only missing insects. The 10,000 or so known endemic species of insects in the Hawaiian Islands are thought to have evolved from 400 immigrant species (Wilson 2010). The number of avian colonisations has been estimated to be in the order of 20; 81 per cent of Hawaiian bird species are endemic (Whittaker and Fernández-Palacios 2007). Prominent among these endemics is a group of finch-like birds, the Hawaiian honeycreepers, which are typically assigned to their own tribe, the *Drepanidini*, within the finch subfamily *Carduelinae* (goldfinches and their relatives).

Darwin's finches are the poster children of adaptive radiation, but the honeycreepers are more deserving of that status. There are more species of honeycreepers: 57, of which at least 25 have been driven to extinction since the arrival of humans on Hawai'i (James 2004); the honeycreepers show more marked and more extreme adaptive diversification, especially with respect to bill morphology; the honeycreepers are spectacularly coloured birds, as opposed to the drab grey-black-brown Darwin's finches; the Galápagos radiation is still in progress to a limited extent as there is evidence of continuing gene flow among species caused by hybridisation among species. Despite their distinctness, the honeycreeper radiation is relatively recent: the oldest extant island, Kauai, dates back some six million years, and the radiation originated at around the same time (Lerner et al. 2011). We thus see extensive morphological divergence in this group coupled with relatively little genetic divergence: the classic evolutionary footprint of adaptive radiation.

The diversity of form enshrined in the honeycreeper bill is extensive (see Figure 4.9). Species in the genus *Psittacirostra* have beefy parrot-like bills that are ideal for cracking large, tough seeds; species of *Drepanis* and *Hemignathus* have delicate, probing bills for extracting nectar from flowers and picking up small insects; *Hemignathus wilsoni* has taken over the woodpecker niche on Hawai'i (Hawai'i's disharmonic fauna is predictably bereft of true woodpeckers) with its weirdly asymmetrical 'overbite' bill. The short, stabbing lower mandible is used for hammering and chiselling wood and the delicate, curved, and much longer upper mandible is used to probe for and capture insects in the resulting cavity (Wilson 2010).

Figure 4.9   Hawaiian honeycreepers. A glimpse of one aspect of the radiation: males of three species of *Hemignathus*. The top exemplar, *H. wilsoni*, is an extant species; the bottom two, *H. hanapepe* and *H. vorpalis*, are extinct (James and Olson 2003).

With kind permission of Helen James, Smithsonian National Museum of Natural History, Washington DC, USA and The Auk.

Drosophila and other insects and spiders, Hawaiian Islands

The diversification of *Drosophila* (Diptera) is one of the most spectacular exemplars of adaptive radiation. Yet, it tends to be overlooked because they are, well, just flies (see Figure 4.10). There are over 800 endemic *Drosophila* species in the Hawaian Islands. One of the more remarkable aspects of this radiation is that it is considerably older than the extant above-sea islands (O'Grady and DeSalle 2008): Kauai, the oldest, is around six million years old while molecular analyses have dated the common ancestor of the *Drosophila* radiation to about 30 million years ago. The

Figure 4.10   *Drosophila conspicua*, an endemic Hawaiian picture-winged *Drosophila*. A (relatively) minuscule
Drosophila melanogaster (the familiar laboratory fruit fly, on left) provides scale.

Source: Photo by William P. Mull. © Bishop Museum, Honolulu HI, USA. Reproduced with permission.

key lies in the chain of sea-mounts – submerged, eroded, ex-islands – stretching away to the
northwest of the current islands. These are members of the archipelago that were formed earlier
than the current islands during the Pacific plate's slow march over the geological hotspot. *Drosophila* presumably have progressively colonised (and radiated upon) new islands as they emerged
from the ocean (and the old ones sank beneath the waves). Phylogenetic studies based on shifts
in chromosome structure have revealed that there is indeed an 'old island > new island' polar-
ity in the evolution of *Drosophila*, with the forms on the older (northwestern) islands typically
ancestral to those on the newer (southeastern) islands (Carson 1983). For *Drosophila*, the geological history of the islands has presented a series of emerging-then-submerging stepping stones.

A number of other insect and spider lineages have developed into massive species swarms
in the Hawaiian islands, including the moth genus *Hyposmocoma* (Haines et al. 2014) with over
350 species characterised by all sorts of different cases in which the caterpillars live. Other large
radiations include leafhoppers (Bennett and O'Grady 2012), psyllids (Percy 2017), and crickets
(Shaw and Gillespie 2016), as well as several lineages of spiders (Gillespie 2016). Comparative analysis of biological diversification across these and many other lineages reveals that the
evolution-tracking-geology pattern is a common feature of adaptive radiations on hot-spot
archipelagos (Shaw and Gillespie 2016). Hormiga et al. (2003) present a clear example of this
process in a spider group (see Figure 4.11). This stepping stone pattern can be accompanied by
repeated spurts of adaptive radiation within each island of the archipelago.

One important reason for the hyper-diversity of Hawaiian arthropods is their facility for
finding islands within islands. In particular, the frequent lava flows on the geologically active
(that is, youngest) island create what are locally known as *kipuka*. These are islands of undisturbed

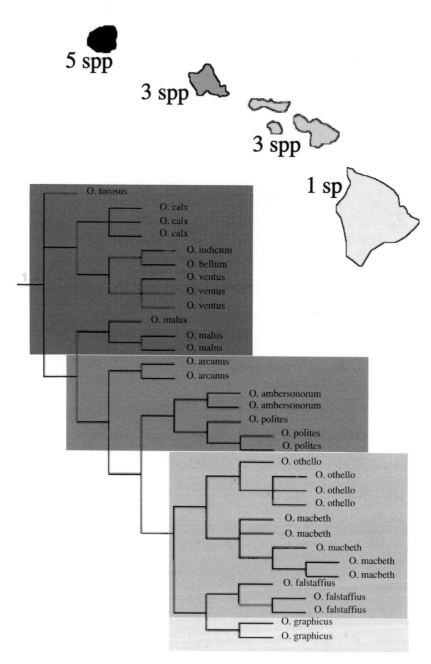

Figure 4.11 Phylogeny of sheet web spiders, *Orsonwelles*, with shading to indicate each species' home island. The Hawaiian islands have been generated by a volcanic hot-spot travelling southeast, so the spiders' evolution tracks the evolution of the archipelago.

Adapted by Rosemary Gillespie from Hormiga et al. (2003).

Figure 4.12   Hawaiian silversword (*Argyroxiphium sandwicense* ssp *macrocephalum*) on Mount Haleakala, Maui, Hawaiian islands.

Source: Photo by Steve Goldsmith, Austin College TX, USA. Reproduced with permission.

forest cut off from each other by inhospitable, sterile tracts of lava. From the point of view of an individual fly, Hawai'i is a pair of nested archipelagos: the islands themselves and then the complex of 'habitat islands' within each island. Each type of island – whether a 'true' geographical island or a 'habitat' one – has the potential to spawn and harbour an independent and genetically isolated population that may ultimately diverge to become a new species. Given the complexity of this island-rich world, it is not surprising that arthropod diversity is so remarkably pronounced.

Tarweeds and other plant radiations, Hawai'i

The botanical equivalent of Hawai'i's honeycreepers is the endemic radiation of tarweeds, the *Madiinae* (*Asteraceae*) (see Figure 4.12). Following the arrival of an ancestral species from California in Hawai'i some five million years ago, this group has diversified into three endemic genera of the so-called silversword alliance, *Argyroxiphium*, *Dubautia*, and *Wilkesia*. There are 50 species, of which 25 are single island endemics (Carlquist et al. 2003). The radiation of lobelioid plants in the Hawaiian Islands is a good deal larger (125 species), the largest on any island archipelago. Like the tarweeds, these plants show a tremendous diversity of form (Givnish et al. 2008).

Anolis lizards, Caribbean

For many organisms, the Caribbean Islands are not particularly isolated. However, for *Anolis* lizards, they have imposed sufficient isolation to allow adaptive radiation (Losos 2009). One striking aspect of this radiation is the role of convergent evolution in similar habitats, leading to parallel evolution of similar ecological forms on the different islands. Each species can be recognised as an 'ecomorph', occupying a characteristic microhabitat (e.g., tree twigs, grass, tree trunk) with corresponding morphological and behavioural attributes named for the part of the habitat they occupy. Most remarkably, similar sets of ecomorphs are found on each island and have generally arisen through convergent evolution on each island, showing that similar communities on the different islands have evolved independently (see Figure 4.13).

Cichlids, African Rift Lakes

Lakes are typically islands from the perspective of the fish that inhabit them. The African Rift Lakes host vast numbers of endemic species (Brawand et al. 2014) (see Figure 4.14):

Lake Victoria	500 species
Lake Malawi	500 species
Lake Tanganyika	250 species

Figure 4.13   Anolis cristatellus in Dominica.

Source: Photo by Postdlf, Wikimedia Commons, https://commons.wikimedia.org/wiki/File:Anolis_cristatellus_in_Picard,_Dominica-2012_02_15_0357.jpg.

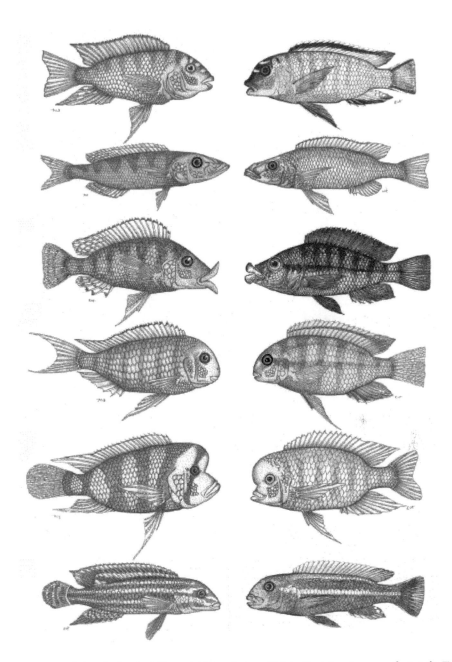

Figure 4.14 Parallel evolution in African cichlid radiations. Fish in the left column are from Lake Tanganyika; those in the right column are from Lake Malawi. Those in each column are more closely related to each other than they are to the superficially similar fish in the other column.

Source: Drawing by R. Craig Albertson, University of Massachusetts, USA. Reproduced with permission.

In a manner that parallels adaptive evolution in bill morphology in birds, the primary feature of cichlids that is subject to adaptive change is the mouthparts – or, to be precise, their pharyngeal jaws. Face and jaw morphology vary widely, even among closely related species, according to diet. One of the most striking feeding adaptations is scale feeding in which a predator approaches its prey by stealth and rips a mouthful of scales out of its side. The entire jaw apparatus is oriented to one side, right or left, depending on which side the fish makes its approach on.

Perhaps the most remarkable of these parallel radiations is the one in Lake Victoria, where all 500 species have evolved within the past 15,000 years. Meier et al. (2017) have recently provided evidence of an unusual genetic mechanism behind this, suggesting that an ancient hybridisation event was crucial. This occurred some 150,000 years ago, between two distantly related cichlid lineages, one from the Congo basin and the other from the Upper Nile. The resulting hybrid, with its two sets of divergent alleles – the genetic raw material for future rounds of natural selection – seeded the evolutionary frenzy that produced all Lake Victoria's cichlids.

Understand adaptive radiation and you understand evolution. All the key processes of evolution – natural selection, divergence, competition, speciation – are integral to adaptive radiation. In fact, the island-based examples laid out above can be regarded as examples in microcosm of the process whereby life took over our planet. The 'Cambrian Explosion' is the term applied to the sudden appearance of many major animal groups in one seemingly frenzied burst of evolutionary innovation. The earliest known metazoan animals date from about 620 to 550 million years ago (mya) but these were unimpressive, two-dimensional creatures. Then, with the Lower Cambrian, at about 530 mya, the explosion occurred: a quantum leap in the biological colonisation of the planet. Animals – in just the sense we think of them today – had arrived. Not only was a wide taxonomic range represented – priapulids, annelids, arthropods, even chordates – but there was considerable morphological diversity within some of these groups. And the world had suddenly become ecologically diverse, too: animals were no longer inert filter feeders but had branched out into more ambitious trophic domains, including predation (Gould 1989). How did this ecological and evolutionary revolution occur? There are plenty of contentious ideas; but one factor is clear: that the Cambrian Explosion was an adaptive radiation on a grand scale. In this case, the island was Planet Earth. This, like the Galápagos when the ancestral finches first arrived there, was ecologically 'empty': an invitation to wholesale evolutionary diversification. The Cambrian Explosion was a massive and worldwide instance of adaptive radiation.

Convergent evolution: the island effect

A striking aspect of evolution on islands is that it often tracks well-worn trajectories: what happens on one island also happens on other islands. Some aspects common to all islands result in shared evolutionary forces on islands the world over. Table 4.2 details four examples of these repeated trends.

Attempting to explain these patterns is a popular parlour game for evolutionary ecologists, but in each case there is a single explanation that probably best fits the facts.

- *Flightlessness in birds.* As we have seen, islands are often too small to sustain populations of terrestrial mammalian predators, meaning that island endemics are released from the requirement that they be able to escape – and fly away from – predators.
- *Flightlessness in insects.* Darwin suggested that insects lose their wings on islands because flying is a dangerous business in an environment where any puff of wind can blow you out to sea and a watery grave. While this notion clearly is true in some cases, it is not the only important factor because there are plenty of low-lying, exposed islands whose insect faunas

Table 4.2 Shared evolutionary process on islands. (Data from: Carlquist 1965, Whittaker and Fernández-Palacios 2007, Williamson 1981).

Process at Work	Taxa and location

Flightlessness in birds	Galápagos: Cormorant, *Phalacrocorax harrisi*
	Campbell and Auckland Islands (New Zealand): flightless duck, *Anas aucklandica*
	Réunion: solitaire, *Ornithaptera solitaria* (extinct)
	Flightless rails:
	  Inaccessible Island: Rail, *Atlantisia rogersi*
	  Aldabra: Rail, *Dryolimnas aldabranus*
	  Takahe (New Zealand): Rail, *Porphyrio mantelli*
	New Zealand: Kakapo, *Strigops habroptilus*
	Cuba: Giant flightless owl, *Ornimegalonyx oteroi*, (extinct)
	Hawai'i: Moa-nalos, *Chelychelynechen quassus, Thambetochen xanion, T. chauliodous, Ptaiochen pau* (all extinct)
	New Caledonia: Kagu, *Rhynochetos jubatus*
	Mauritius: Dodo, *Raphus cucullatus* (extinct)
	Peru/Bolivia (i.e. habitat island): Titicaca Grebe, *Centropelma micropterum*

Flightlessness in insects	Madeira: flightless components of the beetle fauna: e.g. *Eurygnathus latreillei* (carabid); *Loricera wollastoni* (carabid); *Atlantis vespertinus* (weevil); *Echinosoma porcellus* (weevil). Hawai'i: carabid beetle fauna: 184 spp., only 20 winged Hawai'i: lacewings: *Pseudopsectra swezeyi, P. usingeri, Nesomicromus drepanoides* Tristan da Cunha: Moth, *Dimorphinoctua cunhaensis* Amsterdam Islands: Moth, *Brachypteragrotis patricei* Tristan da Cunha: Fruit flies without wings: e.g. *Scaptomyza frustulifera*
Dwarfism in large vertebrates	Mediterranean (Malta): Pygmy elephant *Palaeoloxodon falconeri* (extinct) Channel Islands: Dwarf mammoth *Mammuthus exilis* Madagascar: pigmy hippo *Hippopotamus madagascariensis*
Gigantism	**In vertebrates** Komodo (Indonesia): Komodo dragon, *Varanus komodoensis* Galápagos and Aladabra: Tortoises (*Chelonoidis nigra* and *Aldabrachelys gigantea*) California night lizard: *Xantusia riversiana* on the Channel Islands compared to the smaller mainland form, *X. henshawi* Gargano (in the past, an island off the Italian coast): predatory giant hedgehog, *Deinogalerix koenigswaldi* (see Figure 4.15) Gough Island: House mouse, *Mus musculus domesticus* (see Figure 4.16) Madagascar: Elephant bird, *Aepyornis maximus* (extinct) **In plants (increase in arboreal habit and woodiness)** Galápagos: Prickly pear, *Opuntia echios* Hawai'i: *Lobelia* Plantains, *Plantago*. St Helena: *P. robusta*; Hawai'i: *P. princeps*; Juan Fernandez: *P. fernandeziana*; Canary Islands: *P. arborescens* St Helena: Composites attain tree-like form: *Aster burchelli, Psiadia rotundifolia, Senecio leucadendron*

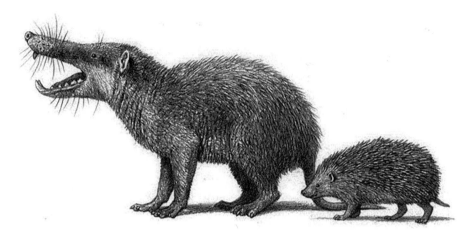

Figure 4.15   Island giantism. A reconstructed life appearance of the giant hedgehog *Deinogalerix koenig-swaldi* (with a modern hedgehog, *Erinaceus europaeus*, for scale).

Source: Artwork by Mauricio Antón. Wikimedia Commons, https://commons.wikimedia.org/wiki/File:Deinogalerix_Gargano_fauna.jpg.

Figure 4.16   More island gigantism. A house mouse from Gough Island, South Atlantic Ocean (left) with a mainland conspecific (right). These are both *Mus musculus domesticus*, the same species as the laboratory mouse (Gray et al. 2015).

Source: Photo courtesy of Michelle Parmenter, Dept. of Genetics, University of Wisconsin-Madison, USA. Reproduced with permission.

are winged. Flightlessness, then, is often a response to the specific demands of particular island habitats. For example, a species' preferred habitat may be very small in area on an island and there is therefore no requirement for winged dispersal within it (Darlington 1943, Gillespie et al. 2012).

- *Dwarfism in vertebrates*. The lack of terrestrial mammalian predators on islands permits herbivore populations initially to boom. This in turn places marked stress upon plant resources, and selection acts on herbivore populations to reduce their food requirements. One obvious way of doing this is through a reduction in size.

- *Gigantism.* How can islands promote dwarfism on one hand, and gigantism on the other? The simple answer is that dwarfism stems from large species becoming smaller, and gigantism from small species becoming larger. Again here the absence of top predators may be a factor: the onus, in their absence, is no longer on growing up as fast as possible. More leisurely development may result in greater adult size. In plants, gigantism takes the form of shrubs evolving to become tree-like and woody. Partly, this may be stimulated by the absence of normal tree species – the ex-shrubs are, in good island evolution fashion, taking advantage of an available ecological opportunity. It has also been suggested that woodiness is a response to insular environmental conditions that permit year-round growth (Runemark et al. 2015).

Invasion! Introduced species and the decline of island biodiversity

To conclude, a note of warning: island ecosystems are under a merciless siege. Habitat destruction and the introduction of invasive species are wreaking a terrible havoc on islands the world over (Quammen 1997). Since they are home to so many endemics, courtesy of the evolutionary processes discussed here, and their populations are necessarily limited in extent (and therefore especially fragile), islands are at the epicentre of the ongoing extinction crisis. Whittaker and Fernández-Palacios (2007, p. 293) estimate that, of the 855 extinctions that have been documented since 1600, 504 (59 per cent) have occurred on islands.

Many of evolution's most wonderful creations reside on islands. Conservation is vital. One story will suffice. The Brown Treesnake, *Boiga irregularis* (see Figure 4.17), a native of east Asia and Australia, was inadvertently introduced to the Pacific island of Guam some time shortly after

Figure 4.17 Guam's nemesis: *Boiga irregularis*, the Brown Treesnake.

Source: Photograph by Gordon H. Rodda, U.S. Fish Wildlife Service. Wikimedia Commons, https://commons.wiki media.org/wiki/File:Brown_tree_snake_picture.jpg.

the Second World War. Finding itself in an environment free of natural enemies and pathogens, the snake has established populations of unprecedented density. Within a mere 50 years, it had driven to extinction ten of Guam's 12 native forest bird species (Rodda and Savidge 2007). Rogers et al. (2017) have shown that the absence of forest birds in Guam is having a severe knock-on effect on the Guam forest ecosystem: plants traditionally dependent on birds for the dispersal of their seeds are struggling, with seedling recruitment rates down by as much as 90 per cent. The long-term impact of *B. irregularis* is thus likely to be even more catastrophic than the short-term one. For control, the idea of introducing to Guam a natural predator of *B. irregularis* – for example, the red-bellied black snake, *Pseudechis porphyriacus* – may seem initially attractive, but what havoc would *P. porphyriacus*, a toxic generalist predator, additionally wreak on Guam's remaining native fauna? In general, the elimination of island invasives has been achieved using brute force methods, such as targeted saturation poisoning campaigns (Nicholls 2013). *B. irregularis*, unfortunately, has proved hard to eradicate, though curiously it is sensitive to even small doses of acetaminophen (Tylenol), and the snakes have been successfully baited using dead mice tainted with acetaminophen. But even this strategy comes with a potential headache: the passage of acetaminophen through the ecosystem as scavengers consume the dead snakes. Happily, despite the abundance of cautionary tales typified by Guam, there are eradication success stories. There have been some 900 recorded successful eliminations of mammalian island invasives (DIISE 2015), albeit typically from very small islands. With respect to rather larger islands, a concerted effort eliminated rats and mice by 2011, and rabbits by 2014, from Macquarie, a 130 km^2 island to the south of New Zealand (DIISE 2015).

References

Albertson, R.C., and Kocher, T.D. (2006) Genetic and developmental basis of cichlid trophic diversity. *Heredity*, 97, 211–221.

Baldursson, S., and Ingadottir, A. (eds.) (2006) *Nomination of Surtsey for the UNESCO World Heritage List*, Reykjavik, Iceland, Icelandic Institute of Natural History.

Bennett, G.M., and O'Grady, P. (2012) Host – plants shape insect diversity: phylogeny, origin, and species diversity of native Hawaiian leafhoppers (Cicadellidae: Nesophrosyne). *Molecular Phylogenetics and Evolution*, 65(2), 705–717.

Brawand, D., Wagner, C.E., Li, Y.I., Malinsky, M., Keller, I., Fan, S., . . . Di Palma, F. (2014) The genomic substrate for adaptive radiation in African cichlid fish. *Nature*, 513, 375–381.

Carlquist, S. (1965) *Island life: a natural history of the islands of the world*. New York, Natural History Press.

Carlquist, S., Baldwin, B.G., and Carr, G.D. (eds.) (2003) *Tarweeds and silverswords: evolution of the Madiinae (Asteraceae)*. St Louis, MI, Missouri Botanical Garden Press.

Carson, H.L. (1983) Chromosomal sequences and inter-island colonisations in Hawaiian Drosophilidae. *Genetics*, 103(3), 465–482.

Coyne, J.A., and Price, T.D. (2000) Little evidence for sympatric speciation in island birds. *Evolution*, 54(6), 2166–2171.

Daniel, M.J. (1990) Bats: Order Chiroptera. In C.M. King (ed.), *Handbook of New Zealand Mammals*. Auckland, Oxford University Press, pp. 114–137.

Darlington, P.J. Jr. (1943) Carabidae of mountains and islands: data on the evolution of isolated faunas, and on atrophy of wings. *Ecological Monographs*, 13(1), 37–61.

Darwin, C. (1859/1985) *The origin of species*, Harmondsworth, Penguin.

Diamond, J.M. (1972) Biogeographic kinetics: estimation of relaxation times for avifaunas of southwest Pacific islands. *Proceedings of the National Academy of Sciences of the United States of America*, 69, 3199–3203.

Diamond, J.M. (1984) 'Normal' extinctions of isolated populations. In M. Nitecki (ed.), *Extinctions*. Chicago, IL, University of Chicago Press, pp. 191–246.

DIISE (2015) The database of island invasive species eradications, developed by island conservation, coastal conservation action laboratory UCSC, IUCN SSC invasive species specialist group, University of Auckland and Landcare Research New Zealand. Available from: http://diise.islandconservation.org.

Gibbs, G. (2001) *The New Zealand weta*. Auckland, New Zealand, Reed Publishing.

Gillespie, R.G. (2007) Oceanic islands: models of diversity. In S.A. Levin (ed.), *Encyclopedia of biodiversity*, Oxford, Elsevier, pp. 1–13.

Gillespie, R.G. (2016) Island time and the interplay between ecology and evolution in species diversification. *Evolutionary Applications*, 9(1), 53–73.

Gillespie, R.G., Baldwin, B.G., Waters, J.M., Fraser, C.I., Nikula, R., and Roderick, G.K. (2012) Long-distance dispersal: a framework for hypothesis testing. *Trends in Ecology and Evolution*, 27, 47–56.

Gillespie, R.G., and Clague, D.A. (eds.) (2009) *Encyclopedia of islands*. Berkeley, CA, University of California Press.

Givnish, T.J., Millam, K.C., Mast, A.R., Patterson, T.B., Theim, T.J., Hipp, A.L., Henss, J.M., Smith, J.F., Wood, K.R., and Sytsma, K.J. (2008) Origin, adaptive radiation and diversification of the Hawaiian lobeliads (Asterales: Campanulaceae). *Proceedings of the Royal Society B. Royal Society*, 276, 407–416.

Gould, S.J. (1989) *Wonderful life: the Burgess shale and the nature of history*. New York, W.W. Norton and Company.

Grant, P.R. (1999) *The ecology and evolution of Darwin's finches*. Princeton, NJ, Princeton University Press.

Grant, P.R., and Grant, B.R. (1996) Speciation and hybridisation in island birds. *Philosophical Transactions of the Royal Society: Biological Sciences*, 351, 765–772.

Grant, P.R., and Grant, B.R. (2015) Introgressive hybridisation and natural selection in Darwin's finches. *Biological Journal of the Linnean Society.*, 117, 812–822.

Grant, P.R., Grant, B.R., and Petren, K. (2005) Hybridization in the Recent Past. *The American Naturalist*, 166(1), 56–67.

Gray, M.M., Parmenter, M.D., Hogan, C.A., Ford, I., Cuthbert, R.J., Ryan, P.G., Broman, K.W., and Payseur, B.A. (2015) Genetics of rapid and extreme size evolution in island mice. *Genetics*, 201, 213–228.

Haines, W.P., Schmitz, P., and Rubinoff, D. (2014) Ancient diversification of *Hyposmocoma* moths in Hawaii. *Nature Communications*, 5, 3502–3509.

Hormiga, G., Arnedo, M., and Gillespie, R.G. (2003) Speciation on a conveyor belt: sequential colonisation of the Hawaiian islands by *Orsonwelles* spiders (Araneae, Linyphiidae). *Systematic Biology*, 52, 70–88.

James, H.F. (2004) The osteology and phylogeny of the Hawaiian finch radiation (Fringillidae: Drepanidini), including extinct taxa. *Zoological Journal of the Linnean Society*, 141, 207–255.

James, H.F., and Olson, S.L. (2003) A giant new species of Nukupuu (Fringillidae: Drepanidini: Hemignathus) from the Island of Hawaii, *Auk*, 120(4), 970–981.

Krushelnycky, P.D., and Gillespie, R.G. (2008) Compositional and functional stability of arthropod communities in the face of ant invasions. *Ecological Applications*, 18, 1547–1562.

Lack, D. (1947) *Darwin's finches*. Cambridge, Cambridge University Press.

Lamichhaney, S., Berglund, J., Almén, M.S., Maqbool, K., Grabherr, M., Martinez-Barrio, A., Promerová, M., Rubin, C.J., Wang, C., Zamani, N., Grant, B.R., Grant, P.R., Webster, M.T., and Andersson, L (2015) Evolution of Darwin's finches and their beaks revealed by genome sequencing. *Nature*, 518, 371–375.

Lerner, H.R.L., Meyer, M., James, H.F., and Fleischer, R.C. (2011) Multilocus resolution of phylogeny and timescale in the extant adaptive radiation of Hawaiian Honeycreepers. *Current Biology*, 21, 1838–1844.

Losos, J. (2009) *Lizards in an evolutionary tree: ecology and adaptive radiation of anoles*. Berkeley CA, University of California Press.

MacArthur, R.H., and Wilson, E.O. (1967) *The theory of island biogeography*. Princeton, NJ, Princeton University Press.

Meier, J.I., Marques, D.A., Mwaiko, S., Wagner, C.E., Excoffier, L., and Seehausen, O. (2017) Ancient hybridisation fuels rapid cichlid fish adaptive radiations. *Nature Communications*, 8, Article No. 14363. Available from: www.nature.com/articles/ncomms14363.

Millien, V. (2006) Morphological evolution is accelerated among island mammals. *PLoS Biology*, 4, e321.

Montgomery, S.L. (1982) Biogeography of the moth genus Eupithecia in Oceania and the evolution of ambush predation in Hawaiian caterpillars (Lepidoptera: Geometridae). *Entomologia Generalis*, 8(1), 27–34.

Newman, D. (1987) *Tuatara*, endangered New Zealand wildlife series, Dunedin, New Zealand, John McIndoe.

Nicholls, H. (2013) Invasive species: the 18-km^2 rat trap. *Nature*, 497, 306–308.

O'Grady, P., and DeSalle, R. (2008) Out of Hawaii: the origin and biogeography of the genus Scaptomyza (Diptera: Drosophilidae). *Biology Letters*, 4(2), 195–199.

Parent, C.E., Caccone, A., and Petren, K. (2008) Colonisation and diversification of Galápagos terrestrial fauna: a phylogenetic and biogeographical synthesis. *Philosophical Transactions of the Royal Society B: Biological Sciences*, 363(1508), 3347–3361.

Percy, D.M. (2017) Making the most of your host: the Metrosideros-feeding psyllids (Hemiptera, Psylloidea) of the Hawaiian islands. *ZooKeys*, 649, 1–163.

Petren, K., Grant, B.R., Grant, P.R., and Keller, L.F. (2005) Comparative landscape genetics and the adaptive radiation of Darwin's finches: the role of peripheral isolation. *Molecular Ecology*, 14, 2943–2957.

Quammen, D. (1997) *The song of the dodo: island biogeography in an age of extinctions*. New York, Scribner Reprint.

Robinson, W.D. (1999) Long-term changes in the avifauna of Barro Colorado Island, Panama, a tropical forest isolate. *Conservation Biology*, 13, 85–97.

Rodda, G.H., and Savidge, J.A. (2007) Biology and impacts of Pacific island invasive species. 2. Boiga irregularis: the brown tree snake (Reptilia: Colubridae). *Pacific Science*, 61, 307–324.

Rogers, H.S., Buhle, E.R., Hille Ris Lambers, J., Fricke, E.C., Miller, R.H., and Tewksbury, J.J. (2017) Effects of an invasive predator cascade to plants via mutualism disruption. *Nature Communications*, 8, 14557, 1–8.

Runemark, A., Kostas S., and Erik I. Svensson, E.I. (2015) Ecological explanations to island gigantism: dietary niche divergence, predation, and size in an endemic lizard. *Ecology*, 96, 2077–2092.

Schluter, D. (2000) *The ecology of adaptive radiation*. Oxford, Oxford University Press.

Schluter, D., Price, T.D., and Grant, P.R. (1985) Ecological character displacement in Darwin's finches. *Science*, 227, 1056–1059.

Shaw, K.L.R., and Gillespie, R.G. (2016) Comparative phylogeography of oceanic archipelagos: hotspots for inferences of evolutionary process. *Proceedings of the National Academy of Science*, 113, 7986–7993.

Simberloff, D.S., and Wilson, E.O. (1969) Experimental zoogeography of islands: the colonisation of empty islands. *Ecology*, 50, 278–296.

Simkin, T., and Fiske, R.S. (1983) *Krakatau 1883: The volcanic eruption and its effects*. Washington, DC, Smithsonian Institution Press.

Stuart, Y.E., Campbell, T.S., Hohenlohe, P.A., Reynolds, R.G., Revell, L.J., and Losos, J.B. (2014) Rapid evolution of a native species following invasion by a congener. *Science*, 346, 463–466.

Thornton, I.W.B., Zann, R.A., and Van Balen, S. (1993) Colonisation of Rakata (Krakatau island) by non-migrant land birds from 1883 to 1992 and implications for the value of island equilibrium theory. *Journal of Biogeography*, 37, 441–452.

Tilman, D., May, R.M, Lehman, C.L., and Nowak, M.A. (1994) Habitat destruction and extinction debt. *Nature*, 371, 65–66.

Wallace, A.R. (1857) On the natural history of the Aru Islands. *Annals and Magazine of Natural History*, 20 (2nd s.), 473–485.

Wallace, A.R. (1880) *Island life*. London, Palgrave Macmillan.

Weiner, J. (1880/1995) *The beak of the finch*. New York, Vintage Reprint.

Whittaker, R.J., and Fernández-Palacios, J.M. (2007) *Island biogeography: ecology, evolution and conservation*, 2nd edn. Oxford, Oxford University Press.

Williamson, M.H. (1981) *Island populations*. Oxford, Oxford University Press.

Wilson, E.O. (1990) *Success and dominance in ecosystems: the case of the social insects*, Oldendorf/Luhe, Germany, Ecology Institute.

Wilson, E.O. (2010) *The diversity of life*, 2nd edn. New York, Norton.

5

FLORA

Diana M. Percy, Quentin C. B. Cronk and Stephen Blackmore

Introduction: plants as common features of island life

As primary producers, plants form the basis of island ecosystems. The particular plants that arrive on an oceanic island will therefore determine the evolutionary and ecological trajectory to be taken by other organisms on the island. Within the subject of island biology, plants are either the primary unit or key component in ecosystem functions and therefore are particularly interesting objects of study. Almost all islands provide amenable examples of plant-based evolutionary and ecological processes, yet many of the available studies have focused on a few 'model' islands or archipelagos.

Favourable geographical and political circumstances have led to some islands becoming well known scientifically, while others remain relatively little known. To what extent do we need to study the plant life of every island? Can the systems that have been studied in detail provide us with all the information to generalise about the ecology and evolution of islands as a whole? The answer is both yes and no. No, because each island system has a unique natural history and a unique assemblage of plants that colonised, established and evolved on a particular island. But the answer is also a limited yes, because there are definite generalities between islands, at least between oceanic islands.

This chapter reviews some of the generalities of island plants, and how their interaction with other living organisms, notably but not exclusively humans, is played out on island habitats. An extended case study of the Indian Ocean atoll of Aldabra serves to exemplify the discussion in a focussed, single island, context through time.

Floral traits on islands

In his classic studies of islands, Carlquist (1965, 1974) pointed out morphological traits with parallels in unrelated plant groups across many islands. Notable among these are the presence of island woodiness, reduced dispersal capabilities, and changes in breeding systems. Island woodiness refers to the woody growth-form found in island members of otherwise herbaceous groups. It is well exemplified in the Hawaiian silverswords (*Asteraceae: Madiinae*), a woody group for which the ancestral state has been reconstructed as herbaceous; as are its near relatives, *Deinandra fasciculata*, the tarweeds of California (Baldwin and Sanderson 1998) (see Figure 5.1).

Colonising plants, such as those that successfully undergo long-distance dispersal to islands, must have effective seed dispersal. However, once established on an island, the selection for

Figure 5.1   The flower and buds of Clustered Tarweed (*Deinandra fasciculata*).

Source: Photo taken in Santa Monica Mountains National Recreation Area, California, USA, Wikimedia Commons, https://commons.wikimedia.org/wiki/File:Deinandrafasciculata.jpg.

dispersal ability may plausibly become relaxed, and they may even be selected for reduced dispersal (Carlquist 1996, Gillespie et al. 2012). Wild lettuce plants (*Lactuca*) that reach the Barkley Sound islands in British Columbia seem to have experienced strong selection for reduced dispersal within a decade of initial colonisation (Cody and Overton 1996). Specifically, they develop lower ratios of the amount of dispersal hair (pappus) to fruit (achene). Generalities in plant-breeding systems are more arguable. Plants have a greater chance of colonising if they are bisexual and capable of self-fertilisation: an observation known as 'Baker's law' (Baker 1955, 1967). Thus, island plants with self-compatible breeding systems should be more numerously represented in island floras. Once established on islands, however, there may be selection for outbreeding to maintain genetic diversity in small populations. This may give rise to novel mating systems such as dioecy. Dioecious plants are those that have separate male and female individuals, therefore eliminating the possibility of self-fertilisation, and many islands do seem to have numerous examples of dioecious plants, some of which may have evolved *in situ* (Sakai et al. 1995a, 1995b; Weller et al. 1995). However, it cannot be ruled out that on many islands dioecious plants will have arrived from dioecious source populations (Baker and Cox 1984).

There are ecological commonalities between islands too. An example of the generality of island plants is found in the Indo–Pacific strand flora. Walking along a beach on any island in the tropical Pacific or Indian oceans, one is likely to encounter certain plants, regardless of the particular island. These are the 'supertramp' species with very effective dispersal by sea currents. The coconut (*Cocos nucifera*) is one of these (see Figure 5.2a): this is the species most readily associated with island life. The coconut is not a tree but, like all palms, a large modified herb. Another is the goatfoot creeper (*Ipomoea pes-caprae*) which is a coastal vine that protects sand dunes and the shape of whose leaf resembles the footprint of a goat (see Figure 5.2b). Similar species are the portia tree (*Thespesia populnea*), beach naupaka (*Scaevola taccada*) and tree heliotrope (*Tournefortia argentea*). What these plants have in common is fruit with high floatability and survivorship in seawater, and an ecological suitability for the strand-line habitat. Plants such as these may form the majority of species on the low islands (i.e. raised coral islands) of the Indo–Pacific (Parnell et al. 1989). In contrast, species of the mountain interiors of much more recent volcanic 'high

Figures 5.2a and 5.2b The coconut (*Cocos nucifera*).

Source: Photo by Philip Gabrielsen, Wikimedia Commons, https://commons.wikimedia.org/wiki/Cocos_nuci
fera#/media/File:Palme_cuba.jpg. The goatfoot creeper (*Ipomoea pes-caprae*). *Source*: Photo by Marion Schneider
and Christoph Aistleitner, Wikimedia Commons, https://commons.wikimedia.org/wiki/File:Beach_Morning_
Glory_-_Ipomoea_pes-caprae.jpg.

islands' are more likely to have arrived by bird dispersal and to have berry fruits, like *Cyrtandra* (Cronk et al. 2005) and some island lobelioids (Givnish et al. 1996). These events are rarer and such mountain floras are much more likely to be individualistic.

Island uniqueness

What then can be derived from the botanical uniqueness of each island? The geographic structure and prevailing climate determines the diversity of habitats and microclimates on an island, and island environments are in large part generated by differences in areas and height above sea-level, characteristics related to island age and origin. Given enough time, an island may eventually contain a representative breadth of the source flora(s): however, many ocean islands are of relatively recent origin and at various distances from a given source (either continental, or other islands/archipelagos). An island flora therefore often reflects only a skewed representation of these source floras, and because of the vagaries of dispersal and local environmental peculiarities, island plant communities do not necessarily reflect the community structure of source floras, as one might be tempted to predict.

In the Pacific, the most island-rich ocean body in the world, we continue to puzzle over why plant groups are present on some islands but not others. One example is the absence of *Weinmannia* (a shrub in the Cunoniaceae) and *Cyathea* (a tree fern) in the Hawaiian Islands, but both are common in other Pacific islands. Both *Cyathea* and *Weinmannia* are wind-dispersed taxa like *Metrosideros* (a shrub in the Myrtaceae, called ohia in Hawai'i and pohutukawa in New Zealand), but *Metrosideros* has dispersed throughout the Pacific, apparently making it to the Hawaiian archipelago from the Marquesas (Wright et al. 2000), where both *Cyathea* and *Weinmannia* also occur. If prevailing climatic conditions explain the close relationship between Marquesan, Hawaiian (eastern Pacific) and New Zealand (western Pacific) *Metrosideros* taxa (Percy et al. 2008), then why have not similar means of dispersal and distribution put *Cyathea* and *Weinmannia* in the Hawaiian Islands? There are also puzzling cases of disjunct plant taxa, such as *Ascarina* (a genus in the 'primitive' family, *Chloranthaceae*) in the west and eastern Pacific, but apparently lacking from many central Pacific islands. In addition to colonisation barriers and filters, different evolutionary trajectories after colonisation are influential contributors to the uniqueness of individual island biotas. These may reveal both how unique morphologies arise as well as how equivalent morphologies arise independently through convergence. Some of these are discussed below.

Evolutionary radiation of plants on islands: neo-endemics

A dispersing plant propagule faces tremendous odds in order to reach and become established on an island, leading to relatively few species and small gene pools, which in turn contribute to unique adaptive radiations and island individuality. The fact that evolution has converged repeatedly on 'island phenotypes' including derived woodiness (e.g. the tree cucumber of Socotra, the only woody member of this plant family – see Figure 5.3) and columnar and monocarpic growth forms (e.g. *Echium* species from the Canary Islands, which are commonly cultivated in gardens), implies some predictability of trait evolution on islands, possibly as a result of similar selection/relaxation of selection. Insular woodiness on islands may be an adaptive response to a lack of competitors in the woody plant niche or the availability of permissive oceanic climatic conditions. Both these features consistently differentiate island from continental habitats.

Recent decades have heralded an age of molecular phylogenetic studies on an unparalleled scale. Island plants have not been neglected, and the origin and diversification of island groups has been the focus of a number of studies, particularly in the Hawaiian and Canary Islands (e.g. Lindqvist and Albert 2002, Percy and Cronk 2002). Many of the molecular phylogenetic studies have looked at recently derived groups often exhibiting spectacular adaptive radiations as in the tree sowthistles of the Canary Islands and the Hawaiian silverswords (Kim et al. 1996, Baldwin and Sanderson 1998). Fewer studies have looked at older groups that show possibly relictual distributions such as distributions in the Mediterranean and Macaronesian region (which includes Atlantic islands such as the Canary Islands and Madeira) of *Lauraceae* and other families with tertiary fossil records (Cronk 1992). Molecular systematics has also been useful in identifying the origins of phenotypically unique island forms, such as endemic island genera. These can often be placed within more widely dispersed groups, even though their affinities were previously obscured by the development of unusual morphologies *in situ*, such as the Hawaiian silverswords evolution from a Californian tarweed (Barrier et al. 1999).

Figure 5.3 Tree Cucumber (*Dendrosicyos socotrana*) from the Island of Socotra, Yemen, the only tree in its family.

Source: Photo by Gerry and Bonni. Wikimedia Commons, https://commons.wikimedia.org/wiki/File:Cucumber_tree_(6407165121).jpg.

As island plants exhibit evolutionary radiation in growth forms, they provide potential model systems for developmental biology, using the techniques of evolutionary molecular developmental genetics (dubbed 'evodevo'). The adaptive radiation of the Hawaiian silverswords has been used in this connection (Robichaux and Purugganan 2005). The silverswords are a group of tarweed-related *Compositae*, which include the magnificent silverswords, *Argyroxiphium*. These are the large arborescent rosette plants famously found in the crater of the Haleakala Volcano on the island of Maui (see Berry and Gillespie, this volume). However the group also includes closely related plants with very different growth forms, such as *Dubautia* and *Wilkesia*. The evolution of genes that may be involved in this radiation have been examined. As might be expected, the results so far are mixed. Certain floral regulatory genes show an elevated relative rate of evolution of the proteins they encode (Barrier et al. 2001), whereas certain growth regulatory genes show no such elevation of protein evolution (Remington and Purugganan 2002).

Island relicts: palaeo-endemics

The volcanic origin and age of many islands can be determined by potassium-argon dating. However, dating the origin of an island's plant life is not as simple as assigning a maximum age correlated to island emergence. Studies of several plant groups using methods of dating diversification based on calibrating molecular phylogenies independently of island age indicate clearly asynchronous relationships between the age of diversification in the plant group and island age, with certain groups proving much older than the islands on which they are currently found. This is explained by the assumption of a 'stepping-stone' hypothesis whereby plant groups previously existed on former islands that are now submerged (Kim et al. 1998). A particularly persuasive case is the genus *Hillebrandia* that apparently diverged from the related genus *Begonia* long before its current home in the Hawaiian islands existed (Clement et al. 2004) (see Figure 5.4).

In the Pacific, oceanic islands sink rapidly, eventually disappearing completely to become seamounts or coral atolls. Most of the unique plants are therefore, understandably, recently diverged or 'neo-endemic'. However, in other oceans, islands may be longer lived. St Helena in the Atlantic Ocean has had a continuous subaerial existence for some 14.5 million years. During

Figure 5.4   Hillebrandia sandwicensis, from Maui, Hawai'i.

Source: Photo by Forest and Kim Starr, Wikimedia Commons, https://commons.wikimedia.org/wiki/File:Hillebrandia_sandwicensis.jpg.

this time, there have been considerable changes in the southern African source flora. Some of the early recruits to the island therefore represent 'time capsules' of a past African flora. The divergence of these 'time capsule' species can even be used to date events on continents. Richardson et al. (2001) used the method to date the radiation of *Phylica* in the Cape flora of South Africa, by calibrating with the divergence of the St Helena endemic *Nesiota*. A north African example is the dragon's blood tree, so named for the Canary Island and Socotran island species (*Draceana draco* and *D. cinnabari*) which have blood-red resin reputedly used in ancient Roman times and in medieval alchemy and, more recently, in the making of musical instruments and the decoration of ceramic pots on Socotra (see Figure 5.5). Each new branch of *Draeceana* takes some 10–20 years to flower, then branches again beneath the flowers. These two island species are closely related but one occurs in the Atlantic Ocean and one in the Indian Ocean. There are small relictual populations in the Moroccan Anti-Atlas Mountains and the Red Sea Hills and these disjunct distributions together with Pliocene fossil remains in southern Europe suggest there was once a much wider occurrence of these trees, but the inter-continental distribution became fragmented with the aridification of North Africa.

Some of the lineages present on St Helena are represented by nine-million-year-old fossils, whereas others seem to have reached the island more recently. Thus, the endemics have a series of different ages (Cronk 1987, 1990). Pollen very similar to that of trees like the St Helena redwood (*Trochetiopsis erythroxylon*) and ebony (*Trochetiopsis ebenus*) have been found in the nine-million-year-old deposit (Cronk 1990). However, the dryland succulent *Hydrodea cryptantha* (for example) is very similar to a Namib desert species and that lineage probably arrived on St Helena more recently.

Figure 5.5 The Dragon's Blood Tree (*Dracaena draco*).

Source: Photograph by Frank Vincentz, Wikimedia Commons, https://commons.wikimedia.org/wiki/File:Ses_Salines_-_Botanicactus_-_Dracaena_draco_01_ies.jpg.

Plant conservation: rare plants versus invasive plants

Due to the small size of islands, island plants are represented by a restricted number of individuals, and limited spatial area, compared to continental species. As a result, human disturbance can have a catastrophic effect on species survival. Many island plants have become totally extinct: one example is the St Helena endemic genus *Nesiota elliptica* (the St Helena wild olive). All attempts by the St Helena government to conserve the species failed and it became extinct in the wild in 1994 and totally extinct in 2003 (see Figure 5.6). Many other plants are reduced to minute world populations: of under ten individuals say, or even to a single individual. Such plants present tremendous conservation challenges as they may show the serious effects of prolonged inbreeding. To take a pessimistic view, the extinction of these species may be inevitable, in which case they are 'living dead'.

There is another aspect of plant life on islands that has great significance for conservation. Human introductions of invasive plants on islands do astounding damage to native habitats

Figure 5.6   The leaves and flower of the St Helena wild olive (*Nesiota elliptica*), now extinct.
Source: Melliss (1875, p. 337).

(Macdonald et al. 1991, Meyer and Florence, 1996). *Miconia calvescens* from central America has been introduced into Pacific islands, originally for ornament. In Tahiti, French Polynesia, its large leaves with dark purple undersides cast a deep shade which prevents the growth of native seedlings and converts diverse natural vegetation to pure stands of this exotic. It spreads fast and is causing an unprecedented ecological catastrophe (see Figure 5.7). It is now the dominant plant in over 65 per cent of the forest on Tahiti, and infestations in some Hawaiian islands are already too widespread for effective chemical/mechanical control (Le Roux et al. 2008).

Similarly, on the island of Mauritius, the Brazilian fruit tree, strawberry guava (*Psidium cattleianum*) is an introduced species. It too casts a dense shade and spreads rapidly. It has wiped out vast tracts of native forest. The strawberry guava is 'facilitated' by other introduced organisms,

Figure 5.7 The central American plant *Miconia calvescens* introduced in Pacific islands.

Source: Photo by Forest and Kim Starr, Wikimedia Commons, https://commons.wikimedia.org/wiki/File:Starr_031118-0097_Miconia_calvescens.jpg.

as it is pollinated by the introduced honeybee (*Apis mellifera*) and the fruits are dispersed by an introduced bird, the red-throated bulbul (*Pycnonotus jocosus*) (Cronk and Fuller 2001).

Mechanical or chemical control of invasive plants, let alone total eradication, is very difficult due to the vast, often inaccessible, areas affected and the tough physical labour involved in such programmes. Biological control is therefore often the only practical solution. However, the procuring and testing of biological control organisms is a lengthy and expensive undertaking, and the risks of unforeseen consequences can never be completely eliminated (Messing and Wright 2006, Seastedt 2015).

Interactions between plants and other organisms on islands

There are occasions when interacting organisms become extinct before the plants, and we are left with shadows of past interactions. The spines on certain Hawaiian lobelioids (*Cyanea*) may result from defence against now extinct herbivorous birds, perhaps geese driven to extinction by early Polynesian colonists (Givnish et al. 1994). Something similar has been suggested for the extinct moa in New Zealand which may have left 'ghosts' in the branching habits of New Zealand plants (Greenwood and Atkinson 1977, Bond et al. 2004), although other hypotheses have been suggested (Howell et al. 2002, Lusk 2002).

Island plants can also play an important part in promoting and shaping the diversification of an interacting animal group. One example is the spectacular radiation of Hawaiian honey-creepers (see Berry and Gillespie, this volume). Originally derived from a finch-like ancestor, these birds have diversified most dramatically in bill morphology (Lovette et al. 2002, Lerner et al. 2011). The nectar-feeding honeycreepers have long, distinctly curved bills that are highly adapted to feeding from the flowers of several members of the Hawaiian endemic flora, while the insectivorous and seed-eating honeycreepers have short fine beaks, or larger seed-cracking beaks.

As with birds like the moa, herbivory by other large vertebrates also creates plant–animal evolutionary interactions. Although mammals (with the exception of bats) are absent from oceanic islands, the large herbivore niche may be taken by more dispersible reptile lineages. The land iguana (*Conolophus subcristatus*) and giant tortoises (*Geochelone elephantopus*) of the Galápagos islands have shaped plant evolution there, as evidenced by the adaptations to avoid grazing. The prickly pear cacti (*Opuntia* spp.) of the Galápagos tend to be taller and spinier on the islands with abundant reptile herbivores than on other islands (Dawson 1966). On the Galápagos islands of Marchena and Genovesa, which were never colonised by grazing reptiles, the prickly pears are low growing, and less spiny (Stewart 1911).

The diversity of insect herbivores is often directly determined by the extent of diversification in the plant community, particularly when insects are specialised on certain plants (Roderick and Percy 2008, Percy 2009). An increase in insect speciation can be promoted by diversification in the preferred host-plant group. Conversely, insect diversity can be limited if the preferred host plant is monotypic or rare, and alternative host plants have been filtered out by barriers to island colonisation (Percy 2003).

The decline of pollinators and seed dispersers has important conservation consequences for plants as they mediate very important ecosystem processes. Individual pollinators or dispersers may be at a premium on islands where diversity is low (Cox et al. 1991, Sekercioglu et al. 2004) and their extinction can spell disaster for plants. As well as the loss of coevolved animals, the gain of introduced animals can be equally disastrous for plants. This is particularly true of mammals. As noted above, oceanic islands are generally free of all terrestrial mammals except bats, since mammals tend to be poor transoceanic dispersers. However, herbivorous domesticated mammals are frequently introduced to islands by humans. The goat was introduced to St Helena in 1511 and by 1588 existed as flocks "a mile long" (the island is only 16 km [10 miles] long) (Cronk 2000). They destroyed about half the native vegetation of St Helena; and, without vegetation, the soil washed into the sea, causing an ecological catastrophe (Cronk 1989). These areas of St Helena are still largely barren of vegetation, even though wild goats have been almost completely eradicated.

Island ethnobotany

Many of the Indian Ocean and Mid-Atlantic islands were not colonised before European and Arab traders encountered them in the Middle Ages. In contrast, the history of human colonisation in the Pacific is a very different story because the Polynesian peoples of Oceania had such exceptional navigational skills and reached every corner of the Pacific. Many Polynesian sea voyages were exploratory, with the anticipation of colonising new lands. With this in mind, provisions were taken on these voyages, including stock cargos of plants and animals. The stock animals were pigs, dogs and chickens. The stock plants included important food crops, including bananas, coconuts, taro, sweet potatoes and breadfruit. In some cases, the precise historical pattern of island migration for these plants can be traced using molecular data (Zerega et al. 2004). Two interesting examples of Polynesian plant introductions which have become widely naturalised in the Pacific are the Polynesian chestnut (*Inocarpus fagifera*), a nutritional food source; and the candlenut tree (*Aleurites moluccana*), the oily nut of which was used, candle-like, as a source of illumination (see Figure 5.8). Other plants that had practical or ceremonial use, such as flax, cordylines and bull rushes, may also have been widely introduced by Polynesians.

Figure 5.8 The candle-nut tree (*Aleurites moluccana*).

Source: Photo by Forest and Kim Starr, Wikimedia Commons, https://commons.wikimedia.org/wiki/File:Starr_070215-4559_Aleurites_moluccana.jpg.

Aldabra

We now move on to exemplify most of the above arguments on the atoll of Aldabra, one of the few 'model' islands for which relatively extensive biotic data is available over time.

An introduction

Aldabra has many claims to fame. As the largest of the four islands that make up the Aldabra Group (the others being Assumption, Astove and Cosmoledo) it is the most extensive raised atoll in the western Indian Ocean and home to a remarkable assemblage of species, many of them endemic (Fryer 1911, Seaton et al. 1991, Fosberg and Renvoize 1980; see Figure 5.9a). The giant land tortoise *Geochelone gigantea* is its most celebrated and well-documented inhabitant (Bourne and Coe 1978, Bourne et al. 1999, and references therein) (see Figure 5.9b).

In common with many islands, Aldabra has faced numerous threats as a result of the actions of humans, even though it is but a few short centuries since people first visited (Stoddart 1971). This brief period of human-influenced history could, perhaps it is not too fanciful to suggest, have been encompassed within the life span of a single one of Aldabra's giant land tortoises. A full appreciation of Aldabra and its unique biota requires consideration over geological time scales. A series of expeditions, between 1967 and 1969, organised by the Royal Society (Westoll and Stoddart 1971, Stoddart and Westoll 1979) and the construction of the Royal Society Aldabra Research Station which opened in 1971 (Griffin 1974) have added greatly to our knowledge of the terrestrial ecology of Aldabra. Today, this low-lying coral atoll is threatened by rapid changes in sea level and global climate change. Aldabra has been submerged in the past; but

Figure 5.9a   Satellite image of the largest island in the Aldabra group.

Source: Photo by Simisa, Wikimedia Commons, https://commons.wikimedia.org/wiki/File:Seychelles_outer_islands_25.08.2009_10-20-09.jpg

Figure 5.9b Aldabra Giant Tortoise (*Geochelone gigantean*).

Source: Photo by Mohammed Abdi Karim, Wikimedia Commons, https://commons.wikimedia.org/wiki/File:Aldabra_Giant_Tortoise_Geochelone_gigantea_edit1.jpg.

now there are no natural refugia or remote populations for many of its most important species to escape to, or recolonise from.

This brief account considers the origins of the terrestrial biota (both flora and fauna) of Aldabra, the impact of human visitation and settlement on this biota, and its prospects for the future. Since their discovery, the economy of the islands of the western Indian Ocean has followed a widespread pattern with a pioneer stage founded on the exploitation of natural resources, a settled stage with a mixed economy based on agricultural products and natural resources and most recently an economy in which the largest sector is tourism.

Origins and significance of Aldabra's terrestrial biota

Aldabra is not a biodiversity hotspot. The number of species reported from its 97 km² of land is quite small. Yet, Aldabra's distinctive terrestrial biota serves as a reminder that, whilst global hotspots offer the best prospects for conserving the majority of the Earth's species, it would be remiss to neglect less species diverse places, especially islands with their elevated levels of endemism.

Given that Aldabra has been completely submerged at least three times during Pleistocene interglacials (Braithwaite et al. 1973), the last time being about 90,000 years ago (Fosberg and Renvoize 1980), the current biota is only the latest in a sequence of changing biotas as sea levels changed. A surprisingly rich fossil record for many of the vertebrates shows clear evidence of repeated cycles of invasion and extinction from a variety of source populations on the African mainland or Indian Ocean islands (Taylor et al. 1979); yet, knowledge of the vegetation history during the same period is lacking because sediments suitable for preserving a record of the pollen have not been found.

Complex terrestrial ecosystems can only develop on an atoll when the coral reef limestone emerges above sea level for an extended period of time allowing the process of colonisation and establishment to begin. Whilst the establishment of a land flora is not a necessary prerequisite for some nesting seabirds or animals with marine larval stages, it is necessary for Aldabra's primary herbivore, the giant tortoise (see Figure 5.9b). Vegetation is also necessary for the 15 native land birds (Benson and Penny 1971), of which two are considered full species: *Dicrurus aldabranus*, the Aldabran drongo or '*moulanba*' and the Aldabra Brush Warbler, *Nessilus aldabranus*. The latter species was only discovered in the 1960s, was never abundant and has scarcely been seen since the mid-1980s (Hambler et al. 1985). The most celebrated land bird is *Dryolimnas cuvieri aldabranus*, the Aldabran flightless rail, known locally as '*tyomityo*'. The rail is an endemic subspecies (Wanless 2003), as are a further ten other land birds. Many formerly widespread but now rare species of plants and animals have their last stronghold on Aldabra (Seaton et al. 1991). The robber crab (*Birgus latro*) provides a good example of a formally widespread animal that was eaten to extinction by humans in most of its former western Indian Ocean localities.

The origins of the land flora have been considered in detail by several authors. From a total flora of 175 species known at that time, Renvoize (1975) analysed the distribution of those that also occur elsewhere, using the chorological divisions of Meusel et al. (1965). He noted that the Aldabran flora comprised species characteristic of coastal bushland or sea shore habitats and concluded that, in common with the other coral islands of the western Indian Ocean, Aldabra was open to colonisation by plants from all directions. Indeed, Renvoize noted that 33 Aldabran plant species were widespread palaeo-tropical plants ranging from tropical Africa to the Far East, including Madagascar and the high islands of the western Indian Ocean. Ten plant species were of Deccan-Malaysian/Madagascan affinities, occurring from the Far East to the islands of the western Indian Ocean, including Madagascar. A total of 30 Aldabran plant species were restricted to the western Indian Ocean: nine of them shared with Madagascar, five with Seychelles, one with Mauritius and a further 15 that are also known from Madagascar, Seychelles, Mauritius, Réunion and Rodrigues. Finally, 25 east African/Madagascan plant species found on Aldabra also occur in coastal east Africa and Madagascar. Renvoize pointed out that the prevailing winds and ocean currents did not favour transportation to Aldabra from the African coast and suggested that birds played a part in the arrival of many of this latter group of species. He considered Aldabra to be unique among the coral islands of the western Indian Ocean because it is a centre of endemism for plants, a feature it has in common with each of the high, granitic islands. The Flora of Aldabra (Fosberg and Renvoize 1980) describes 34 endemic species out of a total which had increased to 274 known species (87 of which Fosberg and Renvoize considered had probably been introduced by humans). The proportion of introduced plant species is high on Aldabra, as it is on many islands, but fortunately very few have become aggressively invasive.

Island ecosystems can include some fascinating examples of plant and animal interactions. Since Aldabra's terrestrial ecosystem is dominated by a large, vegetarian reptile, a vegetation type that Grubb (1971) called "tortoise turf" (*Pemphis acidula*) has developed in response to heavy grazing. Tortoise turf is an assemblage of small, salt-tolerant grass and sedge species mixed with prostrate herbs and averaging about 2 cm in height (or up to 15 cm when less heavily grazed). An endemic grass, *Sporobolus testudinum*, described as a new species from the tortoise turf by Renvoize (1971, 1972), has narrow, contracted inflorescences interpreted as an adaptive response to protect the developing flowers from intense grazing. The giant tortoise plays a part in dispersing the species, since the seeds are known to survive passage through its gut system (Hnatiuk 1978).

Human settlement

The first human visits to Aldabra are unrecorded, although it is generally assumed that it would have been well known to early Arab sailors with their dhows. The name of the atoll is said to be derived from 'al-khadra' (*the green* in Arabic), a likely reference to the green colour of the vegetation that is reflected in the clouds above the island and is visible from a considerable distance (Lionnet 1995). In 1503, Vasco de Gama was one of the earliest European explorers to visit the region, passing by the Amirantes, named in his honour, and returning to land there a year later. The atoll appeared with the name Aldhadra on a Portuguese map of 1511 (Skerrett and Mole 1995). The earliest recorded landing on Aldabra was made in 1742 by Captaine Lazare Picault who was unable to find fresh water but was impressed by the large size and quality of meat of the giant tortoises (ibid.).

For centuries, people have largely passed Aldabra by. Ships paused only to fish, hunt turtles and exchange ballast from their holds for a living cargo of giant land tortoises for meat. The hostile terrain, with difficult landings and absence of a ready source of year-round fresh water, meant that Aldabra escaped permanent human settlement during the first centuries of its encounter with the human species. However, this was a period of introductions when rats, cats, dogs and goats all entered an ecosystem that was, at that time, devoid of mammals (except for a few species of bat and the dugong) and dominated by a giant vegetarian reptile (the tortoise). These invasive animals and a number of introduced plants have had the biggest impact in Aldabra. Relative to all other relatively species-rich islands in the region, Aldabra was unscathed by human colonists. The demise of the dodo of Mauritius has come to stand as the emblem of anthropogenic extinctions, but it was not the only casualty in the western Indian Ocean. Giant land tortoises, robber crabs, green snail, endemic birds, including other flightless land birds, also shared the same fate. In all, humans account for the extinction of at least 60 terrestrial species in the Indian Ocean islands (Balouet 1990). The initial economy of the Seychelles and other islands of the western Indian Ocean relied heavily upon the unsustainable exploitation of natural resources until an economy based on introduced products such as vanilla and cinnamon could be established. These remained important through to the 1960s and 1970s when they were replaced by tourism as the economic mainstay.

Despite escaping the major terrestrial extinctions found on other islands, Aldabra attracted the attention and concern of Charles Darwin and others when, in the late 19th century, it was feared that the giant land tortoises there were finally following the fate of others into extinction. An expedition to Aldabra in 1878 landed and found only a single specimen (Skerrett and Mole 1995). By the end of the 19th century, a small settlement was established on Île Picard where coconut trees were planted for copra production and this, together with the production of dried fish, followed the way of life common throughout the islands of the western Indian Ocean. This economy, based on the exploitation of coconuts and natural resources such as green turtle, dried fish and green snails, continued well into the 1970s (e.g. Travis 1959, Thomas 1968). On neighbouring Assumption Island, which lacks a lagoon, guano extraction provided a viable local industry that endured until the late 1970s when stocks of guano were depleted. This way of life based around isolated island communities has now almost disappeared; and, once again, tourism has become the new foundation of the economy. The settlement on Île Picard was small and its impact on the terrestrial ecosystem of Aldabra was limited. Exploitation of some marine species, such as the Green Turtle (*Chelonia mydas*) led to a sharp decline in numbers, from 6,000 to 8,000 nesting turtles per annum at the end of the 19th century to below 1,000 by the 1960s and 1970s (Mortimer 1988). But again, relative to the exploitation of this species elsewhere in the Indian Ocean, Aldabra remained a stronghold and has recovered since the low point of the 1970s.

In the 1960s, a major threat arose when Aldabra was selected for development as a military base (Beamish 1970, Skerrett and Mole 1995). The lagoon was to have been deepened to provide a deep-water anchorage and an airstrip constructed on Grande Terre, home to the largest numbers of giant tortoises. For a while the future of Aldabra hung in the balance (Stoddart 1974). The resulting outcry eventually reached the highest levels and instead the military base was constructed on Diego Garcia where the local population was summarily deported. Finally, Aldabra was saved for science and for humanity, by being declared a UNESCO World Heritage Site in 1982.

Prospects for the future

In 1979, responsibility for the Aldabra Research Station passed to the Seychelles Islands Foundation (SIF) which has since successfully managed the atoll (Blackmore 2001). Today, the research station has been rebuilt and has undergone a second phase of reconstruction. Active conservation projects include monitoring and research on the marine and terrestrial ecosystems. Steps have been taken to reduce a number of invasive species. Feral goats have so far proved impossible to eradicate completely but are greatly reduced in numbers. Sisal plants have been removed from Île Picard (the part of the atoll where the research station and former settlement are situated) although they remain on some islets in the lagoon. Flightless rails have been successfully reintroduced onto Île Picard. Green turtle numbers have continued to increase and the giant tortoise populations remain the largest of any island system in the world, although the total number has declined since the 1960s (Bourne et al. 1999).

It is satisfying to see that Aldabra remains one of the least spoilt and important atolls in the world; its future now appears secure. However, the prospects for the future depend on a series of global problems that not even a remote island system can escape. Global warming raises the prospect of dramatic changes in sea level, if there is significant melting of the polar ice caps. At its worst this would inundate the terrestrial ecosystems of Aldabra. Elevated sea surface temperatures, such as those experienced in 1998, have already caused coral bleaching throughout the western Indian Ocean, including Aldabra (Stobart et al. 2005). Concern is growing about pollution of the world's seas and oceans and fishing is now a globalised activity, with fish from the western Indian Ocean being exploited for international use. These global trends make coral reefs more vulnerable than ever, however remote they are. As Hodgson (1999) points out in an excellent review of the threats to reefs, a fundamental problem is the lack of baseline information for understanding changes in reef ecosystems. Changes within a human life span are noticed and reported, at least anecdotally; but information on the pristine baseline would require knowledge going back over centuries. Such knowledge is rarely available, but the detailed scientific exploration and monitoring of Aldabra Atoll make it one of the few places with detailed records for almost 50 years. All the more reason for scientists to see Aldabra as a natural laboratory during the decades ahead: a period of change unprecedented in human history.

Conclusion: amplification by compression

In conclusion, it appears that almost all ecological and evolutionary processes concerning plants are amplified on islands; generally speaking, the smaller the island, the more amplified these processes are. Small geographic area and low diversity seem to be the main factors. With populations existing in miniature, they are prone to stochastic, or random, processes. Climatic zonation is compressed too, since islands have stark zonations of rainfall with altitude over short distances: rainfall of oceanic islands (unlike continents) is often largely a function of relief. Such a mosaic of habitats in a tiny area promotes evolutionary radiation. Conversely, the small size of islands means that they are extremely vulnerable to biological invasion and disturbance, particularly human mediated, as there are few distance barriers to dispersal by humans, and few areas are immune to disturbance by inaccessibility. On the plus side, 'amplification by compression' makes islands particularly useful to biologists: on islands, evolutionary patterns are often more clearly evident and processes that may be subtle on continents tend to be more clearly exposed.

References

Baker, H.G. (1955) Self-compatibility and establishment after 'long-distance' dispersal. *Evolution*, 9(3), 347–349.

Baker, H.G. (1967) Support for Baker's law: as a rule. *Evolution*, 21(4), 853–856.

Baker, H.G., and Cox, P.A. (1984) Further thoughts on dioecism and islands. *Annals of the Missouri Botanical Garden*, 71(1), 244–253.

Baldwin, B.G., and Sanderson, M.J. (1998) Age and rate of diversification of the Hawaiian silversword alliance (Compositae). *Proceedings of the National Academy of Sciences USA*, 95(16), 9402–9406.

Balouet, J.-C. (1990) *Extinct species of the world*. London, Letts.

Barrier, M., Baldwin, B.G., Robichaux, R.H., and Purugganan, M.D. (1999) Interspecific hybrid ancestry of a plant adaptive radiation: allopolyploidy of the Hawaiian silversword alliance (Asteraceae) inferred from floral homeotic gene duplications. *Molecular Biology and Evolution*, 16(8), 1105–1113.

Barrier, M., Robichaux, R.H., and Purugganan, M.D. (2001) Accelerated regulatory gene evolution in a plant adaptive radiation. *Proceedings of the National Academy of Sciences USA*, 98(18), 10208–10213.

Beamish, T. (1970) *Aldabra alone*. London, George Allen & Unwin.

Benson, C.W., and Penny, M.J. (1971) The land birds of Aldabra. *Philosophical Transactions of the Royal Society of London, Series B*, 260(836), 529–548.

Blackmore, S. (2001) Proceedings of the Aldabra science and conservation workshop held on Aldabra 8–18 December 2000. *Phelsuma*, 9 (Supplement A), 1–36.

Bond, W.J., Lee, W.G., and Craine, J.M. (2004) Plant structural defences against browsing birds: a legacy of New Zealand's extinct moas *Oikos*, 104(3), 500–508.

Bourne, D., and Coe, M. (1978) The size, structure and distribution of the giant tortoise population on Aldabra. *Philosophical Transactions of the Royal Society of London, Series B*, 282(988), 139–175.

Bourne, D., Gibson, C., Augeri, D., Wilson, C.J., Church, J., and Hay, S.I. (1999) The rise and fall of the Aldabran giant tortoise population. *Philosophical Transactions of the Royal Society of London, Series B*, 266(1424), 1091–1100.

Braithwaite, C. J., Taylor, J. D., and Kennedy, W. J. (1973) The evolution of an atoll: the depositional and erosional history of Aldabra. *Philosophical Transactions of the Royal Society of London, Series B*, 266(878), 307–340.

Carlquist, S. (1965) *Island life: a natural history of the islands of the world*. Garden City, NY, Natural History Press.

Carlquist, S. (1974) *Island biology*. New York, Columbia University Press.

Carlquist, S. (1996) Plant dispersal and the origin of Pacific island floras. In A. Keast and S.E. Miller (eds.), *The origin and evolution of Pacific Island biotas, New Guinea to Eastern Polynesia: patterns and processes*, Amsterdam, SPB Academic Publishing, pp. 153–164.

Clement, W.L., Tebbitt, M.C., Forrest, L.L., Blair, J.E., Brouillet, L., Eriksson, T., and Swensen, S.M. (2004) Phylogenetic position and biogeography of *Hillebrandia sandwicensis* (Begoniaceae): a rare Hawaiian relict. *American Journal of Botany*, 91(6), 905–917.

Cody, M.L., and Overton, J.M. (1996) Short-term evolution of reduced dispersal in island plant populations *Journal of Ecology*, 84(1), 53–61.

Cox, P.A., Elmqvist, T., Pierson, E.D., and Rainey, W.E. (1991) Flying foxes as strong interactors in South-Pacific island ecosystems: a conservation hypothesis. *Conservation Biology*, 5(4), 448–454.

Cronk, Q.C.B. (1987) The history of endemic flora of St Helena: a relictual series. *New Phytologist*, 105(4), 509–520.

Cronk, Q.C.B. (1989) The past and present vegetation of St Helena. *Journal of Biogeography*, 16(1), 47–64.

Cronk, Q.C.B. (1990) The history of the endemic flora of St Helena: late Miocene *Trochetiopsis*-like pollen from St Helena and the origin of *Trochetiopsis*. *New Phytologist*, 114(1), 159–165.

Cronk, Q.C.B. (1992) Relict floras of Atlantic islands: patterns assessed. *Biological Journal of the Linnean Society*, 46(1–2), 91–103.

Cronk, Q.C.B. (2000) *The endemic flora of St Helena*. Oswestry, UK, Anthony Nelson.

Cronk, Q.C.B., and Fuller, J.L. (2001) *Plant invaders: the threat to natural ecosystems*. London, Earthscan.

Cronk, Q.C.B., Kiehn, M., Wagner, W.L., and Smith, J.E. (2005) Evolution of *Cyrtandra* (Gesneriaceae) in the Pacific Ocean: the origin of a supertramp clade. *American Journal of Botany*, 92(6), 1017–1024.

Dawson, E.Y. (1966) Cacti in the Galápagos islands, with special reference to the relationship with tortoises. In R.L. Bowman (ed.), *The Galápagos*, Berkeley CA, University of California Press, pp. 204–209.

Fosberg, F.R., and Renvoize, S.A. (1980) *The flora of Aldabra and neighbouring islands*, Kew Bulletin Additional Series No. 7, London, HMSO.

Fryer, J.C.F. (1911) The structure and formation of Aldabra and neighbouring islands, with notes on their fauna and flora. *Zoological Journal of the Linnean Society of London*, 14(3), 397–422.

Gillespie, R.G., Baldwin, B.G., Waters, J.M., Fraser, C.I., Nikula, R., and Roderick, G.K. (2012) Long-distance dispersal: a framework for hypothesis testing. *Trends in Ecology and Evolution*, 27(1), 47–56.

Givnish, T.J., Knox, E., Patterson, T.B., Hapeman, J.R., Palmer, J.D., and Sytsma, K.J. (1996) The Hawaiian Lobelioids are monophyletic and underwent a rapid initial radiation roughly 15 Million years ago. *American Journal of Botany* (Supplement), 83(6).

Givnish, T.J., Sytsma, K.J., Smith, J.F., and Hahn, W.J. (1994) Thorn-like prickles and heterophylly in *Cyanea*: adaptations to extinct avian browsers on Hawai'i. *Proceedings of the National Academy of Sciences USA*, 91, 2810–2814.

Greenwood, R.M., and Atkinson, I.A.E. (1977) Evolution of the divaricating plants in New Zealand in relation to moa browsing. *Proceedings of the New Zealand Ecological Society*, 24(1), 21–33.

Griffin, D.J.H. (1974) The Aldabra research station. *Notes and Records of the Royal Society of London*, 29(1), 111–119.

Grubb, P. (1971) The growth, ecology and population structure of giant tortoises on Aldabra. *Philosophical Transactions of the Royal Society of London, Series B*, 260(836), 327–372.

Hambler, C., Hambler, K., and Newing, J.M. (1985) Some observations on *Nesillus aldabranus*, the endangered brush warbler of Aldabra atoll, with hypotheses on its distribution. *Atoll Research Bulletin*, 290, 1–19.

Hemsley, W.B. (1919) Flora of Aldabra. *Bulletin of Miscellaneous Information (Royal Botanic Gardens, Kew*, 3, 108–153.

Hnatiuk, S.H. (1978) Plant dispersal by the Aldabra giant tortoise, *Geochelonia gigantea* (Schweigger). *Oecologia*, 36(3), 345–350.

Hodgson, G. (1999) A global assessment of human effects on coral reefs. *Marine Pollution Bulletin*, 38(5), 345–355.

Howell, C.J., Kelly, D., and Turnbull, M.H. (2002) Moa ghosts exorcised: New Zealand's divaricate shrubs avoid photoinhibition. *Functional Ecology*, 16(2), 232–240.

Kim, H.G., Keeley, S.C., Vroom, P.S., and Jansen, R.K. (1998) Molecular evidence for an African origin of the Hawaiian endemic *Hesperomannia* (Asteraceae). *Proceedings of the National Academy of Sciences USA*, 95(26), 15440–15445.

Kim, S.C., Crawford, D.J., Francisco Ortega, J., and Santos Guerra, A. (1996) A common origin for woody *Sonchus* and five related genera in the Macaronesian islands: molecular evidence for extensive radiation. *Proceedings of the National Academy of Sciences USA*, 93(15), 7743–7748.

Le Roux, J.J., Wieczorek, A.M., and Meyer, J.-Y. (2008) Genetic diversity and structure of the invasive tree *Miconia calvescens* in Pacific islands. *Diversity and Distributions*, 14(6), 935–948.

Lerner H.R.L., Meyer, M. James, H.F., Hofreiter, M., and Fleischer, R.C. (2011) Multilocus resolution of phylogeny and timescale in the extant adaptive radiation of Hawaiian honeycreepers. *Current Biology*, 21(21), 1838–1844.

Lindqvist, C., and Albert, V.A. (2002) Origin of the Hawaiian endemic mints within North American *Stachys* (Lamiaceae). *American Journal of Botany*, 89(10), 1709–1724.

Lionnet, G. (1995) A virgin land. In M. Amin, D. Willets and A. Skerret (eds.), *Aldabra: world heritage site*, Nairobi, Kenya, Camerapix, pp. 19–22.

Lovette, I.J., Bermingham, E., and Ricklefs, R.E. (2002) Clade-specific morphological diversification and adaptive radiation in Hawaiian songbirds. *Proceedings of the Royal Society, Series B*, 269(1486), 37–42.

Lusk, C.H. (2002) Does photoinhibition avoidance explain divarication in the New Zealand flora? *Functional Ecology*, 16(6), 858–869.

Macdonald, I.A.W., Thebaud, C., Strahm, W.A., and Strasberg, D. (1991) Effects of alien plant invasions on native vegetation remnants on La Réunion (Mascarene islands, Indian Ocean). *Environmental Conservation*, 18(1), 51–61.

Melliss, J.C. (1875) *St Helena: a physical, historical, and topographical description of the island* (Illustrated by J.C. Melliss). London, L. Reeve & Co.

Messing, R.H., and Wright, M.G. (2006) Biological control of invasive species: solution or pollution? *Frontiers in Ecology and the Environment*, 4(3), 132–140.

Meusel, H., Jager, E., and Weinert, E. (1965) *Vergleichende chorologie der zentraleuropaischen flora*. (2 vols.) Jena: Veb Gustav Fischer Verlag.

Meyer, J.Y., and Florence, J. (1996) Tahiti's native flora endangered by the invasion of *Miconia calvescens* DC (Melastomataceae). *Journal of Biogeography*, 23(6), 775–781.

Mortimer, J.A. (1988) Green turtle nesting at Aldabra atoll: population estimates and trends. *Bulletin of the Biological Society of Washington*, 8, 116–128.

Parnell, J.A.N., Cronk, Q., Jackson, P.W., and Strahm, W. (1989) A study of the ecological history, vegetation and conservation management of Île aux Aigrettes, Mauritius. *Journal of Tropical Ecology*, 5(4), 355–374.

Percy, D.M. (2003) Radiation, diversity and host plant interactions among island and continental legume-feeding psyllids. *Evolution*, 57(11), 2540–2556.

Percy, D.M. (2009) Insect radiations. In R. Gillespie and D. Clague (eds.), *Encyclopedia of islands*, Berkeley, CA, University of California Press, pp. 460–466.

Percy D.M., and Cronk, Q.C.B. (2002) Different fates of island brooms: contrasting evolution in *Adenocarpus*, *Genista*, and *Téline* (Genisteae, Fabaceae) in the Canary Islands and Madeira. *American Journal of Botany*, 89(5), 854–864.

Percy, D.M., Garver, A.M., Wagner, W.L., James, H.F., Cunningham, C.W., Miller, S.E., and Fleischer, R.C. (2008) Progressive island colonisation and ancient origin of Hawaiian *Metrosideros* (Myrtaceae). *Proceedings of the Royal Society, B-Biological Sciences*, 275(1642), 1479–1490.

Remington, D.L., and Purugganan, M.D. (2002) GAI homologues in the Hawaiian silversword alliance: molecular evolution of growth regulators in a rapidly diversifying plant lineage. *Molecular Biology and Evolution*, 19(9), 1563–1574.

Renvoize, S.A. (1971) Miscellaneous notes on the flora of Aldabra and neighbouring islands: I. *Kew Bulletin*, 25(3), 417–422.

Renvoize, S.A. (1972) Miscellaneous notes on the flora of Aldabra and neighbouring islands: II. A new species of *Dichrostachys* (Leguminosae) from Aldabra. *Kew Bulletin*, 26(3), 433–438.

Renvoize, S.A. (1975) A floristic analysis of the Western Indian Ocean coral islands. *Kew Bulletin*, 30(1), 133–152.

Richardson, J.E., Weitz, F.M., Fay, M.F., Cronk, Q.C.B., Linder, H.P., Reeves, G., and Chase, M.W. (2001) Rapid and recent origin of species richness in the Cape Flora of South Africa. *Nature*, 412(6843), 181–183.

Robichaux, R.H., and Purugganan, M.D. (2005) Adaptive radiation and regulatory gene evolution in the Hawaiian silversword alliance (Asteraceae). *Annals of the Missouri Botanical Garden*, 92(1), 28–35.

Roderick, G.K., and Percy, D.M. (2008) Insect-plant interactions, diversification, and coevolution: Insights from remote oceanic islands. In K.J. Tilmon (ed.), *Specialization, speciation and radiation: the evolutionary biology of herbivorous insects*, Berkeley, CA, University of California Press, pp. 151–161.

Sakai, A.K., Wagner, W.L., Ferguson, D.M., and Herbst, D.R. (1995a) Origins of dioecy in the Hawaiian flora. *Ecology*, 76(8), 2517–2529.

Sakai, A.K., Wagner, W.L., Ferguson, D.M., and Herbst, D.R. (1995b) Biogeographical and ecological correlates of dioecy in the Hawaiian flora. *Ecology*, 76(8), 2530–2543.

Seastedt, T.R. (2015) Biological control of invasive plant species: a reassessment for the Anthropocene. *New Phytologist*, 205(2), 490–502.

Seaton, A.J., Beaver, K., and Afim, M. (1991) *A focus on Aldabra: conserving the environment*, Vol. 3. Victoria, Seychelles, Government of Seychelles.

Sekercioglu, C.H., Daily, G.C., and Ehrlich, P.R. (2004) Ecosystem consequences of bird declines. *Proceedings of the National Academy of Sciences USA*, 101(52), 18042–18047.

Skerrett, J., and Mole, L.U. (1995) Somewhere in an empty ocean. In M. Amin, D. Willets and A. Skerrett (eds.), *Aldabra: world heritage site*. Nairobi, Kenya, Camerapix, pp. 23–48.

Stewart, A. (1911) A botanical survey of the Galápagos islands. *Proceedings of the California Academy of Science*, 4(1), 7–288.

Stobart, B., Teleki, K., Buckely, R., Downing, N., and Callow, M. (2005) Coral recovery at Aldabra atoll, Seychelles: five years after the 1998 bleaching event. *Philosophical Transactions: Mathematical, Engineering and Physical Sciences*, 363(1826), 251–255.

Stoddart, D.R. (1971) Settlement, development and conservation of Aldabra. *Geographical Journal*, 134(4), 471–485.

Stoddart, D.R. (1974) The Aldabra affair. *Biological Conservation*, 1(1), 63–69.

Stoddart, D.R., and Westoll, T.S. (1979) *The terrestrial ecology of Aldabra: a Royal Society discussion*. London, Royal Society.

Taylor, J.D., Braithwaite, C.J.R., Peake, J.F., and Arnold, E.N. (1979) Terrestrial faunas and habitats of Aldabra during the Late Pleistocene. *Philosophical Transactions of the Royal Society of London, Series B*, 286(1011), 47–66.

Thomas, A. (1968) *Forgotten Eden*. London, Longman Group.

Travis, W. (1959) *Beyond the reefs*. London, George Allen & Unwin.

Wanless, R.M. (2003) Can the Aldabra white-throated rail *Dryolimnas cuvieri aldabranus* fly? *Atoll Research Bulletin*, 508, 1–7.

Weller, S.G., Wagner, W.L., and Sakai, A.K. (1995) A phylogenetic analysis of *Schiedea* and *Alsinidendron* (Caryophyllaceae, Alsinoideae): implications for the evolution of breeding systems. *Systematic Botany*, 20(3), 315–337.

Westoll, T.S., and Stoddart, D.R. (eds.) (1971) A discussion on the results of the Royal Society expeditions to Aldabra, 1967–68. *Philosophical Transactions of the Royal Society of London, Series B*, 260(836), 1–654.

Wright, S.D., Yong, C.G., Dawson, J.W., Whittaker, D.J., and Gardner, R.C. (2000) Riding the ice age El Nino? Pacific biogeography and evolution of *Metrosideros* subg. *Metrosideros* (Myrtaceae) inferred from nrDNA. *Proceedings of the National Academy of Sciences USA*, 97(8), 4118–4123.

Zerega N.J.C., Ragone, D., and Motley, T.J. (2004) Complex origins of breadfruit (*Artocarpus altilis*, Moraceae): implications for human migrations in Oceania. *American Journal of Botany*, 91(5), 760–766.

6

FAUNA

R. J. 'Sam' Berry and Adrian M. Lister

Introduction: beyond Noah's Ark

In 1684, Thomas Burnet (1635–1715) published *The sacred theory of the Earth* in London. It was highly idiosyncratic; Burnet might have become Archbishop of Canterbury if he had not written it. Still, it was extremely influential. Burnet began with the almost unquestioned assumption of his time that God had created the world as a series of concentric layers, with "no Rocks or Mountains, no hollow caves, nor gaping Channels, but even and uniform all Over". Rivers ran from the poles to the tropics, where they dissipated. This primitive order was devastated by the biblical Flood. For Burnet, the Earth's current surface was a ruin of the 'perfect' situation which existed until Noah's flood; for him, the oceans were gaping holes and the mountains upturned fragments of the old Edenic crust.

Burnet's ideas were widely criticised as 'poetic fiction'; but his recognition that earthquakes and other phenomena (apart from the Flood) could "contribute something to the increase of rudeness and inequalities of the earth in certain places" was eventually to lead to the sciences of geomorphology and biogeography. Until Burnet's assumptions were disposed of, the study of the plants and animals on islands had no interest. Put in another way, island biology can be dated to the increasing difficulty in the 18th century of maintaining a literal interpretation of the story of Noah's Ark, with its implication that the colonisation of the whole Earth sprang from a single source in the Middle East (Browne 1983, Rudwick 2005). As the exploration of the globe became more complete, the restriction of some species to very limited areas became more and more apparent.

The challenge to views such as Burnet's came in the context of the European Enlightenment, which also stimulated improvements in taxonomy and exploration. This made possible, for the first time, the ability to define the precise limits of distribution of different species (Cain 1984). Taken together, these developments gave birth to the science of biogeography; and the recognition of the importance of island studies.

Zoogeography can be regarded as advancing through three phases. The first was defined by the Comte de Buffon (1707–1788) and his 1778 essay on the differences between the mammals of the Old and New Worlds, including the presence of the opossum and the absence of the native horse in the Americas. The second phase dealt with the studies of Humboldt (1769–1859), Cuvier (1769–1832) and Lyell (1797–1875) (Mayr 1982, Wulff 2015). The third phase was a period of consolidation, geological support and inspiration from the travels of Charles Darwin (1809–1882) and Alfred Russel Wallace (1823–1913), supplemented by the work of Edward Forbes (1815–1854), Hewett Cottrell Watson (1804–1881), Thomas Vernon Wollaston (1822–1878), and Joseph Hooker (1817–1911). It culminated in the geographical chapters of the *Origin of species* (1859) and Wallace's paper on the Malay archipelago (1860), followed by his two great books on the *Geographical distribution of animals* (1876) and *Island life* (1880).

It was Darwin, Wallace and Hooker who introduced island biology into science; but Wollaston, an English naturalist, was perhaps the effective founder of island zoogeography. In *Insecta Maderensia*, Wollaston (1854, 1878) described vast numbers of endemic species on the Canary Islands, the Madeira Group (see Figure 6.1) and the Azores. He was so intrigued with the invertebrates of St Helena that he and his wife spent six months there; he described three-quarters of the more than 150 endemic species of beetles on the island.

Loss of flight ability is common in island invertebrates. Tristan da Cunha has 20 endemic species of beetles, 18 with reduced wings; while on Hawai'i, 184 out of the 200 endemic species of carabid beetle are flightless, and restricted to single islands or volcanoes (Hawaii's Comprehensive Wildlife Conservation Strategy 2005). Darwin sent Wollaston an advance copy of the *Origin*. Wollaston did not like it at all, damning it in an anonymous review (Anon 1860). To him, the fauna of the islands were relics of a submerged land mass. He refused to accept 'transformism' on theological grounds (Cook 1995).

The flightlessness of birds on islands is well-known. Many of these are rails (e.g. coots and moorhens): at least 23 species are flightless on islands, while another eight species are poor fliers. In fact all rails, including those on mainlands, are poor or reluctant fliers. A better example is probably the best known flightless bird of all: the dodo (*Raphus cucullatus*), one of a group of flightless birds related to pigeons once found on the Indian Ocean islands of Mauritius, Réunion and Rodrigues. They were very easy to catch and formed an easy prey for sailors; while their eggs, laid on the forest floor, were easy prey for the rats that arrived with the sailors. The dodo has been extinct for around 350 years (see Figure 6.2).

The moa were species of flightless birds, endemic to New Zealand. They were the dominant herbivores in New Zealand's forest, shrubland and sub-alpine ecosystems for thousands of years, but were then hunted by the Maori when these people arrived in New Zealand. The moa went extinct around AD 1400 (see Figure 6.3). The moa's closest cousin is probably the tinamou, species of ground-dwelling turkey-like South American birds which are able to fly, and still exist. By Darwin's time, it was becoming obvious that something more than climate and soil was needed to explain observed plant and animal distributions, and there were clear signs of biogeographical 'provinces'. Lyell was arguing that historical processes involved in determining such distributions were at least a possibility. In the second volume of the *Principles of geology* (1832), he proposed multiple "centres of creation" with different species being created (and becoming extinct) *in situ*. Part of his reasoning for this was the situation on oceanic islands, where

> the original isle was the primitive focus or centre of a certain type of vegetation, yet all belonging to the same group, giving the appearance of centres of foci of creation ... as if there were favourable points where the creative energy has been in greater action than in others.
>
> *(Lyell 1832, p. 126)*

He referred particularly to "the Madeiras and Canaries as types of oceanic archipelagos", confident because he had been there himself.

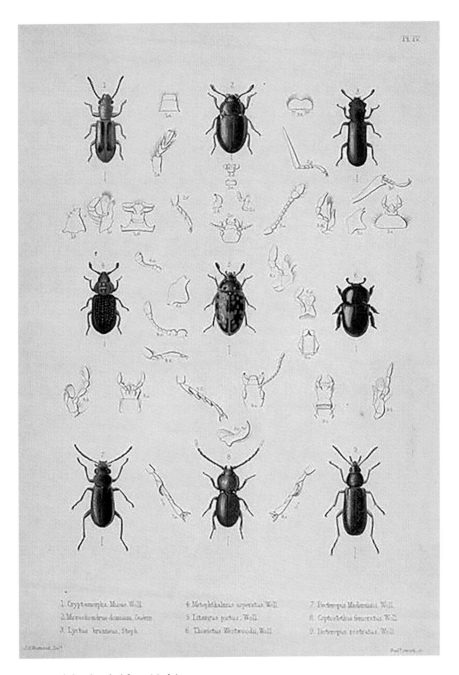

Figure 6.1   'Fightless beetles' from Madeira.

Source: Wollaston (1854). Picture ID: 9614548. Wikimedia Commons and Charles Darwin's Library. www.biodiver sitylibrary.org/pageimage/9614548.

Figure 6.2 Early 20th century representation of the dodo (*Raphus cucullatus*).

Source: © The Trustees of the Natural History Museum, London. Reproduced with permission.

Key data on islands was collected by Joseph Hooker, who soon after Darwin had sailed round the world on the *Beagle* (during which he visited over 30 islands) voyaged on the ship *Erebus* to the Antarctic and round the world (1838–1843). Hooker was influenced by his observations as the voyage proceeded. His notes quickly expanded from the minutiae of collecting to questions of geographical distribution. In a letter to his father written during the journey through the South Atlantic, he clearly thought that the island biotas he would find would be determined by temperature (Huxley 1918, pp. i, 82). By the time the *Erebus* reached Kerguelen Island, in the Southern Indian Ocean, he had begun to ask deeper questions about the relation between the biotas of islands and continents.

At the time, most geological processes were effectively unknown: it was only in 1837 that Louis Agassiz had propounded the notion of a 'Great Ice Age'. Indeed, the Yale zoologist George Baur went as far as to argue that the Galápagos Islands, nearly 1,000 km from the coast of South America, were 'continental islands', isolated by the subsidence of a land bridge (Baur 1897).

In contrast, Darwin believed that long–distance dispersal over water was more important. He carried out a series of experiments to test the survival of various seeds and fruits in water.

Figure 6.3 An early view of a Maori hunting a moa.

Source: Hutchinson (1896, pp. 232–233) and Wikimedia Commons.

Hooker set out his preferred scheme in a Linnean Society paper, read in 1860 (Hooker 1862), but his definitive conclusions were set out during the 1866 meeting of the British Association for the Advancement of Science in Nottingham (Hooker 1866; reprinted in Williamson 1984, Berry 2009). He took a lot of trouble with its preparation. Hooker wrote to Darwin:

> You must not suppose me to be a champion of Continental Connection, because I am not agreeable to trans-oceanic migration. I have no fixed opinion on the subject . . . Either hypothesis appears to me to well cover the facts of Oceanic Floras, but there are grave objections to both. Botanical to yours, geological to Forbes.

The fossil record showed that extinction was a natural process and not one dependent on either the Flood or human depredations. Some islands seemed to have arisen *de novo* from the bottom of the sea, and must have acquired their terrestrial biota from some other land.

It is often said that the *Origin of species* convinced people of evolution because it provided an easily understood mechanism (natural selection) for evolution. But the consensus of articles and books published in 1909, 50 years after the *Origin*, show clearly that it was principally the facts of geographical distribution that had convinced the majority. Darwin himself was very grateful for Hooker's support, and particularly Hooker's 1866 paper; it was one of the first significant post-*Origin* supports for Darwin's ideas. Moreover, it was the first systematic statement of the importance of islands for evolutionary studies. Hooker's identification of the main characteristics of island biotas still stands:

- they contain a high proportion of forms found nowhere else (endemics), although these endemics are usually similar to species found on the nearest continental mass;

- islands are impoverished in comparison with comparable continental areas; that is, there are fewer species on islands than on mainlands;
- dispersal must play a part in the colonisation and establishment of islands, unless the island has been cut off from a neighbouring area and therefore carries a relict of a former continuous fauna and flora; and
- the relative proportions of different taxonomic groups on islands tend to be different from non-island biota, i.e. there is taxonomic 'disharmony'. Hooker does not use the word 'disharmony', but it is implicit in his analysis. He described the "unequal dispersion" of species, "the most singular feature" of the flora.

Island inhabitants

What animals would we find on any particular island? The answer to that question would be a largely indigestible list – a triumph for those who compiled it and useful for visitors – but out of date, even before it was published, because new records appear and local populations commonly go extinct. The dynamics of colonisation and extinction will differ between continental and oceanic islands, but the principles are the same.

The avifauna illustrates the process of colonisation and extinction well. The presence or absence of a particular bird is more likely to be noted than for any other group; local lists are continually being revised. A pioneer in this area was the British ornithologist David Lack (1910–1973). During war service in the Orkney archipelago (for this and other locations in the British isles, consult Figure 1.1 on Page 5), he collected a mass of information on the local birds, which was published in two long papers in *Ibis* (Lack 1942a, 1943). This inspired him to develop more general ideas by listing the breeding birds of Caithness (at the northeast corner of Scotland), Orkney, Shetland and the Faroes, and comparing them with previously published lists (Lack 1942b). He noted that, at the time of his survey, 26 bird species had established themselves on Orkney since 1800, while at least another eight species (and perhaps as many as 20) bred occasionally. Thirteen of the 26 'new' species could be attributed to the planting of woods, gardens, trees and bushes in the 19th century, assisted by an increase in land under cultivation. A further eight species resident in Orkney in 1800 had extended their range to other island groups; in contrast, six species no longer bred on Orkney. Lack found that 61 (84 per cent) of the inland breeders and 15 (60 per cent) of the sea and shore birds changed significantly in density over his study period. This turnover in the avifauna is one of the most striking features of his survey.

Widening the focus: Ireland has only three-fifths the number of breeding bird species found in Britain, although it is only 80 km from Wales, which might be thought to be too short a distance to serve as a barrier for most birds. Indeed, all but six of the regular British breeders have been recorded in Ireland on one or more occasions; even the great spotted woodpecker (*Dendrocopos major*), the British race of which is sedentary, has been recorded over 50 times (see Figure 6.4). A clue to the missing Irish breeders is that most of them do not breed in western Britain, including Wales, opposite the island of Ireland. Lack (1969b) recorded 122 regular breeders in Wales compared to 171 in Britain as a whole. Most of the species not found in Ireland are ones that breed only in south or southeast England or the Scottish Highlands. He concluded that their absence from Wales and Ireland must be due to ecological factors rather than mere isolation. The same phenomenon occurs in flowering plants: nearly all those widespread in Britain occur in Ireland, but those least well represented are the ones restricted to eastern England (Praeger 1950).

Figure 6.4   Great Spotted Woodpecker (*Dendrocopos major*).

Source: Photo by Ron Knight, Wikimedia Commons, https://commons.wikimedia.org/wiki/File:Great_Spotted_Woodpecker_(Dendrocopos_major)_(13667857655).jpg.

The details of these findings have changed since Lack's surveys, but his general conclusions stand: there is a remarkably large amount of turnover going on in island bird populations (e.g. Lack 1976, p. 5). He reinforced the point with a study (Lack 1969a) of the birds observed and breeding on the 98 hectare Welsh island of Skokholm from 1928 – when Ronald Lockley began his systematic recording – to 1967 (Lockley 1969); Mark Williamson (1983) extended this to 1979. There were nine land birds which bred regularly throughout the period; while other species failed to breed in some years (see Figure 6.5).

Tim Reed (1980, 1981) expanded Lack's work with data from 73 British islands. He found that the best predictor of species number on any island was the number of habitats on the island, which usually, but not inevitably, increased with the land area of the island and its altitude range: hilly islands will have more area for colonisation than 'flat' ones, as well as scope for a diversity of ecosystems. This conclusion held up in a more intensive study of the birds of the Inner Hebrides, off the west coast of Scotland (Reed et al. 1983), with an additional recognition that any factors which affect habitats on a small island (such as marsh drainage or tree planting) may have a marked effect on the number of breeding species. On the smaller isles (Rum, Muck, Eigg and Canna), the number of woodland species (tits, finches and warblers) was significantly greater in larger and more diverse woods than in smaller, more uniform ones. Stuart Pimm and his colleagues extended these findings, collecting data on 67 species at 16 bird observatories on islands around the British and Irish coasts. They showed that the likelihood of extinction is linked not just to population size (and therefore island area), but also to body size: large birds (which have a longer life span than small ones) are less susceptible to extinction at low numbers, but are at a greater risk at higher ones (above seven pairs); presumably because of their demand for resources (Pimm et al. 1988).

Figure 6.5 Skokholm. The white building at the bottom right corner of the island is the lighthouse and provides a sense of scale.

Source: Wikimapia, http://photos.wikimapia.org/p/00/02/16/98/08_big.jpg.

Land area is also important in another and more indirect way: larger islands will probably have a greater area of suitable habitat for any particular species. A species may reach an island; but, if there is too little suitable habitat, it is unlikely to breed successfully. In practice there must be enough habitat to support a population greater than a threshold size, which will differ for each species. Raptors need a large area for their hunting and therefore always occur at a relatively low density; peregrine falcons disappeared from all but a few British islands in the late 1950s as their numbers fell nationally. Chaffinches are infrequent breeders on the small Welsh island of Bardsey because woody cover there is restricted to small clumps of scrub; annual recruitment is always low and periodic extinctions occur, often with a break of several years before a population is re-established. Species with simple habitat requirements (such as oystercatchers or rock pipits) are found on almost all the British islands, whilst wood warblers (which depend on tall and broad leaved cover) occur rarely (Pimm et al. 1988).

Another factor is that, for any species to establish itself on an island (as opposed to merely getting there), it must be able to reproduce and achieve a population sufficient in size to maintain itself (Hengeveld 1989). This places a premium on asexual methods of reproduction. The citrine forktail, *Ischnura hastata*, on the Azores is the only known population of parthenogenetic damselflies, consisting only of females (see Figure 6.6). The species is common in the Americas, but it is always a traditional sexual breeder there. In plants, hermaphroditism and self-compatibility are valuable after long-term dispersal because they enable successful reproduction even when animal pollinators are absent: as they may well be if the island is sufficiently isolated. For example, only 15 per cent of insect orders are present in the Hawaiian islands and few species seem

Figure 6.6   Citrine forktail, *Ischnura hastata*, the only known species of damselfly with a population consisting only of females.

Source: Photo by Pondhawk in public domain, www.flickr.com/photos/38686613@N08/4780864194/.

to be pollinators; there are only six native species of hawkmoths, two species of butterflies, and
no bumble bees.

Persistence of a species on an island is affected by its distance from the nearest mainland (and
source of immigrants to 'top up' the population); but, as Lack found, persistence depends on
more than physical distance. Species near the edge of their range will be relatively rare immigrants, even if apparently suitable habitat is available; small changes in any of the factors affecting
breeding are likely to affect such species in particular. O'Connor (1986) distinguished between
sporadic breeders, re-colonising species and invaders. He pointed out that the number of bird
species breeding in woodland decreases from southeast to northwest Britain; this might be
linked to the availability of invertebrate food. Scotland has 19 species which breed nowhere else
in Britain; most of them are probably periodically 'topped up' by incomers from Scandinavia.
These are all factors to be taken into account when assessing the likelihood of new recruits to
any island.

Different considerations and conclusions apply to other groups. The past distribution data
for beetles is better known than for most animal groups, because their presence or absence can
be traced reasonably accurately in sub-fossil deposits. A few species have their earliest records
in archaeological sites and can be regarded as human introductions, but many Atlantic species
now have a wide range, from the British islands to the Faroes, Iceland and Greenland. Russell
Coope (1986) has argued that most of the present North Atlantic beetle fauna originated on
ice rafts, dispersed precariously during a fairly short period after the ice sheets finally retreated
15–10,000 years ago. Part of the reason for this conclusion is that, of all the species that could

tolerate the climate on each island group, only 54 per cent occur in Shetland, 26 per cent in the Faroes, 19 per cent in Iceland, and a mere 4 per cent in Greenland. Coope (1986, p. 632) says, "The only satisfactory explanation of these figures is that the faunas were derived from the east and that there were progressive losses in the transatlantic 'sweepstake'." Spiders are different. They are more uniformly distributed than virtually any other animal group. Their eggs are so light that they are swept into high altitudes and may come to land anywhere. This means that individual species are limited by climate or habitat, but not the opportunity to colonise. The Faroe Islands have only two-thirds of the spider species found in the Shetland Islands (off NE Scotland); but a third of the species in the Faroes are absent from Shetland. The difference is probably due to the higher latitude and higher hills of the Faroes. Iceland has almost as many species as Shetland, but more than half of them are different. Since over 80 per cent of the Icelandic species are found in mainland Scotland, their absence in Shetland probably reflects the greater range of habitats on Iceland. Spiders found in Iceland, but not in Britain, are mainly arctic forms. In spite of this, the spiders of the eastern North Atlantic form a common faunal area and are not very useful markers for island faunas (Ashmole 1979).

In general, the colonisation and extinction patterns of flying insects are not unlike those of birds, with the same principles of dispersal and settlement applying. Dinnin (1996) concluded that all the Outer Hebridean insects can be regarded as post-glacial colonisers, and that some species have disappeared as woodland declined on the islands. However, successful colonisation may be exceedingly rare. One of the most studied cases of inter-island movement of animals is that of the 'picture-winged' *Drosophila* of the Hawaiian archipelago, traced using chromosome markers. Even between islands mainly in sight of each other, only one successful colonisation takes place on average every 25,000 years (Williamson 1981, p. 219).

With the exception of marine mammals and bats, mammals have not reached oceanic islands except by human agency. But islands close enough to the mainland have been colonised by terrestrial mammals, which often develop into endemic species, especially when colonisation is a rare, 'sweepstakes' event. Reviewing the Pleistocene mammal faunas of islands in the Mediterranean and southeast Asia, Paul Sondaar (1977) noted that they usually comprise elephants, deer and hippos – all species which are strong swimmers or, in the case of hippos, 'underwater walkers'. The Thylacine or Tasmanian Tiger (*Thylacinus cynocephalus*) was a top marsupial predator native to Australia, which colonised Tasmania and New Guinea thanks to lowered Pleistocene sea levels. The last-known living exemplar died in Hobart Zoo in 1936 (see Figure 6.7). It is officially classified as extinct, although sightings are occasionally reported.

(Re)colonisation following volcanic eruptions

Colonisation is easier and turnover more rapid when empty habitats are available. Some wide-ranging species quickly establish themselves whenever a possibility occurs. They have been described as 'super-tramps': good colonisers but usually poor competitors. The island of Krakatau in Indonesia exploded in 1883. In the first phase of re-establishment of biological life, most of the colonisers away from the shore (an average of two species a year) were wind-borne, although there were also nearly as many species (1.64 per year) that colonised the shore line. In the early years, only an average of one animal species every seven years penetrated into the interior. During the next phase (1897–1919), wind-borne pioneers lost ground and animal penetration of the interior increased ten-fold (to an average of 1.32 species per year) as fruit-bearing plants grew up and provided food for immigrants.

Then in 1930, further volcanic activity produced a new island, Anak Krakatau ('Child of Krakatau'), which has experienced sporadic eruptions ever since (see Figure 6.8).

Figure 6.7   The Thylacine, or Tasmanian Tiger (*Thylacinus cynocephalus*).

Source: The Thylacine Awareness Group of Australia, www.thylacineawarenessgroup.com/.

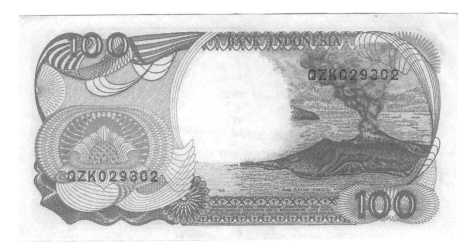

Figure 6.8   Volcanically active Anak Krakatau, depicted on the Indonesian 100 rupiah banknote.

Source: www.summitpost.org/anak-krakatau-on-100-rupiah-banknote-indonesia/769287.

William Syer Bristowe landed on it six months after its birth. He commented:

> Its virgin shore [was] composed entirely of dark grey ash, black cinders and white pumice stone. No plants would grow here until weathering and bacteria had had time to create soil in a year or two, but seeds, along with debris like banana stems and other vegetable matter, were awaiting their time to establish themselves. The only abundant insects were scavengers which could feed on whatever plant life the sea brought them – a springtail, a beetle, a species of ants, a tiny leaf-mining moth and a mosquito.
>
> *(Bristowe 1969, pp. 148–149)*

The continuing process of colonisation on Anak Krakatau has been chronicled by Ian Thornton. The first birds seen were migrants like the Common Sandpiper, Grey Wagtail, Pacific Golden Plover, Mongolian Plover, Whimbrel and Great Knot. Thornton placed sea-water filled plastic containers in places where the lava had flowed and found a constant rain of arthropods, including insects, crustaceans, spiders and scorpions: in ten days in 1985, he collected 72 species. A great increase in bird species occurred in the same year, coinciding with the first fruiting of the island's figs and the consequent opening of a 'habitat window'. As these new birds established themselves, some of the earlier-arrived ground nesting species disappeared. Then raptors came to feed on the fruit-eating species. By the mid-1990s, about 150 plant species inhabited the island (Thornton 1996).

The same process has been occurring on Surtsey, an island produced by a volcanic eruption off the southwest corner of Iceland in 1963–1964 (see Figure 6.9). By 1987, a dense gull colony had developed, and a there was a rapid increase in plants from seeds carried by the birds. The

Figure 6.9 On 14 November 1963, the crew aboard a trawler sailing near Iceland spotted a column of smoke rising from the sea surface. Surtsey, a new island, was being born.

Source: NOAA, www.ngdc.noaa.gov. Available in public domain at Wikipedia, *https://en.wikipedia.org/wiki/Surtsey#/media/File:Surtsey_eruption_2.jpg*.

land surface of Surtsey has been eroding so quickly that the succession has not been as marked as for Krakatau (Fridriksson 1975, 2005).

Old and new arrivals

Volcanoes illustrate the dynamics of species arrival and replacement in particularly vivid ways. But, in fact, the species composition of all communities is similarly affected by turnover resulting from colonisation and extinction. These processes must have been especially marked in the North Atlantic region as the Pleistocene ice receded and gave way to sub-Arctic tundra-steppe and then temperate conditions, a process involving much flux and opportunism.

In Britain, for example, a glacial mammal fauna including mammoth and reindeer first expanded its range after the ice began its retreat some 19,000 years ago, then died out as the climate warmed and woodland became established. By 13,000 years ago, aurochs (*Bos primigenius*) and elk (*Alces alces*) had arrived; by 10,000 years ago, they had been joined by other species such as beaver (*Castor fiber*) and roe deer (*Capreolus capreolus*). At around 8,000 years ago, sea-level rise made Britain an island and there were no further additions to the large mammal fauna except those imported by people (Yalden 1999).

The ecological processes involved in all this have been studied experimentally by Bob Paine (1980) in habitats which are generally assumed to be stable: mature forest or rocky sea-shore. Paine followed the sequence of events after an old tree falls or a patch of shore is artificially cleared of its animals and plants. The dynamics of the re-establishment of the fauna and flora are the same as in the more spectacular circumstances of Krakatau and Surtsey, albeit in a less obvious manner.

Good information on the time of colonisation or extinction is sometimes available for birds, but it is relatively rare for most groups; in most cases, it is usually impossible to know whether a new record is a fresh arrival or merely the result of a more intense search; perhaps simply the arrival of an expert taxonomist.

Some occurrences – and changes in distribution – are surprising. On the islands around Britain, Red Deer (*Cervus elaphus*) are (or were) present on the Outer Hebrides, the Isles of Scilly and the Channel Islands. On the Channel Islands they were presumably relicts from the time when the islands were connected to continental Europe, but elsewhere it seems likely they were humanly introduced in prehistoric times, although it is just conceivable that they managed to reach some of these islands by swimming. There were foxes (*Vulpes vulpes*) on the Isle of Man in the 19th century. They became extinct, almost certainly before the end of the century (Reynolds and Short 2003). Flux and Fullagar (1992) list over 800 islands throughout the world with rabbit (*Oryctolagus cuniculus*), all of them introduced from their original range in Spain.

The Pleistocene fossil record of mammals on islands richly illustrates the dynamic and fluid nature of island communities (Geer et al. 2010). Many caves on Mediterranean islands have been excavated for their fossil content, and they typically show a sequence of differing assemblages, sometimes separated by sterile layers implying extinction of one fauna before immigration and establishment of another. At Ghar Dalam Cave on Malta, for example, a Late Pleistocene sequence starts with a lower layer dominated by remains of hippo (*Hippopotamus*) with some dwarfed straight-tusked elephant (*Palaeoloxodon*) (see Figure 6.10). This is succeeded by a layer without hippo or elephant but dominated by deer, probably derived from mainland red deer (*Cervus elaphus*). All of these species presumably arrived from the adjacent island of Sicily either by crossing a narrow channel or by walking at rare intervals of sufficiently low sea level. Periodic extinctions may reflect climatic fluctuation, but the changing succession of faunas may owe as much to chance colonisation as to the 'suitability' of varying environments to different herbivore species.

Figure 6.10   Skeleton of an adult dwarf elephant, *Palaeoloxodon falconeri*, just 1 metre (3 feet) high, from
the island of Sicily, Italy. Part of an adult human skeleton in the background provides a sense
of scale.

Source: Photo by James St John. Displayed at the State Museum of Natural History, Lincoln NE, USA. Wikimedia
Commons, https://commons.wikimedia.org/wiki/File:Mammuthus_falconeri_(dwarf_mammoth).jpg.

'The theory of island biogeography'

A major advance in the understanding of island biotas was 'the theory of island biogeography'
proposed by Robert MacArthur and Edward Wilson in 1963 and expanded into a book in
1967. The theory was the extension of ideas that had been around for a long time. In the mid-
nineteenth century, H.C. Watson pointed out that one square mile (2.5 km²) in Surrey, England,
held half the plant species found in the county as a whole. Henry Gleason generalised this in
a 1922 paper, 'On the relation between species and area' and Philip Darlington (1957) related
it to islands by pointing out that, in a range of islands, a ten-fold increase in area led only to
a doubling in the number of species. But the species present in any one place are not invari-
able. As we have seen for birds, every community has turnover in its composition: some species
disappear and others appear. Michael Soulé (1983) listed 18 reasons why a population may go
extinct. How can a species be replaced? Isolated communities – and, by definition, all islands
are to some extent isolated – will have fewer potential colonisers: both fewer species and fewer
individuals of any species, and increased potentialities for unique interactions or exploitation of
unusual habitats.

The core of MacArthur and Wilson's thesis is that there is a balance between *immigration* to an
island (which will vary with its distance from the mainland) and *extinction* on it of local populations (which will vary with the island area). Thus, the number of species on an island will be the
difference between those continually reaching it and those which are being lost.

The insight of MacArthur and Wilson was that this is a present dynamic rather than a
simple historical hangover: species are continually going extinct locally; species are continually appearing and establishing themselves. Species turnover on islands is normal; MacArthur
and Wilson suggested that recurrent colonisations and extinctions create an equilibrium in
which the number of species remains relatively constant, although the species concerned will

vary over time. They used data from the re-colonisation of Krakatau to support their thesis. Wilson and Dan Simberloff went on to test the theory by fumigating four small mangrove islands off the coast of Florida so as to kill all the resident animals and then monitoring their re-colonisation over a period of years (Simberloff and Wilson 1969, 1970, Simberloff 1969, 1976).

The insights of MacArthur and Wilson had been anticipated decades earlier. Wallace had expounded much the same idea about balance 80 years earlier, in his book *Island Life* (1880, p. 532):

> The distribution of the various species and groups of living things over the earth's surface and their aggregation in definite assemblages in certain areas is the direct result and outcome of firstly the constant tendency of all organisms to increase in numbers and to occupy a wider area, and their various powers of dispersion and migration through which, when unchecked, they are enabled to spread widely over the globe; and secondly, those laws of evolution and extinction which determine the manner in which groups of organisms arise and grow, reach their maximum, and then dwindle away.

However, it was left to MacArthur and Wilson to give formal expression to these ideas.

Most tests of the MacArthur and Wilson theory have come from birds, and British birds have been particularly significant here: the amount of ornithological data from the British islands is unique in the number of species and the length of time for which it is available. Tim Reed collaborated with US biologists (Manne et al. 1998) to test the generality of his earlier conclusions on British birds and the applicability of the MacArthur and Wilson theory, using data collected from bird observatories. Sufficiently complete data was available from 12 island observatories: Bardsey, Calf of Man, Copeland, Fair Isle, Handa, Hilbre, May, Lundy, Scolt Head, Skokholm, Skomer and Steepholm. Both immigration and local extinction rates increased with the number of breeding species on each island; the number of breeding species also declined, albeit not significantly, in smaller islands and distance from the mainland pool of breeders.

The extinction and re-colonisation data are probably more accurate for birds than for any other group, because it is relatively easy to record a species failing to breed. The reason for such failure may be because numbers are declining generally or through local factors like competition, change in habitat, or even pure chance. A complication is that different species have different mobility, whatever their potential powers of flight. The ancestors of the finches on the Galápagos or the honey-creepers on Hawai'i never 'intended' to settle on the respective islands, and their successful establishment and breeding represent extremely unlikely events. It is impossible to know how many of their relatives perished at sea, although some indication is given by the numbers of 'vagrant' birds from distant parts of the world that appear in the 'wrong' place – mostly to die a lonely death.

MacArthur and Wilson originally described the balance between immigration and extinction as an 'equilibrium', although as Mark Williamson (1981) has pointed out, it is really nothing more than a logical necessity: in ecological time, the number of species on an island can only be increased by two processes – immigration, which in turn will depend on the distance of the island from the source of potential colonisers; and the availability of ecological space for them. They will decrease simply by failing to survive, that is, by extinction. A complication is that immigration and extinction are not independent, since a high immigration rate may 'rescue' an island population of the same species, and hence reduce the likelihood of its extinction (Brown and Kodric-Brown 1977). Moreover, an immigrant might outcompete a resident and lead to its extinction. David Lack castigated the value of the MacArthur and Wilson thesis on the ground that most species turnover is ecologically trivial. Notwithstanding, the MacArthur and Wilson theory works: whilst there is a turnover of species on any island over a period of years, the total number of species tends to remain fairly stable. Examples of its successes are given by David Quammen in *The song of the dodo* (1996). However,

- The theory deals only with the number of species, not with the number of individuals in a particular species.
- The theory considers all species together, and tells us nothing about the functioning and composition of the community. It does not tell us, for example, why there are no rabbits (*Oryctolagus cuniculus*) on Tiree (the most westerly island in the Inner Hebrides of Scotland) or why there are no snakes on Ireland.
- It does not tell us anything about historical factors: why, for example, is the common European vole *Microtus arvalis* found only on Guernsey (in the English Channel) and Orkney (off NE Scotland) while its close relative *M. agrestis* occurs widely in Great Britain and many of the smaller islands, but not in Ireland. The two species co-exist over much of continental western Europe.
- Most importantly, *The theory of island biogeography* is misleading in its discussions of the origins of island endemicity. The authors omit any mention of genetic changes that may occur as a result of a colonising event. Part of the reason for this is historical. MacArthur and Wilson's book was published in 1967. In the previous year, two scientific papers were published which caused a radical rethinking of population biology.

This last point is important. Traditionally, biologists thought of any animal or plant population as genetically rather uniform. Clearly, inherited variation exists (such as bridling [a white eye ring and a thin white line extending behind the eye] in the common guillemot, black coloration in rabbits, or mammalian blood groups), but the proportion of variable gene loci was thought to be very small. Indeed there was a simple calculation that showed too much genetic variation could not be tolerated: it produces a 'genetic load' which reduces fitness and the reproductive potential of the population (Muller 1950). This assumption had to be ditched following the demonstration in 1966 by Harry Harris working in London on human material and Jack Hubby and Dick Lewontin in Chicago on *Drosophila pseudoobscura* that heterozygosity (that is, different alleles inherited from the two parents) was found in 10 per cent or more of genes. (An allele is one of the possible variants at a particular position on a chromosome; different alleles can produce different effects on a character.) This result was rapidly extended to a wide range of organisms.

The consequences of this high heterozygosity are extremely significant in understanding the differentiation of island forms. A small group of individuals drawn from a large population will almost certainly differ from its parental group in the frequency of alleles at a large number of loci. Some alleles will be absent or relatively over-represented in the smaller group. If the small group is either isolated or is a colonising propagule, the daughter group will be immediately different from the source population.

This only became evident after the time that MacArthur and Wilson were developing their ideas, and perhaps explains why they confined themselves to discussing evolutionary changes occurring subsequent to colonisation.

Founder effects

MacArthur and Wilson were not particularly interested in the processes of speciation. However, in the *Theory of island biogeography*, their chapter on evolutionary changes states:

> Since we believe that evolution through natural selection has produced the biotic differences which characterise islands, it is appropriate for us to study how natural selection works on islands. We can think of the evolution of the new population as passing through three overlapping phases. First the population is liable to respond to the effects of its initial small size. This change, if it occurs at all, will take place quickly, perhaps only in a few generations. The second phase, which can begin immediately and must continue indefinitely, is an adjustment to the novel features of the invaded environment. The third phase, an occasional outgrowth of the first two, consists of speciation, secondary emigration and radiation.
>
> *(MacArthur and Wilson 1967, pp. 145, 154)*

MacArthur and Wilson explicitly equate their first phase with 'the founder effect': a concept developed by the Harvard taxonomist and evolutionist, Ernst Mayr in one of the defining works of the neo-Darwinian synthesis, *Systematics and the origin of species* (1942) and described more fully in a later essay, 'Change of genetic environment and evolution' (Mayr 1954). MacArthur and Wilson describe the founder effect as:

> an omnipresent possibility but one easily reduced to insignificance by small increases in propagule size, immigration rate, or selection pressure. The founder principle is actually no more than the observation that a [founding] propagule should contain fewer genes [alleles] than the entire mother population
>
> *(MacArthur and Wilson 1967, pp. 154, 156)*

This is not correct. It would have been a reasonable assumption before the discovery of high levels of heterozygosity, but the post-1966 revolution showing the enormous genetic variability in any group of organisms means that the founder effect will almost certainly change allele frequencies as well as reducing variability to some extent.

Geneticist Sewall Wright, one of the founders of modern evolutionary theory, has written:

> I attribute most significance to the wide random variability of gene frequencies (not fixation or loss) expected to occur simultaneously in tens of thousands of loci at which the leading alleles are nearly neutral, leading to unique combinations of gene frequencies in each of innumerable different local populations. The effects attributed to the "founder effect" by Mayr (gene loss, reduced variability) are the most obvious but the least important of the three I have stressed (letter from Wright to Victor McCusick, quoted in Provine 1989, p. 57). [The third was the commonness of local extinctions and fixation through strong selection].

In other words, Wright believed (*contra* both Mayr and MacArthur and Wilson) that the main effect of the colonisation of an island by a small number of individuals will be a population differing from its parent in the frequencies of a large number of loci, producing instant differentiation. Every colonising event will be a new experiment, exposing to the environment a unique set of genes and their interactions in the alleles carried by the founding group. If members of the group are unable to cope with the environmental conditions to which they are exposed, the group will not survive. Indeed, most colonisations result in rapid extinction because of the failure of the animals or plants to respond effectively to their new situation. If they survive, their response is Phase II in the MacArthur and Wilson scheme; but it will necessarily be limited by the chance collection of alleles present in the founders.

There are many examples of situations where it is difficult to explain differentiation without invoking a founder effect (Berry 1996, 2004). One of the most convincing is that of House Mice (*Mus musculus*) on the Faroe Islands. House Mice are the only small mammals on the islands. They were originally described scientifically by Eagle Clarke as a new subspecies, *faeroensis*, distinguished solely by "immense size" (Clarke 1904). On the same grounds, they were later promoted to specific rank, since the form "differs so conspicuously from all other members of the [species] group" (Miller 1912).

The distinctiveness of these mice is of considerable interest because there is no plausible way that they could have reached the islands, except by human transport. The earliest human inhabitants of the Faroes were 8th-century Irish hermits, but the islands were not settled properly until the Norse began to arrive in the 9th century. A detailed study by Magnus Degerbøl suggested the existence of distinct races of Faroese mice on at least four of the islands. Degerbøl followed the common assumption that they must have originated through isolation for a long space of time, while the big mice of Nolsoy (one of the Faroe Islands), described by Clarke, have then developed by adaptation to the leaping life on the bird cliffs (i.e. as a kind of ecological race) (Degerbøl 1940).

This situation puzzled Julian Huxley. In his definitive exposition of the neo-Darwinian synthesis, he calculated that:

> the normal minimum time for distinct sub-speciation [is 5,000 years] although the facts concerning rats and mice show that sub-speciation may occur much more quickly. In particular, the Faroe house mouse, *Mus musculus faeroensis*, that was introduced to the islands not much more than 250 years ago, is now so distinct that certain modern authorities [including Magnus Degerbøl] have assigned full specific status to it.
>
> *(Huxley 1942, pp. 194–195)*

In fact, differentiation has been even more explosive. The island of Hestur was only colonised by mice in the 1940s when an airport was being built there, but they are as distinct as those on any of the islands in the group (Berry et al. 1978).

It is difficult to explain this in any other way than that their differentiation is the chance consequence of the genes carried by the original colonisers from the large neighbouring island of Streymoy. Similar examples of island differentiation are the field mice (*Apodemus* spp) of the Hebridean and Shetland archipelagos and the voles (*Microtus arvalis*) of Orkney (Berry 1969, Berry and Rose 1975).

Notwithstanding the possibility of post-colonisation evolution, founder changes persist. In his comprehensive overview of island population biology, Mark Williamson (1981, p. 135) objected to explanations based on the uncertain chance of the founder effect, on the grounds that they are entirely speculative. He suggested that the differentiation of the Faroese mice could be explained by "adaptation to the colder and damper situation found on the islands".

To test this, Simon Davis (1983) compared the morphometric characteristics of mice from the mainland of Great Britain with population samples from the Orkney, Shetland and Faroe archipelagos. His null hypothesis was that a major 'maritime effect' would result in the island populations being more like each other than to mainland populations. Using a statistical analysis similar to that employed by Williamson, Davis (1983) found that the mouse population of each island group was distinct and individually more like that from Caithness (northeast Scotland) than to each other. In other words, his data showed a regional geographical effect (which may reasonably be blamed on a Viking influence) rather than an island effect.

Surprisingly, the low heterozygosity found in populations which have gone through a founder effect (Frankham 1996, 2005) is much less than expected. Only if the founder bottleneck persists for a long period or is repeated, does genetic variability remain low (Bryant et al. 1990). Kaneshiro (1995) has suggested that the disruption of a few individuals entering a new habitat may lead to the breakdown of sexual selection and thus involve a higher proportion of males breeding than usual, so increasing the effective gene pool. Populations with little genetic variation are exposed to a range of environmental hazards to which they may have difficulty in adjusting; whilst captive breeding programmes (such as that of endangered species maintained in zoos) properly make great efforts to avoid inbreeding and the potential loss of inherited variation (Soulé 1987). Experimental findings suggest that the risk of heterozygosity loss may be over-emphasised in field (i.e. non-managed) situations (Jamieson 2007). The concept of a minimum viable population is usually more important ecologically than genetically, although clearly the two cannot be entirely separated (Primack 1993).

Experimental islands

MacArthur and Wilson's theory led to questions being asked about the practical significance of islands at a time when natural areas are increasingly becoming 'islands' in seas of developed or despoiled land. Can this sort of island be treated in the same way as a real island in terms of its biological content and diversity? Jared Diamond (1975a) spelt out some of the consequences of applying 'island theory' to 'island reserves':

- a newly isolated area will temporarily hold more species than at equilibrium because it will 'gain' species from the surrounding unsuitable habitat – but the surplus will not persist;
- the rate at which equilibrium occurs will be faster for small reserves than large ones; and
- different species require different minimum areas to support a stable population.

In designing nature reserves, this means we should bear in mind that:

- a large reserve will hold more species at equilibrium than a small one;
- a reserve near other reserves will hold more species than a remote one, because they will behave like a single large one;
- a cluster of reserves will support more species that separated ones, or ones arrayed in a line;
- a round reserve will hold more species than an elongated one, because it will have a lower surface for 'leakage'.

Diamond's arguments led to a debate on SLOSS: is it better to aim for a **S**ingle **L**arge **O**r **S**everal **S**mall reserves? (see Figure 6.11).

This was tested by Dan Simberloff, extending the studies on the experimental islands that he had established with Edward Wilson (see above). He divided the original islands, cutting channels through their roots and canopy so that each island became a mini-archipelago, leaving the overall area only slightly affected. Against expectation, he found four years later that a single island did not always harbour more species than several small ones (Simberloff and Abele 1976).

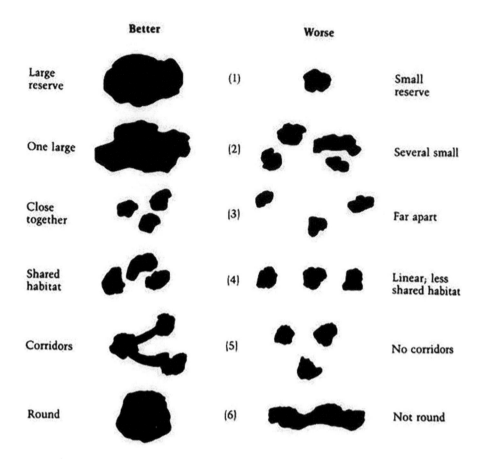

Better **Worse**

Large reserve — (1) — Small reserve

One large — (2) — Several small

Close together — (3) — Far apart

Shared habitat — (4) — Linear; less shared habitat

Corridors — (5) — No corridors

Round — (6) — Not round

Figure 6.11 The predictions of the SLOSS theory: **S**ingle **l**arge **o**r **s**everal **s**mall reserves?
Source: Adapted from Diamond (1975b) and Whittaker (2007).

The SLOSS argument rumbles on (Harris 1984, Gilbert et al. 1998, Lovejoy 2000, Fahrig 2003, Robert 2009). It is an important debate, because it affects not only the biodiversity of particular islands but the survival or extinction of whole species. However, it is only marginal to understanding the natural history of islands.

More on founder effects: is there a genetic revolution?

At the same time that MacArthur and Wilson were exploring island biogeography, a group of evolutionary geneticists were speculating about the genetics of colonising species at a pioneering conference in California (Baker and Stebbins 1965). The discussions were summed up by Ernst Mayr, who emphasised the need for ecologists to know something about the genetics of the organisms they studied, and for geneticists to become familiar with the ecology of their organisms. He had attempted this himself in his description of the 'founder effect' in 1954.

Although his original proposal about the origin of differentiation was similar to that of Wilson and MacArthur – on changes following colonisation – Mayr extended his idea by proposing that animals and plants are faced with a new 'genetic environment' when they are isolated and as a consequence undergo a 'genetic revolution'. As a young man, Mayr had worked on the birds of the islands north of Australia. Time after time, he found that species which differentiated little on larger land masses were represented on the smaller islands by very different forms, and by forms which differed from those on other islands. He wrote that:

> [that] mutation, recombination, selection and isolation are the four corner stones of evolution is now generally acknowledged . . . [However] the role of a sudden change in the genetic environment seems never to have been properly considered.
>
> *(Mayr 1954, p. 161)*

He did not believe that the two factors usually cited for 'the striking dissimilarity' of peripherally isolated populations: differences of physical and biotic environment (MacArthur and Wilson's explanation) or genetic drift (random changes in a small population) were sufficient. He focused on the importance of 'gene-flow or immigration'. He pointed out that gene-related experiments in laboratories go to great length to avoid immigrants, which are regarded as "genetic contaminants" (ibid., p. 157).

Mayr argued that in a widely distributed population continually exposed to individuals moving in from elsewhere, there will be selection for genes which will tolerate combination with 'alien' genes. He cited populations of plants subjected to powerful stresses due to high salinity or desiccation which may develop locally adapted races (or 'ecotypes') but which tend to be very variable due to the inflow of genes from nearby populations. If we think of single genes, we can measure their effects on physiology and reproductive success, but all the genes carried by an individual work together to produce a functioning whole; it is wrong to think of them behaving as discrete entities like beans in a bean bag. This working together means that the genes carried by an individual (or a species) are functionally related to each other in development, producing a 'genetic architecture' where the loss or gain of a particular allele may well affect other gene-loci. Such genetic architecture is often referred to as a 'coadapted' system. For Mayr (ibid., p. 165), "the selective value or viability of a gene (allele) is not an intrinsic property but is the total of its viabilities on all the genetic backgrounds that occur in a population". In other words, there is an internal genetic environment as well as an external one. The implication is that any disharmonious gene or gene-combination introduced into a coadapted complex will be discriminated against by selection. Importantly, as Brakefield (1991, p. 71) has commented:

> the existence of coadaptation between genes and of forms of non-additive genetic contributions to quantitative variation are likely to make variability within populations more resistant to loss than would be expected on the basis of [simple "bean-bag"] theory.

All this led Mayr to suggest that a group isolated on an island will undergo a "genetic revolution" in the sense that they will not have to cope with a constant influx of new alleles and can therefore adapt to the circumstances on the island without the need to compromise due to the disrupting effect of immigrants. It is a nice idea, but there is no direct evidence for it. Although evidences of genetic cohesion or coadaptation are common (e.g. epistasis – the interaction between two or more loci affecting a single phenotype – and other specific gene interactions; complex traits being controlled by many loci; selection favouring the formation of 'balanced' chromosomes with positive and negative traits intermingled), its power remains unknown (Berry 1997). Recent findings from gene-mapping suggest that many higher organisms have fewer genes than previously assumed. This implies many genes act together in development and behaviour and strengthens the idea of the importance of genetic architecture; it is the basis for the interest in epigenetics (Jablonka and Lamb 2014). However, for the moment, the existence of Mayr's 'genetic revolution' remains unproven.

Whatever the ultimate fate of the genetic revolution concept, it is not as important for island differentiation as originally believed by Mayr. He effectively assumed that the genetic changes that take place in the process of colonisation and isolation itself are negligible and can be neglected. As we have seen, this is not so. This does not negate the fact that environmental factors which are likely to be different between islands and mainlands may act selectively. It may be helpful to distinguish the founder principle that occurs through the chance collection of genes carried by founders from founder selection, which occurs following isolation; the founder effect is the result of both processes.

Founder selection

Although there is little doubt that most of the distinctiveness of island forms is due to the genes carried fortuitously by the founder individuals of any population, the species which result will inevitably be exposed to natural selection. We can call this 'founder selection'. Adaptation is everywhere evident in nature, but actual studies of natural selection in operation are not common. (An excellent review is that by John Endler 1986.) The classic study of adaptation through time on islands is the well-documented one of the *Geospiza* finches of the Galápagos archipelago (e.g. Grant and Grant 2008, Weiner 1995).

A particularly good example of such selection taking place after the original founder event is offered by the frog-hopper or spittle-bug, *Philaenus spumarius* (so-called because of the 'cuckoo-spit' which protects the pre-adult life of these insects) on small islands in the Baltic Sea off the coast of Finland. The pattern on the wing-cases (elytra) of these bugs is highly polymorphic, different forms being determined by alleles at a single gene locus. On the mainland of Finland, one form increases in frequency towards the north, whilst another form is commoner in humid areas, implying adaptation to different conditions. Olli Halkka and his colleagues found that some forms seem to be particularly hardy and survive better than others when introduced to new islands, but that the relative advantage of the forms changes as vegetation grows up, with the original colonisers being replaced by another form which has a preference for feeding on newly dominant plants like purple loose-strife (*Lythrum salicaria*) and sea mayweed (*Triplospermum maritimum*).

The island races of *Philaenus* largely retain the variability found in mainland populations. Although several elytral forms are missing on the outer islands, the same forms also disappear at the northern edge of species' range on the mainland, although the bugs there are in breeding contact with a large southern population. In other words, the genetic structure of the island races seems to be adaptive, and not merely the result of ongoing immigration. Direct proof of selection on the islands has come from an experiment in which some 8,000 individuals (three-quarters of the populations) were swapped between two islands with genetically different populations. After three generations, both island populations had reverted to the pre-transfer allele frequencies, although they deviated considerably from another 35 island populations in the area, i.e. the genetic makeup on the islands was not random and must therefore be regulated by natural selection (Halkka et al. 1974, Halkka et al. 1975).

Another example of founder selection is *porphyria variegata*, an inherited defect in humans of porphyrin, part of the haemoglobin molecule. The porphyria allele causes extreme sensitivity to sunburn, so its occurrence is easily recognised. In South Africa, it is carried by about 8,000 individuals, three in every thousand of the white population, but it is very rare outside the country.

All the present day sufferers in South Africa are in 32 family groups, which can be traced back to one of the original 40 pairs of white settlers, an immigrant family from Holland that arrived in 1688. One million of the current population have the same surnames as the original 40: implying an increase of 12,500 times in three centuries. However, the porphyria allele is present in only two-thirds of these. Presumably, its deleterious effects (particularly in the sunny climate of South Africa) have led to selection against porphyria carriers.

Another island example is forceps length in the earwig (*Forficula auricularia*). This is an inherited trait although the actual switch between 'short' and 'long' forceps is determined by larval nutrition. In mainland populations, virtually all the males have short forceps; on many British islands, up to 60 per cent of males have long forceps which are more effective in sexual combat; this may be related to the high density of earwigs on many islands. High frequencies have been found on the British islands of Lundy, the Isles of Scilly and some of the Forth islands, and on the Faroes (Tomkins and Brown 2004).

An intriguing study has been that of the house mouse (*Mus domesticus*) population on Skokholm. The species probably reached there in the 1890s in sacks brought across by rabbit-catchers for transporting their harvest back to the mainland. The Skokholm mice are very odd: more than half of them have a failure of fusion of some of the vertebral neural arches; in effect, this is a very mild form of *spina bifida*. This is very rarely found in wild mice (although it occurs in some laboratory strains), but it is present in just under 10 per cent of the mice from the mainland of Wales opposite Skokholm, from where the rabbit-catchers came. This seems to be a clear case of a founder effect trait in the island mice.

The mice also show founder selection. They breed for approximately six months every year (April to September); for the winter months their chief problem is obtaining enough food to survive low temperatures. Studies of both skeletal and inherited biochemical variations show that gene frequencies change one way during the summer breeding phase and the opposite way during the winter survival phase. In other words, natural selection is operating in different directions during the two phases of the animals' lives. There is also a lesser but continued change evident in the skeletal traits, apparently indicating that the animals are progressively adapting to the island conditions (Berry et al. 1987).

Seasonally varying selection has also been found in mice on three islands in the Antarctic (Macquarie, Marion and South Georgia) (Berry et al. 1979). The details differ, but clearly show natural selection operating on individuals carrying particular genes (or more strictly, carrying the piece of chromosome which carries the genes) in different ways at different times. It is only when longitudinal studies are possible and undertaken (that is, ones repeated over a period and not simple one-off snap-shots taken at a particular time) that it is possible to identify natural selection in operation.

A striking feature of the Pleistocene fossil record of islands is the repeatedly parallel evolution of similar traits, strongly implying a naturally-selected adaptive cause. The best-known example is body-size change of mammals on islands, small mammals generally becoming larger and large mammals smaller, according to the so-called 'Island Rule' as named by Leigh van Valen in 1973. Dwarfing is seen in many species of Pleistocene deer, elephants (including mammoths) and hippos on islands in the Mediterranean, southeast Asia and elsewhere (see above), whereas various rodent and small insectivorous species became larger. The generality and adaptive interpretation of these patterns have been, however, the subject of debate (Lomolino et al. 2013). Dwarfing has most commonly been considered a response to a finite area of available food, together with seasonal shortage where animals cannot migrate as they can on the mainland. Associated adaptive features include the shortening of distal limb bone elements in quadrupedal mammals, considered a locomotory adaptation to the frequently rocky interiors of many islands (Sondaar

1977). For small mammals, size increase has been attributed to factors including lack of terrestrial predators, character release due to lack of competition, and energetic considerations.

Ecosystem frailty

Island animals often lack predators or ecological defence mechanisms and therefore are particularly susceptible to damage by introduced species. Sometimes these introductions are intentional: such as the 'Naturalisation [or 'Acclimatisation'] Societies' established by European settlers in North America, Australia and New Zealand to remind themselves of home; food species like the rat *Rattus exulans* spread through Pacific islands by Polynesian voyagers, or goats by later sailors; or mongoose taken to Hawai'i to control other pests. But often they are inadvertent: rat or mouse species most commonly arrive as commensals, often as a result of ship-wrecks (Baskin 2002).

The frailty of island populations is shown by the fact that 75 per cent of all known Holocene extinctions (i.e. where an identifiable species has become extinct, as distinct from calculated extinction rates based on loss of habitat) have occurred on islands (Turvey 2009). For example, 43 bird species are represented only by sub-fossil bones on the Hawaiian Islands, having disappeared in the centuries between the arrival of the Polynesians and the visit of Captain Cook in 1778. At least eight bird species on Guam have been lost since the introduction of the Brown Snake in the Second World War (see Berry and Gillespie, this volume). Introduced mammal predators are responsible for 42 per cent of historical bird extinctions and the local decline of many others. The Manx Shearwater (*Puffinus puffinus*) is so-called because of the vast numbers that used to nest on the Calf of Man, a teardrop of an island below the Isle of Man between the mainland of Britain and Ireland) proper. According to Njal's Saga, a Viking fleet anchored off the Calf in 1014 prior to a battle near Dublin was 'attacked' by shearwaters, to the extent that the sailors had to protect themselves with swords and shields. In the 17th century, around 10,000 plump young birds were harvested for food each year from the Calf. But in 1780 or thereabouts, rats (*Rattus norvegicus*) reached the island and the colony disappeared within 20 years.

Puffins (*Fratercula arctica*) are especially susceptible. Many islands have been completely cleared by rats. The Puffins on Puffin Island off Anglesey, Wales were likened to "swarms of bees" in 1774 (Harris 1984, p. 47); but rats colonised the island and the colony had disappeared by 1835. In 1892 it was said of Lundy (an island off the west of Britain, its name meaning 'Puffin Isle' in Old Norse) that:"there would not be room for another puffin" (ibid., p. 46). Black Rats were on the island by 1630 and Common Rats sometime before 1877. Rats were cleared by an intensive poisoning programme and a small puffin colony has now become established (Lock 2006). In the Faroes, by far the largest puffin colonies are on the rat-free island of Mykines.

A range of exotic species are a particular menace to island biotas. For example, hedgehogs (*Erinaceus europaeus*) are frequently transported to islands as pets. In most situations they are harmless, but they have become a serious predator of birds' eggs on grassland in the Outer Hebrides where there are few trees and thus a high proportion of ground-nesting birds. Since their arrival, dunlin numbers on the islands have fallen by 65 per cent, ringed plover by 57 per cent, redshank by 40 per cent, snipe by 43 per cent and lapwings by 43 per cent. Within the Outer Hebrides, there are an estimated 5,000 hedgehogs on the Uists and Benbecula, all descended from four individuals released on South Uist in 1974. In one study, excluding Hedgehogs from an area led to a 2.5 times increase in breeding success in wader species within a couple of years (Jackson 2001).

Ascension Island, in the Atlantic Ocean, is a cautionary tale of the serial problems that introductions may cause (Ashmole and Ashmole 2000). The island was colonised by the British in

1815 to deter possible attempts to free Napoleon from St Helena. Cats were introduced by the navy to control the rats which were common (probably coming from the ship of the pirate William Dampier, which foundered on Ascension in 1701). However, within ten years the island was overrun by cats, and dogs were released to control the cats. A few decades later, the Gannets, Boobies, Terns and Frigate Birds which had previously nested on the lower slopes of Ascension were gone: driven to cliff ledges or an offshore islet which remained cat-free. The plants were threatened by goat grazing (from animals left by early sailors; at one time there were several hundred goats, although they have now been cleared) and by other introduced alien species (around 300 have been recorded) – such as the Mexican Thorn (*Prosopis juliflora*) which is spreading rapidly and has the potential to cover 90 per cent of the island's surface (see Figure 6.12).

Determined efforts are now being taken to control or eliminate exotic species in many parts of the world. New Zealand scientists have been at the forefront in this respect, particularly in developing protocols to poison and eliminate rodents (Veitch and Clout 2002).

Conclusion

Islands continue to contribute to biology (Hanski 2016). As already noted, MacArthur and Wilson invigorated the study of biogeography through their landmark 1967 text. Charles Elton, effective founder of animal ecology, was inspired to develop his ideas by the relatively simple ecosystems he encountered on Spitzbergen [Svalbard]. The methods he developed as a consequence helped to resolve the confusion in the early days of ecology of regarding an ecological

Figure 6.12   Mexican Thorn, *Prosopis juliflora*, an invasive species on Ascension Island.

Source: Photo by Forest and Kim Starr. Wikimedia Commons, https://commons.wikimedia.org/wiki/File:Starr_070404-6613_Prosopis_juliflora.jpg.

community as a 'super-organism', which in turn seemed to lend support to the idea of a 'balance of nature', an idea that lives on in the Gaia hypothesis (Lovelock and Lovelock 2000). The diversity and serendipities of the constitution of island populations – what Hooker in his British Association lecture called 'disharmony' – show very clearly the error of regarding a biological community as a self-regulating 'super-organism'. Island populations are triumphs of individual adjustment, not manifestations of some master pattern (Simberloff 1983).

Mayr concluded his 1967 paper with a challenge to recognize the importance of island biotas:

> Islands are an enormously important source of information and an unparalleled testing ground for various scientific theories. Their biota is vulnerable and precious. We must protect it. We have an obligation to hand over these unique faunas and floras with a minimum of loss from generation to generation. What is once lost is lost forever because so much of the island biota is unique. Island faunas offer us a great deal scientifically and aesthetically. Let us do our share to live up to our obligations for their permanent preservation.

(Mayr 1967, p. 374)

References

Anon [Wollaston,T.V.] (1860) Bibliographical notice: on the origin of species by means of natural selection; or the preservation of favoured races in the struggle for life, by Charles Darwin, M.A., F.R.S., F.G.S., etc. *Annals and Magazine of Natural History, Series* 3(5), 132–143.

Ashmole, N.P. (1979) The spider fauna of Shetland and its zoogeographic context. *Proceedings of the Royal Society of Edinburgh*, 78(1–2), 63–122.

Ashmole, P., and Ashmole, M. (2000) *St Helena and Ascension Island: a natural history,* Oswestry, Anthony Nelson.

Baker, H.G., and Stebbins, G.L. (1965) *The genetics of colonising species*, New York, Academic Press.

Baskin,Y. (2002) *A plague of rats and rubber-vines*, Washington, DC, Island Press.

Baur, G. (1897) New observations on the origin of the Galápagos Islands, with remarks on the geological age of the Pacific Ocean. *American Naturalist*, 31(370), 661–680.

Berry, R.J. (1969) History in the evolution of *Apodemus sylvaticus* (Mammalia) at one edge of its range. *Journal of Zoology*, 159(3), 311–328.

Berry, R.J. (1989) Ecology: where genes and geography meet. *Journal of Animal Ecology*, 58(3), 733–759.

Berry, R.J. (1996) Small mammal differentiation on islands. *Philosophical Transactions of the Royal Society of London, Series B*, 351(1341), 753–764.

Berry, R.J. (1997) The history and importance of conservation genetics: one person's perspective. In T.E. Tew,T.J. Crawford, J. Spencer, D. Stevens, M.B. Usher and J.Warren (eds.), *The role of genetics in conserving small populations*, Peterborough, ON, JNCC, pp. 26–32.

Berry, R.J. (2003) *God's book of works: the nature and theology of nature,* London, T&T Clark.

Berry, R.J. (2004) Island differentiation muddied by island biogeographers. *Environmental Archaeology*, 9(1), 117–121.

Berry, R.J. (2009) Hooker and islands. *Biological Journal of the Linnean Society*, 96(2), 462–481.

Berry, R.J., Bonner, N., and Peters, J. (1979) Natural selection in house mice from South Georgia (South Atlantic Ocean). *Journal of Zoology*, 189(3), 385–398.

Berry, R.J., Jakobson, M.E., and Peters, J. (1978) The house mice of the Faroe islands: a study of micro-differentiation. *Journal of Zoology*, 101(185), 73–92.

Berry, R.J., Jakobson, M.E., and Peters, J. (1987) Inherited differences within an island population of the house mouse. *Journal of Zoology*, 211(4), 605–618.

Berry, R.J., and Rose, F.E.N. (1975) Islands and the evolution of *Microtus arvalis* (Microtinae). *Journal of Zoology*, 177(3), 395–407.

Brakefield, P.M. (1991) Genetics and the conservation of invertebrates. In I.F. Spellerberg, F.B. Goldsmith and M.G. Morris (eds.), *The scientific management of temperate communities for conservation,* Oxford, Blackwell Scientific, pp. 45–79.

Bristowe, W.S. (1969) *A book of islands*, London, Bell.Brown, J.H., and Kodric-Brown, A. (1977) Turnover rates in island biogeography: effect of immigration or extinction. *Ecology*, 58(2), 445–449.

Browne, J. (1983) *The secular ark: studies in the history of biogeography*, New Haven, CT, Yale University Press.

Bryant, E.H., Meffert, L.M., and McCommas, S.A. (1990) Fitness rebound in serially bottlenecked populations of the house fly. *American Naturalist*, 136(4), 542–549.

Burnet, J. (1684–1691) *The sacred theory of the Earth: containing an account of the original of the Earth and of all the general changes which it hath already undergone or is to undergo, till the consummation of all things*, 2 vols. London, Walter Kettilby.

Cain, A.J. (1984) Islands and evolution. *Biological Journal of the Linnean Society*, 21(1–2), 5–27.

Clarke, W. Eagle (1904). On some forms of Mus musculus, Linn., with description of a new subspecies from the Faeroe Islands. *Proceedings of the Royal Society of Edinburgh*. 15, 160–167.

Cook, L.M. (1995) T. Vernon Wollaston and the 'monstrous doctrine'. *Archives of Natural History*, 22(1), 33–48.

Coope, G.R. (1986) The invasion and colonisation of the North Atlantic islands: a palaeoecological solution to a biogeographic problem. *Philosophical Transactions of the Royal Society, Series B*, 314(1167), 619–635.

Darlington, P.J. (1957) *Zoogeography*, New York, Wiley.

Darwin, C.R. (1859) *On the origin of species by means of natural selection, or the preservation of favoured races in the struggle for life*, London, John Murray.

Degerbøl, M. (1940) Mammalia. In Jensen, A. et al. (rds.) *Zoology of the Faroes*, 3(2), 1–133. Copenhagen, Horst.

Diamond, J.M. (1975a) Assembly of species communities. In M.L. Cody and J.M. Diamond (eds.), *Ecology and evolution of communities*, Cambridge, MA, Harvard University Press, pp. 324–444.

Diamond, J.M. (1975b) The island dilemma: lessons of modern biogeographic studies for the design of natural reserves. *Biological Conservation*, 7(2), 129–146.

Dinnin, M. (1996) The development of the Outer Hebridean entomofauna: a fossil perspective. In D. Gilbertson, M. Kent and J. Grattan (eds.), *The outer hebrides: the last 40,000 years*, Sheffield, Sheffield University Press, pp. 163–183.

Endler, J.A. (1986) *Natural selection in the wild*. Princeton, NJ, Princeton University Press.

Fahrig, L. (2003) Effects of habitat fragmentation on biodiversity. *Annual Review of Ecology and Systematics*, 34, 487–515.

Flux, J., and Fullagar, P. (1992) World distribution of the rabbit *Oryctolagus cuniculus* on islands. *Mammal Review*, 22(3–4), 151–205.

Forbes (1846) On the connexion between the distribution of the existing fauna and flora of the British Isles, and the geological changes which have affected their area, especially during the epoch of the northern drift. *Memoirs of the Geological Survey of Great Britain*, 1, 336–432.

Frankham, R. (1996) Do island populations have less genetic variation than mainland populations? *Heredity*, 78(3), 311–327.

Frankham, R. (2005) Genetics and extinction. *Biological Conservation*, 126(2), 131–140.

Fridriksson, S. (1975) *Surtsey: evolution of life on a volcanic island*. London, Butterworth.

Fridriksson, S. (2005) *Surtsey: ecosystems formed*, Reykjavik, Iceland, Surtsey Research Society.

Geer, A. van der, Lyras, G., Vos, J. de, and Dermitzakis, M. (2010) *Evolution of island mammals: adaptation and extinction of placental mammals on islands*, Hoboken, NJ, Wiley.

Gilbert, F., Gonzalez, A., and Evans-Freke, I. (1998) Corridors maintain species richness in the fragmented landscapes of a micro-ecosystem. *Proceedings of the Royal Society of London, Series B*, 265(1396), 577–582.

Gleason, H.A. (1922) On the relation between species and area. *Ecology*, 3(2), 158–162.

Grant, P.R., and Grant B.R. (2008) *How and why species multiply: the radiation of Darwin's finches*. Princeton, NJ, Princeton University Press.

Halkka, O., Halkka, L., and Raatikainen, M. (1975) Transfer of individuals as a means of investigating natural selection in operation. *Hereditas*, 80(1), 27–34.

Halkka, O., Raatikainen, M., and Halkka, L. (1974) The founder principle, founder selection, and evolutionary divergence and convergence in natural populations of *Philaenus*. *Hereditas*, 78(1), 73–84.

Hanski, I. (2016) *Messages from islands*, Chicago, IL, Chicago University Press.

Harris, L.D. (1984) *The fragmented forest: island biogeography theory and the preservation of biotic diversity*, Chicago, IL, Chicago University Press. Harris, M.P. (1984) *The puffin*, Calton, T & A.D. Poyser.

Hawaii's Comprehensive Wildlife Conservation Strategy (2005) Terrestrial Invertebrates: Beetles. Order: Coleoptera. Available from: http://dlnr.hawaii.gov/wildlife/files/2013/09/Fact-Sheet-Coleoptera-beetles.pdf.

Hengeveld, R. (1989) *Dynamics of biological invasions,* London, Chapman & Hall.

Hooker, J.D. (1862) Outlines of the distribution of Arctic plants. *Transactions of the Linnean Society,* 23(2), 251–348.

Hooker, J.D. (1866) Insular Floras, Lecture delivered at the British Association for the Advancement of Science Meeting, Nottingham, UK, 27 August 1866. Published in the *Gardeners' Chronicle,* January 1867. Reprinted With an introduction by Mark Williamson in *Biological Journal of the Linnean Society,* 22(1), 55–77.

Hutchinson, H.N. (1896) *Extinct monsters* (with illustrations by Joseph Smit and others), 4th edn. London, Chapman and Hall.

Huxley, L. (1918) *Life and letters of Sir Joseph Dalton Hooker, OM, GCSI,* 2 vols. London, John Murray.

Jablonka, E., and Lamb, M.J. (2014) *Evolution in four dimensions: genetic, epigenetic, behavioural and symbolic variation in the history of life.* Boston, MA, MIT Press.

Jackson, D.B. (2001) Experimental removal of introduced hedgehogs improves wader nest success in the Western Isles, Scotland. *Journal of Applied Ecology,* 38(4), 802–812.

Jamieson, I.G. (2007) Has the debate over genetics and island endemics truly been resolved? *Animal Conservation,* 10(2), 139–144.

Kaneshiro, K.Y. (1995) Evolution, speciation and the genetic structure of island populations. In P.M. Vitousek, L.L. Loope and H. Adsersen (eds.), *Islands: biological diversity and ecosystem function,* Berlin, Springer-Verlag, pp. 22–33.

Lack, D. (1942a) The breeding birds of Orkney. *Ibis,* 84, 461–484.

Lack, D. (1942b) Ecological features of the bird faunas of British small islands. *Journal of Animal Ecology,* 11(1), 9–36.

Lack, D. (1943) The breeding birds of Orkney. *Ibis,* 85, 1–27.

Lack, D. (1945) The Galápagos finches (Geospinizinae): a study of variation. *Occasional Paper of the California Academy of Sciences,* No. 21.

Lack, D. (1947) *Darwin's finches.* Cambridge, Cambridge University Press.

Lack, D. (1969a) Population changes in the land-birds of a small island. *Journal of Animal Ecology,* 38(1), 211–218.

Lack, D. (1969b) The numbers of bird species on islands. *Bird Study,* 16(4), 193–209.

Lack, D. (1976) *Island birds, illustrated by the land-birds of Jamaica,* Oxford, Blackwell Scientific.

Lawton, J.H., and Brown, K.C. (1986) The population and community ecology of invading insects. *Philosophical Transactions of the Royal Society, Series B,* 314(1167), 607–617.

Lock, J. (2006) Eradication of brown rats (*Rattus norvegicus*) and black rats (*Rattus rattus*) to restore breeding seabird populations on Lundy island, Devon, England. *Conservation Evidence,* 3(1), 111–113.

Lockley, R.M. (1969) *The island.* London, Andre Deutsch.

Lomolino, M.V., Geer, A.A. van der, Lyras, G.A., Palombo, M.R., Sax, D.F., and Rozzi, R. (2013) Of mice and mammoths: generality and antiquity of the island rule. *Journal of Biogeography,* 40(8), 1427–1439.

Lovejoy, T. (2000) Biodiversity. In C. Patten, T. Lovejoy, J. Browne, G. Brundtland, V. Shiva and The Prince of Wales (eds.), *Respect for the earth,* London, Profile Books, pp. 22–35.

Lovelock, J., and Lovelock, J.E. (2000) *Gaia: a new look at life on earth.* Oxford, Oxford Paperbacks.

Lyell, C. (1832) *The principles of geology, being an attempt to explain the former changes in the Earth's surface by reference to causes now in operation,* Vol. II. London, John Murray.

MacArthur, R.H., and Wilson, E.O. (1963) An equilibrium theory of island biogeography. *Evolution,* 17(4), 373–387.

MacArthur, R., and Wilson, E.O. (1967) *The theory of island biogeography.* Princeton, NJ, Princeton University Press.

Manne, L.L., Pimm, S.L., Diamond, J.M., and Reed, T.M. (1998) The form of the curves: a direct evaluation of MacArthur and Wilson's classical theory. *Journal of Animal Ecology,* 67(5), 784–794.

Mayr, E. (1942) *Systematics and the origin of species.* New York, Columbia University Press.

Mayr, E. (1954) Change of genetic environment and evolution. In J.S. Huxley, A.C. Hardy and E.B. Ford (eds.), *Evolution as a process,* London, Allen and Unwin, pp. 157–180.

Mayr, E. (1967) The challenge of island faunas. *Australian Natural History,* 15(2), 359–374.

Mayr, E. (1982) *The history of biological thought.* Cambridge MA, Harvard University Press.

McIntosh, R.P. (1998) The myth of community as organism. *Perspectives in Biology and Medicine,* 41(3), 426–438.

Miller, G. S. (1912) *Catalogue of the mammals of western Europe (Europe exclusive of Russia) in the collection of the British Museum.* London, British Museum (Natural History).

Muller, H.J. (1950) Our load of mutations. *American Journal of Human Genetics*, 2(l), 111–176.

O'Connor, R.J. (1986) Biological characteristics of invaders among bird species in Britain. *Philosophical Transactions of the Royal Society of London, Series B*, 314(1167), 583–598.

Owen, R. (1865) Description of the skeleton of the great auk or garefowl. *Transactions of the Zoological Society of London*, 5(4), 317–335.

Paine, R.T. (1980) Food webs: linkage interaction strength and community infrastructure. *Journal of Animal Ecology*, 49(3), 667–685.

Pimm, S.L., Jones, H.L., and Diamond, J. (1988) On the risk of extinction. *American Naturalist*, 132(6), 757–785.

Porter, D.M. (1984) Relationships of the Galápagos flora. *Biological Journal of the Linnean Society*, 21(1–2), 243–251.

Praeger, R.L. (1950) *Natural history of Ireland: a sketch of its flora and fauna*. London, Collins.

Primack, R.B. (1993) *Essentials of conservation biology*. Sunderland, MA, Sinauer.

Quammen, D. (1996) *The song of the dodo: island biogeography in an age of extinctions*. London, Hutchinson.

Reed, T.M. (1980) Turnover frequency in island birds. *Journal of Biogeography*, 7(4), 329–335.

Reed, T.M. (1981) The number of breeding land-bird species on British Islands. *Journal of Animal Ecology*, 50(2), 613–624.

Reed, T.M., Currie, A., and Love, J.A. (1983) Birds of the Inner Hebrides. *Proceedings of the Royal Society of Edinburgh*, 83B, 449–472.

Robert, A (2009) The effects of spatially correlated perturbations and habitat configuration on metapopulation persistence. *Oikos*, 118(10), 1590–1600.

Rudwick, M.I.S. (2005) *Bursting the limits of time*. Chicago, IL, University of Chicago Press.

Simberloff, D. (1969) Experimental zoogeography of islands: a model for insularcolonisation. *Ecology*, 50(2), 296–314.

Simberloff, D. (1976) Species turnover and equilibrium island biogeography. *Science*, 194(4265), 572–578.

Simberloff, D. (1983) When is an island community in equilibrium?. *Science*, 220(4603), 1275–1277.

Simberloff, D., and Abele, L.G. (1982) Refuge design and island biogeographic theory: effects of fragmentation. *American Naturalist*, 120(1), 41–50.

Simberloff, D., and Wilson, E.O. (1969) Experimental zoogeography of islands: the colonisation of empty islands. *Ecology*, 50(2), 278–296.

Simberloff, D., and Wilson, E.O. (1970) Experimental zoogeography of islands: a two-year record of colonisation. *Ecology*, 51(5), 934–937.

Sondaar, P.Y. (1977) Insularity and its effect on mammal evolution. In M.N. Hecht, P.C. Goody and B.M. Hecht (eds.), *Major patterns in vertebrate evolution*, Plenum, New York, pp. 671–707.

Soulé, M.E. (1983) What do we really know about extinction? In C.M. Schonewald-Cox, S.M. Chambers, B. MacBryde and W.L. Thomas (eds.), *Genetics and conservation*, Menlo Park CA, Benjamin/Cummings, pp. 111–124.

Soulé, M.E. (ed.) (1987) *Viable populations for conservation*. Cambridge, Cambridge University Press.

Strickland, H.E., and Melville, A.G. (1848) *The dodo and its kindred: or, the history, affinities, and osteology of the dodo, solitaire, and other extinct birds of the islands Mauritius, Rodriguez and Bourbon*, London, Reeve, Benham and Reeve.

Thornton, I. (1996) *Krakatau: the destruction and reassembly of an island ecosystem*. Cambridge, MA, Harvard University Press.

Tomkins, J.L., and Brown, G.S. (2004) Population density drives the local evolution of a threshold dimorphism. *Nature*, 431(7012), 1099–1103.

Turvey, S.T. (2009) *Holocene extinctions*. Oxford, Oxford University Press.

Veitch, C.R., and Clout, M.N. (eds.) (2002) *Turning the tide: the eradication of invasive species*. Cambridge, International Union for the Conservation of Nature.

Wallace, A.R. (1860) On the zoological geography of the Malay peninsula. *Journal of the Proceedings of the Linnean Society*, 4(16), 172–184.

Wallace, A.R. (1876) *The geographical distribution of animals, with a study of the relations of living and extinct faunas as elucidating the past changes of the Earth's surface*, 2 Vols. London, Palgrave Macmillan.

Wallace, A.R. (1880) *Island life, or the phenomena and causes of insular faunas and floras, including a revision and attempted solution of the problems of geographical climates*. London, Palgrave Macmillan.

Weiner, J. (1995) *The beak of the finch: a story of evolution in our time*. New York, Vintage.

Whittaker, R.J. (2007) *Island biogeography: ecology, evolution, and conservation*. Oxford, Oxford University Press.

Williamson, M.H. (1981) *Island populations*. Oxford, Oxford University Press.

Williamson, M.H. (1983) The land-bird community of Skokholm: ordination and turnover. *Oikos*, 41(3), 378–384.

Williamson, M.H. (1984) Sir Joseph Hooker's lecture on insular floras. *Biological Journal of the Linnean Society*, 22(1), 55–77.

Wollaston, T.V. (1854) *Insects maderensia*. London, van Voorst.

Wollaston, T.V. (1878) *Testacea Atlantica or the land and freshwater shells of the Azores, Madeiras, Salvages, Canaries, Cape Verdes and Saint Helena*. London, Reeve.

Wulff, A. (2015) *The invention of nature*. New York, Knopf.

Yalden, D. (1999) *The history of British mammals*. London, Calton Poyser.

PART II

The human world of islands

7

HISTORY AND COLONISATION

Robert Aldrich and Miranda Johnson

Introduction

Since ancient times, and around the world, islands have fascinated dreamers, thinkers and adventurers, but also commanded the attention of settlers, planters, traders, military strategists and missionaries: those involved in imperial conquest and rule. Islands and archipelagos are home to rich and diverse, as well as interconnected, cultures, language and kin groups, many of which have suffered invasion and occupation by the forces of a power from distant shores (Conrad 2001, Salmond 2009). Islands in the Atlantic, Caribbean, Indian Ocean and Pacific provide important examples of various stages of imperialist expansion, in particular from the 1400s through the 1800s of the Christian Era. Islands were some of the first overseas places to be brought under colonial rule by European powers (and other expanding states), and they were also some of the last territories to be decolonised. Despite enormous differences between islands – those lying off-shore of continents and remotely located in 'blue-water' expanses, large and small in size, volcanic and coral in geology, resource-rich and barren, ones with and without indigenous populations – imperialism has left a lasting imprint on the environment, social structures, political institutions and cultures in islands scattered across all of the 'seven seas'. Enduring island nations continue to draw on deep indigenous histories, as well as engaging more recent legacies of imperialism, in renewed struggles for survival as many face the dire challenges of climate change and globalisation (Conrad 2009).

The colonisation of islands

'Colonisation' – derived from a Latin word for an ancient Roman farming community of former soldiers, a 'colonia', in newly conquered territory – means both the settling of a new population in a place and the takeover, often by conquest, of one country by another. This is well shown by the case of Ireland, with waves of settlement and centuries of rule by British overlords. In both senses, the story is complicated. In regards to the first meaning of the word, the colonisation of islands was a very long process, roughly synonymous with the spread of humankind around the globe. By the time of the ancient Greeks, many islands of the Mediterranean – to use the sole example of that part of the world – had been settled, and the sea-faring Greeks and Phoenicians created new 'colonies' around the Aegean, in Malta, Sicily, Corsica and the Balearics. Other islands, however, were populated only much later: the Carib peoples moved into the islands of the sea bearing their name (but displacing earlier migrants) only around 1200 CE, and Polynesians did not settle in New Zealand until around 1200–1300 CE. Later centuries would see the arrival of more disparate populations in all of the world's islands, though even today some small and isolated ones, those with few resources and ones located in especially challenging climatic zones, remain underpopulated or host no permanent population at all.

Similarly, 'colonisation' in the sense of conquest has a long and complex history. The spread of the Roman Empire brought most of the islands of the Mediterranean under the rule of Rome, which exported the Latin language, its system of laws, and the famous Roman roads and garrisons of legionnaires to the new territories. In the early Middle Ages, Vikings invaded and occupied the British isles and other places as far away as Sicily. Crusaders, waging holy war to wrest the holy land from the Turks, established a French dynasty in Cyprus, while the Knights Hospitaller of St John took over the island of Malta. Invasion and occupation were also the fates of islands far away from Europe, with the expansion and contraction, for instance, of the empires of southeast Asia. In the case of the Srivijaya empire, there was the conquest of a wide-ranging continental empire from a base on the island of Sumatra with hegemonic powers in the region from the seventh to the eleventh centuries, succeeded by the thalassocratic empire of the Majapahits centred on the island of Java in the 14th and 15th centuries. Acquisition of land and natural resources, mastery of trade routes, spread of dynastic power, and affirmation of religion explained the takeover of island territories in both the Mediterranean and the so-called 'Asiatic Mediterranean' of southeast Asia.

In the early modern age (according to Western chronology), there was a new stage of island colonisation by Europeans. Though Mediterranean islands continued to change hands, especially with the expansion of the maritime empire of Venice and the conflicts between Christians and Turks, the voyages of exploration and the actions of the conquistadors focussed European attentions on the Americas and the Indian Ocean. The Treaty of Tordesillas of 1494 divided unknown territories between Spain to the west and Portugal to the east of a meridian passing through today's Brazil. Columbus's landfall in the Caribbean initiated Spanish takeover of West Indian islands, and Spain soon developed plantation economies on Cuba and Hispaniola (see Figure 7.1) Portugal established trading outposts on islands (and mainland enclaves) along the African coast, on Ceylon, in India and further east in Malacca and the Moluccas, where it found the coveted 'spice islands'. Portuguese traders enjoyed a monopoly among Europeans in the east in the 1500s, merchants and missionaries spreading out to locations ranging from Zanzibar to Japan, and the Portuguese flag was raised over even unpopulated islands such as St Helena and Mauritius.

The rising political and commercial powers of Europe, however, did not accept the monopolistic clams of the Iberians. The Dutch, French and British all challenged the Spanish in the Caribbean, eventually adding islands to their own growing imperial portfolios (Palmié and Scarano 2011). For example, the French took over western Hispaniola (Saint-Domingue), Martinique and Guadeloupe; Britain acquired Barbados, Jamaica and a host of other islands as well as Bermuda in the Atlantic; and the Netherlands picked up a smaller share of the spoils. European nations competed sometimes by gentlemanly agreement. Claims on the small island of St Martin (previously claimed by the Spanish) in 1648 reputedly were resolved when a Dutch and a French official set off back-to-back and walked around the island in opposite directions in order to divide it; that division endures today under continued administration of the island by the Netherlands and France. More commonly, warfare was involved, with loss of life to both European troops and especially the indigenous populations, which were often reduced to remnants of their pre-colonial mass by warfare and diseases. By the by, sugar became king in most of the islands, the main cash crop produced by plantations owned by European settlers, and the back-breaking labour carried out by enslaved Africans brought to the West Indies. It was the

Figure 7.1   Eurocentric view of the colonial encounter: An engraving by Theodore de Bry from 1592, part of his 'America-series', showing Christopher Columbus landing on the Caribbean island of Hispaniola in 1492 and meeting the Taino natives.

Source: https://publicdomainreview.org/2014/04/16/1592-coining-columbus/.

consumers of sweets in Europe who were probably the happiest beneficiaries of island colonialism in the New World.

The Dutch meanwhile challenged Portugal in the Indian Ocean world: the powerful United East India Company (VOC), chartered in 1602, wrested Ceylon, Malacca and other ports in the East Indies from the Portuguese. The Dutch thus for a time became the leading colonial power of the Indian Ocean, though having to negotiate with local kings, sultans and other potentates (Pearson 2007, Vickers 2013) (see Figure 7.2).

The 1600s and 1700s also saw the arrival of the French and British, determined to establish their own colonies in the region. The French had a triumph with the taking of Mauritius (renamed the Isle de France) from the Dutch in the early 1700s, but were frustrated elsewhere. The British, under the chartered East India Company, after establishing important and rapidly expanding outposts on the Indian subcontinent from the 1600s, took over Penang island in the 1780s. The Napoleonic Wars then provided a golden opportunity to capture Mauritius from the French and take Dutch holdings in Ceylon (though almost 20 years elapsed before the British defeated the King of Kandy and controlled the whole island of Ceylon by 1815). In 1819, the British extended their reach to Singapore, though the Dutch retained nominal or effective overlordship of most of the 17,000 islands in the East Indies (Holland 2011, Palmié and Scarano 2011, Pearson 2007, Sivasundaram 2013).

Figure 7.2   Logo of the Dutch East India Company (Vereenigde Oost-indische Compagnie; or VOC).

Source: Public Domain/Wikipedia, https://en.wikipedia.org/wiki/Dutch_East_India_Company.

At the start of the 19th century, a map of the island world showed the British triumphant in the Indian Ocean and with a substantial share of the Caribbean; the Dutch increasingly entrenched in the East Indies; a cohort of imperial powers (even including Denmark and Sweden) claiming islands in the West Indies; and Portugal clinging to some of its older conquests, from Madeira and the Azores in the northern Atlantic to Cape Verde and São Tomé off the west African coast. Spain retained control of the Philippines, which it had conquered in the 1500s, though some islands in Asia and most in Oceania remained under the authority of indigenous sultans, rajas and chieftains. By this time, too, the Japanese archipelago had come until the unified rule of the Tokugawa shogun, and the mainland Chinese had established hegemony over Taiwan. The British were securely established on the Australian mainland, Tasmania and Norfolk Island (see Figure 7.3).

The scramble for islands continued during the 1800s, with the British and French the leading players until the 1880s. The French were eager to gain ports for the storage of provisions and access to fresh water, rebuild an overseas empire (having lost most of their colonies by the end of the Napoleonic wars), gain new trading posts, and challenge the British. They thus took over tiny Mayotte on the east coast of Africa in 1841. In the 1890s, an agreement between Paris and London allowed France to colonise the 'Great Island', near-by Madagascar, and left Britain with Zanzibar. In the Pacific, there was a tit-for-tat between the two European powers: Britain claimed New Zealand in 1840, and France raised the Tricolour in the Society Islands (including Tahiti) and the Marquesas Islands in 1842. The French claimed New Caledonia in 1853, and the British took possession of the Fiji Islands in 1874. By the last decades of the 1800s, there was a free-for-all in Oceania, complete with new entrants to the imperial game – Germany and the USA – and by the end of the 1800s, every island had come under at least nominal foreign control. Germany acquired northeastern New Guinea, the western part of the archipelago of Samoa, and several island groups in Micronesia. The USA took over other Samoan islands, and

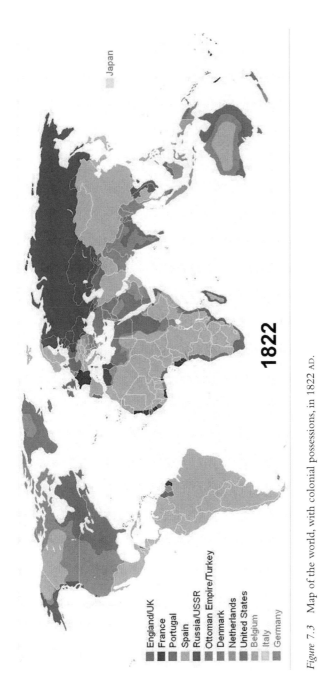

Figure 7.3 Map of the world, with colonial possessions, in 1822 AD.

Source: EarthDirect, Creative Commons, www.youtube.com/watch?v=ihD3_Nm8qA.

in the northern Pacific the USA in 1867 occupied the Aleutian Islands following the purchase of Alaska from Russia. After a war with Spain, the USA acquired the Philippines and Guam in 1898, and American intervention overturned an independent monarchy in the Hawaiian islands. There was also a 'sub-imperialism' in the island Pacific, with the expansion of Australian interests in New Guinea and the New Hebrides, French New Caledonians contesting 'Anglo-Saxon' involvement in the latter case. In one of the odder colonial arrangements, the New Hebrides became a 'condominium' with two resident foreign commissioners, representing France and Britain, two flags and two currencies; since the French and British judges on the local tribunal were likely to disagree, a Spanish judge was appointed to the court on the basis that a Spanish explorer had first 'discovered' the island group. The Melanesian inhabitants of the New Hebrides, paradoxically, were subjects of neither Britain nor France, and enjoyed no political rights, a situation that endured with few changes until the independence of Vanuatu in 1980 (McIntyre 2016). In eastern Asia in the nineteenth century, Japan was also expanding, taking over the formerly independent kingdom of the Ryūkyū islands in 1879, consolidating rule over the northern Japanese island of Hokkaido (and its indigenous population of Ainu), and with victory in the Sino-Japanese War of 1895, winning Taiwan.

By the outbreak of the First World War, few islands thus had not been claimed by continental and colonial powers, though the territory of Sarawak (on the island of Borneo) had been acquired as a personal colony from the Sultan of Brunei by an ambitious Briton: Charles Brooke and his heirs ruled as the 'white rajas' until 1945. Metropolitan governors, colonial lobbyists and adventurers devised coherent or wild plans for island development, and deftly deployed rhetoric and propaganda to advance outsized promises of the potential of even forsaken island groups. For instance, on the Kerguelen Islands of the Southern Ocean, a Frenchman tried, with inglorious results, to establish sheep-rearing, while a small band of French settlers were landed on Clipperton in the eastern Pacific, and then were promptly forgotten by Paris and only rescued, near starving and half-mad, by Mexican ships years later. (In another of the oddities of the colonial era, a conflict over claims to Clipperton by France and Mexico was submitted to the arbitration of the king of Italy, who decided in favour of France.) Some of the new island colonies, however, fared better, with expanding export economies. Fiji, for example, flourished commercially with the production of sugar (much of it sold in Australia), and Hawai'i supplied Americans with pineapples (Okihiro 2009).

Colonisers and the use and abuse of islands

Individuals, groups and nations involved in colonisation found many uses for islands (and engaged in many abuses of the islands and their inhabitants in pursuing their objectives). Islands offered stepping-stones between continents and beachheads for further colonial expansion. One of the first overseas conquests came with the takeover of the Canary Islands, off the African coast, by Castilian Spaniards in 1402, 90 years before Columbus set sail. Columbus first 'discovered' the New World by landing on islands in the Caribbean Sea, and the Spanish then sailed on to imperial conquests on the American continents, as did later the British and French. The Portuguese made their way along the west African islands and coastline before the conquest of mainland Angola, and moved onwards into the Indian Ocean. It should be remembered that such great sites of colonial occupation as Manhattan, Bombay (a set of islands before land reclamation connected the city to the mainland), Singapore and Hong Kong are islands immediately off continents: these insular holdings were not only invaluable in themselves but expanded into mainland possessions. Colonisers also hoped that the islands would lead them elsewhere: the islands of Oceania would provide a chain of transit points across the Pacific from the Americas to

China, promoted in the 19th century as a new axis of world trade with plans, from the 1830s, for the construction of a canal across the Central American isthmus (though the Panama Canal was not opened until 1914) and the 'opening' of China. The earlier Suez Canal, inaugurated in 1869, and creating a passage from the Mediterranean to the Red Sea and thence to Asia, enhanced the value of island ports in Malta and Cyprus.

Furthermore, islands provided vital places for ships to moor and to take on provisions of food and water; tiny St Helena in the southern Atlantic had been a crucial crossroads for sailing ships in transit from Europe to Asia before the Suez Canal cut off 7,000 km in the voyage from Britain to India. In the age of steamships, island and continental ports served as coaling stations and, at a later time, they stocked supplies of fuel for aeroplanes. Island cities such as Singapore served for transhipment of goods in the developing network of global commerce, becoming bustling entrepôts of commerce and finance, as well as for the rough-and-tumble life of the docklands with legendary watering-holes and pleasure quarters for wayfarers.

Trade and the flag accompanied each other around the world, and *prises de possession* of islands gave geopolitical advantage to established great powers and their aspirational rivals. Islands represented new territories to colour with the appropriate hues on imperial maps, sites to garrison troops and build fortifications, and launching-pads for conquering raids and military conflicts. The Napoleonic Wars between France and its opponents (in many ways, the first world war) were fought in part around and over the islands of the Mediterranean, Caribbean and Indian Ocean, victories at sea transferring 'ownership' of islands from one colonial master to another. By the late 1800s, Europeans engaged in a kind of 'preventive' imperialism to take over whichever islands had not yet been claimed in Oceania and the Southern Ocean, even the most remote and commercially unpromising, in large measure to forestall moves by eager competitors.

Though there were only a relatively few military skirmishes in and around islands during the First World War, the Pacific theatre became a major battlefield in the Second World War. The Japanese capture of Singapore represented an enormous blow to the military might and the pride of the British empire. The attack on Pearl Harbour, Hawai'i, brought the USA into the war in Asia, and battles raged on many islands, notably Guadalcanal. The American fire-bombing of Tokyo, and the dropping of atomic bombs on Hiroshima and Nagasaki, caused unparalleled loss of life and destruction to Japanese victims. (Islands elsewhere also played a role in the war, with Greenland serving as an intelligence base for Allied monitoring of the Germans, and King George VI awarded the prestigious George Cross to Malta and its people for their bravery and heroism.) In more recent times, the islands of Oceania served for American and French nuclear testing, and the islands and reefs of the South China Sea in the early 21st century present a domain for China's neo-imperialist ambitions and the making of new islands into militarised colonies. The building of airstrips, accommodation for troops and even potential tourist resorts by the Chinese is effectively transforming reefs into islands. The remaining overseas territories of old colonial powers underline the continuing strategic significance of islands, including those still administered by metropolitan powers; Diego Garcia in the Indian Ocean, is rented by the British to the USA for use as a military base (see Figure 7.4) with the native Ilois having been evicted in the process (Vine 2011). Claims to 'exclusive economic zones' off islands and reefs around the world further underline the strategic and commercial value of islands.

Throughout history, islands have offered a treasure-trove of coveted commodities (Grove 1995, Walvin 1997, Higgs 2012). The 'Spice Islands' of the East Indies produced precious pepper, cloves, nutmeg and other spices that constituted some of the most expensive imports into early modern Europe, essential for culinary and medicinal uses, and Ceylon was the world's leading source of much desired cinnamon. In the 1600s, as mentioned, sugar became the key colonial crop in tropical islands conquered by Europeans, grown on plantations in the West

Figure 7.4 Diego Garcia, part of the British Indian Ocean Territory, and leased by the UK to the USA. It is the site of a US military base, with two 3.7 km runways and facilities visible in this photo.

Source: NASA/Wikimedia Commons, https://commons.wikimedia.org/wiki/File:Nasa_diego_garcia.jpg.

Indies and Indian Ocean, and subsequently on the Dutch islands of southeast Asia and in the Pacific. The islands where sugar was produced were the most prized possessions of Britain and France through the 1700s, though cane sugar later had to compete with beet sugar in European markets, and many of the 'sugar islands' never recovered the central place they had occupied as imperial assets. There were nevertheless other profitable tropical island products in addition to spices and sugar; sandalwood harvested in the South Pacific became a valuable trade good with the Chinese (who turned it into incense), and *bêche-de-mer* was enjoyed as a culinary treat in China (Shineberg 2014). Europeans, meantime, developed a taste for tropical fruit, particularly bananas and pineapples that came in large quantities from Caribbean islands (Walvin 1997). The Europeans established coffee plantations around the tropical world, in both islands such as Java (whose name became synonymous with the drink) and on continental territory. By the late 1800s, tea had become the prime export of Ceylon; thanks to Thomas Lipton and other planters, 'Ceylon tea' was the beverage of choice throughout the British empire and beyond. Chocolate became another European addiction, sourced from mainland plantations in South America and Africa and also from the 'chocolate islands' of Portuguese São Tomé (Higgs 2012).

Europeans tried, less successfully, to produce cotton in island colonies (as in Tahiti in the 1860s, when the American Civil War endangered world supplies), but wool became the leading export of British New Zealand. Such other products as indigo, rice and tobacco were shipped from hot-climate islands to Europe. Islands provided important bases, too, for the processing of whales for food, lamp oil, cosmetics and cleansers, and the whalebone used in fashionable corsets. With the second industrial revolution, demand for phosphates for fertiliser stimulated the economic development and led to the environmental degradation of Christmas Island in the Indian Ocean, and Nauru and Makatea in the Pacific; later the Svalbard Islands in the Arctic provided substantial export quantities of coal. The French discovered in New Caledonia one of the world's largest supplies of nickel, a key alloy in the making of steel, and vast quantities of copper were later mined in Bougainville in Papua New Guinea. The islands of Oceania also provided much of the copra that was transformed in European factories into soap, another product with a greatly expanding market in the late 19th century. More glamorously, Ceylon continued to produce the sapphires and other gemstones mined there since antiquity.

Settlement – in the primordial sense of 'colonisation' as noted at the beginning of this chapter – was a most important aspect of imperialism in the islands. Some islands were unoccupied when Europeans and other conquerors claimed possession, but most had indigenous populations, the outcomes of centuries or millennia of migration, sometimes in multiple waves. The 'natives' often paid the price for colonial takeover, with violence, the import of diseases to which they had no immunity, the effects of overwork, anomie and environmental destruction (Moorehead 2000). Some historians have accused colonisers of committing 'genocide' in islands from the West Indies to Tasmania, though the term is legally complex and laden with moral implications. However, it is certain that populations such as the Caribs in the West Indies and the Veddas in Ceylon were gradually pushed back into mountain or jungle enclaves, and indigenous peoples were often economically exploited, politically disenfranchised and even physically confined to reservations. But indigenous islanders also took advantage of imperial networks, acquiring foreign commodities, finding employment in merchant fleets or colonial armies, migrating from one region to another, and using colonisation to expand their horizons and adopt goods and ideas from the outside world.

Wherever they went, Europeans and other colonisers settled new populations. There were the administrators, soldiers, traders and missionaries who served as 'pioneers' (and, it was hoped, counted among the beneficiaries) of the colonial state. Islands also provided a dumping-ground for prisoners, whether those convicted of common crimes, even often relatively minor offences, or political crimes or taken as prisoners of war: St Helena, after all, is most famous as the place of exile of Napoleon. The British established penal colonies in New South Wales and Norfolk Island in 1788, soon followed by Tasmania (see Figure 7.5) and other colonies on the Australian mainland, and continued to transport convicts to various parts of Australia until the 1860s. Among the convicts counted men from an Irish rebellion in the 1790s and a Canadian uprising in the 1830s, as well as far larger cohort of debtors, fraudsters, thieves and violent criminals from the burgeoning cities of Britain.

Just as the British abandoned transportation to the Australian colonies, the French started sending convicts to New Caledonia: those with long-term sentences and recidivist criminals, but also revolutionaries from the Paris Commune of 1871, and rebels from Algeria in the same year; the shipment of prisoners to New Caledonia did not end until the late 1890s. Russia sent shiploads of its convicts to Sakhalin Island, and the British sent Indian prisoners to the Andaman Islands (Vaidik 2010); such other islands as Alcatraz in the USA, Côn Đảo in Vietnam and Robben Island in South Africa became notorious prison establishments.

Slavery represented one of the largest, and most heinous, transfers of populations in the colonial age, and many enslaved Africans ended up in the plantation colonies of the West Indies and the Indian Ocean. (And enslaved East Indians were moved by the Dutch to the Cape Colony, where many of their descendants still live.) In the Caribbean, black slaves became the numerically dominant group in the population, even if white settlers continued to monopolise

Figure 7.5 Port Arthur Penitentiary, Tasmania.

Source: Photograph by Andrew Braithwaite, Melbourne, Australia. Wikimedia Commons, https://commons.wiki
media.org/wiki/File:Port_Arthur_Penitentiary.jpg.

political and economic power. Though slave-trading was formally outlawed in the early 1800s, it continued clandestinely, with the well-known evils that human trafficking produced. Not until 1834 in the British empire and 1848 in the French empire (and later in some countries) were slaves emancipated. In replacement for enslaved Africans, Europeans sought other labourers, and India provided a vast source. Under the British, enormous numbers of Indians were moved as indentured (or contract) labourers to plantation islands – where they became a majority of the population in Mauritius and Fiji, and a substantial minority in Trinidad – and to possessions such as Singapore and the Malay states. (There were also free Indian migrants, such as Gujarati merchants, around the Indian Ocean, where intricate networks of Indian trade had dominated commerce before the arrival of Europeans.) Under the aegis of the British, large numbers of Tamils were taken from southern India to Ceylon to work on tea plantations, swelling the already resident Tamil population. The French imported Japanese, Vietnamese and Javanese into New Caledonia, and after the First World War encouraged a small current of Lebanese and Syrian migration to the Caribbean. In the late 1800s, the phenomenon of 'blackbirding' saw the recruitment of indentured labourers from Melanesia for the cane fields of Queensland. Many of the men, women and children from the southwestern Pacific who were bonded to work in that sugar industry were however uprooted from their Australian homes decades later under racist immigration laws and deported back to island homes that some of them had never known. The massive Chinese diaspora in the nineteenth century saw the establishment of important Chinese communities on many islands. Various other routes in the *longue durée* of island history took willing or forced migrants from island to island, and from mainlands to islands: Omanis and Persians to the Comoros, Zanzibar and other islands off the eastern coast of Africa before the age of large-scale European colonialism, East Indians to Ceylon under the Dutch, Polynesians to

New Caledonia under the French, Portuguese from the Azores and Madeira to Hawai'i. Islanders also travelled to colonial metropoles, some as individual exotic visitors to Europe, such as Aotourou and Omai from Tahiti in the late 1700s, others in more considerable numbers such as freed slaves and seamen in the 1800s. Far larger populations of West Indians moved to Britain, France and the Netherlands after 1945, many recruited to work in subaltern positions in hospitals, post offices and similar enterprises; that migration wave had profound social effects on both the islands and the European host countries.

Though a few islands or archipelagos retained rather homogenous populations (such as Iceland, Greenland and Japan), others came to count among the most socially and culturally cosmopolitan areas of the world. Singapore is a revealing example, with its Chinese, Malay, Indian and European communities, to which must be added the Peranakan population, descended from liaisons between Chinese and Malays. 'Mixed-race' populations became common in islands: Burghers in Ceylon (descendants of Dutch, Portuguese, Sinhalese and Tamil ancestors), Eurasians in the East Indies, and *métis* in the West Indies, where specific words – such as 'octoroon' in English or '*chabin*' and '*mulâtre*' in French – were used to denote racial blends and place individuals in racial hierarchies. The *métis* sometimes found niches for themselves in social structures. In the French West Indies, *métis* came to dominate the professional classes of law, medicine and politics, while in Tahiti, those of mixed European and Polynesian ancestry, revealingly termed *demis* ('halves'), became the dominant political and economic group. Yet in other cases, men and women known as 'half-breeds' in colonial times lived in a no-man's-land, never accepted by whites and not fully at home in indigenous communities. The co-existence of indigenous, migrant and *métis* populations, and the cohabitation of extremely diverse and competing migrant groups, created severe political tensions both under colonial rule and afterwards.

Islands, therefore, presented many uses for imperial powers, and colonial promoters were always on the lookout for new agricultural, subsoil or maritime resources to exploit, economic projects to develop and sources of labour for plantations and mines, shops and commercial entrepôts. As the colonial era moved onwards, new activities emerged. Tourism, made easier by steamship lines and travel agencies in the late 19th century, was one of the nascent industries in the colonial age. Bali by the 1930s, for example, had already become one of the most fabled destinations, renowned for its luxuriant vegetation, beautiful temples, accomplished dance and music, and a population considered gentle and welcoming. Tourism was destined to grow exponentially during the later 20th century with jet travel, paid holidays, package tours and advertising: a veritable cult of the tropical island that was starting even in the pith-helmeted 19th century. From Ibiza and Mykonos in the Mediterranean to Tahiti and Hawai'i in Oceania, islands are synonymous with travellers' dreams of sunny days and wild nights. Other present-day commercial 'uses' for islands – shipping, offshore banking (and occasional money-laundering and other dubious activities) and the sale of internet domains – owe much to the ways that islands were connected to the global market under imperial flags (Lockhart and Drakakis-Smith 1997).

The influence of islands in imperial culture is manifestly related to tourism. A tropical *dolce far niente* became a trope in European views of the Caribbean, complete with verdant forests, sandy beaches and sparkling seas (despite the legacy of the horrors of slavery and the tribulations of other settlers). Voyagers in the 18th-century South Pacific returned with tales that led to utopian descriptions of *bons sauvages* in the commentaries of Enlightenment *philosophes* (Salmond 2009). Contrarily, others saw spectres of primitive and barbaric peoples in the islands: cannibals and headhunters. For many Europeans, the populations of Melanesia (and Aborigines in Australia) were relics of the Stone Age, humans relegated to the bottom of the ladder in the ascent to civilisation, perhaps fated to die out as evolution took its toll. Their plight inspired evangelists who set out to save heathen souls and 'civilise the savages' (Edmond and Smith 2003). The London

Missionary Society, established in 1795, played a major role in the Christianisation of Oceania; Catholics were also involved, and established a virtual theocracy, for example, in the Polynesian islands of Wallis and Futuna (and Spanish Catholics had, in much earlier times, converted the majority of the Philippines' population to Christianity). Stereotypical views of islands, whether positive or negative, continued to circulate in the colonial era and afterwards, owing much to the paintings of Paul Gauguin, whose depictions of lissom women and girls combined the erotic lure of Polynesia and melancholy at the arrival of European civilisation and its perceived destruction of indigenous life (see Figure 7.6).

Islands were particularly rich laboratories for European scientists and social observers, as well as missionaries and artists, in the imperial age. Charles Darwin's ideas of evolution were developed on observations of the fauna of the Galápagos Islands and northern Australia. Naturalists found an infinite variety of plant and animal specimens unknown to Europeans, which they collected, classified and displayed in museums at home. The 'curios' made by island artists and artisans first filled European 'cabinets of curiosity' and then entered art museums; such works as the ceremonial masks of the southwestern Pacific exercised considerable influence on avant-garde artists in Paris, such as the Cubists, at the beginning of the 20th century. Islands also excited scientific imagination. Agronomists tried to 'acclimatise' various species for transplantation to different climes. Captain Cook's first voyage, as is well known, set out to observe the transit of Venus. Vulcanologists had a field day in islands, since many islands are of volcanic origin. Ethnologists and anthropologists went on expedition to study the 'primitive' peoples of Oceania, Bronislaw Malinowski's 1929 book *The sexual life of savages* and Thor Heyerdahl's theory of South American

Figure 7.6 Parau Api (What news): Women in Tahiti, now French Polynesia. Painting by Paul Gauguin, 1892.

Source: Wikimedia Commons, https://commons.wikimedia.org/wiki/File:Paul_Gauguin_144.jpg.

settlement of Polynesia based on the 1947 voyage of the *Kon-Tiki* offering prime examples of the sometimes fanciful theories to which island cultures gave birth in the imperial age (Malinowksi 1929, Heyerdahl 1996). The toppled stone statues of Rapa Nui (Easter Island) similarly inspired various theories, even including notions that there were created by extraterrestrials.

Fictional works about islands, including novels by Daniel Defoe, Bernardin de Saint-Pierre, Joseph Conrad, Pierre Loti and Robert Louis Stevenson, among others, still figure among classics of colonial-era literature (see McMahon and André, this volume). New cultural movements grew up in the colonial age as well. One of particular importance is the *négritude* movement of the early 20th century, a leading figure in which was the Martinican poet and politician Aimé Césaire who also became an incisive critic of colonialism (Césaire 2000). *Négritude* gathered social commentators, poets, novelists and political activists from around the French and British West Indies and Africa, as well as African Africans, who sought to revalorise an African history and culture depreciated by slavers, colonial conquerors and missionaries (Césaire 1950). Cultural blending begun in colonial times also produced many original forms of music and its exemplars, from *séga* in Mauritius and *zouk* in the French West Indies to the reggae of Bob Marley in Jamaica and the bittersweet Portuguese creole ballads of Cesária Évora in Cape Verde. Writers such as the novelist V.S. Naipaul from Trinidad and the poet Derek Walcott from St Lucia (both Nobel Prize-winners), Michael Ondaatje from Sri Lanka, Jamaica Kincaid from Antigua, Patricia Grace from New Zealand, and Albert Wendt from Samoa, among many others, creatively and critically explored the complexities of island history and contemporary life.

Christianity, once a Western colonial religion, had become 'indigenised' in the West Indies, South Pacific and the Philippines. Hinduism, Buddhism, Christianity and Islam were all brought from mainlands to islands and now help differentiate one segment of a population from another. The syncretic Hinduism of Bali sets it apart from mostly Muslim Indonesia (though there are also predominantly Christian islands in the country). Sri Lanka became one of the most important homes of Theravada Buddhism, linked to such mainland countries as Burma and Thailand by a shared faith. Rastafarianism, which has spread widely in the African diaspora, emerged in Jamaica, and Pacific islands gave birth to various 'cargo cults' (Worsley 1967). The domes of Orthodox churches in the Aegean, the steeples of Protestant churches in the South Pacific, the minarets in the Comoros islands and the Buddhist dagobas in Sri Lanka are not only major landmarks but also sites of cultural heritage and signifiers of identity.

These developments are now seen by islanders to have produced ambiguous effects, especially in Oceania. Missionaries brought literacy to oral cultures, which entailed both enrichment and loss. The translation of the Bible into many different vernaculars offered new metaphors and stories to indigenous peoples and maroon communities, as well as ways of imagining and critiquing the excesses of colonialism. Many indigenous peoples in the Pacific and West Indies found affinity with the Israelites in the stories of the biblical exodus. Increasing literacy through the late nineteenth century and the introduction of printing presses helped to create new island publics who consumed news from afar, producing indigenous 'imagined communities' that were also the seedbed for anticolonial resistance. Literacy could have a democratising effect on tribal hierarchies, challenging traditional leadership and inheritance. Dependence on writing nonetheless entailed the loss or curtailment of aspects of oral culture and the memory traditions of some peoples. Yet 'traditional' song, music, dance and art, as well as the novels, political treatises, daily newspapers and internet blogs, are all forms of expression of diverse cultures, shared island belonging and connection to the wider world (Baldacchino 2012, Dawe 2004).

Island ecologies underwent massive transformations (Grove 1995). Colonisers introduced many foreign species to island environments – there were, for instance, no mammals in New Caledonia other than flying-foxes in pre-colonial times. Some new species enriched food

Figure 7.7 Hong Kong Island skyline, 2009.

Source: WiNG, Wikimedia Commons, https://commons.wikimedia.org/wiki/File:Hong_Kong_Island_Skyline_2009.jpg.

supplies and marketable commodities; others, such as feral cats, had more deleterious effects. The most famous animal victim of imperialism must surely be the dodo, the somewhat ungainly flightless bird native to Mauritius that was hunted to extinction by the Dutch; the Tasmanian tiger is another species that disappeared after the arrival of the Europeans. In many ways, colonisers transformed island environments by clearing land, opening roads, digging mines and constructing buildings. The extraordinary skyline of Hong Kong Island offers the most dramatic illustration of the metamorphosis in the natural and built environments over the colonial era (see Figure 7.7). The development of island cities, not only metropolises such as Hong Kong and Singapore, but contemporary urban centres (and sometimes urban ghettos and shantytowns) throughout the island world, owes much to ecological, economic and social changes initiated under colonialism (see Grydehøj and Swaminatham, this volume).

The decolonisation of islands

Islands often passed from one imperial overlord to another. Corfu, for instance, was ruled in turns by Venetian, French and British masters before becoming part of independent Greece: 'decolonisation' was a protracted and complicated affair. The independence of Corsica in 1755, though short-lived, represents one of the first modern examples of island independence, followed by that of Haiti, the former French colony of Saint-Domingue, in 1803. A few other islands became politically independent later in the century, such as the Dominican Republic in the mid-1800s and Cuba in 1898 (though Americans troops then occupied the island). New Zealand was granted responsible government in 1852 but remained steadfastly a part of the British Empire. Ireland, where Britain long denied the 'home rule' demanded by (largely Catholic) nationalists, finally achieved the status of a 'Free State' in 1922, though at the expense of six counties remaining part of the UK, and Éire did not become a fully independent republic until 1949. Iceland, ruled as a Danish colony and then, after the First World War, as an independent realm of the king

166

of Denmark, did not sever its relationship with the Danish crown until the 1940s. When Ceylon became independent in 1948, it too was a dominion of the Commonwealth, keeping the British monarch as head of state until 1972. However, the wave of nationalism and anticolonialism that swept Asia in the 1940s and 1950s, and the subsequent tide of independence in Africa, reached many smaller island nations only tardily.

Nationalist movements did not gain traction for various reasons (e.g. Banivanua-Mar 2016, McIntyre 2016, on the Pacific). Relatively few strong leaders emerged to lead campaigns of island independence, and some of the ideologies that inspired strategies elsewhere, such as Marxism, were generally unpopular. Representative assemblies were set up in the Caribbean and Pacific islands only in the 1950s or later, and thus the avenues for political apprenticeship and mobilisation were limited. Island populations were divided by ethnicity, religion and language (though that was also the case in continental countries). Early rebellions – such as a slave revolt in Jamaica in the 1830s, a Melanesian uprising in New Caledonia in 1878, and the Mau movement in Samoa in the 1920s – were violently suppressed, and an anti-French outbreak in Madagascar in the 1940s was put down with great bloodshed though Madagascar did gain independence, alongside most other French sub-Saharan African colonies, in 1960. The undeveloped nature of an 'infrastructure' of ideological and political engagement – education systems, the press, broadcasting – worked against large-scale campaigns for independence. Many nationalists in the islands (as was more commonly true throughout the colonial world than is often realised) long wanted some amelioration of the colonial order, greater political rights and autonomy rather than outright and full independence.

The colonial powers were also largely opposed to independence for small island colonies. The commercial and geopolitical benefits islands presented remained considerable even in the mid-20th century, and sometimes assumed increased importance: the French government decided to transfer nuclear testing to Polynesia after the independence of Algeria (in 1962), where testing had previously taken place in the Sahara Desert. Most officials in colonial ministries felt that islands were too small in geographical size and population, too resource-poor (despite the value colonisers placed on their resources) and too underdeveloped to sustain independence. White officials in Australian-administered Papua New Guinea, for instance, held firmly to paternalist ideas in the 1950s, believing that it might be generations or a century before that nation had the maturity to rule itself. In the context of the Cold War, metropolitan governments worried about the contagion of Communism, especially after the establishment of Fidel Castro's Communist regime in Cuba at the end of the 1950s. (Though the island-based Castro regime became one of the major exponents of socialist revolution in the 1960s, his ideas found more fertile ground in continental liberation movements than in other islands.) They also worried about the fate of substantial settler and diasporic populations in colonies that became independent.

The trajectories taken by colonies moving towards the postcolonial era present some unexpected developments. In 1956, the electorate of Malta, for instance, voted for integration into the UK, though British lawmakers turned down the overture. In Fiji in the 1960s, indigenous Melanesian chiefs argued against independence on the grounds that the 'native' population would be electorally overwhelmed by the Indian majority; similarly, in Mauritius, descendants of African and Malagasy slaves feared for their future in an independent state where Indians would form a majority. When a referendum on independence was held in the Comoros Islands in the 1970s, three of the islands voted for independence, while the fourth, Mayotte, chose to remain under French administration. In New Caledonia, in the 1980s, the descendants of French settlers, and more recently arrived French residents, campaigned most strongly against independence, allied with migrants from French Polynesia and Wallis and Futuna. Meanwhile, efforts in French Polynesia to gain independence also came to nought.

Several island colonies were assimilated into metropolitan states. Earlier examples of the union of islands and mainland territories, of course, can be seen in the binding of the Balearic Islands to Spain, the foundation of a unified Italy encompassing Sicily and Sardinia, and the gradual incorporation of Aegean islands into independent Greece from 1830 to the end of the 19th century. Tasmania became part of the Commonwealth of Australia in 1901 (and the Cocos/Keeling Islands, Christmas Island and Norfolk Island were incorporated later), and Newfoundland was confederated with Canada in 1948. In 1945, the *vieilles colonies* of Martinique, Guadeloupe and La Réunion (and French Guiana in South America) were fully integrated into the French Republic, as with Mayotte more recently (Childers 2016). France continues to administer the island territories of Saint-Pierre and Miquelon, Mayotte, French Polynesia, New Caledonia, Wallis and Futuna and the Southern and Antarctic Territories. All of their residents are fully-fledged French citizens, with access to French social services, right of abode in France and living standards generally higher than in neighbouring independent countries, though protests against 'colonialism' are episodically heard. Hawai'i became the 50th state of the USA in 1960, and American possessions from Guam to Puerto Rico remain a part of the USA, their residents also having full citizenship but only limited rights of Congressional representation. Denmark continues to administer largely autonomous Greenland and the Faroe Islands, and the Kingdom of the Netherlands includes six islands (actually, five and a half) in the Caribbean. The remaining British territories stretch from the North Atlantic (Bermuda) to the South Atlantic (Falklands), and from the British Indian Ocean Territory to minuscule Pitcairn Island. The latter is inhabited by several dozen descendants of HMS *Bounty* mutineers and Polynesian women who chose or were forced to accompany them in the South Pacific (see Figure 7.8).

New Zealand administers Niue and Tokelau, and, through a treaty of 'free association', the Cook Islands, while Easter Island comes under the rule of Chile. Some 50 overseas polities,

Figure 7.8   Pitcairn Island, the last British dependency in the Pacific.

Source: Photograph by Jens Bludau. Wikimedia Commons, https://commons.wikimedia.org/wiki/File:Pitcairn_Longboats.jpg.

almost all of them islands, remain constitutionally attached to the old colonial powers, and some have been more recently occupied – as when the then Soviet Union occupied four of the Kuril islands, claimed by Japan, at the end of the Second World War. Whether these territories are still 'colonies' is a complex constitutional and political issue. There have been episodic campaigns in favour of independence, occasionally marked with violence, in several territories distant to mainlands (as well as in proximate French Corsica), but none has achieved independence. Most of the pro-independence movements have now lost support, though there are occasional reports of interest in sovereignty even in such veteran British islands as Bermuda. The winds have seemingly blown away from independence in most of the remaining island 'colonies' (Banivanua-Mar 2016).

If not necessarily seeking full political independence, islanders, especially those who have borne the pain of dispossession, have formed other kinds of solidarity and connection across colonial and settler colonial domains. Political movements have championed greater rights, self-determination and return of lands, from 'nativist' groups in Hawai'i to the 'National Liberation Front' in Corsica. Less stridently, in the 1980s, a cultural movement of '*Créolité*' sought to create cultural links and political solidarity among islands in the Caribbean and Indian Ocean where a French-based Creole is the *lingua franca*. The Commonwealth, and international Francophone, Hispanophone and Lusophone associations, help to join together islands and connect them to mainland states with shared colonial heritages (as do sporting ties, such as taking part in the 'Island Games' and the playing of cricket in the islands of the old British empire). An interconnected 'Black Pacific' identity developed in the 1970s, drawing on the theories and concepts of anticolonial resistance in other parts of the world and expressing itself through shared music and art forms. Resisting continuing attempts to isolate island peoples in conceptual and intellectual terms as well as physical and material ones, indigenous scholars have translated practices of traditional navigation, which linked islands over vast areas, into scholarly concepts. They have emphasised the cultural, economic and kin relationships between island peoples, most famously the idea that Oceania itself is best understood as a 'sea of islands' rather than 'islands in the sea' – the concept is equally applicable to other parts of the world (Waddell et al. 1993).

Uniting indigenous peoples across oceans, deserts and polar regions, an indigenous rights movement that gathered momentum in the 1980s culminated in the 2008 United Nations Declaration on the Rights of Indigenous Peoples. It asserts indigenous peoples' rights to self-determination, social and economic equality, and the integrity of their distinctive nations, cultures and languages. These efforts have not only made an impact at the level of international law but have also drawn together indigenous communities into new political and diplomatic relations. Meanwhile, organisations such as the (South) Pacific Forum and the Caribbean Community (CARICOM) seek to develop cooperation among both independent and dependent polities often separated by the legacies of colonialism.

One phenomenon where formal independence remains a live issue, however, has been in a few islands (and other territories) that were incorporated into postcolonial states. In the 1950s, there was a strong if unsuccessful movement for independence from Indonesia in the Moluccas islands, and in more recent decades a similar movement, also violently repressed, in West Papua (Vickers 2013). The western half of the island of New Guinea, with a Melanesian population, remained a colony of the Netherlands until the mid-1960s, almost two decades after Indonesia became independent; when the Dutch withdrew, the Indonesians occupied West Papua. When Portugal left East Timor in the wake of the 1974 'carnation revolution' in Lisbon, ephemeral independence came to an end as Indonesian forces occupied the territory. Indonesia ruled East Timor for a quarter of a century and violently opposed its independence during a war in the 1990s, though East Timor succeeded in becoming a sovereign state in 2002. Other island nations have also seen intense conflict, as was the case in Sri Lanka, where a civil war pitted Sinhalese against Tamils from 1983 until 2009. Episodic violence has flared elsewhere, in part because of the changes wrought by colonialism; diasporic populations (such as the Chinese in Indonesia) have sometimes been the target and victims. The 'troubles' in Northern Island, violent ethno-religious conflict between Catholics and Protestants that found no resolution until the signing of the Good Friday Agreement in 1998, can also be seen as emblematic of historical tensions within formerly colonised islands. And yet another example of intra-island confrontation is the case of Cyprus, where the historical desire of Greeks for union with Greece came head-to-head with opposition from Turkish Cypriots; the Turkish invasion of 1974 and the setting up of a Turkish sponsored state in northern Cyprus has meant the seemingly unresolvable division of that island. Colonisation, in both cases over the *longue durée*, has played a role in such tensions.

Islands followed many pathways to their current political status, not all of which involved independence. Two widely separated cases illustrate the point. Britain established a protectorate over the island of Zanzibar, off the east African coast – famous for its cloves – in 1890. Neighbouring Tanganyika, a British colony, became independent in 1961. Three years later, immediately after the British protectorate over Zanzibar came to an end, the sultan of Zanzibar was overthrown in a revolution and the island merged with Tanganyika to form Tanzania. In 1997, Britain relinquished control of Hong Kong – the island and the mainland territories that Britain had held under long-term lease – and the crown colony was ceded to the People's Republic of China, where it became a 'special administrative region'. Under the principle of 'one country, two systems', Hong Kong retains its own political, legal and commercial system, although dissidents claim that Hong Kong's liberties are now being seriously curtailed by Beijing.

Despite the obstacles on the road to independence, and the different endpoints at which islands arrived, most of the old island colonies are now independent states (although others continue to refuse to consider independence and prefer to linger in the colonial embrace) (Aldrich and Connell 1998). Leading the way were Jamaica, Trinidad and Tobago, and Western Samoa (now officially called just Samoa) in 1962, Malta in 1964, the Maldives in 1965, Barbados in 1966, Mauritius in 1968 and Fiji in 1970. Smaller islands did not gain independence until the 1970s; or, in the case of Vanuatu, Antigua and Barbuda, and St Kitts and Nevis, the 1980s. Some, especially in the Pacific, are microstates with populations of only a few thousand and very limited resources; for low-lying atoll nations such as Kiribati and Tuvalu, climate change presents a very real menace of eventually forcing evacuation of the entire population (Waddell et al 1993, Armitage and Bashford 2014). Nevertheless, some former tropical island colonies have done well: Mauritius is a stand-out example of success among small independent island states, with political stability and a democratic parliamentary system, a diversified economy, and a continually rising standard of living.

The colonial and imperial history of islands bears many similarities with that of mainland countries that were colonised, though with several particularities: the crucial role of islands as stepping-stones across the seas in the era of sailing ships and steamships, the importance of such commodities as sugar, the cosmopolitan and mixed populations colonial-era migrations introduced to islands, and the late dates of independence (and continued status of the 'last colonies') (Aldrich and Connell 1998). All of the islands of the Mediterranean, Atlantic, Caribbean, Indian Ocean and the Pacific, where European (and American and Japanese) imperialism were most evident, still bear the distinct imprint of the age of empire: the languages, architectures and political institutions are several of the most obvious legacies (see Figure 7.9). Today, many of these islands bear the brunt of climate change. Many leaders of island nations are leading powerful moral campaigns to force large continental states, and the international organisations they dominate, to pay attention to issues that particularly affect islands, drawing on centuries of resistance to continental incursions and pride in enduring local traditions.

Figure 7.9   Colonial nostalgia? Barber shop façade in the main square of the town of Zejtun, Malta.

Source: © Anna Baldacchino, 2017, with permission.

References

Aldrich, R., and Connell, J. (1998) *The last colonies*. Cambridge, Cambridge University Press.

Armitage, D., and Bashford, A. (eds.) (2014) *Pacific histories: ocean, land, people*. London, Palgrave Macmillan.

Baldacchino, G. (2011) *Island songs: a global repertoire*, Lanham, MD, Scarecrow Press.

Banivanua-Mar, T. (2016) *Decolonisation and the Pacific: indigenous globalisation and the ends of empire*. Cambridge, Cambridge University Press.

Césaire, A. (2000 [1950]) *Discourse on colonialism*. Translated by Joan Pinkham. New York, Monthly Review Press.

Childers, K.S. (2016) *Seeking imperialism's embrace: national identity, decolonisation and assimilation in the French Caribbean*. New York, Oxford University Press.

Conrad, P. (2009) *Islands: a trip through time and space*. London, Thames & Hudson.

Dawe, K. (2004) *Island musics*, New York, Berg.

Edmond, R., and Smith, V. (2003) *Islands in history and representation*, New York, Psychology Press.

Grove, R.H. (1995) *Green imperialism: colonial expansion, tropical island edens and the origins of environmentalism, 1600–1800*. Cambridge, Cambridge University Press.

Heyerdahl, T. (1996) *The Kon-Tiki expedition: by raft across the South Seas*. London, Flamingo.

Higgs, C. (2012) *Chocolate islands: cocoa, slavery and colonial Africa*. Athens, OH, Ohio University Press.

Holland, R. (2011) *Blue-water empire: the British in the Mediterranean since 1800*. London, Allen Lane.

Lockhart, D., and Drakakis-Smith, D. W. (1997) *Island tourism: trends and prospects*, London, Thomson Learning.

Malinowski, B. (2006[1929]). *The sexual life of savages in North-Western Melanesia*. Guilford, Genesis Publishing.

McIntyre, W.D. (2016) *Winding up the British Empire in the Pacific islands*. Oxford, Oxford University Press.

Moorehead, A. (2000) *The fatal impact: an account of the invasion of the South Pacific, 1767–1840*. London, Penguin.

Okihiro, G. Y. (2009). *Island world: a history of Hawai'i and the United States*, Berkeley, CA, University of California Press.

Palmié, S., and Scarano, F.A. (2011) *The Caribbean: a history of the region and its peoples*. Chicago, IL, University of Chicago Press.

Pearson, M. (2007) *The Indian Ocean*. London, Routledge.

Salmond, A. (2009) *Aphrodite's island: the European discovery of Tahiti*. Berkeley, CA, University of California Press.

Shineberg, D. (2014) *They came for sandalwood: A study of the sandalwood trade in the South-West Pacific 1830–1865*. University of Queensland Press.

Sivasundaram, S. (2013) *Islanded: Britain, Sri Lanka and the bounds of an Indian Ocean colony*. Chicago IL, University of Chicago Press.

Vickers, A. (2013) *A history of modern Indonesia*. Cambridge, Cambridge University Press.

Vaidik, A. (2010) *Imperial Andamans: colonial encounter and island history*. Basingstoke, Palgrave Macmillan.

Vine, D. (2011) *Island of shame: the secret history of the US military base on Diego Garcia*, Princeton, NJ, Princeton University Press.

Waddell, E., Naidu, V., and Hau'ofa, E. (1993) *A new Oceania: rediscovering our sea of islands*. Suva, Fiji, School of Social and Economic Development, University of the South Pacific.

Walvin, J. (1997) *Fruits of empire: exotic produce and British taste, 1660–1800*. New York, New York University Press.

Worsley, P. (1968). *The trumpet shall sound: a study of 'cargo' cults in Melanesia*, New York, Schocken Books.

8

GOVERNANCE

Edward Warrington and David Milne

Introduction

Treasure islands . . . beleaguered gardens . . . convict settlements . . . Islands serve as malleable material for the construction of clashing images (Pinet 2002): precisely because of this treacherous serviceability, scholars can become victims of their own images, shaping what is actually a variegated, complex subject into one-dimensional categories. In our view, this kind of reductionism is the principal, characteristic failing in commentaries on island governance.

Starting from this premise, this chapter begins by reviewing the literature on island governance, highlights weaknesses in substance and methodology, addresses the nature of islandness, and argues for an ecological framework for understanding the subject. The second section picks up this theme and illustrates its relevance for islands operating within the intricate, multi-layered ecology of governance of contemporary Europe. Finally, after synthesising evidence from several disciplines and numerous islands, the authors propose seven patterns (or species) of island governance that have developed ecologically out of the confluence of geography and history, and they outline the characteristic elements of governance in each.

Perspectives on island governance

The conventional approach to 'island governance' assumes or attributes supposed effects of 'islandness' (such as peripherality, isolation, small scale) on politics and government. This is one furrow in a well-ploughed field of political science, namely the attempt to establish a relationship of cause and effect between a polity's *assumed distinguishing characteristic* – such as scale, a belief system or a socio-economic system – and a corresponding pattern of governance. This tendency reappears in an important literary genre where, as 'laboratories of governance', islands are depicted as utopias in treatises or political novels such as *Utopia, Treasure island* and *The Swiss family Robinson* (More 1516, Stevenson 1888, Wyss 1812) or as dystopias, such as the island settings of *Lord of the flies, The leopard* and *Knowledge of angels* (Golding 1954, Tommasi di Lampedusa 1991, Walsh 1994). While such exercises may produce compelling *images* of islands, they do not *analyse* 'islandness' or governance particularly well.

The meaning of islandness is not single or self-evident (Dommen 1980, Ronström 2009). The deceptively straightforward geographical definition of an island – a tract of land surrounded by water – produces numerous exemplars from just three variables: land area, boundedness and distance. It says nothing about an island's geography, geology or natural ecology; still less does it disclose information about inhabitants, institutions or relations with neighbouring lands.

Philosophy, literature, anthropology and sociology draw attention to 'islandness' that is a state of mind, or a human condition of relative isolation and distinctiveness, expressed across almost the entire range of human experience, from economic activity to speech patterns, from belief systems to genetics (O'Crohan 1977, 1986, Gillis 2004, Bedggood 2004, Brinklow 2016, Crane and Fletcher 2016).

Furthermore, the intense reaction of islanders confronted with the prospect of terminating their 'isolation' through a fixed link to a mainland, or through accession to another polity, testifies to the power of island landscapes to identify communities: witness, for example, the contested connection of Skye and Prince Edward Island by bridges to the mainland (Baldacchino 2006a, Begley 1993, McCabe 1998, Royle 1999) or Malta's bitterly contested accession to the European Union (Mitchell 2002). Clearly, islandness constitutes a peculiarly salient example of landscape identity attachment (Gibbons 2010, Weale, 1991), and of geography in the service of polity-building (Bartmann 1996, 2000, Connell, 1988, Milne 2001, Ackren 2006, Hay 2006). Yet, island distinctiveness may ironically stimulate separatism and fragmentation in archipelagos (Anckar 2007, Baldacchino 2010, Stratford 2013). In still other circumstances, as in Cyprus or Malta, 'islandness' may constitute a weaker pole of attraction compared to the pull of mainland motherlands (Baldacchino 2002a, 2002b).

These ambiguities and contradictions alert us to the danger of making easy generalisations about islands and island governance. Indeed, doubts have been raised about whether islands are an analytically significant category for political science (e.g. Selwyn 1980). One consequence of this uncertainty has been a tendency to conflate 'islandness' and 'small scale', thereby obliterating the distinctiveness of the subject matter (Baker 1992). Not only does this intellectual sleight of hand do little justice to small islands as *islands*: it eliminates from consideration larger islands (New Guinea, Madagascar) and/or small but densely populated islands (Manhattan, Singapore).

Broadly speaking, then, 'islandness' can be approached from two perspectives. The *external* perspective, that of a detached analyst, seeks verifiable patterns and regards island governance as the product of 'foreign' historical and geographical 'facts' penetrating and impinging on 'domestic' matters. The *internal* perspective directs us towards the islanders: it seeks to understand how distinctive 'island' identities develop, and what effects they have on habits of thought and action, on socio-economic structures and political processes, and the way that these engage with the externally determined facts of geography and history. Together, these constitute an 'ecology' within which island governance is played out – the social science equivalent of a naturalist ecology, complete with its language of *tendencies, environment* and *interdependence*. This is not reducible to one-dimensional biological, sociological or political categories: islands as ecological fields transcend and contradict Selwyn's strictures.

The ecology of island governance

Commentaries about the human ecology of islands single out the *insularity* of island communities. However, the apparent clarity of this claim is belied by the wide variety of cultural, political and economic phenomena attributed to insularity – some of which are clearly contradictory. Thus, islanders are thought to be deeply attached to their home, but also presented as seafarers, explorers or migrants. Island communities are lauded for their resourcefulness and resilience, but fears are also expressed for their vulnerability (Commonwealth Advisory Group 1997, European Commission 1994, Fairbairn 2007, Azzopardi 2015, Gitay and Jones 2017). Similarly, island communities are commonly presented as being inherently democratic, though politics is just as often excoriated for traditionalism, parochialism and autocratic tendencies (Richards 1982, Dommen and Hein, 1985, Srebrnik 2000b, 2004, Sutton 1987, Mitchell 2002). Islands are natural polities, self-contained and distinctive, but are also thought to be handicapped by small populations and remoteness. A further source of ambiguity is the undoubted autonomy that island communities generally enjoy, coupled, paradoxically, with subjection that is no less real or common, to a neighbouring or distant metropolis (Al Jazeera English 2013, Holmén 2014).

Singly, these baffling contradictions appear to confirm Selwyn's strictures about transposing biological categories into social science. However, taken together, the tensions and ambiguities disclose the very stuff of 'islandness'. After all, must not geographers and naturalists take account of both isolation and contact in understanding island eco-systems? Sociologists go to the heart of island communities when they address the tension between the separateness granted by geography and the impact of contact whose terms islanders seek to negotiate. The ability to balance tradition and change, and selectively to assimilate alien values or patterns of behaviour, both moulds and preserves islander identity.

A parallel dynamic is at play in politics and economics. Whether they are merely subsistence economies or influential global traders, many island economies negotiate the paradox of their undoubted vulnerability coupled with self-evident affluence (Karides 2013). Island economies as a class do *not* appear to stand out as problem cases from those of other small states; indeed, notwithstanding the phenomenon of the MIRAB economy (Bertram and Watters 1985, Watters 1987), many rank among the most affluent (Armstrong et al. 1998). Island politics gravitates around the tension between autonomy and dependence, whether jurisdictional or economic (Le Rendu 2004); between indigenous and expatriate or immigrant elites, (e.g. Pirotta 1996, Georghalides 1985, Lawson 1996); and between centre and periphery (e.g. Howe 1988). Policy reflects the three-way pull of islander interests: engagement with the outside world more or less on their own terms, subordination to an 'imperial' power, or defiant withdrawal (Cooper and Shaw 2009). These patterns are as evident in large as in small islands.

To sum up, islands and 'islandness' may best be understood in terms of a characteristic set of tensions and ambiguities, opportunities and constraints arising from the interplay of geography and history. *Geography tends towards isolation*, permitting or favouring autarchy, distinctiveness, stability and evolution, propelled endogenously. *History*, on the other hand, *tends towards contact*, permitting or favouring dependence (or inter-dependence), assimilation, change and evolution propelled exogenously. An island's character develops from the interplay of geography and history, evasions and invasions, the indigenous and the exotic.

Geography includes elements that critically influence an island's community, political history and destiny: the 'carrying capacity' of the land and adjacent waters; whether the 'island' is solitary or an archipelago; its location and relation to trade routes; resource endowments; proximity to land masses, organised states and centres of power. Hence, geography conditions an island's *historical experience*. One thread of this is the frequency, duration and strength of contact with other territories, a feature which political scientists tend to overlook, although natural scientists and economists document the ease with which an island's human and other 'ecologies' can be penetrated and overwhelmed – the matter of vulnerability. Attention has recently shifted to 'resilience', that is, the ability of an island's systems (including its government) to preserve their distinctive character by either resisting or assimilating alien penetration (Baldacchino 2000, Briguglio 1995, 2004). In short, the interplay of geography and history moulds an island's political economy, and with that the incidence, depth and pace of changes in the constitutional order, economic activity, social institutions and cultural traditions (e.g. Bayliss-Smith et al. 1988, Beckles 1990, Lee 1989, Anckar 2006). Governance is forged from these raw materials.

Consider what is possibly the most important element of an island's political economy, namely its *geo-strategic profile*, that is, whether it is central or peripheral in relation to the forces and powers at play in its neighbourhood. Of course, the architecture, meaning and burdens of 'centre' and 'periphery' are inherently socially constructed, relational not absolute, and constantly recast in response to shifts in international political economy and power (Milne 1997). It is *political agency*, the ability to act in a changing environment, that determines how peripherality, so often cited as a limitation inherent in islandness (Amaral Fortuna 2001), and epitomised by lonely St Helena (Royle 2001), is experienced and whether it is treated as constraint or opportunity.

Another way of looking at the geo-strategic profile is to examine an island's paramount external relationship: what might be described as 'the imperial connection'. At one extreme, it may be subjugated by another power, deprived of jurisdiction and, indeed, of a political 'personality'. Conversely, it may itself be the dominant element in a large polity, as medieval Venice, Tonga, Britain and Japan once exercised jurisdiction over dependencies on terms which they themselves largely determined. An island may enjoy such dominance as the centre of a traditional empire, or of an internally diverse state such as an archipelago, as Java does within Indonesia (e.g., Anckar 2006, 2007, Stratford 2013). In their turn, peripheral islands may powerfully influence national identity building in the metropole (Farinelli 2017)

Certain elements of the imperial connection profoundly affect island society, economy, political culture, governing institutions and policy-making. Consider, for example, the question whether an imperial power's overriding interest in its dependency is commercial or military; this influences the character and strategic interests of elites: whether bureaucratic or land-owning or entrepreneurial (e.g. Meleisea 1987, Tordoff 1997). That, in turn, shapes political culture, the scope of politics and socio-economic strategies (e.g. Vella 1994).

Furthermore, ever since European powers began claiming jurisdiction over distant islands, the question whether or not, and on whose terms, large-scale immigration was permitted, determined the social and ethnic composition of island polities. This had lasting consequences for the welfare of indigenous islanders, for class, racial and ethnic cleavages, and the identity of island 'nations', as demonstrated in Fiji, Mauritius, or Trinidad (Srebrnik 2000a; 2002a; 2002b). Paradoxically, mass immigration, no longer driven by imperial policy, has become a potent political question in the islands on Europe's periphery in the 21st century (King 2009, Beramendi et al. 2014).

The character of the regime governing relations between centre and periphery is another aspect of the imperial connection: it may be likened to a political apprenticeship, benignly or malignantly influencing political culture, shaping the aspirations, political agenda and strategic options available to islanders (Anckar 2006). Historically, constitutional, law-based regimes were more likely to permit island politics and jurisdiction, as in the English-speaking Caribbean islands. In contrast, restricted autonomy and authoritarian tendencies, as in the Lusophone African islands, are more likely to be legacies of coercive imperial regimes.

Only by having law/policy-making and administrative capacity can islanders hope to affect the terms of transactions with their neighbourhood, and to become actors in their community's destiny (Baldacchino and Milne 2000, Karlsson 2009). Although the literature on vulnerability implies that there is little scope for this kind of agency, islands that have made themselves centres of empire and civilisation flatly contradict this model; many smaller islands, by carving out credible positions of wealth and independence, also defy the stereotype (Betermier 2004, Karides 2013).

In fact, because of their distinctness, islands tend to enjoy higher levels of constitutional recognition and jurisdiction than mainland territories, another testament to opportunities arising from the conjuncture of geography and history (Bartmann 1996, Anckar 2007). No less than

43 (22 per cent) of the world's sovereign states are islands (Baldacchino and Milne 2006). Furthermore, islands are overwhelmingly represented in *special* federal relationships, particularly in federacies (Stevens 1977, Elazar 1987, Watts 1999, Rezvani 2004, Ackren 2006). In these 'asymmetrical federal relationships', a smaller unit is linked to a larger polity, often a former colonial power, but the small unit retains considerable autonomy, has a minimal role in the government of the larger, and the relationship can be dissolved only by mutual consent (Watts 2000): examples include the Åland and Faroe Islands, Bermuda, Puerto Rico, and the Isle of Man. As islands account for more than 90 per cent of federacies (Watts 1999), it might fairly be said that this constitutional relationship has been devised principally to meet the purposes and circumstances of small island territories. The least integrated federacies, such as the Channel Islands, use their jurisdictional advantage to build vibrant economies that seek maximum shelter from the tax and regulatory intrusions of the UK and the EU (Hampton and Abbott 1999, Le Rendu 2004, Foreign and Commonwealth Office 2012).

This relative autonomy and jurisdictional muscle permits the development of entrepôt economies in some island jurisdictions, both sovereign and non-sovereign, which deploy jurisdiction and policy-making capacity as springboards for economic activity (Baldacchino and Milne 2000, 2006, Commonwealth Advisory Group 1997). Arguably, the international environment at the start of the 21st century, characterised by globalisation, regionalisation and multi-level governance (Hooghe and Marks 2001), enhanced these opportunities for achieving autonomy and prosperity (Keating 1999a). Within this malleable environment, even sub-national governments of varying types found little difficulty in conducting foreign relations in the service of their own strategies for economic prosperity (Aldecoa and Keating 1999, Bartmann 2006, Corbin 2001, Karides 2013). Whether this favourable climate will subsist in the future constitutes a profound governance challenge for islands.

In an effort to systematise and explain the extraordinary variety of island experiences, and the bewildering inconsistencies, ambiguities, uncertainties and contradictions disclosed by research, a three-way taxonomy of small-island, socio-economic formations has emerged (Bertram 2006). The PROFIT model emphasises the use of law-making jurisdiction in economic development strategies, particularly in relation to **P**ersonnel, **R**esource management, **O**verseas representation, **FI**nance and **T**ransportation (Baldacchino 2006b). This contrasts with the pioneering MIRAB model of dependent development (based on **MI**gration, **R**emittances, **A**id and **B**ureaucracy), conceived to describe the less developed island states in the Pacific and Caribbean (Bertram and Watters 1985). The third is the so-called SITE model (**S**mall **I**sland **T**ourist **E**conomies) where tourism promises more resilience (McElroy 2006a). With all these models, a constitutional status that guarantees the exercise of certain law-making rights and powers is critical to the success of political economies, though surprisingly full sovereignty has often proved a less attractive option (McElroy 2006b, Armstrong and Read 2000).

Just as critical is the 'ecology' of globalisation, with its densely inter-connected systems of trade, aid, movement of capital and labour, and communications, operating in global or regional contexts (Karides 2013). These conditions, epitomised in the European Union, permit even small islands to act for themselves with greater confidence, although international rules are often set with the interests of larger, populated, continental states in mind.

Patterns, issues and agendas in the governance of European islands

Options for governance and development depend crucially upon the character and proximity of the neighbourhoods in which islands find themselves. For European islands, their neighbourhood, quite unlike other geopolitical spaces, has since the Second World War proven to be especially benign. Here, islands have been able to pursue their own idiosyncratic paths within the continent's complex evolving constitutional geometry. This has been a remarkable part of post-war Europe's story, where older centralised national states, 'recognised' regions and nations within their states, including offshore islands, and devolved power to them. (Judt 2005, Keating 1999b).

Looking back, the socio-economic profile and prospects of Europe's peripheral island communities shifted decisively from a former dismal pattern of emigration and subsistence in crofting or fishing, to newer patterns of inward and internal migration, and economic and political diversification. While the volume, rate, duration, direction and composition of these flows varied from island to island and over time, the patterns were similar in the Baltic, Atlantic and Mediterranean theatres (European Commission 1994, Adler-Nissen and Gad 2013, Beramendi et al. 2014).

The benefits of jurisdiction for declining island communities were not fully appreciated until the onset of Europe's post-war reconstruction. From the first tentative steps, when Italy's constitution-makers sought to remedy peripherality and recognise distinct communities by granting limited legislative and executive jurisdiction to regions like Sardinia and Sicily (Desideri et al. 1996), the process gathered momentum. Comparable initiatives occurred in countries similarly undergoing democratisation, although only in Spain can it be said that this experiment led to a genuine constitutional revolution in governance (Agranoff et al. 1997, Borzel 2000). Furthermore, reorganisations of central administration, specifically to facilitate regional government, accompanied constitutional recognition (Corte-Real 1999, Bar Cendón 1999, pp. 137–140). While *some* mainland territories also benefited from the initiatives, *every* island group acquired jurisdiction to a greater or lesser extent. The powers conferred were specifically oriented towards fostering economic development in 'maritime islands' perceived as 'extreme examples of peripheral regions' (Council of Europe 1986, Karlsson 2009). Three *strategic* islands – Iceland, Cyprus and Malta – deliberately chose sovereignty for that purpose. Simultaneously, decolonisation emancipated dozens of island dependencies of European powers in the Caribbean, the Indian Ocean and the Pacific.

With the decisive emergence of the European Union in the final quarter of the 20th century, the continent's peripheral islands all confronted the fundamental question of their terms of association with it. Two island states – Britain and Ireland – acceded (with Denmark) to the European Economic Community, as it then was, in the first enlargement of 1973.

Their accession in turn altered the balance of power and interests within the EEC. While Ireland sought to maximise the 'draw-down' of EU funds (Kelly 1999, p. 202), Britain, by contrast, was consistently more ambivalent about the integrating Union. So, too, its island dependencies, which established a trend in the development of Europe's 'variable geometry'. The British Crown Dependencies, hosting lucrative offshore financial services centres, "are part of the Customs Union and are essentially within the Single Market for the purposes of trade in goods, but are 'third countries' in all other respects" (Channel Islands Brussels Office 2017). The Faroe Islands, "an autonomous nation with home-rule arrangements within the Danish Kingdom" negotiated similar arrangements (Langset 1999, p. 49). Greenland, whose status is analogous to that of the Faroes, and which has valuable fisheries to protect from intra-Union competitors, was the only territory to have withdrawn from the European Union. Britain's impending withdrawal, in the wake of the 2016 'Brexit' referendum result, constitutes an event of altogether different magnitude, about which more is said below.

Every enlargement has added sovereign or dependent islands to the Union. Along with the EU's effort to strengthen 'economic, social and territorial cohesion', this has added to the scope and variety of regimes governing relations between peripheral islands and the EU. Today,

Europe's islands constitute a multi-layered tissue of constitutional, taxation, incentive, trading, administrative, planning, transportation, communication, judicial and cultural regimes interwoven between island, regional, national and supra-national spheres of governance (Moncada et al. 2010). This opens a fresh, perhaps unintended, perspective on the term 'deepening' that is the goal of European integrationists. Here are some examples.

To start with, each island member state negotiated special arrangements safeguarding national jurisdiction over matters that were perceived to be of vital importance. Thus, Britain and Ireland reserved their positions over border controls, immigration, asylum, police cooperation and judicial cooperation in civil matters. Britain also declined to adopt the euro, in order to preserve an independent monetary policy and London's innovative financial markets. Malta, the Union's smallest member, negotiated the largest number of derogations and transitional arrangements ever conceded to an acceding state (*The Economist* 2004). Alongside these special arrangements, the island members enjoy full shares of the jurisdiction and voice accorded by the treaties to all member states. Some other states, such as Iceland, which decided against accession to protect vital national interests, nonetheless enjoy access to the Single Market.

Non-sovereign island territories are also favoured in jurisdictional terms (Adler Nissen and Gad 2013). As the Single Market developed, the British Crown Dependencies sought to protect their tax and legal regimes from regulatory incursions originating in London and Brussels. Indeed, internal and external pressures have prompted constitutional changes within Jersey, as well as debate about re-defining Jersey's relationship with Britain (Le Rendu 2004). The Åland Islands, established under Finnish legislation as "an autonomous, demilitarised and neutral region guaranteed under international law" Lindstrom 2000, Mäki-Lohiluoma 1999, p. 345), joined the Union along with Finland, after securing permanent concessions permitting duty free transactions on Åland ships within EU waters and confirming restrictions on the right of non-residents to own property(Lyck, 1996). Particularistic arrangements devised in respect of Greenland's fisheries, imports into the EU of petroleum products refined in the Netherlands Antilles, and the Pacific Financial Community formed by France's *départments d'oûtre mer*, are embodied in the treaties.

The treaties articulate a legitimising doctrine for both the constitutional arrangements and the developmental policies formulated by the islands or by the EU in the islands' interests (Kristjánsdottír 2010), synthesising the principles established incrementally over three decadess, as the Union progressively established a three-cornered arena in which EU institutions, member states and non-sovereign island governments negotiate and accommodate converging and diverging interests, continually adjusting the terms of relationships that are mutually but unequally beneficial, and 'especially generous' to islands and micro-states (Fagerlund 1996).

European islands which have not secured particularistic arrangements – such as the Italian and Greek islands – nonetheless benefit from EU programmes devised for disadvantaged regions. These policies merit attention because of their vital importance to islanders and their prominence in Union politics. They were legitimated by the Treaty of Amsterdam (Article 158), and are now enshrined in Article 174 of the Lisbon Treaty.

Two aspects are noteworthy. Firstly, the Union's interest in 'harmonious development' and 'economic, social and territorial cohesion' motivates its relationship with 'the least favoured regions': the relationship is unequal in both law and the power available to the parties, but is mutually beneficial and envisages 'better' integration into the internal market 'on fair conditions' (Amsterdam Treaty: Declaration No. 30 on island regions). Secondly, the Union's policies are grounded in the idea that islandness implies 'backwardness' and 'severe and permanent handicap'; and that 'peripherality' imposes permanent costs (Amaral Fortuna 2001). This outlook has fundamentally affected the Union's strategies and policies for islands, permitting it to accept non-reciprocal arrangements with many islands, without compromising the larger goal of integration.

The Union's relations with *European* islands are embodied chiefly in the so-called 'regional' and 'cohesion' policies, which afford islands (and other supposedly disadvantaged territories) voice, jurisdiction and administrative capacity by which they leverage the terms of their transactions with the world beyond their shores (Baldacchino and Pleijel 2010, Moncada et al. 2010).

'Voice' implies both the right and the ability to participate in decision-making, by lobbying, making alliances, negotiating and voting in appropriate decision-making or advisory institutions. The Committee of the Regions is among the most important European institutions from the perspective of sub-national island jurisdictions designated as 'regions' in their respective member states (Kennedy 1997). Island interests are also articulated through a Conference of Presidents of the outermost regions, a Conference of Peripheral Maritime Regions and a network of island chambers of commerce. The influential Economic and Social Committee, the European Parliament and the Commission, also address island concerns frequently.

The jurisdiction accorded to islands that are constitutionally integrated with member states and with the EU is much more limited. The Union has permitted *selective* derogations from rules governing the single market, principally in relation to taxation and customs duties 'to offset the handicaps arising from remoteness' in the 'outermost territories'. It also permits the *national governments* to establish tax regimes adapted to the islands' developmental needs (European Commission 2004, pp. 18–20). Such policies have helped foster offshore centres and asymmetrical governance on islands like Madeira and the Azores.

Capacity-building programmes are devised especially to carry forward the Union's model of sustainable, integrating development. An overview of the measures applying to the 'outermost territories' reveals the density, size and scope of these regimes (Armstrong et al. 2012). Here too, European law and practice have created, through the regional and cohesion policies, a three-cornered arena in which the European institutions, member states and regions (including island regions) negotiate and accommodate their respective interests and create mutually beneficial, though unequal, relations.

This discussion illustrates the complex, dynamic, asymmetrical arrangements applying to islands within the European Union. Until the onset of a succession of crises in the first decade of the 21st century, Europe was a case study of the extraordinary flexibility of policy-makers in accommodating islander interests within the emerging world order. Two concerns intrude on this benign scenario.

The first is that the performance of comparable islands differs markedly (European Commission 2004). The evidence from much of Europe's southern periphery, which includes numerous islands, suggests that an enabling legal framework, jurisdiction and sound policy design are necessary but not sufficient to overcome peripherality:

> What seems to matter, besides and above higher levels of institutionalisation and cultural identity, [is] the interest of the regional political classes in using European development funds to promote the economic development of their region . . . In other words, formal institutional capacities may lie dormant in the absence of a clear developmental will, while a strong political will to develop a region may not only activate existing capacities . . . but even create them from scratch.
>
> *(Piattoni and Smyrl 2003, pp. 133–135)*

This observation resonates with a recurring theme in the literature on the economic development of non-European island states (e.g. Jacobs 1989, McKee and Tisdell 1990, Ramkissoon

2002, Sutton 1991, Watters 1987, Baldacchino 2006c). The role played by institutions, political elites and policy orientations (in all states) is also a principal concern of donor countries and agencies (Rodrik et al. 2002).

The second concern arises from the long-term consequences of three successive crises which shook the EU: first, beginning in 2007, the financial crisis and the subsequent fiscal crises in Greece, Ireland, Portugal, Spain and Italy (Karyotis and Gerodimos 2015, Stiglitz 2016a); second, the influx of migrants into the Mediterranean member states, especially the 'frontier' islands (Crawley et al. 2016); third, the Brexit referendum in 2016, and the impending withdrawal of the UK from the Union (Clarke, Goodwin and Whiteley 2017). Islands, including island member states, seem to be disproportionately implicated in each successive crisis, from tiny Lampedusa, epitome of the migration crisis, to the mighty UK. Each crisis has called into question the Union's four freedoms, challenged the mechanisms of European policy-making, and demonstrated the limits of intra-European solidarity. As the EU is the heart of this neighbourhood that has nurtured a benign environment for the continent's numerous islands, its fate could be crucial for their future prospects. Ironically, it was Europe's most powerful island state, and the EU's second largest member state, the UK, that precipitated the crisis in earnest with its invocation of Article 50 of the Lisbon Treaty in March 2017, a consequence of seismic shifts in Britain's economy, society and politics (Luce 2017).

In summary, Europe's experience, unfolding over six decades of hope and crisis, underscores the importance of governance *within an island's neighbourhood*, as well as the fallacy of thinking in terms of any single pattern of island governance. In short, the evidence confirms that, as complex systems, islands should be examined within a broad ecological and historical framework that admits of anomalies, accidents and purposeful human intervention. The watchwords are complexity, not simplicity; diversity, not uniformity; contingency, not predictability. It is to this question that our article now turns.

Patterns of island governance

'Governance' emerges from "the governing activities and interaction of a variety of social, political and administrative actors" (Kooiman 1993, p. 1). It extends beyond the framework of governing institutions to embrace elements of civil society that share the role and authority of formal governing institutions. Our review draws attention to four elements of governance that are particularly prominent in and important for islands, namely:

(1) the geo-strategic role and 'imperial' connection: island governance arises from a distinctive ecology that is shaped decisively by the design of either a regional paramount power or, increasingly, by regional and global institutions;
(2) the constitutional order and political system operating within an island and its neighbourhood; along with patterns of leadership, the organisation, alignment and interests of political elites, and the political culture;
(3) dominant policy concerns, including the threats of peripherality and vulnerability, economic and social dislocation; the choice of development strategy; the government's role and administrative capacity; and
(4) islander identity, worldview and social organisation.

Although these elements are common to island governance everywhere, the evidence demonstrates conclusively that islands are not governed according to a single pattern (e.g. Anckar 2012). The Commonwealth, which embraces island members from four distinct maritime regions – the Mediterranean and Caribbean Seas, the Indian and Pacific Oceans – offers unmistakable evidence of the diversity of governance systems, notwithstanding the shared colonial history, constitutional and administrative framework, and developmental trajectories. Thus, for example, whereas Pacific Island studies probe the reasons why the constitutional order is chronically contested or has broken down in multi-ethnic states such as Fiji (e.g. Larmour 1994), corresponding studies on the English-speaking Caribbean investigate the persistence of democracy there, despite ethnic and class cleavages, recurring economic crises and weaknesses in the rule of law (e.g. Allahar 2001, Dominguez, 1993, Payne and Sutton 1993).

If there is more than one pattern of island governance, then, how many such patterns are identifiable, how did they arise, and how may they be grouped and analysed comparatively? This last section takes up this challenge by proposing a typology of *patterns* of island governance. It attempts to make sense of the under-researched, multidisciplinary field of island governance, by abstracting from and synthesising our present knowledge of the history, politics, economy and society of numerous islands, in the same way that the taxonomies of small island socio-economic development noted by Bertram (2006) bring intellectual order to that field of enquiry. Before outlining the typology, it is necessary to state the following cautions:

(1) Islands that lack formal jurisdiction, constitutional personality or agency cannot be classified.
(2) The typology must be grasped as an intellectual distillation and abstraction of dozens of island histories evolving from the early modern era, when Europe's overseas explorations brought hundreds of islands into the orbit of the continent's imperial powers.
(3) Patterns of island governance are elements of broader systems of political economy.
(4) The governance of any one island may evolve over historical time-scales, so that the island (or archipelago) transitions from one pattern of governance to another, more or less rapidly, more or less serenely, and more or less completely.
(5) While it will conform to one of the patterns postulated here, the governance of any single island will always exhibit features that are peculiar to itself anomalies arising from accidents of geography and history, that may be more or less important in that specific setting.

The mere fact that an island is an island does not and cannot account for all the characteristics of its governance. Nor do the patterns we postulate resolve all questions about island governance. Nevertheless, the typology brings order and intelligibility to the diversity of governmental arrangements that islands exhibit; it identifies potential lines of enquiry that will undoubtedly refine the typology, while providing fuller empirical evidence for the patterns.

The typology consists of *seven* patterns of island governance, identified and labelled as follows: civilisation; fief; fortress; refuge; settlement; plantation; entrepôt. Many of these labels will be recognised as they have often been employed by scholars to describe a distinguishing feature which becomes the *leitmotif* of an island territory's history, and a metaphor embodying critical political, sociological and economic phenomena. For example, accounts of Malta single out its history as an imperial *fortress* well into the twentieth century as the *leitmotif* of its existence (e.g. Pirotta 1996, Fenech 2005). Similarly, neighbouring Sicily's despoliation at the hands of predatory elites is the dominant metaphor in commentaries on that island, which commonly – if not altogether accurately – ascribe this to its feudal status (e.g. Robb 1999); Sicily is depicted as an archetypical *fief*.

The *plantation* is a leitmotif in scholarship about the Caribbean islands (e.g. Monroe and Lewis 1971, Myall and Payne 1991, Parry et al. 1990, Singham 1965) and, in attenuated form, of certain Indian Ocean and Pacific island groups (e.g. Bayliss-Smith et al. 1988, Brookfield 1972, Crocombe 1987, Watters 1969). The term *settlement* has just as usefully and as commonly been applied to the more liberal, more autonomous developmental experience of larger island territories populated as a result of large-scale European migration: New Zealand and Iceland are examples of this category.

The foregoing types of political economy are imperial constructs, models of colonisation and dependency that are challenged by island elites. The term *entrepôt*, however, captures the contemporary imagination and political economy of globalisation, where islands like Singapore manage to enlarge their economic space far beyond their shores and to engage the outside world more selectively and actively.

With the term *civilisation*, however, we distinguish the much rarer case of an island or archipelago – such as Japan or Great Britain – which, for quite different reasons, has an extraordinary and distinctive global culture and impact. Finally, we have coined the term *refuge* to cover very rare instances of islands, like Taiwan or Cuba, whose governance defies adjacent empires and exercises an unusual geo-political or economic importance in the contemporary world. Adding the terms 'civilisation' and 'refuge' to accommodate these rarer exceptions, we thereby arrive at a seven-fold classification

In outlining each pattern of governance, we propose two islands as exemplars, though the categories are in fact generally abstracted from the experiences of dozens of islands. Each outline focuses on the four elements of island governance identified earlier and emphasises the crucial role that the 'imperial connection' and local elites play in permitting or obstructing constitutional change, setting standards of governance, and/or shaping options for development. The typology underscores the fact that island jurisdictions confront the challenges of development from quite different starting points and with different mixes of policy instruments and choices.

We begin with the 'civilisation', which is certainly the rarest but the most influential island polity.

The island civilisation

Geo-political profile

In the history of government, civilisations are regarded as originators of great political ideas or institutions, as well as great cultural achievements (Finer 1997). *Island* civilisations are comparatively rare: the modern period has produced just one in the western and eastern hemispheres respectively – Britain and Japan. Through trade or conquest, the islanders propagated their distinctive cultural achievements from the 18th to the end of the 20th centuries, and profoundly altered the contemporary world, although their developmental trajectories differed. Acting through the channels provided by their military, financial or industrial power, Britain and Japan generally took the initiative in engaging the world beyond their shores and set the terms of that engagement. At the height of their power, their foreign policy sought especially to command sea lanes and to maintain a balance of power wherever their interests were at stake (Bartlett 1993, Tarling 1996). An island civilisation thus constitutes the plenitude of insular autonomy, the antithesis of the vulnerability so commonly attributed to islands.

Political system

An island civilisation's comparative security seems to favour constitutional stability. Security and stability in turn encourage the evolution of a distinctive political system whose principal tenets, values and norms are embodied in an enduring founding myth, exemplified by the British doctrine of parliamentary sovereignty (Bagehot 1993, Judge 1993), or the supposedly divine origin of Japan's imperial line. Reinterpretation permits necessary political changes, such as an extension of the franchise in 19th century Britain, or the renunciation of the Japanese Emperor's divine status after the Second World War, while preserving essential elements of the old order as stabilising influences in the new.

Founding myths may obscure the complexity and sophistication of these polities; yet, they underscore the fact that the political system is domestically generated. However, the Meiji Restoration and the post-war Constitution of Japan demonstrate that an island civilisation may borrow institutional forms from abroad or, *in extremis*, have them imposed by a victorious rival power. Even when changes originate externally or under some form of compulsion, they are assimilated under the firm direction of the island's elites, a phenomenon most dramatically evident in the post-war 're-invention' of Japan (Bix 2000). While the trajectory of an island civilisation's political development does not necessarily tend towards democracy – witness Japan's experience – power is distributed among governing institutions and elites in a way that is widely legitimated.

The elites themselves are reinvigorated by the timely assimilation of newly prominent groups, such as the en-nobling and enfranchisement of Britain's entrepreneurs and empire-builders in the 18th and 19th centuries.

Dominant policy concern

The history of the two island civilisations of the modern age suggests that the government is less concerned with state-building or even the aggrandisement of the state at home and abroad. Rather, it is directed towards safeguarding the conditions – domestic and international – that permit domestic production and external trade to flourish. This means that the professional bureaucracy must share power with other groups, such as landowners, merchants and financiers, and may have a lower social status (Chapman 1970, Hennessy 1990). Island civilisations produce models of public administration (Britain) and enterprise management (Britain, Japan) that are widely admired.

Identity

Some elements of the political system become incorporated into the islanders' collective identity: the association between Empire and the English character, so confidently asserted by British imperialists *and* by the Empire's colonial subjects and enemies, exemplifies this pattern (Tidrick 1990). The island civilisation projects a well-rounded, vigorous sense of nationhood; but traces of a lingering insecurity may be discerned in the tendency to assert the islanders' identity over and against the identities of large neighbouring peoples (e.g. McGregor, 2017, French 2017, Peng, 2017, Lewis 1988, p. 56; Brexit at the time of writing).

Three important questions await further study: why are island civilisations so rare; what conjunction of factors generates them; to what extent do governance arrangements account for a civilisation's emergence and success?

The fief

Geo-political profile

As the antithesis of the civilisation, a fief experiences to an extreme degree the peripherality, vulnerability and dependence commonly attributed to islands, compounded by neglect, repression and exploitation at the hands of a rapacious external power or by the design of its own elites. In the modern period, Sicily and Haiti may be regarded as archetypical fiefs, while Irish history is presented as the Catholic experience of subjugation and dispossession by the neighbouring Protestant English and their Presbyterian agents in the north (de Paor 1985).

Political system

Though incorporated into the 'imperial' constitution, pronounced asymmetries characterise the constitutional order – the centre of power versus the fief; imperial agents versus indigenous elites; officials versus ordinary folk. Patron–client relations serve both to institutionalise these asymmetries and to moderate their effects, as a Sicilian aristocrat vividly demonstrated in his classic, semi-autobiographical novel, *The leopard* (Tommasi di Lampedusa 1991). Occasional rebellion may disrupt government without necessarily weakening the empire's grip, as Ireland repeatedly experienced until 1921 (Grierson 1973, pp. 187–195); resistance may then turn to quietly subverting the repressive alien state, as in Sicily (Robb 1999).

The weak, illegitimate constitutional order deforms political life. Politics revolves around a narrow range of imperial or factional interests. The fragmented polity comprises expatriate agents of the imperial power; a 'collaborationist' indigenous elite; and 'oppositional' groups whose activities and claims may place them outside the pale of legitimate politics and economic activity (e.g. Fatton 2002, Lee 1989, Robb 1999).

The co-existence and inter-penetration of the official hierarchy, patron–client networks and criminalised groups, none of which enjoys legitimacy, means that conspiracy and violence feature prominently in politics, government and, indeed, in everyday affairs. Violence may be used to challenge or to uphold the constitutional order, and to secure resources for and from public and private domains. *In extremis*, a mafia may develop – a parasitic class operating at the interstices of lawful and subversive politics, lawful and criminal economic activity. Yet, ironically, the prevailing mistrust, insecurity, official ineptitude and corruption will frequently induce the mafia's victims to call on its aid.

Dominant policy concern

As the public domain is progressively impoverished, people of all classes and conditions rely increasingly on patron–client relations to gain access to public services or resources, aggravating administrative corruption and incompetence (Fatton 2002). Ultimately, the government's control over the territory weakens. In fact, whether or not it is openly challenged, the centre's control over its fief is the dominant policy concern, as Robb argues in scrutinising the connections between Italy's politicians and the Sicilian *Cosa Nostra* (Robb 1999). The government's resources are largely deployed to protect its agents and collaborators.

Profound, defensive traditionalism characterises life on fiefs. Any change which could disturb the status quo, however uneasy, however unsatisfactory that may be, is viewed with suspicion: by the impoverished mass because change threatens further impoverishment; and by the powerful, because they might thereby be displaced (e.g. Tommasi di Lampedusa 1991). Thus, change tends to be precipitated by external contingencies such as war (Norwich 2015).

Identity

Two strains run through the fief's political culture: a dominant but defensive and traditionalist strain complementing the islanders' traditionalism; and the utopian radicalism of a small minority. Both strains subvert governmental authority and, together with repression and poverty, inhibit the emergence of assertive, unifying nationalism. In fact, parochialism stifles and displaces nationalism: expatriates and indigenous elites find their cultural frame of reference in the metropolis; while the impoverished and unorganised masses are unlikely to seek salvation in an abstraction as remote from everyday concerns as 'the nation'. Against that, Ireland's extraordinary cultural renaissance in the 19th century, resting on the paradoxical assimilation of the English language as well as the revival of Gaelic, disseminated at least a partial, though a profound sense of nationhood: the Irish Catholic identity (Lee 1989, pp. 658–687).

Territories such as Sicily and East Timor testify to the continuing existence of fiefs, ironically subjugated by kindred nations. Haiti however shows the anomaly of a rapacious, native-born elite itself perpetuating a predatory regime without the overt presence of an imperial apparatus (Fatton 2002). Against that, the Irish Republic epitomises emancipation from both foreign 'enfeudation' as well as from the violent elements in the body politic (Lee 1989), arguably exchanging one imperial connection for another (the EU), precisely to sustain emancipation. These and other experiences, and comparison with civilisations, draw attention to the determining influence of a fief's indigenous elites, the islanders' identity, the imperial connection and the international order.

The next two types of island polity – the fortress and the refuge – are intimately related to the military ambitions of empires, though in contrasting ways.

The fortress

Geo-political profile

A bulwark and a base for projecting power, the fortress is generally a small dependency commanding trade and communications routes (Buttigieg and Phillips 2013). A fortuitous combination of location and technology ordains that role, but its indigenous community is too small or weak to exploit these *independently* for its own military or commercial designs. Imperial rivalries, conflicts and balances of power particularly affect a fortress's geo-political profile (Gregory 1996). Malta (1530–1979) perhaps epitomises the fortress, but St Lucia during the 18th century (Morris 1968, pp. 157–173), Singapore during the inter-war period and Hong Kong until the transfer of power to China (Morris 1993) also present similar profiles.

Political system

Devising a constitution for a fortress is especially difficult (Cremona 1994, Frendo 1979). The empire will wish to limit the scope of domestic politics. The community's aspirations may be ambivalent, swinging between contradictory platforms: in Malta, for example, this found expression – not for the first time, nor indeed for the last, in the 1950s in the bitterly contending visions of 'integration' with Britain versus sovereignty, in an effort to secure voice, jurisdiction and acceptable living standards (Pirotta 1987). The fortress's value to the empire determines which option triumphs. Even where sovereignty is admitted, the empire may retain residual military facilities or oversee external relations.

The greater the empire's interest, the sharper the distinction between expatriate and indigenous elites: the expatriates wield governmental authority, but disenfranchised local elites dominate politics (e.g. Pirotta 1996). In Malta, the local elites' limited leverage over decision-making encouraged a style of politics based on agitation and mobilisation, rather than discourse and engagement, which persisted following independence (Warrington 1997). The empire's reaction will include accommodation, assimilation or repression, as circumstances dictate. Consequently, constitutional development is retarded or stunted by the irreconcilable imperial and domestic interests, with long-term consequences for governmental legitimacy.

As the examples of Malta, St Lucia and Hong Kong demonstrate, fortress government is predominantly bureaucratic, tending towards autocracy and nurturing an influential, indigenous bureaucratic class.

Dominant policy concern

The empire's interest in the fortress encourages competent, paternalistic, interventionist government that benefits the civilian population. The civilian economy becomes a closely-regulated appendage of the fortress economy, with long-term developmental consequences. The restricted scope of non-military economic activity may induce outward migration, especially if employment in fortress establishments contracts.

While paternalistic government is generally benign, it may cause problems for a newly-independent ex-fortress. Against the benefits of having a large, experienced bureaucracy, it may count the costs of an over-extended, perhaps heavy-handed administration. The large public workforce constitutes a financial burden, as well as an influential player in local politics (Warrington 1995).

Identity

The principal social characteristic of a fortress is the expatriate garrison's presence, a vehicle for the importation of alien life-styles, faiths and practices, around which economic life and political discourse revolves. Numerous accounts testify to the disruption of island life by this alien social presence, as well as by large-scale outward migration. They also underscore the ways in which islanders adapt to, adopt and resist alien influences (e.g. Aldrich and Connell 1998, pp. 188–190, Mallia-Milanes 1988, Sultana and Baldacchino 1994, Mitchell 2002).

Social dislocation stunts 'national' identity, especially where this is complicated by the presence of contending sovereignties. Cyprus, a Western listening post in the Middle East since 1878, has experienced this phenomenon most tragically with competing 'motherlands'. Since 1974, this has produced a divided, 'double fortress' island whose green line is patrolled by UN troops. Here, the aspirations and worldview of an island's native-born elites may never converge, subverting any sense of common purpose and common good (Diez 2002, Georghalides 1985).

All things considered, this pattern of governance is replete with paradoxes: economic affluence and vulnerability; competent, benign administration but fractious politics; a cockpit of imperial politics, but a weak sense of domestic interests.

The refuge

Geo-political profile

A refuge is forced into that role by a rare conjunction of historical circumstances: specifically, a contested emancipation that precipitates prolonged confrontation with the dominant regional power. Cuba and Taiwan are quintessential refuges. Initially plantation colonies of Spain and Manchu China respectively, absorbing the culture, institutions and economic system of the mother country, their contemporary status is complicated by the residual pretensions of the USA and China (Bickers, 2017, Bartmann 2008, Rose 2016, French 2017).

Political system

Authoritarian rule tends to characterise the refuge. The crisis out of which it is born – the Cuban Revolution of 1959 and the Kuomintang's defeat in the Chinese civil war – and the prolonged state of emergency, present serious obstacles in the path of both a constitutional settlement and democratisation. Yet, Taiwan's democratisation, the thaw in US–Cuba relations under Obama, and some liberalisation after Castro's death in 2016 demonstrate that authoritarian rule is neither inevitable nor everlasting. Similarly, the politics of the refuge is always torn by the debate over how far it should engage or resist the adjacent empire.

Dominant policy concern

The geo-political profile of these refuges inclined their governments towards economic and social *dirigisme* to an even greater degree than the fortress. They are also absorbed by foreign affairs, intervening overseas under guises such as aid, humanitarian assistance, military advice, in an effort to legitimise their regimes internationally (Erisman 1993). For decades, both regimes relied on economic and military protection from a Great Power strong enough to counter the threats emanating from the adjacent regional empire. These global dimensions give world prominence to these island refuges. The presence and defence of a sea barrier between the defiant refuge and its menacing neighbour is central, since survival literally depends on maintaining this physical separation and distinction. Sustaining its power through a militarised state bureaucracy, a refuge's ruling elite favours the administrative efficiency in emergency conditions, though it must often accept heavy-handed, corrupt administration in practice (e.g. Erisman 1993 on Cuba).

Identity

The refuge defines itself over and against the empire as its alter ego: hence, Cuban socialism against American capitalism (Valdés 2001); and Taiwanese democracy against Chinese autocracy (Brown 2004, reviewed in Courtenay 2006, Miles 2005, Bartmann 2008). This turns insularity inside out: it is at once intensely nationalistic in the traditional sense, as well as seeking international affirmation. Like the fief, the refuge experiences chronic insecurity, but its elites galvanise the islanders using that very sentiment; unlike the fief, however, they walk boldly on the world stage.

The patterns of island governance reviewed thus far are unquestionably rare. Their very exceptionalism, however, simulates the carefully controlled conditions of a laboratory, offering opportunities to study political phenomena 'uncontaminated' by other elements, although that also requires caution in drawing general conclusions. The rarity of the patterns should not blind observers to the vital importance that all of them have – the fief included – in world or regional affairs. Whereas the civilisation and the refuge are assertive players on the world stage, and the fortress is a base for projecting power, the humble fief, deprived of conventional instruments for exercising jurisdiction, may nonetheless exercise a nefarious influence through criminal networks or, more positively, by embodying the metropole's national identity (Farinelli 2017).

The three remaining patterns are more common. They too are shaped by international economic forces, and virtually all are relics of imperial design, established through European expansion overseas in search of living space and valuable commodities. Thus, migration is more prominent in the ecology of governance here than it is in the foregoing patterns, raising, for example, the profile of racial, ethnic and religious issues in political discourse. Furthermore, the *settlement, plantation* and *entrepôt* muster what leverage they can chiefly through the strategic resources which they hold, the commodities which they produce, or the trade which they generate.

The settlement

Geo-political profile

'Settlement colonies' constituted a principal thrust of Europe's multi-faceted imperial enter-prise: formed predominantly by voluntary migrants, they generally acquired a large, thinly-populated territory for the purpose of absorbing surplus population in the homeland. Thus, Britain's 'white dominions' came into being, among them the settlement islands of New Zea-land and, on a continental scale, Australia (James 1995). Iceland predates both of these, albeit on a smaller scale. These settlements became incorporated into alliances and trade relations centering on the homeland (e.g. Grierson 1973, pp. 282–283), though regional influence is certainly not beyond them (e.g. O'Connell 2005).

Political system

A settlement created by a culturally homogeneous group of migrants will quickly acquire appre-ciable internal autonomy. It develops a robust political tradition modelled on the homeland's constitutional order and political system, exemplified by New Zealand's emancipation as Brit-ain's first 'Dominion', but displaying stronger democratic tendencies associated with the rugged individualism of the migrants – an extended franchise, for example, or an assertive legislature (e.g. Asgeirsson 1994). Against that, descendants of the original settlers, who will generally be the largest landowners, may exercise an oligarchic influence on settlement politics (James 1995, pp. 146–150), promoting a conservative agenda and sharing power reluctantly with later immi-grants. Subsequent waves of immigrants, such as the Chartists in Australia and New Zealand, may introduce novel forms of political and community organisation reflecting partly the coun-try and circumstances of their origin as well as the latitude afforded to them in the settlement's mainstream politics. Trade unions and social democratic parties will be among the most impor-tant innovations. These processes were clearly at work in Australia throughout the 19th and 20th centuries (James 1995, pp. 307–318). Nevertheless, steady constitutional development and governmental stability characterise island settlements.

Dominant policy concern

Both under 'imperial tutelage' and even more under the direction of local politicians, the gov-ernment's central concern is the orderly, rapid and effective settlement of the territory. In a com-munity of pioneers and immigrants where the autonomy of individuals and groups is valued, administrative decentralisation complements constitutional devolution and co-exists with an active, socially-conscious administration. Settlements tend to establish themselves as models of good, innovative governance.

Identity

The identity of settlement communities probably exerts a determining influence on their poli-tics and economies. To begin with, the motherland is likely to be better disposed to granting internal autonomy to, and to trading equitably with, communities who identify with it cultur-ally and are recognised as kin. Furthermore, constraining social conventions and class distinctions will weaken in the new setting, permitting the development of a broader sense of citizenship exemplified in archetypical heroes such as immigrants, pioneers and explorers. The settlement presents an image of self-confidence to its citizens as well as to the outside world: as masters of their destinies, they see no reason for deference in treating with greater powers.

Settlements are generally peaceable, stable, well-governed polities, although these achievements, as in the case of Iceland, may have been won only slowly over centuries of sometimes precarious order under chieftains with no real central government (Karlsson 2000, Asgeirsson 1994). Where island settlements arise on lands inhabited by aboriginal communities, as in New Zealand with the Maori people, the generally good governance record has been tarnished.

The plantation

Geo-political profile

Plantation islands, another pillar of European imperialism, originated as subordinate elements of mercantilist economic systems, to which they supplied a narrow range of commodities: precious metals, gems, timber, sugar, cotton, tea, coffee or spices (Dunn 1973). This specialisation persists, though it now assumes additional forms, such asextraction of hydrocarbons, mass production of clothing, mass tourism or, arguably, offshore finance (Hampton 1999). Many of the island states and dependent territories in the Caribbean Sea, and the Atlantic, Indian and Pacific Oceans conform to this political economy (Aldrich and Connell 1998, Couper 1989, Kaminarides et al. 1989).

Plantations are price-takers: thus, though many enjoy affluence, they remain vulnerable to shifts in the terms of trade. The heavy concentration of ownership of the means of production, either in the hands of a local 'plantocracy' or in multi-national enterprises, accentuates this vulnerability. They are also at risk from depletion of mineral or forest reserves, from disease or storms, which ravage mono-crop agriculture. In the second half of the 20th century, aid, debt, remittances from abroad and government expenditure have had to supplement comparatively scanty private capital formation (Watters 1987).

This could equally be true of settlements, at least at the onset of colonisation. What, then, distinguishes the plantation? Specifically, it is *involuntary migration*: the population originates disproportionately through penal servitude, indenture and slavery, which disenfranchise it, disrupt social organisation and debase its culture. These obstruct the development of a community of equal citizens (e.g. Payne 1993).

Political system

Instead, a plantation's geo-strategic profile and social composition favour the development of a low-trust, clientelistic polity in which land-holding is the basis of power. This is not only true of 'classical' plantocracies in the Caribbean: it is also evident in the Pacific Island states. There, especially in Fiji, well-meaning imperial administrators seeking to avoid the worst features of plantation societies respected the communal land ownership of indigenous communities and consolidated the political ascendancy of a chiefly class (Lal 1990, Lawson 1996). The combination of low trust, clientelism and pronounced socio-economic inequalities disposes the plantation towards personalised politics, populism and conflict (Payne and Sutton 1993, Ramos and Rivera 2001).

Late in the 20th century, numerous plantation islands acquired sovereignty, via a comparatively peaceful process that endowed them with democratic constitutions. They proved, however, to be uneasy democracies, liable to fall under the sway of strongmen, political dynasties and ethnic political groupings. Almost all have faced political crises arising from class or racial grievances, or weak economic performance. Enduring imperial influence and of regional trends is notably evident in plantation politics. Thus, though most of them have experienced some crisis of constitutionalism, the English-speaking Caribbean islands have a free press, lively legislatures and independent judiciaries. French and Dutch-speaking plantation islands were constitutionally incorporated into the mother country. With the notable exception of Mauritius, Indian Ocean states followed the African pattern of personal or one-party rule (Srebrnik 2002a). The larger Pacific island states, which were lightly colonised in the final stage of imperial expansion, experience ethnic and tribal conflict, or the contention of traditional and modern elites for power (Lawson 1996).

Dominant policy concern

Governments find their developmental ambitions constrained by economic dependence, vulnerability and social cleavages. Domestic demands, reliance on external capital and aid flows, and the overseas training typically received by their elites induce them to adopt the current developmental orthodoxy. Administrative performance is mixed, despite – or perhaps because of – on-going, externally designed reforms (Mills 1974, Murray 1985, UNEP 1998). Thus, while plantations provide schools, health care and housing to the general population, their standards, as measured by the Human Development Index, tend to be lower than those of the civilisation or the settlement (UNDP 2016). To borrow a phrase applied to African developing states, the island plantation state is both ambitious and 'soft' (Hyden 1983).

Identity

Social cleavages inhibit nation-building. Consequently, although island society acquires distinguishing folkways, these are variants of a regional identity rather than elements of a national identity that can be harnessed politically for development.

In summary, the plantation experiences *comprehensively* the tensions and ambiguities, opportunities and constraints constituting the ecology of island governance. It is not subjugated as the fief is, nor is it entirely free. It may be wealthy, but remains vulnerable to economic dislocation. It produces valuable commodities, but is peripheral in world affairs. Though capable of sustaining democracy, autocratic strains taint its politics. Though it presents an idyllic image, profound social cleavages fracture its identity. The plantation attracts innumerable visitors and tourists to its welcoming shores, while its own people seek escape from its constraining environment. Discounting the potential of their own institutions and resources, the plantation's leaders look abroad for models of good governance and economic prosperity – and the star by which they chart their course is the entrepôt.

The entrepôt

Geo-political profile

An entrepôt is a marketplace enjoying the fortress's advantages of location and centrality, but generating *internally* the conditions fostering its success: specifically, investment finance, entrepreneurial flair as well as a legal, regulatory and dispute-resolution regime that facilitates market transactions and innovation by minimising cost and risk. Singapore, the Channel Islands, Åland and Mauritius are examples of this political economy; other islands, such as Malta, Barbados and Ireland charted their emancipation from the political economy of the fortress, plantation and fief, respectively, by this star.

It is possible for an entrepôt to accumulate the critical mass needed to gain leverage in world markets. Nonetheless, its commanding position remains vulnerable to contingencies such as regional wars, natural disasters, accidents, epidemics, market shocks or technological developments. Furthermore, lacking a hinterland under its own control, it cannot count on the reserves available to a civilisation or a settlement or even, perhaps, to a sizeable plantation. Therefore, unless an entrepôt tirelessly innovates, re-positions itself and enhances its competitiveness, it will unquestionably decline (World Economic Forum 2017). Mauritius exemplifies both the problem and the strategy (Chernoff and Warner 2002). Regional security arrangements are an unavoidable necessity for this attractive 'prize'.

Political system

An entrepôt requires constitutional and political stability to perpetuate and re-define market-friendly conditions; policy communities that are technically competent and enterprising; public services that enhance competitiveness; a skilled, adaptable workforce. Capital is the key to political power, though not necessarily the key to public office. These factors place their stamp on politics: an entrepôt is formally democratic; the rule of law is secure; the judiciary and public administration untainted by corruption: but freedom of expression and association, and even privacy, may be curtailed officially or informally. Singapore's archetypical experience suggests that the entrepôt inclines to conservative rather than liberal democracy, emphasising consensual politics and suspicion of dissent.

Dominant policy concern

Since government and public administration exist to support economic activity, the entrepôt values efficiency. Far from favouring minimalist government, however, the entrepôt permits ambitious social engineering and welfare schemes that purport to enhance competitiveness.

Identity

All the economic and political features of an entrepôt tend to produce a somewhat paradoxical society having both a conservative ethos and a modern lifestyle, valuing individual well-being as well as social conformism, safeguarding democratic formalities while promoting strong, hierarchical leadership. Though in many respects vastly different, Singapore and Jersey typify this pattern.

The entrepôt's necessary openness to the outside world often attracts immigrants and their various cultures; but, if so, it is also a melting-pot that favours at least outward assimilation. In contrast to the other island polities, the *entrepôt* tends to exploit externally induced change, subordinating it to local direction. The entrepôt's economic success, springing from a deliberate internationalist or regionalist engagement, rather than sovereignty or democracy *per se*, makes it an attractive model for the fortress, settlement, plantation and fief.

Island governance in the 21st century: prospects and a research agenda

This final profile draws attention to the ambitions of island policy-makers who have inherited different starting points, and confronted both domestic and external developmental challenges on very different terms. The virtue of the typology, grounded in an island's distinctive history and experience, is to prepare ambitious reformers more realistically to meet these challenges. Similarly, our treatment of island 'neighbourhood' in the EU case study alerts island policy-makers to the opportunities and constraints that operate in their own geo-political circumstances. Just as individual islands evolve differently in geographical space and time, so too do their ecological fields or 'neighbourhoods'.

In summary, by the onset of the 21st century, international law and practice evolving incrementally during the previous half-century provided an environment broadly supportive of island jurisdiction. The pessimistic outlook so often encountered among social scientists and policy-makers, who tended to emphasise the constraints of island governance, was challenged by a welcome return to earlier perceptions, when philosophers, novelists and ordinary people seeking to make a fresh start saw islands as laboratories for exciting, new and often successful ventures. In this connection, Europe illustrated the vital importance of a nurturing and adaptive policy environment. The consequences for political economy, development prospects and governance in numerous islands were profound.

Since then, the comparatively benign global environment has been shaken by crises that are challenging the international rule of law, dissolving the consensus underpinning politics in the Western world's so-called stable democracies, and presenting daunting new problems for policy-makers (Luce 2017). The long-term consequences of this altered environment for islands and island governance are not yet clear: they may include separatism (the UK and its component nations); the gradual erosion of support and funding for islands development (post-Brexit suspicion of constitutional asymmetry); the assertion of religious fundamentalism in politics (Maldives; Indonesia); existential threats arising from climate change (Tuvalu and other atoll states); unchecked immigration, detention and people trafficking (Lesvos, Lampedusa, Nauru); acute land hunger (Malta, Singapore); increasing reliance on shadowy economic transactions and concomitant pressures from the international community to comply with legitimate business practice (offshore tax havens); new flashpoints for Great Power conflict (the South and East China Seas).

Across the world, the 21st century policy agenda is likely to be dominated by four overarching concerns, namely, environmental distress; migration and social cohesion (King and Connell, 1999) challenges to the neo-classical model of economic growth and redistribution (Piketty 2014); and the weak legitimacy of governance arrangements in their local, state, multi-lateral and supra-national dimensions (Beramendi et al. 2015).

From the point of view of island governance, the altered circumstances of the 21st century may entail one or, more likely, a combination of the following.

First, decision-making systems are required, that are both efficient and legitimate, robust yet adaptable, integrated yet decentralised, if islanders are to address the increasingly complex policy challenges and scenarios unfolding within and beyond their shores.

Second, numerous islands, both sovereign and non-sovereign, will be compelled to reappraise and re-negotiate the terms of their relationships with the unstable geo-politics of their neighbourhoods,(Kaplan 2015) although admittedly the constitutional aspirations of islands, and the jurisdiction that the neighbouring 'imperial powers' are prepared to permit, do not all point in the same direction (Bartmann 2006). Sovereignty may not be a common destination, particularly where a stronger case can be made for a status short of that (Baldacchino and Milne 2009, Karlsson 2009, Ramos and Rivera 2001). On the one hand, by eliminating distances electronically, globalisation permits some islands to escape peripherality. On the other hand, by scaling up and standardising economic transactions, it also threatens livelihoods, forcing islanders and governments into more shadowy economic activity. Indeed, it is these rules and constraints of the 'level playing field' among sovereign states that induce a number of island territories to seek escape and shelter, if they can, as sub-national territories with their own special asymmetrical arrangements. The ensuing consequences for governance are profound. In this connection, even Europe, formerly a model neighbourhood for adjacent islands, must wrestle with compound geo-political crises that may make the continent less stable and secure than in the past half century.

Third, seismic political shifts, the occult nexus of political and business leaders, and the growing incidence of scandals such as 'the Panama Papers' emphatically signal the need for robust measures to increase the legitimacy of governance in the eyes of their inhabitants – native-born, expatriate or immigrant – as well as before the international community (Milbank and Pabst 2016). This will ease political stresses and permit bolder policy-making. Reforms that secure the rule of law, affirm the virtues inherent in constitutionalism, enlarge the protection afforded to fundamental rights and freedoms, and the (inescapable) political emancipation of disenfranchised groups (chiefly migrants), will all contribute to greater governmental legitimacy.

The legal and policy regimes already operating in islands deserve systematic exploration because they provide practical lessons of the ways in which a supportive regime – whether international or national – can shelter and foster island life, *with reciprocal advantages for mainlands*. Other areas of governance also deserve study, among them predominantly administrative issues, such as administrative capacity, transport, communication, disaster preparedness, security, and the distinctive challenges presented by archipelagos.

Individual islands offer numerous lessons of good governance that may be transposed to other islands and mainland states in matters such as constitutional design (Anckar 2012); development strategy (Baldacchino 2006c; 2010); community welfare; heritage protection (Baldacchino 2011); the management of diversity; mechanisms of social change and the transmission of social capital; the efficacy of policy and administrative instruments; democratisation and risk management. All these policy challenges remind us of the critical role that must be played by island leaders and publics intent on charting their futures. Of course, the way will not be easy, as islands in different positions in our typology seek to move forward in an increasingly challenging international environment. Suitably grounded in the specific conditions of individual islands, the typology may help reformers to diagnose systemic obstacles to democratisation, constitutional innovation, competitiveness, social modernisation or capacity-building. Yet, in the final analysis, the wisdom, integrity and judgement of island leadership will play a determining role in securing the future integrity and effectiveness of island governance.

References

Ackrén, M. (2006) The Faroe islands: options for independence. *Island Studies Journal*, 1(2), 223–238.

Adler-Nissen, R., and Gad, U.P. (eds.) (2013) *European integration and post-colonial sovereignty games: the EU Overseas Countries and Territories,* Abingdon, Routledge.

Agranoff, R., and Banon, R. (eds.) (1997) Toward federal democracy in Spain: an examination of Inter-governmental relations. *Publius: The Journal of Federalism*, 27(4), 1–38.

Al Jazeera English (2013) *101 East: The Fight for Rapa Nui* (online documentary).

Aldecoa, F., and Keating, M. (1999) *Paradiplomacy in action: the foreign relations of sub-national governments*, special issue of *Regional and Federal Studies*, 9(1), London, Frank Cass.

Aldrich, R., and Connell, J. (1998) *The last colonies*. Cambridge, Cambridge University Press.

Allahar, A. (ed.) (2001) *Caribbean charisma: reflections on leadership, legitimacy and populist politics*. London, Lynne Rienner.

Amaral Fortuna, M.J. (2001) *The costs of peripherality* (Abridged Multilingual Edition). Luxembourg, European Parliament.

Anckar, D. (2006). Islandness or smallness? A comparative look at political institutions in small island states. *Island Studies Journal*, 1(1), 43–54.

Anckar, D. (2007). Archipelagos and political engineering: the impact of non-contiguity on devolution in small states. *Island Studies Journal*, 2(2), 193–208.

Anckar, D. (2012) Constitutional amendment methods in twenty-one small island democracies. *Island Studies Journal*, 7(2), 259–270.

Anckar, D., and Anckar, C. (1995) Size, insularity and democracy. *Scandinavian Political Studies*, 18(4), 211–229.

Armstrong, H.W., Giordano, B., Kizos, T., Macleod, C., Smed Olsen, L., and Spilanis, I. (2012) The European regional development fund and island regions: an evaluation of the 2000–2006 and 2007–2013 programs. *Island Studies Journal*, 7(2), 177–198.

Armstrong, H.W., Jouan de Kervenoael, R., Li, X., and Read, R. (1998) A comparison of the economic performance of different microstates and between microstates and larger countries. *World Development*, 26(4), 539–556.

Armstrong, H.W., and Read, R. (2000) Comparing the economic performance of dependent territories and sovereign microstates. *Economic Development and Cultural Change*, 48(2), 285–306.

Asgeirsson, J. (1994) The impact of national myth on the foundations of democracy in Iceland: an historical perspective. *Asian Journal of Public Administration*, 16(1), 14–40.

Azzopardi, J. (2015) Solving problems the island way: human resourcefulness in action among the islanders of Gozo. *Island Studies Journal*, 10(1), 71–90.

Bagehot, W. (1993) *The English constitution, with an Introduction by Richard Crossman.* London, Fontana Press.

Baker, R.A. (ed.) (1992) *Public administration in small and island states.* West Hartford, CT, Kumarian Press.

Baldacchino, G. (2000) The challenge of hypothermia: a six-proposition manifesto for small island territories. *The Round Table*, 89(353), 65–79.

Baldacchino, G. (2002a) Jurisdictional self-reliance for small island territories: considering the partition of Cyprus. *The Round Table*, 91(365), 349–360.

Baldacchino, G. (2002b) A nationless state? Malta, national identity and the EU. *West European Politics*, 25(4), 191–206.

Baldacchino, G. (ed.) (2006a) *Bridging islands: the impact of fixed links.* Charlottetown, Canada, Acorn Press.

Baldacchino, G. (2006b) Managing the Hinterland beyond: two, ideal-type strategies of economic development for small island territories. *Asia-Pacific Viewpoint*, 47(1), 45–60.

Baldacchino, G. (2006c) Innovative development strategies from non-sovereign island jurisdictions? A global review of economic policy and governance practices. *World Development*, 34(5), 852–867.

Baldacchino, G. (2010) *Island enclaves: offshoring strategies, creative governance and subnational island jurisdictions.* Montreal and Kingston, Canada, McGill-Queens University Press.

Baldacchino, G. (ed.) (2011) *Extreme heritage management: the practices and policies of densely populated islands.* New York, Berghahn Press.

Baldacchino, G., and Milne, D. (eds.) (2000) *Lessons from the political economy of small islands: the resourcefulness of jurisdiction.* Basingstoke, Palgrave Macmillan.

Baldacchino, G., and Milne, D. (eds.) (2006) Exploring sub-national island jurisdictions. *The Round Table*, 95(386), 1–15.

Baldacchino, G., and Milne, D. (eds.) (2009) *The case for non-sovereignty: lessons from sub-national island jurisdictions.* Abingdon, Routledge.

Baldacchino, G., and Pleijel, C. (2010) European islands, development and the cohesion policy: a case study of Kokar, Aland Islands. *Island Studies Journal*, 5(1), 89–110.

Bar Cendón, A. (1999) The intermediate level of government in Spain. In T. Larsson, K. Nomden and F. Petiteville (eds.), *The intermediate level of government in European states: complexity versus democracy?* Maastricht, The Netherlands, European Institute of Public Administration, pp. 127–174.

Bartlett, C.J. (1993) *Defence and diplomacy: Britain and the great powers, 1815–1914.* Manchester, Manchester University Press.

Bartmann, B. (1996) Saltwater Frontiers: Jurisdiction as a Resource for Small Islands. Canada, Institute of Island Studies, University of Prince Edward Island. Available from: www.upei.ca/islandstudies/art_bb_1.htm.

Bartmann, B. (2000) Patterns of localism in a changing global system. In G. Baldacchino and D. Milne (eds.), *Lessons from the political economy of small islands: the resourcefulness of jurisdiction,* Basingstoke, Palgrave Macmillan, pp. 38–55.

Bartmann, B. (2006) In or out: sub-national island jurisdictions and the antechamber of paradiplomacy. *The Round Table*, 95(386), 541–559.

Bartmann, B. (2008) Between *de Jure* and *de Facto* statehood: revisiting the status issue for Taiwan. *Island Studies Journal*, 3(1), 113–128.

Bayliss-Smith T., Bedford, R., and Latham, M. (eds.) (1988) *Islands, islanders and the world: the colonial and post-colonial experience of Eastern Fiji*. Cambridge, Cambridge University Press.

Beckles, H. (1990) *A short history of Barbados from Amerindian settlement to nation-state*. Cambridge, Cambridge University Press.

Bedggood, D. (2004) Regarding islands: a review of Rod Edmond and Vanessa Smith (eds.), *Islands in history and representation. Australian Humanities Review*, 31–32, April. Available from: www.lib.latrobe.edu.au/AHR/archive/Issue-April-2004/bedggood.html.

Begley, L. (ed.) (1993) *Crossing that bridge: a critical look at the PEI fixed link*. Charlottetown, Canada, Ragweed Press.

Beramendi, P., Bernardie-Tahir, N., and Schmoll, C. (2014) The use of islands in the production of the Southern European migration border. *Island Studies Journal*, 9(1), 3–6.

Beramendi, P., Häusserman, S., Kitschelt, H., and Kriesi, H. (2015) *The poltics of advanced capitalism*. Cambridge: Cambridge University Press.

Bertram, G. (2006) Introduction: the MIRAB model in the twenty-first century. *Asia Pacific Viewpoint*, 47(1), 1–21.

Bertram, G., and Watters, R.F. (1985) The MIRAB economy in South Pacific microstates. *Pacific Viewpoint*, 26(3), 497–519.

Betermier, S. (2004) Selectivity and the economics of independence for today's overseas territories. *Explorations: An Undergraduate Research Journal*, 63–85. Available from: http://explorations.ucdavis.edu/docs/2004/betermier.pdf.

Bickers, Robert (2017) Out of China, Cambridge, Massachusetts, Harvard University Press.

Bix, H.P. (2000) *Hirohito and the making of modern Japan*. New York, Harper Collins.

Borzel, T A. (2000) From competitive regionalism to cooperative federalism: the Europeanisation of the Spanish state of the autonomies. *Publius: The Journal of Federalism*, 30(2), 17–42.

Borzel, T.A. (2002) *States and regions in the European Union*. Cambridge, Cambridge University Press.

Briguglio, L. (1995) Small island developing states and their economic vulnerabilities. *World Development*, 23(9), 1615–1632.

Briguglio, L., and Kisanga, E.J. (2004) *Economic vulnerability and resilience of small states*. Sliema, Malta, Formatek.

Brinklow, L. (2016) A man and his island: the island mirror in Michael Crummey's *Sweetland*. *Island Studies Journal*, 11(1), 133–144.

Brookfield, H.C. (1972) *Colonialism, development and independence: the case of the Melanesian islands in the South Pacific*. Cambridge, Cambridge University Press.

Buttigieg, E., and Phillips, S. (eds.) (2013) *Islands and military orders, c. 1291–c. 1798*. London, Routledge.

Chapman, R.A. (1970) *The higher Civil Service in Britain*. London, Constable.

Chernoff, B., and Warner, A. (2002) *Sources of fast growth in Mauritius, 1960–2000*, Paper prepared for a conference on *Iceland in the World Economy: Small Island Economies in the Era of Globalisation*, Harvard University.

Clarke, C., and Payne, T. (eds.) (1987) *Politics, security and development in small states*. London, Allen and Unwin.

Clarke, H.D., Goodwin, M., and Whiteley, P. (2017) *Brexit: why Britain voted to leave the European Union*. Cambridge, Cambridge University Press.

Commonwealth Advisory Group (1997) *A future for small states: overcoming vulnerability*. London, Commonwealth Secretariat.

Cooper, A.F., and Shaw, T.M. (eds.) (2009) *The diplomacies of small states: between vulnerability and resilience*. Basingstoke, Palgrave Macmillan.

Corbin, C. (2001) Direct participation of non-independent Caribbean countries in the United Nations: a method for self-determination. In A.G. Ramos and A.I. Rivera (eds.), *Islands at the crossroads: politics in the non-independent Caribbean*, London, Lynne Rienner, pp. 136–159.

Corte-Real, I. (1999) The regions in Portugal: a challenging theme for citizens, administrators and politicians. In T. Larsson, K. Nomden and F. Petiteville (eds.), *The intermediate level of government in European states: complexity versus democracy?* Maastricht, The Netherlands, European Institute of Public Administration, pp. 315–328.

Council of Europe (1986) *The development of maritime islands as extreme examples of peripheral regions*, Strasbourg. Council of Europe.

Couper, A.D. (ed.) (1989) *Development and social change in the Pacific islands*. London, Routledge.

Courtenay, P. (2006) Anthropologist asks provocative question: A Review of M. Brown (2004) *Is Taiwan Chinese?* University of California Press. In *Taiwan Journal*, 23(30), 7.

Crane, R., and Fletcher, L. (2016) The genre of islands: popular fiction and performative geographies. *Island Studies Journal*, 11(2), 637–650.

Crawley, H, Duvell, F, Jones, K, McMahon, S., and Signona, N. (2016) *Destination Europe? Unravelling the Mediterranean migration crisis*, Final Report, Medmig Project.

Cremona, J.J. (1994) *The Maltese constitution and constitutional history since 1813*. Valletta, Malta, PEG.

Crocombe, R. (1987) *The South Pacific: an introduction*. New Zealand, Longman Paul.

De Paor, L. (ed.) (1985) *Milestones in Irish history*. Cork, Mercier Press.

Desideri, C., and Santantonio, V. (1996) Building a third level in Europe: prospects and difficulties in Italy. *Regional and Federal Studies*, 6(2), 96–116.

Diez, T. (ed.) (2002) *The European Union and the Cyprus conflict: modern conflict, postmodern union*. Manchester, Manchester University Press.

Dominguez, J.I. (1993) The Caribbean Question: Why has Liberal Democracy (Surprisingly) Flourished? In J.I. Dominguez, R.A. Pastor and R.D. Worrell (eds.), *Democracy in the Caribbean*, Baltimore MD, Johns Hopkins University Press, pp. 1–25.

Dommen, E.C. (1980) Some distinguishing characteristics of island states. *World Development*, 8(12), 931–943.

Dommen, E.C. (ed.) (1980) *Islands*. Oxford, Pergamon Press.

Dunn, R.S. (1973) *Sugar and slaves: the rise of the planter class in the English West Indies, 1624–1713*. New York, The Norton Library.

The *Economist* (2003) On the world's rich list. *London, The Economist Magazine*, 17 May, p. 33.

The *Economist* (2004) European union enlargement: smallness pays. *London, The Economist Magazine*, 26 February.

Elazar, D.J. (1987) *Exploring federalism*. Tuscaloosa, AB, University of Alabama Press.

Erisman, H.M. (1993) The Odyssey of revolution in Cuba. In A. Payne and P. Sutton (eds.), *Modern Caribbean politics*, Baltimore, MD, Johns Hopkins University Press, pp. 212–237.

European Commission (1994) *Portrait of the islands*. Luxembourg, Office for Official Publications of the European Communities.

European Commission (2004) Annex to the communication from the Commission on a stronger partnership strengthened for the outermost regions: assessment and prospects. (Communication of the Commission COM (2004) 343 of 26 May 2004) COM(2004)543, Brussels, European Commission.

European Economic and Social Committee (2003) *Trans-European networks and islands*. Luxembourg, Office for Official Publications of the European Communities.

Fagerlund, N. (1996) Autonomous European Island Regions enjoying a special relationship with the European Union. In L. Lyck (ed.), *Constitutional and economic space of the small Nordic jurisdictions*, Copenhagen, NordREFO, pp. 90–121.

Fairbairn, T.I.J. (2007) Economic vulnerability and resilience of small island states. *Island Studies Journal*, 2(1), 133–140.

Farinelli, M. (2017) Island societies and mainland nation-building in the mediterranean: Sardinia and Corsica in Italian, French and Catalan Nationalism. *Island Studies Journal*, 12(1), 21–34.

Fatton, R. (2002) *Haiti's predatory republic: the unending transition to democracy*. London, Lynne Rienner.

Fenech, D. (2005) *Responsibility and power in inter-war Malta. Book One: Endemic democracy, 1919–1930*. Valletta, Malta, BDL Books.

Finer, S.E. (1997) *The history of government from the earliest times*. Oxford, Oxford University Press (3 volumes).

Foreign and Commonwealth Office (2012) *The Overseas Territories: security, success and sustainability*. White Paper Cm 8374. London, FCO.

French, H. (2017) *Everything under the heavens*. New York, Knopf.

Frendo, H. (1979) *Party politics in a fortress colony: the Maltese experience*. Valletta, Malta, Midsea Books.

Georghalides, G.S. (1985) *Cyprus and the governorship of Sir Ronald Storrs*. Nicosia, Cyprus, Cyprus Research Centre.

Gibbons, M.S. (2010) Islanders in community: identity negotiation through sites of conflict and transcripts of power. *Island Studies Journal*, 5(2), 165–192.

Gillis, J.R. (2004) *Islands on our mind*. London, Palgrave Macmillan.

Gitay, H., and Jones, N. (2017) *Small island states resilience initiative: small islands, big challenges*. Thematic Note, Washington, DC, World Bank Group.

Golding, W.G. (1954/1978) *Lord of the flies: a novel*. New York, Putnam.

Gregory, D. (1996) *Malta, Britain and the European powers, 1793–1815*. London, Fairleigh Dickenson University Press.

Grierson, E. (1973) *The imperial dream*. Newton Abbot, Readers Union.

Hampton, M.P., and Abbott, J.P. (eds.) (1999) *Offshore finance centres and tax havens: the rise of global capital*. London, Palgrave Macmillan.

Hay, Peter A. (2006) A phenomenology of islands. *Island Studies Journal*, 1(1), 19–42.

Hennessy, P. (1990) *Whitehall*. London, Fontana.

Holmen, J. (2014) A small separate fatherland of our own: regional history writing and regional identity on islands in the Baltic Sea. *Island Studies Journal*, 9(1), 139–154.

Hooghe, L., and Marks, G. (2001) *Multi-level governance and European integration*. Lanham, MD, Rowman and Littlefield.

Howe, S. (1988) British decolonisation and malta's imperial role. In V. Mallia-Milanes (ed.), *The British colonial experience, 1800–1964: the impact on Maltese society*, Msida, Malta, Mireva, pp. 329–364.

Hyden, G. (1983) *No shortcuts to progress: African development management in perspective*. Berkeley, CA, University of California Press.

Jacobs, J. (1989) The economic development of small countries: some reflections of a non-economist. In J. Kaminarides, L. Briguglio and H.N. Hoogendonk (eds.), *The economic development of small countries: problems, strategies and policies*, Delft, The Netherlands, Eburon, pp. 83–90.

James, L. (1995) *The rise and fall of the British Empire*. London, Abacus.

Judge, D. (1993) *The parliamentary state*. London, Sage.

Judt, T. (2005) *Postwar: a history of Europe since 1945*. London, Heinemann.

Kaminarides, J., Briguglio, L., and Hoogendonk, H.N. (eds.) (1989) *The development of small countries: problems, strategies and policies*. Delft, The Netherlands, Eburon.

Kaplan, R.D. (2015) *Asia's cauldron: the South China Sea and the end of a stable Pacific*. New York, Random House.

Karides, M. (2013) Riding the globalisation wave (1974–2004): islandness and strategies of economic development in two post-colonial states. *Island Studies Journal*, 8(2), 299–320.

Karlsson, G. (2000) *Iceland's 1100 Years: history of a marginal society*. London, Hurst.

Karlsson, G. (2009) Sub-national island jurisdictions as configurations of jurisdictional powers and economic capacity: nordic experiences from Åland, Faroes and Greenland. *Island Studies Journal*, 4(2), 139–162.

Karyotis, G., and Gerodimos, R. (eds.) (2015) *The politics of extreme austerity: Greece in the Eurozone crisis*. Basingstoke, Palgrave Macmillan.

Keating, M. (1999a) Regions and international affairs: motives, opportunities and strategies. In F. Aldecoa and M. Keating (eds.), *Paradiplomacy in action: the foreign relations of subnational governments*, special issue of *Regional and Federal Studies*, 9(1), 1–16.

Keating, M. (1999b) Asymmetrical government: multinational states in an integrating Europe. *Publius: The Journal of Federalism*, 29(1), 71–86.

Kelly, M. (1999) The intermediate level of government in Ireland: local government with a regional overlay. In T. Larsson, K. Nomden and F. Petiteville (eds.), *The intermediate level of government in European states: complexity versus democracy?* Maastricht, The Netherlands, European Institute of Public Administration, pp. 197–209.

Kennedy, D. (1997) The European Union's committee of the regions. *Regional and Federal Studies*, 7(1) (whole issue).

King, R. (2009) Geography, islands and migration in the era of global mobility. *Island Studies Journal*, 4(1), 53–84.

King, R., and Connell, J. (1999) *Small worlds, global lives: islands and migration*. London, Cassell.

Kooiman, J. (ed.) (1993) *Modern governance: new government-society interactions*. London, Sage.

Kristjánsdóttir, A. (2010) *Small island differentiation in EU law*. Unpublished thesis, University of Iceland, Faculty of Law.

Lal, V. (1990) Fiji: coups in paradise - Race, politics and military intervention. London, Zed Books.

Langset, M. (1999) Intermediate government in Denmark: structure and developments. In T. Larsson, K. Nomden and F. Petiteville (eds.), *The intermediate level of government in European states: complexity versus democracy?* Maastricht, The Netherlands, European Institute of Public Administration, pp. 49–72.

Larmour, P. (1994) 'A foreign flower': Democracy in the South Pacific. Pacific Studies, 17(1), 45-77.

Lawson, S. (1996) *Tradition versus democracy in the South Pacific: Fiji, Tonga and Western Samoa.* Cambridge, Cambridge University Press.

Le Rendu, L. (2004) *Jersey: independent dependency? The survival strategies of a micro-state.* UK, Ex Libris Press.

Lee, J.J. (1989) *Ireland, 1912–1985: politics and society.* Cambridge, Cambridge University Press.

Lewis, F. (1988) *Europe: a tapestry of nations.* London, Unwin Hyman.

Lindstrom, B. (2000) Culture and economic development in Åland. In G. Baldacchino and D. Milne (eds.), *Lessons from the political economy of small islands: the resourcefulness of jurisdiction,* Basingstoke, Palgrave Macmillan, pp. 107–120.

Luce, E. (2017) *The retreat of western liberalism.* New York, Atlantic Monthly Press.

Lyck, L. (ed.) (1996) *Constitutional and economic space of the small Nordic jurisdictions.* Copenhagen, NordREFO.

Mallia-Milanes, V. (ed.) (1988) *The British colonial experience, 1800–1964: the impact on Maltese society.* Msida, Malta, Mireva.

Mäki-Lohiluoma, K.-P. (1999) 'Intermediate level of public administration in Finland. In T. Larsson, K. Nomden and F. Petiteville (eds.), *The intermediate level of government in European states: complexity versus democracy?* Maastricht, The Netherlands, European Institute of Public Administration, pp. 329–346.

McCabe, S. (1998) Romantic notions? Representing the space of 'island' in the wake of the fixed link. Vancouver, Canada, Department of Geography, University of British Columbia, 32pp.

McKee, D., and Tisdell, C. (1990) *Developmental issues in small island economies.* New York, Praeger.

McElroy, J.L. (2006a) Small island tourist economies across the lifecycle. *Asia Pacific Viewpoint,* 47(1), 61–77.

McElroy, J.L. (2006b) The advantages of political affiliation: dependent and independent small island profiles' in sub-national island jurisdictions. *The Round Table,* 95(386), 529–539.

McGregor, Robert (2017) Asia's Reckoning: China, Japan, and the Fate of U.S. Power in the Pacific Century, New York, Viking

Meleisea, M. (1987) *The making of modern Samoa: traditional authority and colonial administration in the modern history of Western Samoa.* Suva, Fiji Islands, University of the South Pacific.

Milbank, J., and Pabst, A. (2016) *The politics of virtue: post-liberalism and the human future.* London: Rowan and Littleton International.

Miles, J. (2005) Dancing with the enemy: a survey of Taiwan. *The Economist,* 15 January.

Mills, G.E. (1974) *Issues of public policy and public administration in the Commonwealth Caribbean.* Kingston, Jamaica, University of the West Indies.

Milne, D. (1997) Placeless power: constitutionalism confronts peripherality. *North,* 8(3–4), 32–38.

Milne, D. (2000) Ten lessons for economic development in small jurisdictions: the European perspective. Online library at Institute of Island Studies, UPEI. Available from: www.upei.ca/islandstudies/rep_dm_1.htm.

Milne, D. (2001) Prince Edward Island: politics in a beleaguered garden. In K. Brownsey and M. Howlett (eds.), *The provincial state in Canada: politics in the provinces and territories.* Peterborough, ON, Broadview Press, pp. 111–138.

Mitchell, J.P. (2002) *Ambivalent Europeans: ritual, memory and the public sphere in Malta.* Abingdon, Routledge.

Moncada, S., Camilleri, M., Formosa, S., and Galea, R. (2010) From incremental to comprehensive: towards island-friendly EU policy-making. *Island Studies Journal,* 5(1), 61–88.

Monroe, T., and Lewis, R. (eds.) (1971) *Readings in government and politics of the West Indies.* Kingston, Jamaica, Institute of Social and Economic Research.

More, T. (1516) *Utopia.* Available from: http://etext.lib.virginia.edu/toc/modeng/public/ MorUtop.html.

Morris, J. (1968) *Pax Britannica: the climax of an empire.* New York, Harcourt Brace Jovanovich.

Morris, J. (1993) *Hong Kong: epilogue to an empire.* London, Penguin.

Murray, F. (2001) The EU and member state island territories. Online library, Institute of Island Studies, University of Prince Edward Island, Canada. Available from: www.upei.ca/islandstudies/art_fm_1.htm.

Myall, J., and Payne, A. (eds.) (1991) *The fallacies of hope: the post-colonial record of the commonwealth third world.* Manchester, Manchester University Press.

Norwich, J.J., (2015), *Sicily: an island at the crossroads of history.* New York, Random House.

O'Connell, D. (2005) Vikings in suits invade Crawley. *The Sunday Times,* 27 February.

O'Crohan, T. (1977) *The islandman.* Oxford, Oxford University Press.

O'Crohan, T. (1986) *Island cross-talk: pages from a Blasket Island diary.* Oxford, Oxford University Press.

Parry J.H., Sherlock, P., and Maingot, A. (1990) *A short history of the West Indies.* London, Palgrave Macmillan.

Payne, A. (1993) Westminster adapted: the political order of the commonwealth Caribbean. In J.I. Dominguez, R.A. Pastor and R.D. Worrell, (eds.), *Democracy in the Caribbean: political, economic and social perspectives,* Baltimore, MD, Johns Hopkins University Press, pp. 57–73.

Payne, A., and Sutton, P.K. (eds.) (1993) *Modern Caribbean politics*. Baltimore, MD, Johns Hopkins University Press.

Piattoni, S., and Smyrl, M. (2003) Building effective institutions: Italian regions and the EU structural funds. In J. Bukowski, S. Piattoni and M. Smyrl (eds.), *Between Europeanisation and local societies: the space for territorial governance*, Lanham, MD: Rowman and Littlefield, pp. 133–156.

Piketty, T. (2014) *Capital in the twenty-first century*. Cambridge, MA, Belknap Press of Harvard University Press.

Pinet, S. (2002) *Archipelagos: insularity and fiction in medieval and early modern Spain*. Unpublished dissertation. Cambridge, MA, Harvard University.

Pirotta, G.A. (1996) *The Maltese public service 1800–1940: the administrative politics of a micro-state*. Msida, Malta, Mireva.

Pirotta, J.M. (1987) *Fortress colony: the final act, 1945–1964*. Rabat, Malta, Studia Editions.

Ramkissoon, R. (2002) Explaining differences in economic performance in Caribbean economies. Paper presented at international conference on *Iceland and the world economy: small island economies in the era of globalisation*, Center for International Development (CID). Cambridge, MA, Harvard University.

Ramos, A.G., and Rivera, A.I. (2001) *Islands at the crossroads: politics in the non-independent Caribbean*. Kingston, Jamaica, Ian Randle.

Rezvani, D.A. (2004) On the emergence and utility of 'federacy' in comparative politics. Working Paper, Harvard University, pp. 1–30.

Richards, J. (1982) Politics in small, independent communities: conflict or consensus? *Journal of Commonwealth and Comparative Politics*, 20(2), 155–171.

Robb, P. (1999) *Midnight in Sicily*. London, Harvill Press.

Rodrik, D., Subramaniam, A., and F. Trebbi (2002) *Institutions rule: the primacy of institutions over geography and integration in economic development*, Harvard, MA, Center for International Development, CID Working Paper No 97.

Ronström, O. (2009) Island words, island worlds: the origins and meanings of words for 'Islands' in North-West Europe. *Island Studies Journal*, 4(2), 163–182.

Rose, G. (ed.) (2016) *Cuba libre*? US–Cuba relations from revolution to rapprochement. *Foreign Affairs*, Spring Issue.

Royle, S.A. (1999) Bridging the gap: Prince Edward Island and the Confederation Bridge. *British Journal of Canadian Studies*, 14(2), 242–255.

Royle, S.A. (2001) *A geography of islands: small island insularity*. London, Routledge.

Selwyn, P. (1980) Smallness and islandness. *World Development*, 8(12), 945–951.

Singham, A.W. (1965) *The hero and the crowd in a colonial polity*. New Haven, CT, Yale University Press.

Srebrnik, H.F. (2000a) Can an ethnically-based civil society succeed? The case of Mauritius. *Journal of Contemporary African Studies*, 18(1), 7–20.

Srebrnik, H.F. (2000b) Islands and governance. *Archipelago*, 5(1), 5–8.

Srebrnik, H.F. (2002a) 'Full of sound and fury': three decades of parliamentary politics in Mauritius. *Journal of Southern African Studies*, 28(2), 277–289.

Srebrnik, H.F. (2002b) Ethnicity, religion and the issue of aboriginality in a small island state: why does Fiji flounder? *The Round Table*, 91(364), 187–210.

Srebrnik, H.F. (2004) Small island nations and democratic values. *World Development*, 32(2), 329–341.

Stevens, M.R. (1975) *Federacy: the federal principle in the post-colonial era*. Unpublished Doctoral thesis, Philadelphia, Department of Political Science, Temple University.

Stevenson, R.L. (1888/1998) *Treasure island*. New York, Signet Classic.

Stiglitz, J.E. (2016a) *The Euro: how a common currency threatens the future of Europe*. New York, Norton.

Stiglitz, J.E. (2016b). *The great divide*. New York, Norton.

Stratford, E. (2013) The idea of the archipelago: contemplating island relations. *Island Studies Journal*, 8(1), 3–8.

Sultana, R.G., and Baldacchino, G. (eds.) (1994) *Maltese society: a sociological inquiry*. Msida, Malta, Mireva.

Sutton, P.K. (1987) Political aspects. In C. Clarke and A. Payne (eds.), *Politics, security and development in small states*, London, Allen and Unwin, pp. 3–25.

Sutton, P.K. (1991) Constancy, change and accommodation: the distinct tradition of the Commonwealth Caribbean. In J. Myall and A Payne (eds.), *The fallacies of hope: the post-colonial record of the Commonwealth Third World*, Manchester, Manchester University Press, pp. 106–128.

Tarling, N. (1996) *Britain, Southeast Asia and the onset of the Pacific war*. Cambridge, Cambridge University Press.

Tidrick, K. (1990) *Empire and the English character*. London, I.B. Tauris.

Tommasi di Lampedusa, G. (1958/1991) *Il gattopardo* [The leopard]. New York, Pantheon (in translation).

Tordoff, W. (1997) *Government and politics in Africa*. Basingstoke, Palgrave Macmillan.

UNDP (2016) *Human Development Report 2016: Human development for everyone*. New York, United Nations.

UNEP (1998) Commission on Sustainable Development, Sixth Session, *National institutions and administrative capacity in small island developing states*. Available from: http://islands.unep.ch/dd98-7a6.htm.

Valdés, N.P. (2001) Fidel Castro (b. 1926), charisma and santeria. In A. Allahar (ed.), *Caribbean charisma*, London, Lynne Rienner, pp. 212–241.

Walsh, J.P. (1994) *Knowledge of angels*. New York, Houghton Mifflin.

Vella, M. (1994) 'That favourite dream of the colonies': Industrialisation, dependence and the limits of development discourse in Malta. In R. Sultana and G. Baldacchino (eds), Maltese society: A sociological inquiry, Malta, Mireva, pp. 53-77.

Warrington, E. (1995) Introduction. In Commonwealth Secretariat, *A profile of the public service of Malta*, London, Commonwealth Secretariat, pp. 1–14.

Warrington, E. (1997) *Administering Lilliput: the higher civil service in Malta, Barbados and Fiji*. Unpublished D.Phil thesis, Oxford, University of Oxford.

Watters, R. (1969) *Koro: economic development and social change in Fiji*. Oxford, Clarendon Press.

Watters, R. (1987) MIRAB societies and bureaucratic elites. In A. Hooper (ed.), *Class and culture in the South Pacific*, Suva, Fiji Islands, University of the South Pacific.

Watts, R.L. (1999) *Comparing federal systems*. Kingston ON, Institute of Intergovernmental Relations, Queen's University.

Watts, R.L. (2000) Islands in comparative constitutional perspective. In G. Baldacchino and D. Milne (eds.), *Lessons from the political economy of small islands: the resourcefulness of jurisdiction,* Basingstoke, Palgrave Macmillan, pp. 17–37.

Weale, D. (1991) Islandness. *Island Journal*, Rockland ME, Island Institute, 8, 81–82.

World Economic Forum (2017) *The global competitiveness report, 2016–2017*. Available from: http://reports. weforum.org/global-competitiveness-index/.

Wyss, J. (1812/1967) *The Swiss family Robinson*. London, Blackie.

9

ECONOMICS AND DEVELOPMENT

Geoff Bertram and Bernard Poirine

Introduction

Island economies, and especially small ones (population below one million), exhibit a remarkably wide range of economic structures built on a correspondingly wide range of development strategies, only a few of which fit conventional notions of 'economic development'. Common elements of 'islandness' may serve to define island economies as a general class, but there exist several distinct 'species' within that class, and a corresponding menu of strategic options open to islander communities in relation to the terms of their incorporation into the global economy.

The defining elements of small island economies are three: isolation, small size, and economic openness. Islands are physically accessible only by sea or air, which makes them rather more expensive to invade, occupy and integrate with neighbouring territories to form larger units. They also tend to develop close-knit communities which treasure their common identity and culture; this in turn underpins their exercise of a large degree of economic and political agency in the management of their local affairs, even when a particular island is nominally incorporated into a larger political unit (Baldacchino 2010). The survival of separate small-island jurisdictions reflects the political-economy consequences of being entirely surrounded by water.

From isolation and size follow scale and scope constraints on economic structure. Very small economic and political units, which on continents become submerged as provinces or local regions, take on a different character when bounded by sea. The network economies and strategic exposure to land transport which bind continental communities into large population units are truncated by the constraint of an aquatic boundary. Islands are the laboratory setting for the very small open economy as an ideal-type.

Isolation is related directly to physical distance from larger landmasses, which means that, where islands are clustered tightly in archipelagos close to a mainland, reducing isolation, they can more readily become absorbed into larger political units centred on the mainland and cease to be visible as separate economic units. Such is the situation of the 18,000 small islands within Indonesia, 6,000 of them inhabited. Of that country's 250 million population, the great majority are settled on four large islands (Borneo, Java, Sumatra, Sulawesi), leaving the remainder of the archipelago with an average population of around 5,000 per island. Similarly, there are over 7,000 islands in the Philippines, with most of the population on the two largest (Luzon, Mindanao) and the remainder averaging fewer than 5,000 per island. Because of their statistical invisibility, the small-island constituents of these two countries are excluded from the quantitative material in this chapter, though many of the qualitative observations can be applied to them.

Underpinning any study in political economy must be some consideration of the range of formal political institutions in small islands and the varying degrees of local agency in policy-making. From that starting point, this chapter focuses on the crucial role of the external balance of payments as the key economic constraint, reviews the process of specialisation into divergent economic 'species', and presents a taxonomy built around the available balance of payments statistics.

Size, jurisdiction, strategies and sustainability

Which islands to include?

Any study of small island economies confronts immediately the difficulty of securing meaningfully representative statistical data. Many islands lie at or beyond the outer limit of coverage for the major international statistical yearbooks and databases and, as already noted, the vast majority of the world's inhabited islands are statistically-invisible geographical units within larger countries.

The economies to be analysed here comprise geographical entities that satisfy four criteria:

- they are completely bounded by sea;
- they have populations that do not exceed one million;
- at least some statistical and other useful data on their economies is available;
- they have clear jurisdictional identities, whether as sovereign states or as well-identified and somewhat autonomous sub-national territorial units.

An initial list can be assembled from the *CIA World Factbook* (Central Intelligence Agency 2017). The *Factbook* is a modern almanac whose compilers select places and topics for inclusion on the basis of informal, subjective, strategically-driven criteria which transcend the constraints of conventional statistical reportage. Its mandate is to provide information on territorial locales of potential interest to the US military and intelligence community, and its selection criteria are unencumbered by category limitations such as human occupation, sovereign statehood, minimum size thresholds, availability of reliable data, or membership of international agencies. The inclusion of a large number of non-sovereign jurisdictional units makes the *Factbook* especially suitable as a starting point for assembling a sample of small-island economies.

In 2016 the *Factbook* listed data for 249 locales, of which just over one-third (89 entities) are bounded entirely by sea. (Enclaves which display some 'island' characteristics but which have a land border with a contiguous neighbouring territory – such as Sint Maarten/St Martin, Nunavut (Baffin Island), and East Timor are excluded.) Two of these locales – Australia and Antarctica – are continents rather than islands, and 14 others are uninhabited (or virtually so, such as Pitcairn) or occupied solely by meteorological stations or military bases. Nine more (Indonesia, Japan, Philippines, Taiwan, Singapore, UK, Sri Lanka, Madagascar, and Cuba) are large countries with populations over five million. Another seven (Bahrain, Cyprus, Jamaica, Mauritius, New Zealand, Puerto Rico and Trinidad and Tobago) are between one and five million. This leaves 57 island jurisdictions with fewer than one million inhabitants for which the *Factbook* provides some description of their economies. This list includes all of the 26 small islands that are sovereign states and are full (voting) members in the United Nations General Assembly. The other 31 entities are small, sub-national island jurisdictions (SNIJs) with varying degrees of autonomy from their metropole.

Missing from the *CIA Factbook*'s coverage are some small islands that meet the four criteria listed above. Obvious cases are four departments or collectivities of France – Réunion, Martinique, Guadeloupe, St Barthélemy – and the Portuguese autonomous regions of Madeira and the Azores.

Other potential candidates for inclusion are to be found in the list of sub-national small islands and enclaves in Baldacchino (2010, Appendix, pp. 207–214) and in the online database of his 'Jurisdiction project' (SNIJ database 2017). Applying the four criteria above as a filter on the 132 sub-national entities listed in those two sources has enabled the addition of a further 13 small SNIJs. The final result, set out in Appendix 9.1, is a set of 26 sovereign small island states and 48 SNIJs.

Income and political status

'Sovereign state' is a reasonably clearly-defined political status, equated for our purposes with membership of the United Nations General Assembly. Constitutionally, the sovereign small islands range from constitutional monarchy in Tonga, through various forms of republican government, to the occasional military dictatorship (for example, periodically in Fiji). Non-sovereign, 'sub-national' jurisdictions, however, span a wide range of institutional arrangements, from the near-complete autonomy of freely-associated self-governing states such as the Cook Islands, to the politically integrated status of islands such as the Shetlands and Orkneys, Réunion (a department of France) or Aruba (one of the four 'countries' of the Kingdom of the Netherlands).

Figure 9.1, adapted from Kerr (2005, p. 504, Figure 1) brings together several key elements for the analysis of the political economy of small islands. At the right-hand end of the spectrum lie the territories with nominally the least political autonomy. At the left-hand end are fully independent nation states. The key dividing line between sovereign and non-sovereign entities is shown as membership or non-membership of the United Nations, but in the centre of the picture the group of 'states with limited independence and territories with state-like autonomy' share many characteristics that span the sharp dividing line.

In terms of formal ability to exercise domestic agency in policy-making, there is a steady progression from right to left in terms of increasing autonomy. Two external sets of forces operate however: one to offset this tendency across the spectrum and one to reinforce it for the sub-nationals. Globalisation reduces the economic policy space for all (but most obviously for sovereign nation states); while localisation (the widespread trend towards devolution of authority from the metropole to its peripheral SNJs) has its strongest effects as the sovereignty boundary is approached from the right (Baldacchino 2010, Baldacchino and Milne 2009, Baldacchino and Hepburn 2012).

While 'autonomy' is certainly a key dimension of the distinction between sovereign and non-sovereign economies, it is important not to assume that sovereignty is correlated with economic prosperity. One of the key stylised facts of small-island economics is that non-sovereign island economies exhibit higher incomes, and better scores on other aspects of human development, than sovereign small-island states (Bertram 1999, 2004).

Figure 9.2 sets out the per-capita income of our sample of 74 small island economies as at 2015, using data as close to that year as possible at the time of writing. Wherever available, we use per capita Gross National Income (GNI); but, in most cases, the available data is limited to per capita Gross Domestic Product (GDP). As in Figure 9.1, the 74 economies are separated into sovereign independent nation states on the left and sub-national entities associated more or less closely with metropolitan states on the right. Within each group, a distinction is drawn between islands that are inside the core of the global economy, and those that are not. Among

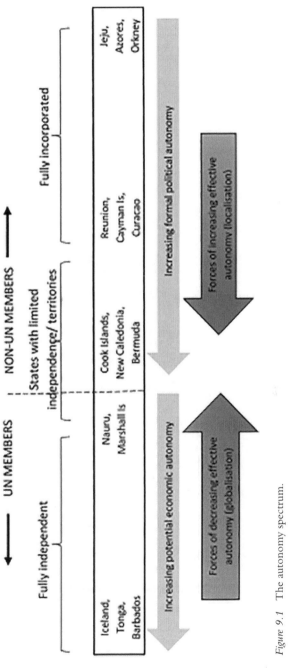

Figure 9.1 The autonomy spectrum.

Source: © Geoff Bertram.

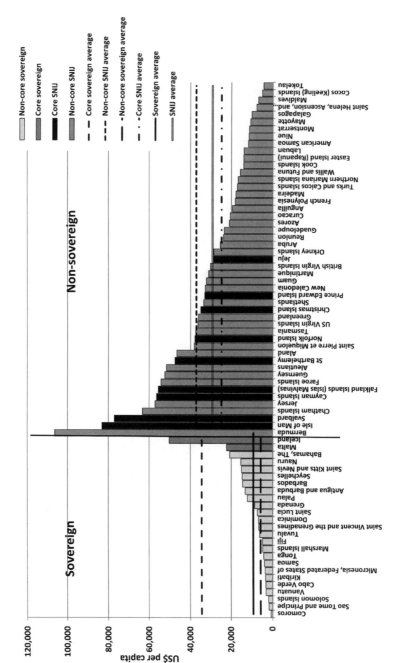

Figure 9.2 Per capita GNI or GDP for 74 small island economies.

Source: Appendix 9.1.

the sovereigns, core status is assigned to two that are actual or potential members of the European Union (EU): Malta and Iceland. For non-sovereign islands, core status is assigned on the basis of being located next to the metropole and tightly integrated into its economy. On both the sovereign and non-sovereign sides of the chart, higher incomes are associated with core status; but, both within the core and outside it, the non-sovereign economies have higher incomes than the sovereigns.

On each side of the chart, horizontal lines show the population-weighted average per capita income for, respectively, core economies, non-core economies and the full set. There is clearly much closer convergence between sovereign and non-sovereign islands within the core of the global economy than on the periphery. It is the big difference between sovereign and non-sovereign economies on the global periphery that gives rise to the substantial income differential across the whole sample.

The range of incomes within each set of island economies is, however, very wide. Amongst sovereign islands on the periphery, the ratio of per capita income of highest (Bahamas) to lowest (Comoros) is 26:1. Amongst peripheral non-sovereigns the ratio is 24:1 (Bermuda relative to Tokelau). (Note that in both cases the high extremes hail from the Caribbean.) With such widely-dispersed values (meaning large standard deviations in the data), average figures have to be treated with caution. Still, at least one clear conclusion can be drawn from Figure 9.2: national independence, whatever its non-economic attractions, is not an automatic recipe for greater prosperity.

Why sub-national islands tend to have higher incomes than sovereign ones remains an intriguing research question (McElroy and Albuquerque 1995, McElroy and Parry 2012, McElroy and Pearce 2006, Feyrer and Sacerdote 2009, Bertram 2004, 2015, 2016). It might be suggested that size makes the difference (that is, that sub-national islands are smaller than sovereign ones), and Figure 9.3 gives some visual support to this, though again with too much variation to allow a statistically significant generalisation.

Bertram (2015) assembled data back to 1900 on per capita imports, as a proxy for income, for 51 small islands, and concluded that the separation of small island economies into better-off present-day non-sovereigns, versus less well-off present-day sovereigns, dates back before their assignment of political status occurred (that is, prior to 1950, after which formal decolonisation got under way). This result contradicted previous work by Bertram and others in which it was argued that the unequal development status of independent versus sub-national small islands was *caused by*, as distinct from merely associated with, different political status. If we assume that economic development was the cause of present political status rather than its consequence, there are two competing theories of how unequal development affected sovereignty in the long run: Demand pulled or Supply pushed sovereignty.

The first "demand side" theory says that the driving factor behind independence was the demand for sovereignty from the island people: since it is easier to contemplate independence when the island is more developed, and its economy can stand on its own feet without foreign assistance, rich islands would have become independent first, leaving the less developed islands no other choice than political dependence to alleviate poverty. If that was the case, the remaining non sovereign islands should be poorer, not richer as we observe now.

The opposite "supply side" theory says that the driving factor was the supply of sovereignty by the colonial power: independence was willingly supplied to poor islands with few economic prospects. In that case, colonial powers would have let go of the poorest islands first (since they cost too much in terms of economic assistance), and would have tried to keep the richest islands as long as possible (since they were an economic asset, rather than a burden, to the metropolitan state). In that case again, the level of development causes the political status, but with the

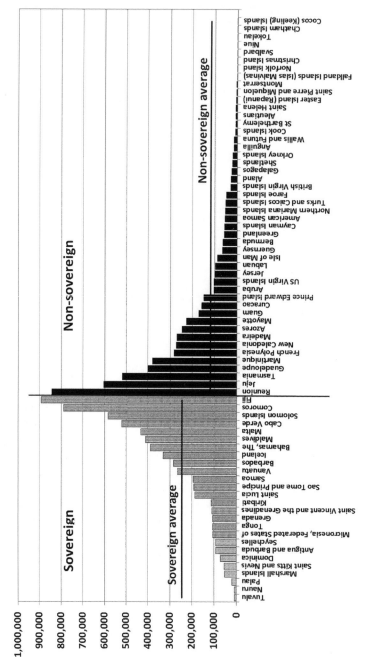

Figure 9.3 24 sovereign and non-sovereign small islands ranked by population at 2016.

Source: Appendix 9.1.

opposite result (that we find statistically) that the remaining non sovereign islands end up being richer than the sovereign ones.

The hypothesis that independence led to slower growth – which should have shown up as increasing divergence in 1950–2008 – was not supported; the data showed parallel paths of per capita imports (and hence implicitly per capita income) for independent and non-sovereign small islands over the six to seven decades since decolonisation. Further research beckons, particularly at regional level.

Size, specialisation and openness

Economies with fewer than one million inhabitants (including, incidentally, landlocked enclaves such as the European microstates of Andorra, Liechtenstein, Monaco and the Vatican, as well as small islands) tend to exhibit extreme specialisation into one or two globally-linked leading sectors which, once selected, determine the character of the economy as a whole. The selection process obeys not so much the orthodox theory of comparative advantage (in which an economy responds in passive fashion to exogenous relative-price signals in a competitive market) as a strategic game-theoretic process of self-selected hyper-specialisation, which Baldacchino and Bertram (2009) label 'speciation' to reflect the way in which the structure of the entire economy is adapted to achieve opportunistic colonisation of one or more niches of opportunity in the global system, on the basis of some absolute (as distinct from comparative) advantage.

'Speciation' refers to the sort of specialisation in which a community takes advantage of a niche of evolutionary opportunity by adopting a particular economic 'personality' with its own distinctive set of institutions, policy imperatives, and mutual understandings amongst the participating population. Economic speciation involves a conscious or quasi-conscious collective decision to embrace the economic phenomenon of crowding-out, with 'Dutch Disease' (the process by which one dominant export sector squeezes other sectors producing tradeable goods and services, by driving up the real exchange rate) treated as an evolutionary opportunity rather than a threat (Matsen and Torvik 2005).

The smaller and more isolated the economy, the greater the need to be open to the world market, and to specialise in a narrow set of income-generating activities in that market. Openness and hyper-specialisation follow from the absence of economies of scale, from high transport costs reducing the scope of trade opportunities, from the lack of a varied pool of mineral resources to draw upon, from the lack of 'agglomeration externalities' associated with the geographical proximity of clients and providers (too many empty cells in the input–output matrix), and from the fact that island residents, as sophisticated consumers, want to choose from a wide variety of goods not made locally.

'Vulnerability' is a red herring

The constraints of small size and geographical separateness are sometimes presumed to render islands particularly economically 'vulnerable' (Briguglio 1995, Streeten 1993, Guillaumont 2010) but this normative categorisation is both conceptually and empirically unhelpful. Conceptually, there are advantages as well as disadvantages of smallness and isolation. Empirically, on balance, island economies appear quite robust in a globalising world. Briguglio's 'vulnerability index' is positively, not negatively, related to per capita income: the more supposedly 'vulnerable' the economy, the higher is its per capita income (Armstrong and Read 2002a). Proponents of the vulnerability hypothesis have implicitly conceded the point by introducing a countervailing concept of 'resilience', placed in a contradictory dialectical relationship to vulnerability to produce indeterminacy of outcomes (Briguglio et al. 2005). While it is true that the more specialised an island economy is, the more 'vulnerable' it is (because of exposure to wide swings in external receipts), the other side of this coin is that the gains from trade, and the realisation of scale economies in the specialised export activity, more than compensate.

The vulnerability-versus-resilience paradigm is also flawed in its underlying assumption that vulnerability is exogenously imposed whereas resilience is endogenously created as a response. A review of the components of the two indices reveals that exogenous and endogenous elements are found on both sides of the supposed dialectic. The image of vulnerability may be instrumentally useful in the rhetoric of political lobbying and aid justification, but lacks solid roots in economic reality.

While the vulnerability/resilience dichotomy gives no analytical leverage, the concepts of speciation and strategic flexibility go to the heart of the economic and geopolitical dynamics of island development.

Most islands – especially those with well-established links to metropolitan patron economies – enjoy external opportunities which are specific to the particular facts of each island's history as well as to the identity of its patron state (Bertram 2004, Bertram and Karagedikli 2004). In the era of decolonisation in the late 20th century, for example, the UK took a fundamentally different approach to the citizens of its island territories than did France and Portugal; the USA was different again (Hintjens and Newitt 1992). Modern island economic structures are path-dependent (Hampton and Christensen 2002, pp. 1668–1669) – outcomes of specific historical paths, not necessarily able to be imitated or reproduced by others, and commonly representing the end product of a cumulative series of collective strategic choices made by (or imposed on) the home community as a whole. This renders problematic any unidimensional conception of what 'economic development' means in an island context (Bertram 1986, Baldacchino 1993).

In treating the economic structure of small islands as a matter of strategic behavioural adaptation within the constraints of smallness, isolation and history, rather than of passive competitive response to exogenously-set world market prices, we are implicitly rejecting the idea that there is a simple linear relationship between country size and market power in the global arena. Conventional international economics distinguishes between 'large' countries which carry sufficient weight in global markets to operate as price-makers, and 'small' economies which are price-takers. But the tendency for market power to fall with population size does not extend down to the smallest size categories. At the very small end of the size spectrum, the strategic behaviour that is intrinsic to speciation creates and reproduces its own form of market power: what Baldacchino (2010, p. xxxi) calls "the power of powerlessness". Small islands can 'get away with' economic policies that would not be accepted from larger players in the global arena, because large countries see the possible adverse consequences for themselves as very small. Some examples:

- very high import duties (e.g. French Polynesia and New Caledonia);
- offshore financial services (Channel Islands, Cayman Islands, Bermuda, British Virgin Islands) (see Suss et al. 2002, Christensen and Hampton 2000, Shaxson 2011);
- securing duty free access to export to large markets such as the USA (Northern Marianas) or the EU (Mauritius);
- securing free or preferential immigration rights to selected countries because of past colonial ties (Comoros, Samoa, Cook Islands);
- providing Cyprus (EU) passports to Russian investors in exchange for a € 2 million investment in Cyprus, which amounts to selling to Russian millionaires the right of free circulation in all EU member states and easier access to other western countries.

This is not to say that all attempts at speciation are successful: consider, for example, Vanuatu's attempt in the 1990s to become an offshore financial centre. Nor do all successful mutations prove durable: the Northern Marianas' export manufacturing enclave which was highlighted in Bertram and Poirine (2007 pp. 335–336) collapsed spectacularly in 2006–2009.

External resources: the key to material prosperity

The more external resources that can be secured to fund imports, the higher the per capita income that can be sustained, because the small island economy's import capacity is the key determinant of its sustainable material standard of living; see Figure 9.4. The central strategic economic problem for a small island is not to choose between an outward-oriented development strategy and some inward-directed alternative. It is how to secure external resources to sustain imports. "In a small economy, the constraint imposed on growth by the external sector is a continuing phenomenon" (Demas 1965, p. 48; Hein 1988, p. 35). A small island economy cannot generate self-sustaining growth from its own internal market, because it usually lacks the raw materials and energy sources to develop a competitive industry, and because the small domestic market rules out economies of scale for local industry. In addition, the cost of imported inputs for local industries is increased by high shipping costs, lack of competition among freight service providers, and in many cases high import duties imposed to fund government budgets. The sustainability of a small open economy boils down to being able to finance its import requirements.

There are (at least) two ways of interpreting the tight relationship between external resources and per capita income shown in Figure 9.4. One approach is the Keynesian multiplier model, with external funds treated as the key injection of purchasing power. Each dollar of external funding flows into the local economy as additional demand for goods and services, which has

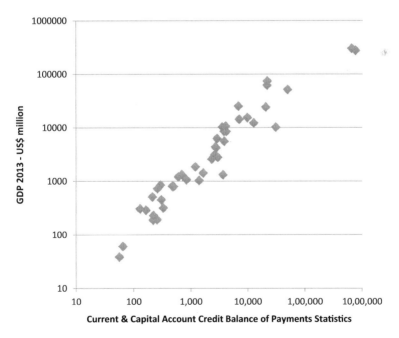

Figure 9.4 Relationship between external resources and per capita income in small island economies: GDP and current and capital account credit (2013) for 45 islands (US$ million, logarithmic scales).

a multiplier effect on local output. As domestic income and output rise, some fraction of the additional income flows back out into demand for imports. That fraction, the 'marginal propensity to import', determines how far the domestic multiplier effect can run before all of any new injection of external funding has been spent on imports. The lower the import propensity, the greater is the amount of domestic output and income that can be sustained by a given amount of external funding.

The other way of thinking about Figure 9.4 is in terms of a balance of payments constraint model, in which domestic demand presses always up against the limited supply of imports (set equal to the available external funding), with all possible import substitution pushed to its economic limit.

Whichever way the issue is framed, the prosperity of a small island population depends ultimately on its ability to secure external funding to pay for imports, and the sustainability of that prosperity depends on the long-term sustainability of the external source of funds. External resources – drawn from what Baldacchino (2006a) has labelled "the hinterland beyond" – are the economic base upon which small islands must build the growth and sustainability of their gross domestic product. To secure those resources without resorting to offshore borrowing, each small island economy must identify and occupy some niche or niches of opportunity in its 'external hinterland'. As will be outlined later in this chapter, a taxonomy of small-island 'species' can be assembled on the basis of their sources of funding for imports.

Comparative advantage submerged by trading costs

It is tempting, and common in the economic development literature, to equate external funding with merchandise export earnings, but in fact very few small islands in the modern world make their way as successful exporters in the traditional sense of producing goods to be transported abroad and sold in external markets. This point is demonstrated clearly by Figure 9.5, which assembles averaged 2010–2015 data for 53 small island economies to show the percentage of their imports of goods and services that is covered (funded) by the export of goods. Of the 53 economies, only five (three non-sovereign and two sovereign) fund more than 50 per cent of imports from this source. Three others have coverage ratios barely over the 40 per cent threshold, marked in the chart by a dashed line, which we use as a rough benchmark to identify export-led economies. Fully half of the sample in Figure 9.5 has coverage ratios lower than 10 per cent.

Iceland, American Samoa and the Faroes have large fisheries exports; Solomon Islands, Fiji and the Marshall Islands export a range of primary commodities (timber, palm oil, fish, copra and cocoa from Solomon Islands; sugar, mineral water, gold and garments from Fiji; fish, copra and coconut oil from the Marshall Islands). The US Virgin Islands' high coverage ratio in Figure 9.5 is dominated by the St Croix oil refinery, which closed in 2012, after which that economy's export coverage ratio dropped to just 21 per cent by 2015. Similarly Aruba's exports have fallen steeply since closure of its refinery in 2009, which brought its coverage ratio down to just 16 per cent by 2015 as its oil exports were reduced to bunkerage. Thus, while the 2010–2015 averages in Figure 9.5 make these two look like export economies, this no longer applies as of 2015.

Bertram and Poirine (2007, Figure 9.4) identified 21 merchandise exporters among their sample of island economies, but more than half of those were economies with more than one million population. Only ten of them qualified to be included in the small-island list compiled for the present chapter and, of these, the Northern Marianas and Malta have dropped out of merchandise exports (Northern Marianas has switched to tourism, while Malta now exports services rather than goods). No major new export economy has emerged in the past decade among the 74 included in Appendix 9.1.

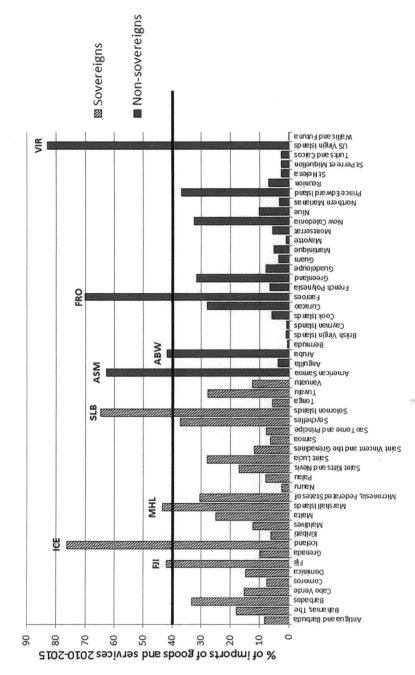

Figure 9.5 Coverage ratio' of merchandise exports relative to imports of goods and services.

Source: Appendix 9.1.

Winters and Martins (2004a, 2004b) and Winters (2005) have shown how the high transport costs associated with isolation, distance and small size (hence small shipments in a global freight transport system characterised by economies of scale) effectively restrict the potential for small islands to exploit the Ricardian comparative advantage that is taught to first-year students of economics. "Comparative advantage does not matter . . . if you do not trade internationally or if you cannot survive (literally) when you do" (Winters and Martins 2004a, pp. 350–351). Instead, small island economies rely on finding sources of absolute advantage that enable them to secure rents of one sort or another in the global economy.

Winters and Martins (2004a) drew two main conclusions from their analysis:

- First, "smallness ... does not usually introduce marginal distortions that need to be acted upon, but an overall constraint on feasibility...". [I]f unviable economies are to be made viable, an additional source of [external] income must be found" (Winters and Martins 2004a, p.376).
- Second, an isolated small island faced with high trade costs "could remain trading [i.e. importing] in two ways. It could receive a non-trading flow of foreign exchange – e.g. from accumulated assets, remittances or aid – which permitted some imports in the absence of exports. Alternatively or additionally, it could receive prices for its exports at preferential prices, which permit exports despite a fundamental un-competitiveness. Both these cases amount to living on rents" (Winters and Martins 2004a, p. 352).

The general point to be drawn from the Winters–Martins model is that any on-island economic activity whose earnings are squeezed by transport and other trading costs, and/or that uses inputs that incur such costs, will generate lower value-added than an identical activity carried on under competitive conditions in a metropolitan economy. The table below, adapted from Poirine (2007, p. 14), compares a hypothetical brewery in France with one in Tahiti that exports to France, assuming freight costs are 20 per cent of the value of exports and imports. All values are in French Pacific francs, roughly equivalent to US cents:

The export sales revenue received by the Tahitian producer has 20 francs (cents) deducted for the costs of getting the product to the French market. In addition, its costs of imported inputs and depreciation (to fund replacement of plant and equipment with imported machinery) are 20 per cent higher than those of the French producer. Value added in Tahiti is then 40 francs (cents) compared with 70 for the French brewery. If the Tahiti brewery pays its workers the same as French workers (Column 2 of Table 9.1) it makes a loss of 10. If it collects a

Table 9.1 Costs of isolation squeeze value added, cutting the return to at least one factor of production.

		(1)	*(2)*	*(3)*
		Brewery in France	*Brewery in Tahiti assuming wage parity*	*Brewery in Tahiti assuming profit parity*
.A	Price at factory gate	120	100	100
B=C+D	Total cost of production	100	110	80
A-B	Profit per unit	20	−10	20
C	Intermediate inputs and depreciation	50	60	60
D	Labour costs	50	50	20
E=A-C	Value added	70	40	40

margin big enough to make its return on capital equal to that of the French brewery (Column 3 of Table 9.1), it must pay its workers 20 francs (cents) rather than 50. The point here is that value-added determines the total income available to be shared among the factors of production – here, capital and labour – and both of these cannot be paid the same as their equivalents in France. The costs of isolation must be borne by one or both of them. (If the rate of return on capital is equalised across countries, then for any non-land-using industry it is the wage rate that must be lower the greater the effect of isolation.)

Of course, in the local Tahitian market, the local brewery has the advantage of being able to charge over 140 per bottle and still compete against beer imported from France. Yet, the small size of the local market rules out economies of scale (which means costs per unit will actually be above those of the French operator), and even selling at the higher price with unchanged costs, the island producer's value added will still be less than that of the French brewery. The higher the trading costs, the greater is the squeeze on value-added in the production of any traded good in a small island location under competitive conditions. Less value added per worker means in the end a lower GDP per capita, for a given ratio of workers to total population. This means that higher transport costs tend to depress GDP per capita as well as the wage rate, *ceteris paribus*.

Absolute advantage and rents

It follows from the preceding analysis that, in order to prosper, a small island economy must overcome the disadvantages of isolation and small size by taking advantage of some special asset that raises value added in its external-resource-earning sector(s) above that attainable by competing suppliers in metropolitan economies. In other words, it must have some source of absolute advantage in the form of a rent-yielding asset. In some cases, the asset may be part of the island's natural endowment; examples familiar from the small-island literature are fisheries, high-value mineral deposits, and desirable tourist destinations:

- A few small island economies such as Iceland and the Aleutians operate their own fishing fleets and fish-processing plants. More often, islands such as Tuvalu, Tokelau and Kiribati simply collect rental income from foreign fishing fleets operating in their exclusive economic zones.
- Since the exhaustion of Pacific island phosphate, mineral deposits are no longer of any economic significance to small island economies, with the sole exception of nickel in New Caledonia. In future, however, oil may become a leading sector for São Tomé and Príncipe.
- Tourism (and its various specialised niches) has been a booming industry in a large number of (mainly tropical) small islands in recent decades, enabling these island economies to collect rents for their climate, beaches, cultures, biodiversity and landscape.

Geographical location can also be a kind of natural resource. Poirine (1999) argues that because of their geographical position, small islands can have an absolute advantage in the production of national defence for the benefit of a large country. Such a supply of strategic services can then be treated as an export, which is of no utility to the islanders but of great utility to the patron country. Missile or nuclear testing facilities and military bases on politically-aligned small islands save the need to maintain expensive aircraft carriers in the region. Small islands similarly gain from exporting other services such as financial services, shipping lines, internet services, and fisheries access to exclusive economic zones. All these can thrive on the basis of absolute advantage arising partly from geographical distance from the rest of the world.

To generalise this in terms of standard trade theory, Figure 9.6 shows a small island economy's 'production possibilities' as the curved line GG' showing combinations of two goods, X and Y, that can be produced with the available resources. We combine all the goods that are of value to the local population – food, clothing, shelter and so on – into a composite good Y and show this up the vertical axis. The potential export good X is plotted along the horizontal axis. Good X could be phosphate, bêche-de-mer, or black pearls (all unvalued locally but prized elsewhere in the world); or the experience of being on the island (free to locals but valued highly by foreign tourists), or the provision of naval port or communication facilities (valued highly by geopolitical patron states). The zero value placed by the local community on the X good is shown by the horizontal lines ('indifference curve') I_1, I_2, and I_3. An indifference curve shows the various combinations of the two goods that leave the consumer (here, the small-island community) equally happy (well off). A horizontal indifference curve means that having more of good X makes no difference to the local community's welfare. More of good Y, in contrast, shifts the community to a higher level of happiness, such as that shown by I_2. In the absence of international trade, the self-sufficient economy would produce OG of good Y and none of good X, achieving the welfare level I_1. The higher welfare contours I_2 and I_3 show that the islanders would be better off if more of the Y good were available.

Suppose that the small island is Nauru, that good X is phosphate, and that good Y is food. Nauruans can grow food or dig phosphate, but they have no use for phosphate. The slope of the straight line PQ_1P shows the rate of exchange between units of food and units of phosphate on the world market. If transport costs are zero, and local resources are reallocated to produce the combination of phosphate (X) and food (Y) shown by point Q_1, then by exporting phosphate in exchange for food the Nauruans can consume the quantity of food OP, greater than the OG that they could get without trade. Their welfare is therefore increased from I_1 to I_3.

If transport costs are high, however, the trade line PQ_1P' is no longer valid (Winters 2005). Less food can be imported for any given amount of phosphate exports, because the price of imported food is augmented by the cost of transport to Nauru while the returns from phosphate sales at world prices are reduced by the need to cover transport costs. Trading possibilities are now the kinked dashed line BQ_2B', and the highest attainable welfare (shown by I_2) is now

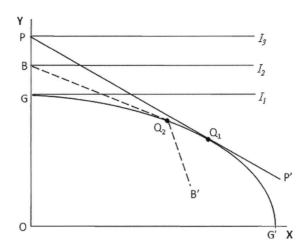

Figure 9.6   Economics of phosphates, nuclear test sites, tourism.

reached by producing at Q_2 and trading phosphate for food to achieve food consumption of OB: still a gain from trade, but much reduced by transport costs.

This tendency for transport costs to reduce the gains from trade, and possibly to negate them altogether, is a large part of the explanation for the small number of successful commodity exporters in Figure 9.5. Trade opportunities become fewer the higher are trading costs (reflecting the distance from trade partners); and per-unit trading costs tend to be higher for smaller traded volumes. This has been documented by an abundance of econometric literature on the (trade) 'gravity equation': trade between any two countries has been repeatedly shown to increase with their respective economic mass (GDP) and to decrease with the distance separating them. Both size and isolation thus work against small island exports.

The gains from trade will be even less if the potential export good is one that the island population themselves value (that is, if the indifference curves in Figure 9.6 were to slope down rather than being horizontal). The discussion above explains why in most small islands only a very few export items make up the bulk of the value of total exports, and why those exports are often things of little value to the islanders, such as phosphate, nickel, copra, pearls, or strategic services. It should be noted that distance is also a handicap for tourism exports. The gravity equation applied to tourism on a sample of 211 countries has shown that when the distance between any country pair doubles, the flow of tourists between them decreases by about two-thirds, *ceteris paribus* (Dropsy et al. 2015).

For small islands whose limited natural resource endowments rule out those sorts of export options, absolute advantage can often be created by exercising, through the agencies of local government (or of transnational firms located in the island): what Baldacchino and Milne (2000) have labelled "the resourcefulness of jurisdiction". Examples are tax havens, offshore banking, company registries and ship registries, even philatelic sales and high level internet domain names. In one or two cases, absolute advantage accrues from the historical location on a small island of some private company that possesses special attributes, such as the international shipping operation based in Åland; but these cases are rare.

Finally, three rent-yielding assets of a different kind must be borne in mind: entitlements to international aid, migrant diasporas that send back remittances to the home economy, and sovereign wealth funds (such as Kiribati's Revenue Equalisation Reserve Fund) which return a stream of dividends. Aid and the returns on sovereign funds mostly flow into the government accounts, expanding the public sector of the economy. Migrant remittances, in contrast, flow into the small-scale household sector and translate into increased private sector consumption and investment.

Income distribution and the lure of industrial development with no comparative advantage

How exactly are the costs of isolation, and the benefits of rent flows, distributed amongst land, labour and capital? This will depend partly on the precise source of rent, and partly on how open are the markets for factors of production (land, labour and capital). Land, obviously, is not footloose and cannot be moved to a more profitable location; hence the rental return to land is the most vulnerable to the Winters–Martins squeeze on value added (which is why primary commodity exports are viable only where high-value natural resources can be appropriated at low or no cost to producers, and/or where the product has very high scarcity value in export markets; commercial fisheries in Iceland and the Aleutian Islands, nickel in New Caledonia).

Turning to labour, the crucial issue is labour mobility: that is, whether workers in the small-island economy have the option of migrating to work overseas. Only in a closed labour market can the wage be pushed down below the international opportunity value of labour – what workers can earn as migrants overseas. International mobility of labour varies greatly across the small-island world, ranging from cases where islanders have shared citizenship and hence free access to the metropolis (for example Cook Islands, Overseas France, Niue, American Samoa, US Virgin Islands, the Azores) to islands where migration outlets are mostly closed. Kiribati is an extreme example of the latter, with only limited access to offshore employment opportunities in, for example, international shipping (Borovnik 2006), and with tight quota restrictions on migration to New Zealand and Australia.

If labour is often footloose, capital is always more so. Hence, depressing the rate of return on capital (as in the second column of the brewery example of Table 9.1) is an incentive for investors to take their money elsewhere. The combination of high trading costs and relatively incompressible wages has spelt doom for numerous attempts to secure 'economic development' for small islands by promoting a modern industrial capitalist sector to imitate those of advanced industrial economies. Even 'rich' islands have very few industries and a large service sector.

Modelling the small-island labour market

In a closed labour market in long-run equilibrium, the real wage will tend to adjust endogenously to bring the demand for labour into line with the supply. This means that wages could be depressed far below the levels prevailing in the wider world (see, for example, the outcome in the right-hand column of Table 9.1). In contrast, in a fully open labour market, the level of wages available in the wider world and accessible to migrants provides an exogenous benchmark, which puts a floor of sorts under the real wage rate in the island economy. In addition, in small islands where the government sector depends to a large extent on external funding, the terms of that funding constitute another exogenous element in the local labour-market equilibrium.

Figure 9.7 shows the economic structure of a small island labour market open to migration. The vertical axes on both sides of the diagram show the real wage rate in local currency terms. The wage available to labour in the outside world (the value of the option of emigrating) is shown by the horizontal line at the wage rate $w^{\star}$ which is equal to the wage rate in the migrant

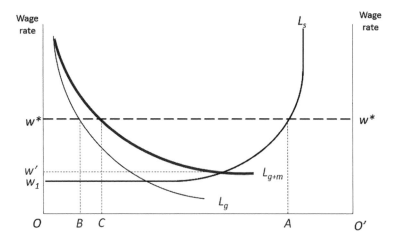

Figure 9.7   Economic model of an open small-island labour market.

Source: Bertram (1986, p. 816).

destination, discounted for the costs and risks of moving out of local employment into the diaspora. The total working population of the islander community (including both island-resident workers and migrants working overseas) is shown by the length of the horizontal axis OO', and this total labour force is allocated across four sectors on the basis of sectoral labour demand.

The line L_s is the demand for labour in subsistence/village activity, drawn with respect to the right-hand origin of the diagram at O'. Because this curve represents the willingness of the village sector to release labour as the wage rate rises, it can be read as a supply curve of village-sourced labour, when read with respect to the left-hand origin at O. Along the horizontal segment at its left-hand end, L_s represents a 'traditional subsistence' minimum income w_1, while at its right-hand end it turns vertical, reflecting the minimum labour input required to sustain the viability of the village economy. Subsidies to 'traditional' life, along with tourist demand for the cultural and heritage values embodied in the village, shift the L_s curve to the left, increasing the proportion of the population sustained in the village sector.

The curve L_g is the government sector's demand for labour. Its position is determined by the size and structure of the government budget (including aid funding); increases in the government budget shift the curve to the right, drawing in a greater proportion of the available labour. In Figure 9.7, the curve is drawn assuming that the government has an exogenously fixed total budget to spend on wages and salaries. (The curve is therefore a rectangular hyperbola, which means that all points on the curve represent the same total spending on wages and salaries in the public sector.)

Finally, the curve L_{g+m} shows the total modern-sector demand for labour, constructed by adding the capitalist private sector's labour demand L_m to that of the government sector.

For a first analysis, assume that the real wage rate in the small island economy is equalised across sectors so that all local wage rates are equalised with the external opportunity wage w^*, and that the labour market is fully open so that all labour not utilised in the village sector, the government, or the capitalist private sector will migrate to work overseas. Under these assumptions, in long run equilibrium the public sector employs OB of the labour force, the capitalist private sector employs BC, the village economy holds $O'A$ of labour, and the migrant diaspora accounts for the remainder, CA.

If one or more of the above assumptions do not hold, the model is easily adjusted to capture particular real-world situations. In the French overseas island departments and collectivities such as French Polynesia and Réunion, for example, the government sector's pay rates are set (and funded) from Paris at levels well above anything available in other sectors. This produces the labour-market structure shown in Figure 9.8. Here the public sector pays a wage rate w_g dictated exogenously (by the metropolitan government in Paris) and employs an exogenously determined number of workers OB. The government demand for labour L_g is therefore a horizontal straight line with length equal to OB. Again the total labour force OO' is allocated across four employing sectors, but the existence of a high public-sector wage holds out an attractive alternative to out-migration so that instead of a diaspora, the economy exhibits unemployment: a pool of workers queuing up in the hope of securing high-paid local employment (see Harris and Todaro 1970 for original analysis). If the private capitalist sector pays a wage of w_2 established (say) by some sort of bargaining, then unemployment is AC (the distance between L_s and L_{g+m} at that wage rate). If the local private sector pays the lower wage w_0 then open unemployment is smaller and some out-migration would be likely.

As a construction drawn from neo-classical economics, the basic diagram abstracts from real-world details (especially in relation to the behavioural reactions of the village sector and the assumed uniform level of skill and aptitudes across the population) but it does capture underlying forces at work in all small island economies, and is readily adapted to different real-world circumstances.

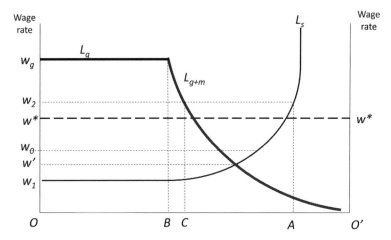

Figure 9.8 Small-island labour market with a high-wage government sector.

Source: Poirine (1993).

Dropping the assumption of an open labour market, for example, involves simply removing the $w\star w\star$ line from Figures 9.8 and 9.9, leaving the island labour market to settle at the lower closed economy equilibrium wage of w' with expanded government and village sectors plus a low-wage private sector – essentially the Kiribati situation.

The model has clear policy implications. Increasing official aid or financial transfers shifts the L_{g+m} curve to the right, drawing labour out of either the diaspora, or unemployment, or both. Increased remittances (and/or NGO-funded aid flowing into the village sector) shift the L_s curve to the left, reducing out-migration and/or unemployment while expanding the village sector, leaving government and private sectors unaffected (to a first approximation). And so on: different outcomes can be modelled by changing the shapes and positions of the curves.

The quest for rents

Aid as a form of trade

Figure 9.9 shows that sovereign islands receive more aid per capita when their populations are smaller, and that non-sovereign islands receive more aid per capita than sovereign island states, whatever their population. Poirine (1999) points out that small islands have an absolute advantage in the export of 'strategic and diplomatic services' to large countries. Many islands are strategic because they can be logistic stepping stones in case of conflict: think of Hawai'i, Guam and most of the Pacific islands during the Second World War; Eniwetok, Bikini, Mururoa and Fangataufa where long-range missiles and nuclear testing took place for a long time; Guam again during the Vietnam war. Others, such as the Falklands in the 1980s, become strategic emblems in their own right.

Islands can make up for the lack of aircraft carriers (China, for example, is building artificial islands to control the South China Sea), and giving aid to a small island population in exchange for the privilege of using a military base saves the very high cost of maintaining an aircraft carrier.

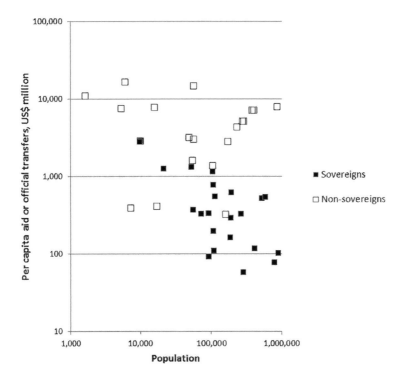

Figure 9.9 Aid and official transfers per capita in 43 small island economies.

Source: Appendix 9.1.

Similarly, islands can serve as secure prisons. Nauru, for example, operates a large detention camp for would-be migrants and asylum seekers who are refused entry to Australia, and is able to negotiate a rent in exchange for such strategic services.

Because the value of a strategic service has nothing to do with the number of people living on the island, it follows that, the less populated the island, the more it will receive per capita in exchange for the rent of strategic services from the large country, and hence the higher the probability that the small island government will accept the deal.

Since sovereign countries may in theory repeal such treaties allowing exclusive military use of their territory and airspace (as happened in the Philippines for Olongapo and Clark US bases) while non-sovereign territories may not do so, it is safer and more valuable for the patron country to put funding into its own island territories rather than sovereign islands. In some cases, however, the required strategic asset has been simply seized by the dominant power at the expense of the local population rather than for their benefit; consider the case of the British Indian Ocean Territory, which hosts the Diego Garcia base but whose indigenous population has been relocated, against its will, to Mauritius (Vine 2011).

The postcolonial transition

In the colonial era prior to 1950, small island economies generally fitted into a standard model driven by the needs of the colonial power. Their primary function was to produce commodity exports, especially of tropical products, as inputs for the industries of the metropolitan economy. A secondary role was geopolitical: to fly the flag of the colonial power in distant corners of the world.

For reasons discussed above, commodity exports from an isolated location with a small resource base usually could not sustain the material living standards of small-island populations at levels aspired to in the post-war world. In the decolonisation era, small island economies shifted towards new rent-yielding leading sectors; and as the field of external opportunities shifted over time, they had to follow those shifts and exploit new niches of opportunity in the global economy. As Baldacchino (2011 p. 236) describes it:

> a "strategic flexibility" approach ... explain[s] how actors practise intersectoral migra-tion: cleverly shifting focus, interest and scope, not just out of necessity (reactively) but in 'smelling' promising opportunities (proactively). In a scenario where change is taken as a given, managing and coping with such change become the hallmarks of economic survival: just like surfers handling the ocean swell.

New economic structures were not always or solely the work of domestic change agents, of course. In several cases, outside agencies drove the processes of speciation and adaptation. The emergence of the Cayman Islands as an international financial centre after 1965, for exam-ple, was driven mainly by the quest of US financiers and businesses for a secure offshore tax haven, encouraged by the Bank of England and the City of London (Shaxson 2011 pp. 90–96, 132–136, 211–215). On a grander scale, as Gay (2012) has described, the ongoing willingness of the French state to provide massive financial transfers to France's overseas departments and collectivities has been the key driver in holding up salaries and wages, government spending, and real exchange rates in those island economies, narrowing the opportunities for sectors such as tourism to compete in world markets (Poirine 2011 Chapters 7 and 8).

Endowment and development strategies: specialisation and speciation

A key requirement for sustainability in a situation of hyper-specialisation is flexibility and rapid response capability. Retention of the ability to mutate, to undertake a rapid shift to a different 'species' in response to shifts in external opportunities, remains a crucial reserve asset in the small island's portfolio of social capital. The greater this evolutionary flexibility, the more extreme can speciation become without endangering the long-run survival chances of the home economy. Cases do emerge from time to time of small islands caught in development culs-de-sac, one example is the struggling economy of São Tomé e Príncipe, which has been unable to carry through its expected switch from cocoa exporter to oil producer due to slow progress in its joint venture with Nigeria to develop the Gulf of Guinea oilfields. Other examples refer to situations where attractive transition opportunities are blocked by externally-imposed constraints: cases in point include the intervention since 1999 of the OECD's Financial Action Task Force to restrict the emergence of unregulated offshore financial centres (Hampton and Christensen 2002, Financial Action Task Force 2005a, 2005b); and the blockage of all except one of the emigration pathways out of Kiribati in the quarter-century following independence in 1979, which for a long time left seafarers as the only outwardly-mobile group in the labour force (Borovnik 2006).

Strategic flexibility in action

A central component of the social capital of islander communities is therefore their flexibility and adjustment capacity. In practice the 'sustainability' of island economies has very little to do with self-sufficiency or environmental protection, with which it is often equated. The basic sustainability requirement is the social capital – people (including diasporas), institutions, and collective willingness to adapt (Baldacchino 2005) – that underpins effective collective response to strategic opportunities, and adaptability in the global arena.

Three examples of transition from one strategic niche to another follow below: offshore finance in the Cayman Islands; the Northern Marianas' transitions from military base to tourism, then to garment manufacturer, and then back to tourist economy; and the Cook Islands' transition from MIRAB to SITE.

Caymans: from MIRAB to offshore bank

The Cayman Islands are today one of the world's major financial centres and tax havens. In 2008, there were more than 93,000 companies registered there, including almost 300 banks, 800 insurers, and 10,000 mutual funds. Foreign assets of over US$4.1 trillion held in the Cayman Islands are 1,500 times the GDP of around $3 billion. 60 per cent of global hedge fund assets are held there (Fichtner 2016, p. 1035, Hampton and Christensen 2002, p. 1659). There are no direct taxes in the Cayman Islands: no income tax, company or corporation tax, inheritance tax, capital gains or gift tax. There are no property taxes or rates, and no controls on the foreign ownership of property and land. The government charges stamp duty of 6 per cent on the value of real estate at sale, with reduced rates available for Caymanians. There is a 1–1.5 per cent per cent fee payable on mortgages. The key revenue source is an annual licensing fee paid to the government by companies registered in the jurisdiction. The government's total revenue runs at over US$800 million per year and there is generally a budget surplus (Cayman Islands Government 2016).

According to legend, the Cayman Islands's tax-free status originated as an 18th-century royal grant in reward for rescuing a shipwrecked member of the British Royal Family (Markoff 2009, Part 1; Brittain-Catlin 2005, p. 14). That was converted into a market niche by three events during the decolonisation period: separation of the islands from the colony of Jamaica (then moving to independence) in 1959; subsequent entrenchment of Crown Colony status in the 1960s; and the expiry of pre-existing tax treaties with the USA in the late 1960s, which opened the Cayman Islands up as a tax haven for US corporations ranging from medical insurers to Enron.

In the mid-1960s the islands had only a single bank, no telephones, and a population of 8,000 (Brittain-Catlin 2005, p. 7). For the preceding half-century, a MIRAB (Bertram and Watters 1985) structure had prevailed, with cash incomes sustained by remittances from seafarers. In 1937, half of the working-age male population was employed in international shipping, and the main post-war employer until the 1960s was a supertanker operator (Brittain-Catlin 2005, p. 17).

The transition from migrant-remittance economy to offshore financial powerhouse took only about a decade (Roberts 1995, Markoff 2009). The accountancy, legal and business skills required to negotiate financial deals and fine-tune Cayman law to the needs of finance capital were acquired or hired, infrastructure investment completed, local legislation passed, and an international reputation for confidentiality and security built up at breakneck pace, even though the new strategic direction was one which had been unforeseen ten years earlier. Dislocation there certainly was, but the Cayman Islands successfully made the transition from one of the poorest to one of the three richest Caribbean island economies (along with Bermuda and the British Virgin Islands).

The main industries are financial services, tourism, and real estate sales and development. In total, services were estimated to account for 93 per cent of GDP in 2016 (CIA 2017). Tourism expenditure (largely by finance-centre customers) of US$500–600 million annually is a mainstay of the balance of payments. Aid flows and migrant remittances are zero, merchandise exports account for not more than US$47 million p.a., and the islands' annual import bill of around US$600 million is almost entirely funded by the offshore finance sector and its tourism appendage.

The Cayman Islands have been among the most diplomatically successful offshore financial centres in confronting and adjusting to the OECD's drive to clamp down on rogue tax-haven and money laundering jurisdictions, and also to the increasing regulatory activities of the US and UK governments with regard to tax havens. In 2000, the Caymans secured early exemption from the Financial Action Task Force's list of 'non-cooperating jurisdictions' (Hampton and Christensen 2002, p. 1670 note 9), by June 2001 they were fully 'delisted', and from June 2002 they were no longer subject to FATF monitoring (Financial Action Task Force 2005, p. 31).

Since 2005, the Cayman Islands have had a fully informative tax information exchange arrangement under the European Union Savings Directive (EUSD) with all EU member states. In 1990, Cayman entered into a transparent all crimes Mutual Legal Assistance Treaty with the USA and, in 2001, a comprehensive US Tax Information Exchange Agreement. In 2013 the OECD found the Cayman Islands to be 'largely compliant' with international standards of tax information exchange (Fichtner 2016, p. 1038).

Northern Marianas

The Northern Marianas are a Commonwealth territory of the USA. In pre-1945 Japanese colonial times, the islands were an agricultural export economy. Following the Second World War, they became a US military base, then from the 1970s a tourist destination for Japanese holidaymakers. Tourism peaked in 1996 and thereafter declined as Japanese recession and the Asian economic crisis took hold. From 727,000 in 1997, visitor arrivals were down to 425,000 by 2002. But as tourism fell, manufactured exports rose.

The US Office of Insular Affairs (2006) recorded that:

> Garments produced or substantially transformed in the CNMI enter into the United States customs territory free of quotas and duties. Under the Covenant, imports into the U.S. from the CNMI receive the same treatment as imports from Guam; however, the CNMI was able to develop a garment assembly industry because it is not subject to U.S. immigration laws, as is Guam. Garment shipments to the United States increased from under US$200 million in 1990 to over US$1 billion in each of 1998 and 1999.

Up to the mid-2000s, the competitiveness of light manufactured exports from the Marianas rested upon the availability of a low-wage immigrant workforce, recruited mainly from the Philippines and China, and concentrated on the main island of Saipan, where the locally-born population was quickly outnumbered by migrant workers. The jurisdictional niche that enabled the CNMI to become a manufactured exporter was highly specific and a product of the negotiations leading to commonwealth status: the treaty-based absence of visa requirements for migrant workers to enter from Asia, combined with duty-free onward access to the US market for manufactured goods (but not migrant workers). Migrant remittances flowed out from the Marianas towards the source countries in East Asia.

Figure 9.10 shows the subsequent events. In 2005 the USA relaxed its quota restrictions on Chinese imports, and major US retailers switched to the new cheaper source of supply. In four years, manufactured exports from the Northern Marianas dropped 75 per cent. Manufacturing employment which had peaked at around 17,000 in 2000, fell to 700 by 2010 (Central Statistics

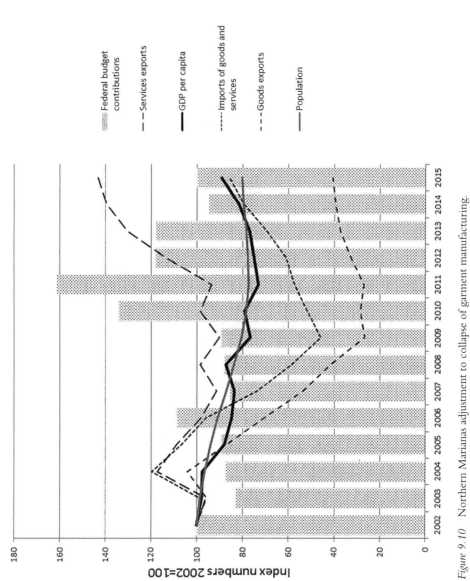

Figure 9.10 Northern Marianas adjustment to collapse of garment manufacturing.

Sources: Exports, imports and GDP from Furlong and Ludlow (2016) and from BEA annual national accounts at www.bea.gov/
national/gdp_territory.htm. Population from World Development Indicators. Federal budget contributions from annual government
financial statements at www.opacnmi.com/sec.asp?secID=4.

Division 2016, p. 58). The manufacturing sector dropped from 34 per cent of GDP in 2003 to 2 per cent by 2010 and 1 per cent by 2013 (Bureau of Economic Analysis 2012, Table 2.2) while the total manufacturing wage bill dropped from $74 million in 2007 to just $8 million by 2010 and $7 million in 2013 before stabilising (BEA 2016 p. 12 Table 2.6). Per capita GDP fell 20 per cent for a brief period, but then recovered fast from 2012 on, driven by three processes: outward migration of the now-redundant temporary manufacturing labour force, a new surge of tourism, and a four-year boost in federal transfers to support the government budget.

As many of the migrant labour force returned home, the population of the Northern Marianas dropped from 69,000 in 2001 to 53,000 in 2011. There was then a sharp upturn in tourism from Korea and China, making up for the earlier drop in tourists from Japan. The resulting surge of services exports meant that imports and per capita income quickly recovered from the collapse of manufactured exports, though as of 2016 they had not quite returned to 2002 levels.

Cook Islands

In the mid-1980s, the Cook Islands was one of the originally-identified MIRAB economies, with imports of US$19 million funded by remittances and aid of US$10 million p.a., commodity exports of US$3.6 million and philatelic and tourism earnings of US$6 million (Bertram 1986: 815 Table 9.4, converted at US$0.70=NZ$1). The Cook Islands diaspora in New Zealand had grown from under 1,000 in 1951 to 14,000 by 1981, approaching parity with the home-resident population. By 1996, there were 47,000 Cook Islanders resident in New Zealand, compared with about 20,000 home residents. Following a financial crisis in the mid-1990s, aid from New Zealand dropped by about half between 1995 and 2002 while, at the same time, tourism earnings roughly doubled. Meantime although around one-third of the resident population emigrated after the crisis, remittances stagnated, then fell away.

The long-run transitions of the Cook Islands from colonial export economy in 1892–1945, to MIRAB in the mid-1980s, then to massive borrowing and financial adventures in the early 1990s, and finally to complete domination by tourism by the 2010s, is traced in Figure 9.11. By 2014, tourism earnings had reached US$175 million compared with imports of goods and services of US$164 million, aid of US$22 million, and remittances of (probably) around US$1 million. The Cook Islands had made a full transition out of MIRAB and to SITE status.

Development of small-island taxonomy

MIRABs, SITEs and PROFITs

Over the past three decades, a widely-used taxonomy of small island economies has emerged, built around a three-way classification according to whether a particular island funds its import needs primarily by

(1) securing financial flows of primary and secondary income (aid, remittances, dividends and interest) which represent (implicitly at least) the return on external assets of some kind;
(2) becoming a tourist economy on the basis of its natural resource endowments of landscape, culture and climate;
(3) colonising niches of opportunity in the global economy in which it can exercise some absolute advantage over competitors, often on the basis of jurisdictional or institutional features.

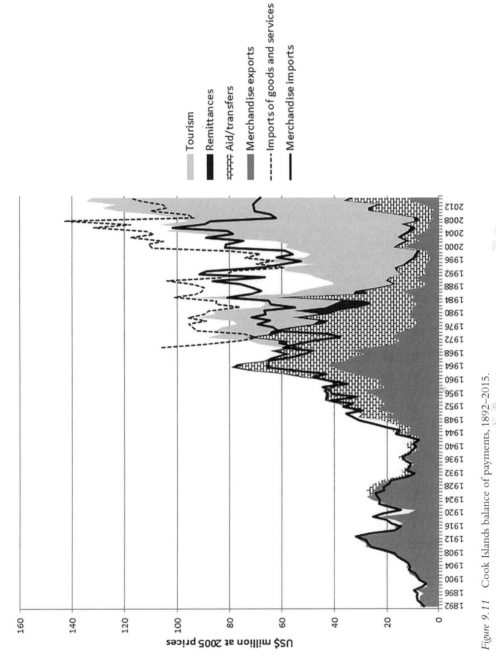

Figure 9.11 Cook Islands balance of payments, 1892–2015.

Sources: UN, WTO, New Zealand Government, OECD official sources, plus unofficial remittance estimates.

The first of these strategic options was described by the MIRAB model, developed in the mid-1980s (Bertram and Watters 1985, 1986) to describe economies funded by migrant remittances (MIR) or by aid which was used to fund local bureaucracies (AB). As pointed out by Poirine (1994) and Tisdell (2016), the MIRAB acronym conflated what were really two quite separate funding processes: the migration-remittance process driven by the decentralised behaviour of myriad individuals completely outside the ambit of government, and the aid-bureaucracy process which is predominantly a public-sector phenomenon. The developmental role of the state is quite different in the two cases: sidelined in a pure migration-remittance economy, but central in an aid-bureaucracy one – epitomised by the case of French Polynesia in the nuclear testing era which was described as 'ARAB' (atomic-rent-aid-bureaucracy) in Poirine (1994, p. 1998).

The second strategic option is encapsulated by the SITE (small island tourism economies) model of McElroy (2006, Oberst and McElroy 2007), and the third by the PROFIT model of Baldacchino (2006a). In Bertram (2006) and Baldacchino and Bertram (2009), a large number of small islands were classified under the MIRAB-SITE-PROFIT schema. The underlying purpose was to demonstrate how few were the cases of recent economic success based on traditional commodity exporting, while in the process categorising alternatives to the export-economy model.

A more complex and nuanced classification was used in Bertram and Poirine (2007 p. 363, Figure 9.11) to group 68 island economies into nine clusters in roughly increasing order of economic success: primary exporters with aid or remittance support, MIRABs, tourism plus exports, geostrategic aid (which can be viewed as a trade specialisation in export of geostrategic services – see Poirine 1999), moderate-impact tourism, geostrategic rent with exports, high-value exports, high-impact tourism, and offshore finance plus tourism. The next section updates that earlier work.

Classifying our 74 small island economies

To see how individual small island economies finance their import requirements, and hence to lay down a quantitative basis for classifying them into species, we use the concept of 'coverage ratios' already seen in Figure 9.5. The issue is how to cover the funding of an economy's total imports of goods and services – in other words, everything that has to be bought from external suppliers.

For the purposes of our analysis, we write the balance of payments identity in the following form:

$$M \equiv X + T + R + G + Z + F + B + O$$

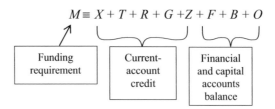

| Funding requirement | Current-account credit | Financial and capital accounts balance |

where M is total imports of goods and services
   X is merchandise exports valued FOB
   T is total spending in the economy by overseas tourists
   R is private remittances
   G is government transfers, comprising both official aid and budgetary support grants

Z is a residual balancing item showing all other flows of funds within the current account, including any debt servicing and dividend flows

F is foreign direct investment net inflows

B is net offshore borrowing

O is other net funding on capital and financial account.

By using the import funding requirement as our basic unit of account ('numeraire'), we can express all the other elements in the balance of payments as percentages of the required import funding – that is, as 'coverage ratios' relative to the total funding need. The sum of the items X, T, R, G and Z minus total imports M will be equal to the current account balance (with Z calculated to produce this result). The sum of the items F, B and O is the negative of the current account balance (with O calculated to ensure this).

The aim of this exercise is to break down the external resources being accessed by each small island economy into a few key categories, in so far as it is possible to locate the necessary statistical information.

Appendix 9.1 shows, for the 53 of our 74 small islands with sufficient data, a statistical break-down of the balance of payments in the period 2010–2015 (or for those years within this period for which data was available) into five sources of external funding to pay for imports of goods and services: merchandise exports, tourism, remittances, aid, and a residual (which is equal to Z in the equation above whenever a figure for the current account balance was available, or else $(Z+F+B+O)$ where no overall current account balance was available).

Data on imports of goods and services, merchandise exports, current account. and some other balance of payments items, come primarily from the World Bank's *World Development Indicators*, the WTO database of merchandise trade statistics (WTO 2017), and the United Nations national accounts database (UN 2017), supplemented by balance of payments statistics produced by the Institut d'Emissions d'Outre-mer for New Caledonia and French Polynesia, the Centrale Bank van Curaçao en Sint Maarten for Curaçao, the East Caribbean Central Bank for Anguilla and Montserrat, the Cook Islands Statistics Office, the Faroes Statbank, and the Aland Statistical Yearbook. For economies that did not have full current account balances available, less complete data on current funding of imports was obtained, especially from the web publications of CEROM, IEDOM and FEDOM for Overseas France (FEDOM 2016), from US Bureau of Economic Analysis national accounts data on US territories, Eurostat NUTS2 and NUTS3 data on EU-linked island regions, the Statistical Yearbooks of American Samoa, Northern Marianas, and Guam, and the statistics offices of Azores and Madeira.

Figures for tourist expenditure and aid flows are not separately identified in the IMF's standard balance-of-payments statistics. For Appendix 9.1, tourist data come from the UN database and from a range of country-specific sources. Aid and government transfers data come mainly from the UN *Statistical Yearbook* (UN 2016), supplemented by the OECD DAC database, island-specific data from various sources, and the audited government accounts of the US territories. In the case of the French overseas departments and collectivities, the aid estimates are the sum of official transfers and net salaries paid from abroad, as recorded in the Institut d'Emission d'Outre-Mer balance-of-payments statistics for French Polynesia and New Caledonia (IEDOM 2016) and in Cour des Comptes (2013) and FEDOM (2016).

The first column of Appendix 9.1 shows per capita total imports of goods and services, and the subsequent six columns show the sources of the funding to sustain those imports. The final columns show population, income per head, life expectancy, and a crude welfare index to be discussed below. Numbers are annual averages for the period 2010–2015 wherever possible, or else the nearest available equivalent.

Appendix 9.1 also shows population counts taken mostly from the US Census Bureau (2016) and the World Bank (2016). Per capita income is drawn mostly from the *World Development Indicators* with income measured as gross national income where available, and from the UN national accounts database. Life expectancy data is drawn from the above sources, supplemented by data from IndexMundi. For island economies not covered by those sources, country-specific data has been used as available, including information from the *CIA World Factbook* and *Wikipedia*.

Figure 9.5 earlier in this chapter showed the results of this statistical exercise with regard to commodity exports. Figures 9.12 to 9.16 show the coverage ratios for: tourism (Figure 9.12), remittances (Figure 9.13), government transfers and aid (Figure 9.14), remittances and aid combined (Figure 9.15) and the residual (Figure 9.16): either the residual Z in the case of economies for which current account balances were available, or $(Z+F+B+O)$ for the rest. In each chart, a threshold is set at 40 per cent of imports of goods and services covered by the respective source of external funding, and economies that equal or exceed this threshold are identified.

Taking tourism first, Figure 9.12 identifies 15 small-island candidates for inclusion in the SITE category: eight of them sovereigns and seven non-sovereigns. Four of these economies – Cook Islands, Maldives, Northern Marianas and Turks and Caicos – have tourism coverage ratios of 80 per cent or more of their import-funding requirements. Another ten of the 50 small islands have tourism coverage ratios of over 20 per cent.

Comparing Figure 9.12 with the corresponding chart in Bertram and Poirine (2007 p.346, Figure 9.5), 11 of the 12 small islands identified as SITEs in that study reappear ten years on, with the addition of another four: Barbados, Dominica, Vanuatu and Aruba. Only the Cayman Islands has reduced its reliance on tourism below 40 per cent, apparently reflecting a stronger relative role for its financial services sector.

Turning to remittances, Figure 9.13 shows that there are only a limited number of purely remittance-led small island economies, if the World Bank's database for 2010–2015 is correct. Remittances are difficult to track accurately and are incompletely recorded in official statistics. In their detailed study of Tuvalu, Boland and Dollery (2005, pp. 32–33) estimated that counting remittances sent in the form of 'non-commercial imports' would add 5–10 per cent to recorded money transfers (probably more, given under-reporting of valuations), in addition to which substantial transfers in cash go unrecorded in official statistics. Consequently Figure 9.13 is almost certain to understate the true importance of remittances; but, even so, they now clearly dominate only a few small-island economies.

Among the non-sovereigns, the only one with remittances covering over 40 per cent of imports of goods and services is, strangely enough, Bermuda, where the so-called 'remittance' flow is associated with high-income employees rather than the more commonly noted blue-collar migrants who provide the large remittance flows seen in the Comoros, Tonga and Samoa. Apart from these four economies, remittances play only a minor role in import funding. Quite a number of small islands show negative remittances, due to repatriation of the earnings of migrant workers. (Mayotte, for example, is the source of around half of the remittances received by the Comoros.) Several other small island economies have unexplained (unclassifiable) outflows of funds (see Figure 9.16) which are probably outward remittances of surplus cash.

Matters are very different when we turn to the other component of the MIRAB model, official transfers of aid and budgetary support, shown in Figure 9.14. Eleven non-sovereign small islands and four sovereigns derive more than 40 per cent of their import funding from this source, and for seven of the 15 the ratio of transfers to imports is greater than 80 per cent. Overseas France accounts for seven of the 14: French Polynesia, Guadeloupe, Martinique,

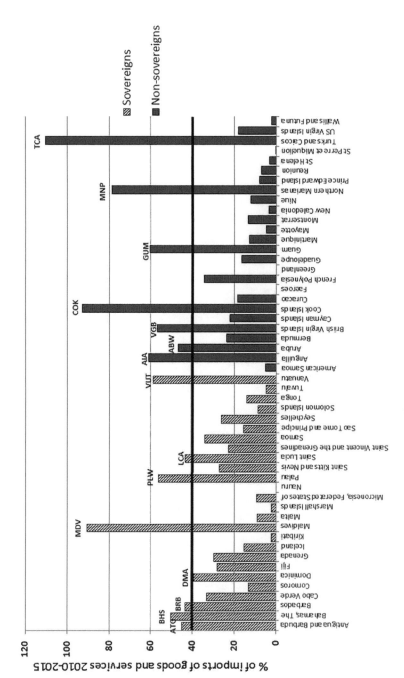

Figure 9.12 Tourism spending relative to imports of goods and services: identifying 'SITES'.

Source: Appendix 9.1.

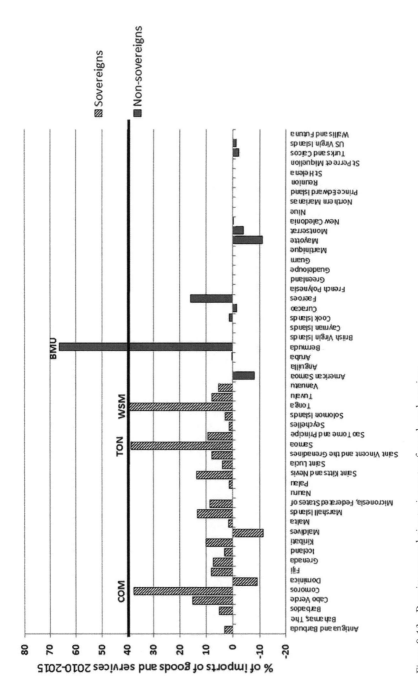

Figure 9.13 Remittances relative to imports of goods and services.

Source: Appendix 9.1.

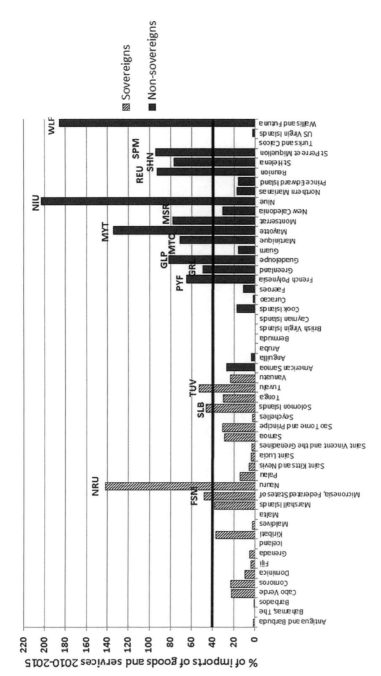

Figure 9.14 Government transfers/aid relative to imports of goods and services.

Source: Appendix 9.1.

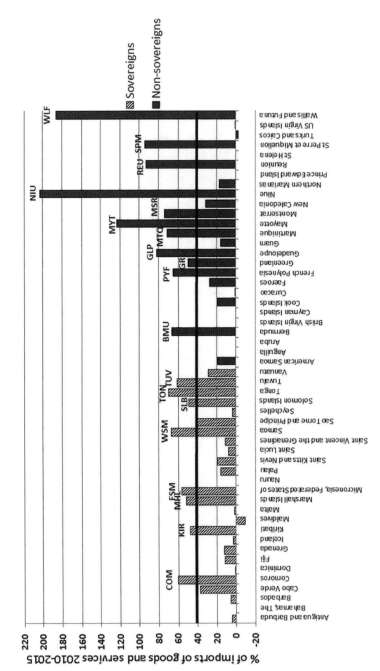

Figure 9.15 Remittances plus official transfers relative to imports of goods and services.

Source: Appendix 9.1.

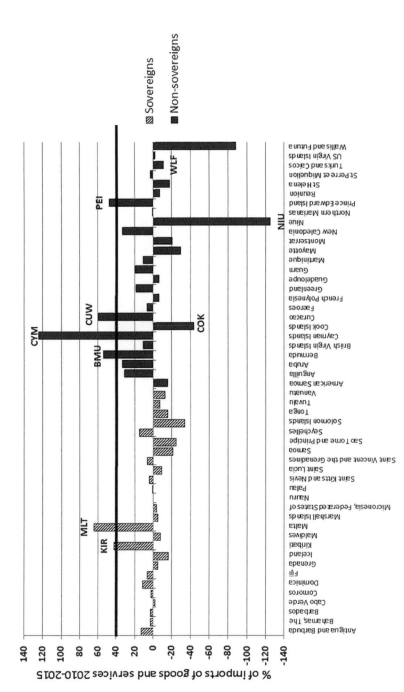

Figure 9.16 Residuals as % of imports of goods and services.

Source: Appendix 9.1.

Mayotte, Réunion, St Pierre et Miquelon, and Wallis et Futuna. Niue, associated with New Zealand under a constitutional guarantee of budgetary funding, heads the bunch with a ratio of around 200 per cent (more than half of which, as Figure 9.16 shows, flows back out again as unrecorded flows). Nauru has largely ceased to be a phosphate export economy – although the recovery of secondary reserves has temporarily revived Nauru's phosphate exports in recent years, as reflected in Figure 9.5 – and now operates an offshore detention centre for asylum seekers and other migrants who have been refused entry by Australia, and is financially rewarded for doing so. Tuvalu, Micronesia, the Solomon Islands and Montserrat make up the rest of the transfer-dependent economies, with Marshall Islands and Kiribati very close to the 40 per cent threshold.

Combining remittances and official transfers produces the picture in Figure 9.15 where 19 of the 50 economies rely on these two rent categories for more than 40 per cent of their import funding; and, for 14 of these, the ratio is over 60 per cent.

Finally, in Figure 9.16, we turn to the residual balancing item, which has to be interpreted on a case-by-case basis. In the case of Kiribati, the residual corresponds to 'income from abroad' in the form of dividends on the country's sovereign wealth fund, the Revenue Equalisation Reserve Fund (established back when the economy was a phosphate exporter prior to 1979); this shows up in the column 'net income from abroad' in Appendix 9.1. For Malta and Prince Edward Island, the residual represents substantial exports of services. For Bermuda, the Caymans and Curaçao, their large positive residuals represent the returns on their offshore finance operations. For Aland, returns from international shipping are the dominant component of the residual. In Niue and Wallis et Futuna, as already noted, the large negative residuals are the overflow from excess coverage by official transfers.

Obviously, the scope of the statistical analysis has been limited by data constraints, with the result that for 21 of our 74 small islands, we have only qualitative descriptive information about their leading sectors. Figure 9.17 classifies the full set of economies, updating and expanding the similar chart in Bertram and Poirine (2007 p.363, Figure 9.12). Ten generalised economic strategies are ranked in roughly ascending order in terms of the level of material welfare they support, as measured by an index of income and life expectancy using the old Human Development Index methodology for those two components. (The third component of the HDI, education, could not be included due to data limitations.)

The strategies are mix-and-match assemblages of six key sectoral foci: simple subsistence, exports, tourism, remittances, aid-official transfers, and financial services. As would be expected, financial services and associated tourism show a clear lead, but high-value exports perform well in small-island appendages of core metropolitan economies or of the EU. MIRAB and subsistence economies are at the lower end of the welfare spectrum.

Conclusion

This chapter has shown that, when they are both small and isolated, islands face a dilemma when choosing a development strategy: with their small domestic market, import substitution is not an option, and competitiveness can be gained only by opening up to trade, that is, by exporting goods or services to the global market in order to gain economies of scale. However, because they are small and isolated, the gains from trade are eaten up by transport costs and the lack of agglomeration economies – which are benefits that accrue when firms and people bunch together, as in cities and industrial clusters – reducing value added per worker compared to less remote countries.

Therefore, development strategies have to cope with what we call the *small island paradox*: islands have to open up (if they are not to stay very poor in autarky), but the gains from trade are harder to come by as the distance from the rest of the world increases. Strategies to get around this dilemma include the search for rents (geostrategic or geopolitical rents generating public transfers, private remittances from the diaspora, fishing rights, and/or very high end tourism) or the export of goods of very high value per unit weight, such as pearls or stamps, or the export of intangible services, such as financial services (and concomitant risks).

Flexible specialisation is the key attribute of island economies, and is more easily achieved the smaller the population involved and the greater the degree of cultural and social cohesion within that population. The actors in and of small island economies are best thought of as entrepreneurs, actively engaged in seeking out external opportunity and deploying their scarce resources to maximise rents and quasi-rents from the exploitation of any market niches they can find and develop (Baldacchino 2015, Connell 2013). The common tendency of observers to treat small island economies as though they are marginal dependents in the world economy is not only demeaning to islanders but profoundly misleading as the basis for economic theorising about their development potential.

A strategic, game-theoretic conceptualisation brings into focus the active role played by island actors in securing their economic place in the world. The statistical record shows a low incidence of poverty; genuinely destitute small-island economies are few, and those which do exist are mostly searching out dynamic escape paths. Fiscal management is generally solid, and democratic institutions are more secure and widely encountered than in continental comparators. Health status and literacy – two indicators of human welfare not analysed in this chapter but implicit in Figure 9.17 – are generally good, endowing islander migrants with a well-grounded start towards employment and success in large host economies.

Resilience and adaptability are long-established traits nurtured by the conditions of island life. Applying these to the economic problem of securing material and non-material welfare in the 21st-century global economy will require new challenges to be overcome – challenges that are likely to prove tougher than those of the 20th century. Global markets for financial services no longer offer the easy pickings that boosted Bermuda and the Cayman Islands to prosperity. Migration and remittance flows face increasing political resistance as the number of migrants and refugees has swelled. Keeping abreast of rapidly-evolving digital technologies in order to hold their own as exporters of services requires small islands to undertake costly investments in physical infrastructure to improve their connectivity and data speeds. Mass tourism has placed new stresses on fragile small-island ecosystems. And for many low-lying small islands, the threat of climate change and rising sea levels overshadows their economic futures.

Not all the emerging trends of the 21st century are negative, however. Global tourism demand continues to grow, especially from the middle classes of newly-prosperous economies such as China and India, bringing with it increased incentives for aid and capital flows to improve infrastructure in small-island destinations. Geopolitical tensions and rivalries are again on the rise, in a multipolar world in which small-island members of the UN General Assembly wield voting power out of all proportion to their population size, and in which the politics and economics of climate change must eventually move to the top of the policy agenda. After all, the smallest 11 Pacific island nations have one UN General Assembly seat (and vote) per 210,000 people. The USA has one vote for 320 million people. India has one vote for 1.1 billion. China has one vote for 1.3 billion.

Baldacchino and Milne's (2000) "resourcefulness of jurisdiction" still has plenty of life in it.

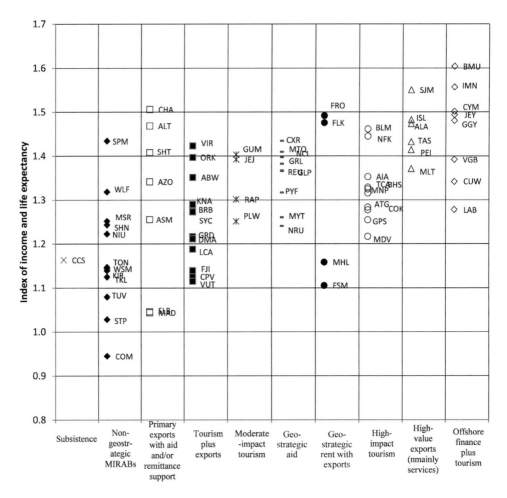

Figure 9.17 Welfare levels and economic strategies for 74 small island economies.

Source: Appendix 9.1.

Appendix 9.1

BASIC DATA FOR 74 SMALL ISLAND ECONOMIES

Island	Abbreviation	Political status	Imports of goods and services per capita 2010–2015 US$	% of imports of goods and services (2010–2015)										2015 population	2015 per capita income US$	Life expectancy	Index of income and life expectancy	Strategic classification
				Merchandise exports	Tourism spending	Remittances	Aid/government transfers	Remittances + aid	Residual	Net income from abroad	Current account balance	Net FDI inflow	Other funding of current account					
Åland	ALA	SNJ												28,916	46,549	82.50	1.473	High-value exports (shipping services) plus some tourism
Aleutians	ALT	SNJ												9,043	51,476	78.70	1.468	Primary exports with federal financial support
American Samoa	ASM	SNJ	11,176	62.8	5.2	−8.0	27.2	19.1	11.0	0.0	−1.8	0.0	0.0	55,538	11,809	74.40	1.255	Primary exports with federal financial support
Anguilla	AIA	SNJ	11,480	4.0	60.9	0.0	3.6	3.6	3.2	0.0	−28.3	13.5	14.8	16,752	19,474	81.40	1.353	High impact tourism with capital inflow
Antigua and Barbuda	ATG	SOV	7,679	8.3	45.1	3.0	1.2	4.2	12.6	−5.6	−29.8	17.2	12.6	91,818	13,270	75.94	1.277	High impact tourism with capital inflow
Aruba	ABW	SNJ	29,868	42.0	46.8	−2.0	0.0	−2.0	12.9	0.0	−0.3	4.3	−3.9	103,889	25,354	75.45	1.351	Tourism plus exports
Azores*	AZO	SNJ	786	70.1			105.3							246,746	21,030	77.8	1.341	Primary exports with large EU regional development funding
Bahamas, The	BHS	SOV	12,003	17.9	50.8	−3.0	0.0	−3.0	5.6	−7.0	−28.8	4.2	24.6	388,019	20,740	75.23	1.325	High impact tourism with capital inflow
Barbados	BRB	SOV	8,083	34.9	46.1	3.3	0.7	4.0	1.9	−8.7	−13.1	11.7	1.4	284,215	14,510	75.50	1.285	Tourism with exports
Bermuda	BMU	SNJ	29,204	0.7	23.8	66.4	0.0	66.4	54.0	75.4	44.9	−1.0	−43.9	65,235	106,140	80.80	1.604	Offshore finance plus tourism
British Virgin Islands	VGB	SNJ	23,218	3.5	59.8	0.0	0.0	0.0	59.0		22.3	−22.3	0.0	30,117	30,144	78.60	1.392	Offshore finance plus tourism

(Continued)

Appendix 9.1 (Continued)

| Island | Abbreviation | Political status | Imports of goods and services per capita 2010–2015 US$ | % of imports of goods and services (2010–2015) | | | | | | Net income from abroad | Current account balance | Net FDI inflow | Other funding of current account | 2015 population | 2015 per capita income US$ | Life expectancy | Index of income and life expectancy | Strategic classification |
				Merchandise exports	Tourism spending	Remittances	Aid/ government transfers	Remittances + aid	Residual									
Cabo Verde	CPV	SOV	2,250	15.4	37.2	15.1	22.2	37.3	-5.8	-6.2	-16.1	8.7	7.4	520,502	3,280	73.15	1.126	Tourism with exports, remittances and aid
Cayman Islands	CYM	SNJ	37,637	1.0	22.1	0.0	0.0	0.0	45.7		-31.2			57,268	56,282	82.30	1.501	Offshore finance plus tourism
Chatham Islands	CHA	SNJ												600	63,264	79.60	1.506	Primary exports with aid and/or remittance support
Christmas Island	CXR	SNJ												2,072	34,964	82.80	1.434	Geostrategic aid
Cocos (Keeling) Islands	CCR	SNJ												596	5,125	69.00	1.164	Subsistence
Comoros	COM	SOV	394	3.4	12.5	36.0	21.9	58.0	11.0	0.1	-15.1	3.1	12.0	788,474	798	63.26	0.945	Non-geostrategic MIRAB
Cook Islands	COK	SNJ	15,520	5.9	94.1	17.5	17.5	35.0	18.1	4.5	53.1	0.0	0.0	9,838	14,119	75.80	1.283	High impact tourism
Curaçao	CUW	SNJ	18,711	26.4	23.3	0.7	-0.9	-0.1	23.8	0.0	-25.7	1.9	23.8	155,909	20,547	78.30	1.341	Offshore finance plus tourism and exports
Dominica	DMA	SOV	3,631	13.3	36.3	8.7	9.1	17.8	41.3	-5.6	8.7	-29.9	21.2	72,680	6,800	77.05	1.217	Tourism plus exports with capital inflow
Easter Island (Rapanui)	RAP	SNJ												6,600	14,100	78.80	1.301	Moderate impact tourism
Falkland Islands (Islas Malvinas)	FLK	SNJ												2,918	55,400	77.90	1.476	Primary exports with aid
Faroe Islands	FRO	SNJ	18,820	70.2		15.8	11.4	27.2	8.5	8.1	5.8	-31.5	25.6	48,199	54,118	81.69	1.492	Primary exports with aid and remittances
Fiji Islands	FJI	SOV	2,858	42.1	37.3	7.6	3.6	11.2	-2.3	-5.3	-11.7	13.5	-1.9	892,145	4,830	70.09	1.139	Tourism plus exports
French Polynesia	PYF	SNJ	8,088	6.8	34.2		55.0	55.0	3.9	0.0	5.4	3.6	-9.0	282,764	18,161	76.54	1.317	Geostrategic aid
Galapagos	GPS	SNJ												25,124	10,200	76.80	1.255	High impact tourism
Greenland	GRL	SNJ	29,489	27.4			46.1	46.1	26.5					57,728	36,111	72.40	1.382	Geostrategic rent with exports
Grenada	GRD	SOV	3,244	8.2	29.5	7.1	5.2	12.3	-1.1	-8.3	-51.1	12.5	38.7	106,825	8,650	73.37	1.217	Tourism plus exports
Guadeloupe	GLP	SNJ	8,791	7.9	16.9		83.1	83.1	-7.9					400,132	23,416	79.75	1.366	Geostrategic aid
Guam	GUM	SNJ	18,695	3.7	56.9		14.1	14.1	25.3					169,885	32,122	79.13	1.403	Moderate impact tourism plus aid
Guernsey	GGY	SNJ												66,297	52,300	80.44	1.480	Offshore finance plus tourism
Iceland	ISL	SOV	18,647	76.3	15.3	1.7	0.0	1.7	8.5	-3.8	1.8	7.1	-8.8	330,823	50,140	82.06	1.482	High-value exports
Isle of Man	IMN	SNJ												88,195	83,100	80.44	1.557	Offshore finance plus tourism
Jeju	JEJ	SNJ												604,771	28,722	79.80	1.392	Moderate impact

Name																		
Jersey	JEY	SNJ												98,069	57,000	80.44	1.493	Offshore finance plus tourism
Kiribati	KIR	SOV	1,466	4.1	3.1	9.6	37.4	47.0	48.9	69.3	3.1	0.2	−3.3	112,423	3,390	65.95	1.079	Non-geostrategic MIRAB
Labuan	LAB	SNJ												96,800	13,932	75.10	1.278	Offshore finance plus tourism
Madeira	MAD	SNJ	608	70.3	1.4		2.3		−3.2					270,000	17,842	74.90	1.044	Primary exports with aid
Maldives	MDV	SOV	5,986	12.3	90.7	−11.4		−9.1		−14.2	−9.3	13.5	−4.1	409,163	6,950	76.77	1.217	High impact tourism
Malta	MLT	SOV	34,263	25.0	9.0	−6.4	0.0	−6.4	72.3	−2.9	1.2	26.7	−27.9	431,333	22,248	81.75	1.371	High-value exports especially services (in residual)
Marshall Islands	MHL	SOV	3,463	43.5	2.4	13.3	38.2	51.5	−3.9	22.9	−6.5	−0.4	6.9	52,993	4,770	73.10	1.158	Geostrategic rent with exports
Martinique	MTQ	SNJ	10,150	13.9	12.3	0.0	72.5	72.5	1.3					378,243	31,058	82.70	1.410	Geostrategic rent (French budget support)
Mayotte	MYT	SNJ	3,356	1.2	3.7	−11.0	134.5	123.6	−28.5					226,900	10,754	76.83	1.260	Geostrategic rent (French budget support)
Micronesia, Federated States of	FSM	SOV	2,390	30.5	9.3	1.5	48.4	49.8	4.6	9.7	−5.8	1.8	4.0	104,460	3,560	69.10	1.106	Geostrategic rent with exports
Montserrat	MSR	SNJ	9,982	5.7	13.4	−4.1	77.8	73.7	−20.4	−4.5	−27.6	20.3		5,267	11,363	74.40	1.251	Non-geostrategic MIRAB
Nauru*	NRU	SOV	7,264	54.3	0.0	0.0	41.3	41.3						10,222	15,420	67.10	1.241	Geostrategic aid
New Caledonia	NCL	SNJ	16,860	32.5	0.0	−0.1	31.1	31.1	2.9	13.3	−33.5	43.4	−9.9	273,000	32,736	77.57	1.397	Geostrategic rent with exports
Niue	NIU	SNJ	11,476	9.1	22.1		180.0	180.0	−111.2					1,612	11,539	69.45	1.222	Non-geostrategic MIRAB
Norfolk Island	NFK	SNJ												2,210	37,828	82.80	1.445	High impact tourism
Northern Mariana Islands	MNP	SNJ	9,072	3.4	78.8	1.2	17.4	17.4	0.4					55,070	16,742	78.00	1.316	High impact tourism
Orkney Islands	ORK	SNJ												21,667	28,550	80.80	1.397	Tourism plus exports
Palau	PLW	SOV	8,704	8.1	58.2		14.5	15.7	18.0	−5.4	−17.2	11.5	5.7	21,291	12,180	73.10	1.251	Moderate impact tourism
Prince Edward Island	PEI	SNJ	25,947	36.8	7.5	0.0	16.3	16.3	39.4					148,649	32,986	80.50	1.414	High-value exports especially services (in residual)
Réunion*	REU	SNJ	7,198	6.3	6.9		94.7	94.7	−7.9	3.5	−40.5			843,529	23,837	79.54	1.367	Geostrategic aid
Saint Helena, Ascension and Tristan da Cunha	SHN	SNJ	4,177	7.0	3.1		76.8	76.8	−13.5					7,776	7,800	79.21	1.243	Non-geostrategic MIRAB
Saint Kitts and Nevis	KNA	SOV	7,486	10.1	24.2	10.7	4.9	15.5	17.8	−6.7	−32.3	28.2	4.1	55,572	15,060	75.70	1.290	Tourism plus exports

(Continued)

Appendix 9.1 (Continued)

Island	Abbreviation	Political status	Imports of goods and services per capita 2010–2015 US$	% of imports of goods and services (2010–2015)										2015 population	2015 per capita income US$	Life expectancy	Index of income and life expectancy	Strategic classification
				Merchandise exports	Tourism spending	Remittances	Aid/government transfers	Remittances + aid	Residual	Net income from abroad	Current account balance	Net FDI inflow	Other funding of current account					
Saint Lucia	LCA	SOV	4,158	23.9	43.8	3.7	3.9	7.5	-5.5	-3.5	-30.3	11.6	18.7	184,999	7,350	75.05	1.211	Tourism plus exports
Saint Vincent and the Grenadines	VCT	SOV	3,632	11.1	22.9	6.2	3.0	9.2	5.7	-1.5	-51.0	27.5	23.5	109,462	6,630	72.94	1.188	Tourism plus exports
Saint Pierre and Miquelon*	SPM	SNJ	17,977	2.8	18.1	0.0	93.6	93.6	-14.5					5,595	38,204	80.50	1.434	Non-geostrategic
Samoa	WSM	SOV	2,057	16.1	35.3	37.2	29.8	67.0	-29.4	-6.9	-10.9	3.5	7.4	193,228	3,930	73.51	1.144	Non-geostrategic MIRAB: remittances and aid plus substantial tourism
São Tomé and Príncipe	STP	SOV	922	7.8	15.4	8.7	31.2	39.9	-17.6	0.9	-54.4	17.3	37.1	190,344	1,760	66.38	1.028	Non-geostrategic MIRAB
Seychelles	SYC	SOV	15,172	37.3	31.1	-2.9	2.2	-0.7	15.2	-5.1	-17.1	14.5	2.6	92,900	14,760	73.23	1.273	Tourism plus exports
Shetland Islands	SHT	SNJ												23,210	33,516	79.00	1.408	Primary exports with financial support
Solomon Islands	SLB	SOV	1,144	64.7	10.0	-5.6	46.3	40.7	-26.7	-30.6	-11.3	10.5	0.8	583,591	1,920	67.93	1.046	Primary exports with aid support
St Barthélemy	BLM	SNJ												9,279	47,388	79.75	1.461	High impact tourism
Svalbard	SJM	SNJ												1,872	77,080	81.80	1.550	High-value service exports
Tasmania	TAS	SNJ												519,128	37,426	80.65	1.432	High-value exports
Tokelau	TKL	SNJ												1,337	4,461	69.00	1.125	Non-geostrategic MIRAB
Tonga	TON	SOV	2,458	5.8	14.9	38.5	31.3	69.7	-13.3	3.6	-22.9	5.8	17.1	106,170	4,280	72.79	1.147	Non-geostrategic MIRAB
Turks and Caicos Islands	TCA	SNJ	8,887	2.6	104.0	-2.3	0.0	-2.3	-4.3					51,430	17,157	79.80	1.329	High impact tourism
Tuvalu	TUV	SOV	4,394	1.4	5.4	7.9	60.2	68.1	25.2	32.9	-8.6	1.2	7.4	9,916	6,230	66.50	1.139	Non-geostrategic MIRAB
US Virgin Islands	VIR	SNJ	58,514	83.0	18.7	-1.3	2.3	1.0	-2.8	0.0	0.0	0.0	0.0	103,574	36,351	79.77	1.423	Tourism plus exports (commodity exports have fallen away sine 2012)
Vanuatu	VUT	SOV	1,655	12.5	65.4	4.5	23.6	28.1	-8.0	-4.9	-2.0	11.6	-9.6	264,652	3,170	71.92	1.115	Tourism plus aid and some exports
Wallis and Futuna*	WLF	SNJ	4,315	0.2	2.2	0.0	186.0	186.0	-88.4					15,664	15,682	79.70	1.318	Non-geostrategic MIRAB

Note: * For these economies, the data show imports of goods only; this is used as the denominator for the coverage ratios.

References

Aldrich, R., and Connell, J. (1998) *The last colonies*. Cambridge, Cambridge University Press.

Armstrong, H.W., De Kervenoael, R.J., Li, X., and Read, R. (1998) A comparison of the economic performance of different micro-states and between micro-states and larger countries. *World Development*, 26(4), 639–656.

Armstrong, H.W., and Read, R. (2000) Comparing the economic performance of dependent territories and sovereign micro-states. *Economic Development and Cultural Change*, 48(2), 285–306.

Armstrong, H.W., and Read, R. (2002a) The phantom of liberty? Economic growth and the vulnerability of small states. *Journal of International Development*, 14(4), 435–458.

Armstrong, H.W., and Read, R. (2002b) Small states, islands and small states that are also islands. *Studies in Regional Science*, 33(1), 1–24.

Baldacchino, G. (1993) Bursting the bubble: the pseudo-development strategies of microstates. *Development & Change*, 24(1), 29–51.

Baldacchino, G. (1998) The other way round: manufacturing as an extension of services in small states. *Asia Pacific Viewpoint*, 39(3), 267–279.

Baldacchino, G. (2005) The contribution of 'social capital' to economic growth: lessons from island jurisdictions. *The Round Table: Commonwealth Journal of International Affairs*, 94(378), 31–46.

Baldacchino, G. (2006a) Managing the hinterland beyond: Two, ideal-type strategies of economic development for small island territories. *Asia-Pacific Viewpoint*, 47(1), 45–60.

Baldacchino, G. (2006b) Innovative development strategies from non-sovereign island jurisdictions: A global review of economic policy and governance practices. *World Development*, 34(5), 852–867.

Baldacchino, G. (2010) *Island enclaves: offshoring strategies, creative governance and subnational island jurisdictions*. Montreal and Kingston, QC, McGill-Queen's University Press.

Baldacchino, G. (2011) Surfers of the ocean waves: change management, intersectoral migration and the economic development of small island states. *Asia Pacific Viewpoint*, 52(3), 236–246.

Baldacchino, G. (2015) *Entrepreneurship in small island states and territories*. London, Routledge.

Baldacchino, G., and Bertram, G. (2009) The beak of the finch: insights into the economic development of small economies. *The Round Table: Commonwealth Journal of International Affairs*, 98(401), 141–160.

Baldacchino, G., and Hepburn, E. (eds.) (2012) Independence, nationalism and subnational island jurisdictions. *Commonwealth and Comparative Politics*, 50(4), 395–568.

Baldacchino, G. and Milne, D. (2000), *Lessons from the political economy of small islands: the resourcefulness of jurisdiction*, New York, St. Martin's Press, in association with Institute of Island Studies, University of Prince Edward Island, Canada

Baldacchino, G., and Milne, D. (eds.) (2009) *The case for non-sovereignty: lessons from sub-national island jurisdictions*. London, Routledge.

Banque de France and Institut d'Emission d'Outre-Mer (2004–2015) *Polynesie Française: Rapport annuel: La balance des paiements*. Available from: www.ieom.fr/polynesie-francaise/publications-71/balance-des-paiements-75/.

Bennell, P., and Oxenham, J. (1983) Skills and qualifications for small island states. *Labour and Society*, 8(1), 3–38.

Bertram, G. (1986) 'Sustainable development' in Pacific micro-economies. *World Development*, 14(7), 809–822.

Bertram, G. (1998) The MIRAB model twelve years on. *The Contemporary Pacific*, 11(1), 105–138.

Bertram, G. (1999) Economy. In M. Rapaport, M. (ed.), *The Pacific islands: environment and society*, Honolulu, HI, Bess Press, pp. 337–352.

Bertram, G. (2004) On the convergence of small island economies with their metropolitan patrons. *World Development*, 32(2), 343–364.

Bertram, G. (2006) Introduction: the MIRAB model in the twenty-first century. In G. Bertram (ed.), *Beyond MIRAB: The political economy of small islands in the 21st century*. Edited special issue of *Asia Pacific Viewpoint*, 47(1), 1–21.

Bertram, G. (2012) Trade and exchange: economic links between the Pacific and New Zealand in the twentieth century. In S. Mallon, K. Māhina-Tuai and D. Salesa (eds.), *Tangata o le Moana: New Zealand and the people of the Pacific*, Wellington, New Zealand, Te Papa Press, pp. 201–219.

Bertram, G. (2015) Is independence good or bad for small island economies? A long-run analysis. *Région et Développement*, 42(1), 31–54.

Bertram, G. (2016) Sovereignty and material welfare. In A. Holtz, M. Kowasch and O. Hasenkamp (eds.), *A region in transition: politics and power in the Pacific island countries*, Saarbrücken, Germany, Saarland University Press, pp. 385–430.

Bertram, G., and Karagedikli, O. (2004) Core-periphery linkages and income in small Pacific island economies. In J. Poot (ed.), *On the edge of the global economy*, Cheltenham, Edward Elgar, pp. 106–122.

Bertram, G., and Poirine, B. (2007) Island political economy. In G. Baldacchino (ed.), *A world of islands: an island studies reader*, Charlottetown, Canada and Luqa, Malta, Institute of Island Studies, University of Prince Edward Island and Agenda Academic, pp. 325–377.

Bertram, G., and Watters, R.F. (1985) The MIRAB economy in South Pacific microstates. *Pacific Viewpoint*, 26(3), 497–519.

Bertram, G., and Watters, R.F. (1986) The MIRAB process: earlier analysis in context, Pacific Viewpoint, 27(1), 47–59.

Boland, S., and Dollery, B. (2005) *The economic significance of migration and remittances in Tuvalu*. Working Paper 2005–10, University of New England School of Economics.

Borovnik, M. (2006) Working overseas: seafarers' remittances and their distribution in Kiribati. *Asia-Pacific Viewpoint*, 47(1), 151–161.

Briguglio, L. (1995) Small island developing states and their economic vulnerabilities. *World Development*, 23(9), 1615–1632.

Briguglio, L., and Cordina, G. (eds.) (2004) *Competitiveness strategies for small states*, Msida, Malta and London, Islands and Small States Institute and Commonwealth Secretariat.

Briguglio, L., Cordina, G., Farrugia, N., and Vella, S. (2005) Conceptualising and measuring economic resilience. In S. Chand (ed.), *Pacific islands regional integration and governance*, Canberra, ANU Press, pp. 26–49.

Brittain-Catlin, W. (2005) *Offshore: the dark side of the global economy*. New York, Farrar, Strauss and Giroux.

Bureau of Economic Analysis, US Department of Commerce (2012) *Estimates of gross domestic product, gross domestic product by industry, compensation by industry, and detailed consumer spending for the Commonwealth of the Northern Mariana* Islands. BEA 12–39. Available from: /www.bea.gov/newsreleases/general/terr/2012/CNMIGDP_091312.pdf.

Bureau of Economic Analysis, US Department of Commerce (2016) *Gross domestic product for the Commonwealth of the Northern Mariana Islands increases for the fourth year in a row*. BEA 12–39. Available from: www.bea.gov/newsreleases/general/terr/2016/CNMIGDP_112916.pdf.

Cayman Islands Government (2016) *About Cayman*. Available from: www.gov.ky/portal/page/portal/cighome/cayman.

Central Intelligence Agency (2017) *CIA world factbook*, Washington, DC, Central Intelligence Agency. Available from: www.cia.gov/library/publications/the-world-factbook/.

Central Statistics Division, Government of the Northern Marianas Islands (2016) *2015 Statistical Yearbook*. Available from: http://i2io42u7ucg3bwn5b3l0fquc.wpengine.netdna-cdn.com/wp-content/uploads/2017/02/2015-Yearbook-02102017-1-FINAL-FOR-PUBLICATION-USE-THIS-ONE.pdf.

Chamon, M. (2005) The importance and determinants of international remittances to Samoa. In *Samoa: selected issues and statistical appendix*, Country Report 05/221, Washington, DC, International Monetary Fund, June.

Christensen, J., and Hampton, M. (2000) The economics of offshore: who wins, who loses? *The Financial Regulator*, 4(4), 24–26.

Connell, J. (2006) Nauru: the first failed Pacific state? *The Round Table: Commonwealth Journal of International Affairs*, 95(383), 47–63.

Connell, J. (2013) *Islands at risk? Environments, economies and contemporary change*. Cheltenham, Edward Elgar Publishing.

Cour des Comptes (2013) *L'autonomie fiscale en outre-mer*. Available from: www.ccomptes.fr/Publications/Publications/L-autonomie-fiscale-en-outre-mer.

Demas, W.G. (1965) *The economics of development in small countries, with special reference to the Caribbean*. Montreal, QC, McGill University Press.

Dropsy, V., Montet C., and Poirine B. (2015) *Tourism, insularity and remoteness: a gravity based approach*. Working Paper 2015/1. Laboratoire Gouvernance et Développement Insulaire, Université de la Polynésie Française.

FEDOM (2016) *Tableau de Bord de FEDOM*. Available from: www.fedom.org/wp-content/uploads/2-15/06/TdB-15-Janv-2016.pdf.

Feyrer, J., and Sacerdote, B. (2009) Colonialism and modern income: islands as natural experiments. *Review of Economics and Statistics*, 91(2), 245–262.

Fichtner, J. (2016) The anatomy of the Cayman Islands offshore financial centre: Anglo-America, Japan, and the role of hedge funds. *Review of International Political Economy*, 23(6), 1034–1063.

Financial Action Task Force (2005a) *Money laundering and terrorist financing typologies 2004–2005*. Paris, OECD.

Financial Action Task Force (2005b) *Annual and overall review of non-cooperative countries or territories June 2005*. Paris, OECD.

Furlong, K., and Ludlow, E.M. (2016) *Results of the first comprehensive revision of the territorial economic accounts – new estimates of GDP for 2014 and revised estimates for 2002–2013*. Available from: www.bea.gov/scb/pdf/2016/04%20April/0416_territorial_economic_accounts.pd.

Gay, J.-C. (2012) Why is tourism doing poorly in Overseas France? *Annals of Tourism Research*, 39(3), 1634–1652.

Guillaumont, P. (2010) Assessing the economic vulnerability of small island developing states and the least developed countries. *Journal of Development Studies*, 46(5), 828–854.

Hampton, M.P. (1996a) Sixties child? The emergence of Jersey as an offshore finance centre: 1955–71. *Accounting, Business and Financial History*, 6(1), 51–71.

Hampton, M.P. (1996b) *The offshore interface: tax havens and offshore finance centres in the global economy*. Basingstoke, Palgrave Macmillan.

Hampton, M., and Christensen, J. (2002) Offshore pariahs? Small island economies, tax havens and the reconfiguration of global finance. *World Development*, 30(2), 1657–1673.

Harris, J.R., and Todaro, M.P. (1970) Migration, unemployment and development: a two-sector analysis. *American Economic Review*, 60(1), 126–142.

Hein, P.C. (1988) Problems of small island economies. In J. Crusol, P. Hein and F. Vellas (eds.), *L'enjeu des petites économies insulaires*, Paris, Economica, pp. 1–45.

Hintjens, H.M., and Newitt, M.D.D. (eds.) (1992) *The political economy of small tropical islands: the importance of being small*. Exeter, University of Exeter Press.

IEDOM (Institut d'Emission d'Outre-Mer) (2006) *Evolution des principaux indicateurs économiques, monetaires et financiers dans les DOM et les Collectivités d'Outre-Mer*. Available from www.iedom.fr/doc/principaux_indicateurs_economiques_DOM2.pdf#search=%22evolution%20des%20principaux%20indicateurs%20economiques%20monetaires%20et%20financiers%20dans%20les%20DOM%22.

IEDOM (2016). Balance-of-payments statistics for French Polynesia and New Caledonia. Available from www.ieom.fr/ieom/balance-des-paiements-46/les-rapports-annuels.html.

Kerr, S.A. (2005) What is small island sustainable development about? *Ocean & Coastal Management*, 48(7–8), 503–524.

McDaniel, C.N., and Gowdy, J.M. (1999) The physical destruction of Nauru: an example of weak sustainability. *Land Economics*, 75(2), 333–338.

McDaniel, C.N., and Gowdy, J.M. (2000) *Paradise for sale: a parable of nature*. Berkeley, CA, University of California Press.

McElroy, J.L. (2006) Small island tourist economies across the life cycle. *Asia Pacific Viewpoint*, 47(1), 61–77.

McElroy, J.L., and de Albuquerque, K. (1995) The social and economic propensity for political dependence in the insular Caribbean. *Social and Economic Studies*, 44(1), 167–193.

McElroy, J.L., and Parry, C. (2010) The characteristics of small island tourist economies. *Tourism and Hospitality Research*, 10(4), 315–328.

McElroy, J.L., and Parry, C. (2012) The long term propensity for political affiliation in island microstates. *Commonwealth and Comparative Politics*, 50(4), 403–421.

McElroy, J.L., and Pearce, K.B. (2006) The advantages of political affiliation: dependent and independent small island profiles. *The Round Table: Commonwealth Journal of International Affairs*, 95(386), 529–539.

Markoff, A, (2009) From obscurity to offshore giant. *Cayman Islands Financial Review*, 5 January, 14 April, 17 April, 7 July, 1 October and 5 October 2009. Available from www.caymanfinancialreview.com/2009/01/05/part-1-the-early-years-1960s-the-cayman-islands/.

Matsen, E., and Torvik, R. (2005) Optimal Dutch disease. *Journal of Development Economics*, 78(2), 494–515.

Milne, S. (2005) *The economic impact of tourism in SPTO member countries: Final Report*, South Pacific Tourism Organisation, August. Available from: http://csrs2.aut.ac.nz/NZTRI/nztrinew/documents/Pacific_Economic_Impact_Report.pdf.

Newitt, M.D.D. (1992) The perils of being a microstate: São Tomé and the Comoros Islands since independence. In H.M. Hintjens and M.D.D. Newitt (eds.), *The political economy of small tropical islands: the importance of being small*. Exeter, University of Exeter Press, pp. 76–92.

Oberst, A., and McElroy, J.L. (2007) Contrasting socio-economic and demographic profiles of two, small island, economic species: MIRAB versus PROFIT/SITE. *Island Studies Journal*, 2(2), 163–176.

Poirine, B. (1993) Le développement par la rente dans les petites économies insulaires. *Revue Economique*, 44(6), 1169–1199.

Poirine, B. (1994) Rent, emigration and unemployment in small islands: the MIRAB model and the French Overseas Departments and Territories. *World Development*, 22(12), 1997–2009, December.

Poirine, B. (1995a) *Les petites economies insulaires: theorie et strategies de developpement*. Paris, Harmattan.

Poirine, B. (1995b) Toujours plus est-il toujours mieux? Refus du développement, emigration et rationalité économique. *Revue d'économie du développement*, 2(1), 29–56.

Poirine, B. (1996) Is more always better? The South Pacific viewpoint on development and cultural change. *Pacific Viewpoint*, 36(1), 47–71.

Poirine, B. (1997) A theory of remittances as an implicit family loan arrangement. *World Development*, 25(4), 589–611.

Poirine, B. (1999) A theory of aid as trade with special reference to small islands. *Economic Development and Cultural Change*, 47(4), 831–852.

Poirine, B. (2006) Remittances sent by a growing altruistic diaspora: how do they grow over time? *Asia-Pacific Viewpoint*, 47(1), 93–108.

Poirine, B. (2007) Éloignement, insularité et competitivité dans les petites économies d'outre-mer: s'ouvrir pour soutenir la croissance? Available from: https://hal.archives-ouvertes.fr/hal-2007.

Poirine, B. (2011) *Tahiti: une economie sous serre*. Paris, l'Harmattan.

Roberts, S. (1995) Small place, big money: the Cayman Islands and the international financial system. *Economic Geography*, 71(3), 237–256.

Shaxson, N. (2011) *Treasure islands: uncovering the damage of offshore banking and tax haven*. New York, Palgrave Macmillan.

SNIJ Database (2017) *Subnational island jurisdictions database*. Charlottetown, Canada: Institute of Island Studies, University of Prince Edward Island. Available from http://projects.upei.ca/iis/island-jurisdiction-database/.

Streeten, P. (1993) The special problems of small countries. *World Development*, 21(2), 197–202.

Suss, E.C., Williams, O.H., and Mendis, C. (2002) *Caribbean offshore financial centers: past, present and possibilities for the future*, Working Paper, WP 02/88, Washington, DC, International Monetary Fund.

Tisdell, C.A. (2016) The MIRAB model of small island economies in the Pacific and their security issues. In A. Holz, M. Kowasch and O. Hasenkamp (eds.), *A region in transition: politics and power in the Pacific Island countries*, Saarbrucken, Germany, Saarland University Press, pp. 431–450.

UN (2016) *Statistical Yearbook 2016*. Available from: https://unstats.un.org/unsd/publications/statistical-yearbook.

UN (2017) *UN national accounts database*. Available from: http://unstats.un.org/unsd/snaama/resQuery.asp.

United Nations Development Programme (UNDP). (2016) *Human Development Report 2016*. New York: UNDP. Available from: http://hdr.undp.org/en/2016-report/download.

US Census Bureau (2016). International database. Available from: www.census.gov/population/international/data/idb/informationGateway.php.

Vine, D. (2011) *Island of shame: the secret history of the US military base on Diego Garcia*. Princeton, NJ, Princeton University Press.

Winters, L.A. (2005) Policy challenges for small economies in a globalising world. In S. Chand (ed.), *Pacific Islands regional integration and governance*. Canberra, ANU Asia Pacific Press. Available from: http://press.anu.edu.au/publications/pacific-islands-regional-integration-and-governance/download.

Winters, L.A., and Martins, P.M.G. (2004a) When comparative advantage is not enough: business costs in small remote economies. *World Trade Review*, 3(3), 347–383.

Winters, L.A., and Martins, P.M.G. (2004b) Beautiful but costly: business costs in small remote economies. *Economic Paper 67*. London, Commonwealth Secretariat.

World Bank (2016) *World development indicators*. Available from: http://databank.worldbank.org/data/reports.aspx?source=world-development-indicators.

WTO (2017) *Database of merchandise trade statistics*. Available from: http://stat.wto.org/StatisticalProgram/WsdbExportZip.aspx?Language=E.

10

TOURISM

Sonya Graci and Patrick T. Maher

Introduction

Islands have clear borders and a coastline that encloses them within a body of water. These sharp boundaries make them easy to conceptualise. As Péron (2004) notes, this condition makes island spaces relatively easy to mentally grasp and understand, which in turn makes humans feel secure once they are thus snugly ensconced. This is what fuels and drives the worldwide fascination of and with islands, often irrespective of their actual material size: after all, some islands are quite big: think Greenland, Madagascar and New Guinea. Islands have what has been described as a particular lure or fascination to visitors (Lockhart 1994, 1997, King 1993, Baum et al. 2000). Such powerful tropes of safe containment and clarity of knowledge lie at the root of the assumption that visitors can cross over to, circumnavigate and/or fully discover an island with relative ease. Increasingly, people are travelling to indulge in 'the island experience' and doing so in a variety of ways, including visiting communities that are isolated and displaying rich and diverse cultures, unique environmental attributes and exotic species, or just lingering carelessly at a sandy beach (Douglas 2006, Graci and Dodds 2010, Baum 1997). Islands present themselves, or find themselves presented, as locales of unfettered desire, as tantalising platforms of paradise, as habitual sites of fascination, emotional offloading or religious pilgrimage. The link with tourism is clear: no surprise, therefore, that many island communities have been looking at tourism as a major economic alternative to such traditional livelihoods as fishing and agriculture, recognising its contribution as an economic diversification tool (Lockhart 1997).

A caveat which has been recently recognised in the literature is the distinction between 'warm water' and 'cold water' islands, a distinction which has implications on both the nature of tourism and its potential side-effects. Islands in tropical and lower temperate latitudes typically enjoy a climate where (mainly beach and coastal) tourism unfolds over an extended season, and where a heavy dose of tourism visitations are not subject to dizzy seasonality spikes. The dominant tourism leitmotifs are those of relaxation and hedonistic enjoyment. In contrast, cold water islands, lying at higher temperate and sub-Arctic/Antarctic latitudes, offer close encounters with a wilder, starker and a more raw and rugged experience of nature for lower numbers of tourists, typically during a narrow window of opportunity (Baldacchino 2006). Tourism activity here tends to be more physical and strenuous. Gössling and Wall (2006, p. 432) describe these distinct characteristics summatively (see Table 10.1).

Table 10.1 Features of warm water versus cold water islands as tourism destinations.

Warm water islands	Cold water islands
Extended season	Short season, often within two months
High air and water temperatures	Low air and water temperatures
Sun	Rain, potentially even snow and hail
White, sandy beaches	Rocks, cliffs, hiking trails
Lush vegetation, colourful flowers	Barren land, little vegetation
Fruits, spices, rich buffets	Few dishes
Relaxation, party fun	Exploration, adventure
Leisure (sun, sea, sand, sex, shopping)	Serious leisure

Environmental and social impacts are likely to be more prevalent on warm water islands. The situation is generally different in cold water islands, as there are more regulations governing the development of tourism and related infrastructure, while tourism numbers tend to be much lower, and so are much more manageable. Cold water islands may have also achieved some degree of sustainability through the establishment of protected areas, waste treatment systems, waste management programmes, environmental regulations and governance (Gössling and Wall 2006).

Three specificities of island tourism

Tourism tends to bring along some distinct challenges to island locales. These impacts will loom larger with a decreasing size of the island territory and/or the size of its resident population.

First, there is a real risk of a very rapid saturation of an island as a tourism destination. Once 'discovered' and placed on the tourist trail, an island may find itself overwhelmed by progressively larger numbers of visitors. Iceland from 2008 to 2016 is an excellent contemporary example of this phenomenon: tourism visitations spiked from 0.5 to 2 million in this period (Icelandic Tourism Board 2017). Hospitality and welcoming behaviour may quickly turn to disgust and irritation, which can threaten the sustainability and future of the tourism industry on any small island. Coastal areas may experience considerable erosion and degradation. Rare and endemic species risk habitat loss or disturbance. Attractive neighbourhoods and beachfront properties become the targets of extra-territorial buyouts, which in turn lead to a gentrification where locals find themselves increasingly tempted to sell their property to outsiders, but then are increasingly unable to afford property on their own island (see Clark and Kjellberg, this volume).

Second, many small island economies can become totally enamoured of and gripped by the tourism industry, to the exclusion or detriment of any other serious alternative productive activity. In some island economies, tourism contributes more than 50 per cent of the GDP and is a key source of direct employment and government revenue (Hampton and Jeyacheya 2013, 2015, Royle 2001). 'Small island tourism economies' represent the highest order of dependence on this one industry (McElroy and Parry 2010): indeed, all the countries where tourism accounted for more than 50 per cent of total exports consisted of small island states (Barrowclough 2007). A high-profile crime, murder or terrorist act can however dangerously send such an economy into a nosedive overnight; a recovery from such a mishap can take years (De Albuquerque and McElroy 1999).

Third, and unless connected by 'fixed links', tourism visitation to islands depends exclusively on sea and/or air connections. Seaports, ferry and cruise ship terminals and airports – all of which are typically located at or near the island's capital city and main urban area – become the arrival and departure points *de rigueur* of all tourists. In such situations, rural and remote areas (in single island units) or outlying islands (in archipelagic units) tend to miss out on the tourism action, and may agitate politically to redress what they consider as an unfair marginality (Baldacchino 2016).

Development and economic importance of islands

The smaller the island, the greater its inevitable engagement with openness and dependence. Islands must face out towards the horizon and await the suite of imports that hail from elsewhere: not just tourists, but cargo, invaders, settlers, missionaries, investment and disease. Historically, islands have served as transit points, as ships moved resources to and from the periphery (colonies), back to the core (centre of empires). This has occurred for centuries across the globe, whether it was shuttling nutmeg from Indonesia to the Netherlands, transporting slaves from São Tomé to Brazil, or ferrying whale blubber from Greenland and Iceland to Denmark. This has changed somewhat with the advent of air travel and the growth of the practice of going 'on holiday' (Löfgren 2002).

Tourism generally produces beneficial economic results for island destinations. However, these can also be accompanied by significant social and environmental impacts. Living standards and quality of life can be raised by income; new employment and educational opportunities can be gained; and improved international understanding can be the result of tourism initiatives (Bramwell and Lane 1993, Clifton and Benson 2006, Eber 1992, Graci and Dodds 2010, Elliott 1997).

Nevertheless, the characteristic complexities of island destinations give rise to many social and environmental issues (Lockhart 1997). Island destinations can be isolated and restricted in dealing with social and environmental challenges due to their location and scale restraints (Graci and Dodds 2010). Many of the services provided by island tourism are resource intensive, resulting in a significant environmental footprint (Briguglio et al. 1996a, 1996b, McElroy and Albuquerque 1998, Selwyn 1975, Graci and Dodds 2010). In small island developing states (SIDS), tourism is touted as the economic saviour for many communities, without understanding how the service intensive industry can destroy marine and land environments while also threatening, instrumentalising and monetising cultures and traditions. In addition, tourism is often associated with colonial and neo-colonial practices that contribute inadequately to local livelihoods: consider expatriate European or North American managers supervising local staff (Britton 1982, Hollinshead 2004, Mowforth and Munt 2008, Tolkach and King 2015). The fast and uncontrolled increase in tourist flows has caused significant negative impacts on the natural and built environment in several island destinations (Mathieson and Wall 1982). In many cases, this results in the degradation of the tourist product and the reduction of profits for host communities and the national economy. This is problematic as a large percentage of GDP in islands may be dependent on tourism.

Due to the rapid and pervasive growth of tourism on many islands, there has been a significant change in the landscape of these destinations as well as a plethora of social and ecological impacts. Impacts include water and soil pollution, excess waste and sewage runoff, loss of natural terrestrial or marine habitats in the face of construction sprees, shoreline erosion, as well as a deterioration of local cultures and a disruption of traditional livelihoods (Graci and Dodds 2010). Any such environmental and social impacts are likely to be more pronounced and detrimental in a small island context.

Island ecosystems are also more vulnerable to global environmental change. Increasing mean temperatures, extreme weather events, sea level rise, and changing rainfall patterns also have an immense effect on the viability of islands and island life (Gössling and Wall 2006). With their high 'coast to land area' ratio, many island populations are sensitive to sea level rise, volcanic eruptions, tsunamis and hurricanes; island communities have had to be relocated elsewhere due to flooding and other dangers. Indeed, a morbid streak of dark tourism is now encouraging many to visit endangered island locales – from the Venice lagoon to the atoll archipelago of Tuvalu – before these succumb to the forces of nature, and become un-visitable (Farbotko 2010, Hindley and Font 2014).

Sustainable island tourism

Butler (1993, p. 29) defined sustainable development in the context of tourism in this way:

> Tourism which is developed and maintained in an area (community, environment) in such a manner and at such a scale that it remains viable over an indefinite period and does not degrade or alter the environment (human and physical) in which it exists to such a degree that it prohibits successful development and well-being of other activities and processes.

Small islands are bedevilled by scarcity and limits. Water shortage is an issue on many island destinations. In addition, growing island populations can apply considerable pressure on limited resources, and sustainable alternatives are difficult to implement due to cost, location and infrastructure issues. It is usually more expensive to provide technical expertise: where the expertise is not locally available, it may have to be flown in (Graci and Dodds 2010); where it is locally available, it may consist of just one person.

Islands are often dependent on foreign investment. Small island businesses that are owned by foreigners may not contribute as much to the local economy (Cole 2006). Foreign investment or ownership can increase an island's dependency on export. In a volatile market, small islands are often hit the hardest as they depend on imports of food, water and raw materials (Graci and Dodds 2010).

Sustainable island living often requires the active participation of and by local people in the development and management of tourism. In addition,, a carefully designed and long-term strategic tourism plan, intensive capacity building, and the training of both national public officials and private sector management are needed to minimise the environmental and social impacts of tourism on small islands (Fennell 2003, Hashimoto 2004). Due to their limited size, isolation, marginalisation and resource limitations, islands can face significant challenges in securing the sustainable development of tourism. Nevertheless, by virtue of being permanently threatened by change, and where necessity is the mother of invention, small island communities can often deploy significant levels of resilience, resourcefulness and social capital; they are very good exemplars of innovative coping mechanisms. But migration and 'ex-isle' are also historical responses to threats to island livelihoods, and these escape hatches are increasingly problematic to resort to in an age where borders are shut and policed in the name of securitisation or of some vague and invented yet powerful sense of national purity and identity (see Connell, this volume).

So, despite the potential environmental and social effects related to tourism, islands can successfully rise to the occasion and somehow contain or manage the negative impacts. The following case studies discuss two islands that have been successful in doing so.

Case studies: Gili Trawangan and Cape Breton Island

As noted earlier, we have situated this chapter within a 'warm water–cold water' division and thus we chose one case study which fits in each of these two general categories. In addition, we felt that it was critical to explore destinations that have some defining similarities, and in doing so we chose to examine Gili Trawangan in Indonesia and Cape Breton in Canada. Sub-national island destinations were chosen, with no autonomous levels of governance, unlike sovereign island states (such as Iceland or Mauritius) or sub-national island jurisdictions (such as South Georgia and French Polynesia). We have also opted for islands within countries known for the large extent of their islands and coastlines (Indonesia and Canada), but still avoiding both the largest island constituent of each (Kalimantan and Baffin Island) as well as their most popular island tourist destinations (Bali and Vancouver Island) in either context. These islands will hopefully offer an important glimpse of the issues and challenges of island tourism.

Case study 1: Gili Trawangan, Indonesia

Gili Trawangan lies at the northwest tip of Lombok in eastern Indonesia. It is the largest of a group of three islands (the others being Gili Meno and Gili Air). Gili Trawangan is dependent on 'sun, sand and sea' mass tourism, with a particular emphasis on dive tourism. Divers are attracted to the island by three main marine species: sea turtles, reef sharks and manta rays. Gili Trawangan is a small island (some 6 km² in land size) with a low-lying topography and experiences a tropical climate (Hampton 1998, Dickerson 2008, Cushnahan 2004). The island has a very small local population: in 2007, it was some 1,900 inhabitants and by 2010 it had risen to some 4,440 (Graci 2007, Willmott and Graci 2013). Infrastructure on the island consists of a limited system of dirt roads and a generator for electricity provision. The island does not have a sewage treatment system. Land transportation is by horse-drawn carts, as motorised vehicles have not been permitted. Gili Trawangan is located within a nationally designated marine park, the Gili Matra Marine Natural Recreation Park, which is preserved for research, science, education, recreation and tourism purposes (Willmott and Graci 2012) (see Figure 10.1).

Evolution of tourism

The island of Gili Trawangan is primarily focused on dive and party tourism and has attracted small-scale tourism since the 1980s (Dodds, Graci and Holmes 2010). Due to its limited access to main tourist areas (such as Bali), it was mainly a destination for backpackers. Much of the development until the late 1990s was small-scale, local establishments and 'developer tourists'; visitors who decided to stay on the island and open a dive shop (Cushnahan 2004, Hampton and Jeyacheya 2014). Starting in 2004, the island went from having very little accommodation other than homestays and one resort, a few dive shops and restaurants to foreigners buying up land to build resorts and villas. More recent developments have included the establishment of up-scale tourism businesses such as high quality restaurants, hotels and spas. These reflect a trend towards attracting more affluent visitors and the growth of Gili Trawangan's expatriate community (Ver Berkmoes et al. 2009, Charlie et al. 2013).

Areas for growing agricultural crops have been diverted to tourism-related development, leaving a highly exposed, tourism-dependent economy. Prior to 2004, development was restricted to a quarter of the island but, as the island rapidly began to develop, a construction boom resulted in buildings that ignored the basic principles of the island's implicit building code: such as not building on the beach or not building higher than the tallest coconut tree. Gili Trawangan had always been a vehicle-free island but recently there is discussion of allowing motorbikes on the island. Due to this rapid and unplanned development, the local population has become increasingly concerned with the state of their environment (Hampton and Jeyacheya 2014).

Figure 10.1 The landscape of Gili Trawangan.

Source: Photo by Midori, Wikimedia Commons, https://commons.wikimedia.org/wiki/File:Sunrise_Gili_Trawan
gan_1.JPG.

The influx of tourism on Gili Trawangan has led to several issues related to water quality, waste and sewage management, marine degradation, beach erosion, unplanned or unauthorised development and illegal fishing (Graci 2013). There are also serious concerns on the island in relation to power and the downstream environmental pressures resulting from waste and sewage. Fresh water does not exist on the island, other than being imported via plastic bottles; and, with the extreme pressure from the building boom, waste management is an issue. Litter is strewn all over the island and, up until recently, there was no organised waste management system in place (Graci 2013, Willmott and Graci 2012). Many stakeholders are resistant to change because of the costs associated with improved environmental management, the potential loss of income through restrictions on their activities, and a general lack of knowledge or disregard about the impact of their actions on the environment (Graci 2007).

Successes and challenges

In order to manage the issues related to tourism development, the local community mobilised to create a not-for-profit organisation to manage the issues on the island. The strength of the community in Gili Trawangan, which is an island entirely dependent on tourism, has been manifested in an organisation that is meant to develop initiatives to protect the island's fragile resource base. The Gili Ecotrust (GET) was established by the island's expatriate-managed dive shops in 2002 as a not-for-profit initiative, prompted by the looming destruction of the coral reefs due to global warming, untreated waste, uncontrolled tourism activities and destructive fishing practices (Dodds et al. 2010, Graci 2007, Charlie et al. 2013). The approach involved the levying of an 'ecotax' on divers and snorkelers by the various dive shops. Initially, the ecotax collected US$3 per diver and US$1 per snorkeler, but it has been increased to US$6 per diver and US$3 per snorkeler. The revenue is used by GET to employ a full-time environmental coordinator, clean up beaches, pay fishers to desist from destructive fishing activities around the coral reefs, and rebuild the coral reefs through the Biorock reef programme (Graci 2007, Charlie et al. 2013). GET collaborates with SATGAS – a law enforcement NGO also set up by the locals – to undertake island patrols by monitoring fishing practices, protecting the reefs and running the Biorock programme (Charlie et al. 2013). In 2005, an environmental audit and sustainable tourism strategy, funded by GET, was developed. In recent years, GET has focused on waste management and recycling, and on environmental awareness, education and training.

The GET is an innovative initiative for dealing with sustainability issues on an island. Due to the influx of tourism on the island, the marine environment has been severely degraded and pressures from overfishing, diving and the increase in boats for the transportation of tourists and supplies has led to an increase in negative impacts. On Gili Trawangan, everything has to be shipped in, from drinking water and food to alcohol and daily essentials. In addition, due to the burgeoning population of tourists and expatriates coming to work in the tourism and allied businesses, the island has effectively exceeded its carrying capacity. The GET is a method to ensure that issues on the island are managed better and, through the eco-tax, tourists can contribute to the protection of the island and ensure that resources are managed more sustainably. The GET ensures that illegal fishing does not occur and manages the waste, sewage and other environmental issues on the island. A waste management system has been developed on the island: the enormous amount of plastic generated from water bottles and other items is separated and recycled, while green (biodegradable) waste is composted. The GET has also been instrumental in regenerating the coral reefs through installing several biorocks around the island. These biorocks are designed to provide low levels of electricity to a structure and coral then grows, attracting fish and other life back to the marine environment. Since Gili Trawangan is primarily a dive island, the biorock has been critical in ensuring the adaptability of the island to overcome the serious challenges to the marine environment that it is facing. The 'GET effect' suggests that people on islands can become resourceful in terms of ensuring that resources are utilised to their full capabilities and yet that these resources are sustained so that the tourism industry is not destroyed. When local island communities mobilise in this and similar ways to protect their resources and sustain their tourism industry, there is hope that challenges can be addressed at the local level rapidly and effectively.

Case study: Cape Breton Island, Canada

Cape Breton Island is located in the North Atlantic Ocean, on the eastern coast of North America. It is part of the Canadian province of Nova Scotia, separated from the 'mainland' peninsula of Nova Scotia by the Strait of Canso. Although it is connected to the mainland by the 1.3 km long Canso Causeway, built in 1955, unlike nearby Prince Edward Island, which now has a permanent bridge, Cape Breton remains a 'true' island because the causeway has a swing bridge to break the link and allow boat traffic to pass. This connection limits the sustainability of the island economy in so much as threatening the distribution of food, fuel and such if this link is compromised.

Cape Breton has a land area of 10,310 km^2; making it the 76th largest island in the world; slightly smaller than the big island of Hawai'i and slightly larger than Cyprus. The island surrounds one of the largest saltwater lakes in the world, Bras D'Or Lake, which is now the centre of a UNESCO Biosphere Reserve. Cape Breton is divided into four administrative counties: Cape Breton, Inverness, Richmond and Victoria.

The population of Cape Breton Island is 132,010, based on the most recent Canadian census (Statistics Canada 2016) and this has been on a steep decline even in just the past two decades (Cape Breton Post 2017). More than 75 per cent of the island's population (98,722; Statistics Canada 2016) lives in the Cape Breton Regional Municipality (CBRM), an entity that resulted from the merger of various small cities and towns in 1995. The CBRM covers the entire County of Cape Breton and is often referred to as Industrial Cape Breton.

Historically, there are three distinct groups in the population of Cape Breton Island: Scottish Gaelic, French Acadian and Mi'kmaq: the local Indigenous population (Destination Cape Breton 2017a). After centuries in relative isolation, these three cultural braids were joined more recently by migrating Black Loyalists escaping the USA, Italians and eastern Europeans. The latter groups all settled primarily in Industrial Cape Breton to work in the coal and steel industries.

Evolution of tourism

The evolution of tourism on Cape Breton Island can be seen largely as an alternative trajectory to extractive resource industries on the island. Through the 19th and 20th centuries, the coal fields of Industrial Cape Breton supported some of the largest private employers in Canada. There were numerous mines employing entire towns, and so the need for additional industries, apart from those that serviced the mines, was simply not there. Coal also went to fuel the strong steel industry in the CBRM. The Cape Breton Development Corporation (DEVCO) operated the coal mines from 1968 to 2001, and the Sydney Steel Corporation (SYSCO) was strong in this same era. Both began to run into economic difficulties in the 1970s/1980s, and by 2001 they were both completely closed. This death of heavy industry, while tragic at the time, is when economic diversification in tourism truly started (Brown and Geddes 2007, Regional Tourism Development Plan 1981).

Since the Second World War, tourism on Cape Breton has been mainly scenic cruising, focused on driving along the spectacular Cabot Trail, a world class highway, completed initially in 1932, that loops 300km around the northern tip of the island. With its recognisable scenic vistas, the Cabot Trail is an attraction on its own, but it also connects other important tourism attractions (Regional Tourism Development Plan 1981). It passes alongside and within Cape Breton Highlands National Park, overlooks the Bras d'Or Lake, rolls next to the Margaree River (a Canadian Heritage River and world-class fly-fishing destination) and bridges many unique Gaelic and Acadian cultural communities along the way (see Figure 10.2).

Often referred to as part of the Cabot Trail, Cape Breton Highlands National Park (CBHNP) is a tourism attraction in its own right (Parks Canada 2017). Many visitors see Cape Breton as the landscape contained here, especially stone-cobbled beaches north of Ingonish or the Skyline Trail. Parks Canada, which is the agency that manages CBHNP, also manages other key attractions linked to Cape Breton's past, such as the Fortress of Louisbourg, the Alexander Graham Bell and Marconi National Historic Sites.

More recently, new attractions have come on stream, and they speak to the strength of sustainability on the island. The Bras d'Or Lake, at the centre of the island, has been designated a UNESCO Biosphere Reserve (BLBRA 2017). That designation, and subsequent marketing exposure, have resulted in a more diverse population of visitors; visitors interested in increased adventure. This has seen a growth in operators (sea kayaking, whale watching) supporting tourism along traditional driving routes (Destination Cape Breton 2017b). A new Marine Protected Area has also recently been designated to the southeast, just off Scatarie Island (St Anne's Bank;

Figure 10.2   The landscape of Cape Breton.

Source: Photo by Michael Sprague, Wikimedia Commons, https://commons.wikimedia.org/wiki/File:2009_Trip_-_Cape_Breton_Island_(3940249474).jpg.

DFO 2017) and there is a clear recognition that this can support commercial marine recreation and tourism uses with the key species for viewing being blue whales and leatherback turtles. These links to the sustainability of the marine ecosystem are critical to Cape Breton as an island tourism destination.

The progress of two other 'human-made' developments has also increased tourism traffic on the island. The Port of Sydney is the ocean gateway to the island, and has been welcoming increased traffic over the past ten years, to the extent that there is now talk of a second cruise ship berth (CTV Atlantic 2017). One specific attraction just outside the Port of Sydney's Joan Harriss Cruise Pavilion is the 60 foot *Fidheal Mhor A' Ceilidh*, or the Big Fiddle of the Ceilidh, that connects to the symbols of Celtic culture on Cape Breton Island. As a cold water island destination, the port of Sydney has also just recently been named as an Arctic gateway for One Ocean Expeditions, and when combined with the attractions already listed, speaks highly to the cold water island feature of exploration and adventure, noted by Gössling and Wall (2006).

One final attraction for Cape Breton is a series of new golf courses developed near Inverness on the western side of the island. Golf has always been popular, and Cape Breton boasted one of the only courses run inside a National Park (Highland Links aligned with Keltic Lodge), while various courses have hosted PGA Canada events. Since 2010, Cape Breton has become home to two of the most highly anticipated and ranked courses in the world: Cabot Links 2012 and Cabot Cliffs 2015. These courses have done wonders for the local economy (Taber 2012, McCarten 2016), and connect to both the local history and the cultural connections of the island and the sport of golf back to Scotland (Cabot Links 2017). World recognition has come alongside the economic benefits. Golf Digest placed Cabot Cliffs as the 19th best course in the world in 2016; while Cabot Links stayed in the top 100 at 93 (down from 42nd the year before) (Golf Digest 2016). Various levels of government appear to be well-supportive of these golf developments, with funding secured for future marketing and promotion (ACOA 2016).

Successes and challenges

Tourism in Cape Breton has always had some level of reliance on other development. When DEVCO and SYSCO were doing well, locals had money via well-paying jobs, so the tourism industry saw visitors coming and going and infrastructure investment, but not really much interest in working in the industry. Now, without those steel and coal industries, there is widespread unemployment and poverty, yet still not much interest in working in the service industry of tourism, due to an expectation of higher wages.

There is also a newly emergent suite of social problems as Cape Bretoners have started to move for work; either on a week-to-week basis with many men (and it is largely men) flying in and out for work in the oil patch of Alberta (a Western Canadian province) for weeks at a time; or moving permanently, which causes the population of the island to further decline. This movement is not sustainable, nor is the reliance on high-paying resource development jobs elsewhere. One positive side of this is an influx of domestic tourists with a personal connection to Cape Breton Island. As Brown (2010) has noted, there is a strong 'Visiting Friends and Relatives' (VFR) sector in the area. Families may have moved permanently to Alberta or British Columbia for work, but they invariably return home to their island each summer for a visit.

Tourism on Cape Breton Island is seen as a valuable component of economic diversification, but is yet to be universally accepted. There is a lingering stigma attached to the typically low-paying service jobs associated with tourism. This lack of respect for tourism as an employer, with below-average pay, low social status and poor recognition, creates a ripple effect. If another exploitable resource quickly appeared, tourism would quickly lose ground, despite the understanding by policy-makers and municipal leaders of the 'boom and bust' nature of resource extraction: whether it is coal, fishing, or another resource; and the consistent sustainability of tourism.

Seasonality is a huge issue for the local industry as there is basically no tourism from mid-October until late May. Even on the number one attraction (the Cabot Trail), everything closes down outside this peak summer season. Even attractions such as the Cape Breton Highlands National Park and Fortress of Louisbourg, which are managed at a higher level by a federal agency (Parks Canada), do not seem willing or able to break out of the seasonality pattern. Various festivals (Kitchen Fest and Celtic Colours) tend to book mark the beginning and end of the season (early July to late October) and this is a huge improvement over the recent past. Sports such as golf are inherently seasonal, and the climatic changes seen in eastern Canada in recent years do not give winter sports much hope.

There is also a lot of fragmentation in the industry. Large attractions appear to do well, especially when they 'piggy back' off the cruise ship growth; but, for the 'Free Independent Traveller' (FIT), there are many missing pieces to the sector. Many in the industry believe that Destination Cape Breton (the regional DMO) only wants to promote the Celtic heritage angle, and there are many other small but significant issues (no shuttle services, poor opening hours of downtown business, and nothing available on Sundays) that plague the sector.

One positive aspect of growth is around Mi'kmaq cultural tourism. Lynch et al. (2011) speak to the sustainability of Mi'kmaq eco-tourism experiences in Kejimkujik National Park and National Historic Site; but on Cape Breton Island there are the award-winning attractions at

Membertou Heritage Park (Doucette 2008), as well as the Eskasoni Cultural Journeys (MacPherson et al. 2016). Combined with individual success is the growing collaborative success secured with the new Unama'ki Tourism Association which, amongst the five Mi'kmaq communities on Cape Breton Island, seems to be more productive than Destination Cape Breton, with regards to community development and shaping a multicultural narrative.

2017 came with increased optimism due to a 'Trump Bump'. In 2016, a local radio personality created a website: 'Cape Breton if Trump Wins'. This started out as a spoof, but really took off as Americans (particularly those who supported the Democratic Party) saw that Donald Trump might win the US election. Hundreds of thousands of inquiries flooded in, and when Trump did win the election many of those inquiries turned to tourist bookings (albeit not permanent moves, as some expected) (Dean 2017). This type of grassroots innovation, even if it started as a joke, showcases what islands need to do: take advantage of every possible situation. Cape Breton has a lot to offer; and its tourism sector has a tremendous reputation upon which to capitalise. Ranked the top island destination in North America by *Travel + Leisure* magazine in 2008–2010, it has seen its reputation slip in recent years; in 2017, it ranked third best in Canada, behind Prince Edward Island and Vancouver Island (Travel and Leisure 2017). In any case, the people of Cape Breton are seen as friendly and welcoming; resilient to changing situations and living in a landscape that is familiar to many (links to Scotland), yet awe-inspiring in its own right. For Cape Breton to truly be sustainable and world-class, it needs to fix a somewhat disjointed industry and match expectation with experience. Keep the 'down home' charm (music and drink; ceilidhs and kitchen parties), yet lose the 'down home' headaches (lack of shops and restaurants on Sunday, poor shuttle services).

Conclusion

Although not necessarily unique to islands, most of the challenges faced by Gili Trawangan and Cape Breton are heightened because of their geographic nature. Due to their size and isolation, historical marginalisation and resource limitations, islands can face significant challenges to the sustainable development of their tourism industry. Yet, there are some extraordinary coping mechanisms in place. The resourcefulness of island people showcases their disposition to be resilient (Graci and Dodds 2010); a necessary adaptation to fast-changing environments.

For many islands, tourism is seen as the key economic base; a diversification of what may have been a single or multiple sector development of the past. If this is the case, then a holistic and integrated approach to tourism development and management is recommended for long-term survival and prosperity. For some island destinations, this is an issue. Despite decades of 'seeing the writing on the wall', it has only been recently that tourism development in Cape Breton has seen a consistent approach and, even now, it is skewed heavily towards a few aspects (driving, Celtic culture/music and more recently golf).

For many islands, especially warm water ones that focus on the sun, sand and sea approach to tourism, there is great competition, low differentiating factors and the product has become commoditised. Long-term strategies need to be put into place to deal with increasing and detrimental environmental impacts but also to changing market trends (Graci and Dodds 2010). For cold water islands, the challenge is very often the simple factor of seasonality. If they cannot capitalise tremendously on a short season, then what can be done to extend that? Islands must constantly innovate to maintain and grow their position in a changing global marketplace. Islands have the opportunity to reinvent themselves and to utilise technology and innovation to assist in inventing and reinventing themselves as sustainable destinations in ways that encompass their limits to growth.

While islands are geographically bounded, given the right kind of people, sound policies and sufficient support, they may know no bounds. Island tourism is volatile, but with the right structures in place it can be resilient and successful in spite of this volatility. The sustainability of island tourism, whether in cold water or warm water locations, has seen many success stories; destinations such as Gili Trawangan and Cape Breton offer both cautionary tales and reasons for cautious optimism.

References

ACOA (2016, 8 November) A hole-in-one for Cape Breton tourism. Available from: www.acoa-apeca. gc.ca/eng/Agency/mediaroom/NewsReleases/Pages/4983.aspx.

Baldacchino, G. (ed.) (2006) *Extreme tourism: lessons from the World's Cold Water Islands*. Oxford, Elsevier.

Baldacchino, G. (ed.) (2016) *Archipelago tourism: policies and practices*. London, Routledge.

Baum, T. (1997). The fascination of islands: a tourist perspective. In D.G. Lockhart and D. Drakakis-Smith (eds.), *Island tourism: trends and prospects*, London, Mansell, pp. 21–35.

Baum, T., Hagen-Grant, L., Jolliffe, L., Lambert, S., and Sigurjonsson, B. (2000) Tourism and cold water islands in the North Atlantic. In G. Baldacchino and D. Milne (eds.), *Lessons in the political economy of small islands: the resourcefulness of jurisdiction*, Basingstoke, Palgrave Macmillan, pp. 214–229.

BLBRA (2017) Welcome to the Bras D'Or Lake Biosphere Reserve. Available from: http://blbra.ca.

Bramwell, B., and Lane, B. (1993) Sustainable tourism: an evolving global approach. *Journal of Sustainable Tourism*, 1(1), 1–4.

Briguglio, L., Archer, B., Jafari, J., and Wall, G. (eds.) (1996a) *Sustainable tourism in islands and small states: issues and policies*. London, Pinter.

Briguglio, L., Butler R, Harrison D., and Filho W.L. (eds.) (1996b) *Sustainable tourism in islands and small states: case studies*. London, Pinter.

Britton, S.G. (1982) The political economy of tourism in the Third World. *Annals of Tourism Research*, 9(2), 331–358.

Brown, K.G. (2010) Come on home: visiting friends and relatives: the Cape Breton experience. *Event Management*, 41(4), 309–318.

Brown, K.G., and Geddes, R. (2007). Resorts, culture, and music: the Cape Breton tourism cluster. *Tourism Economics*, 13(1), 129–141.

Butler, R.W (1993) Tourism: an evolutionary perspective. In J.G. Nelson, R. Butler and G. Wall (eds.), *Tourism and sustainable development: monitoring, planning and managing*, Department of Geography Publication Series No. 37, Waterloo, Canada, University of Waterloo.

Cabot Links (2017). Welcome to Cabot links. Available from: www.cabotlinks.com.

Cape Breton Post (2017, 8 February) Cape Breton population continues decline. Available from: www. capebretonpost.com/news/local/2017/2/8/cape-breton-population-continues-decline.html.

Charlie, C., King, B., and Pearlman, M. (2012) The application of environmental governance networks in small island destinations: evidence from Indonesia and the Coral Triangle. *Tourism Planning & Development*, 10(1), 17–31.

Clifton, J., and Benson, A. (2006) Planning for sustainable ecotourism: the case for research ecotourism in developing country destinations, *Journal of Sustainable Tourism*, 14(3), 238–254.

Cole, S. (2006) Information and empowerment: the keys to achieving sustainable tourism. *Journal of Sustainable Tourism*, 14(6), 629–644.

Cushnahan, G. (2004) Crisis management in small-scale tourism. *Journal of Travel & Tourism Marketing*, 15(4), 323–338.

CTV Atlantic (2017, January 24) Funding for second cruise ship berth announced for Sydney harbour. Available from: http://atlantic.ctvnews.ca/funding-for-second-cruise-ship-berth-announced-for-sydney-harbour-1.3254297.

De Albuquerque, K., and McElroy, J.L. (1999) Tourism and crime in the Caribbean. *Annals of Tourism Research*, 26(4), 968–984.

Dean, F. (2017, 15 February) Cape Breton enjoys 'Trump bump'. Available from: http://o.canada.com/travel/cape-breton-enjoys-trump-bump.

Destination Cape Breton (2017a) Cape Breton island: your heart will never leave. Available from: www. cbisland.com/about-cape-breton/the-island/.

Destination Cape Breton (2017b) Cape Breton connect: industry update. Available from: http://dcba-info.com.

DFO (2017) St Anns Bank marine protected area. Available from: www.dfo-mpo.gc.ca/oceans/mpa-zpm/stanns-sainteanne-eng.html.

Dickerson, H. (2008) Trouble in paradise. *Inside Indonesia*. Available from: www.insideindonesia.org/trouble-in-paradise.

Dodds, R., Graci, S., and Holmes, M. (2010) Does the tourist care? A comparison of visitors to Koh Phi Phi, Thailand and Gili Trawangan, Indonesia. *Journal of Sustainable Tourism*, 19(2), 207–222.

Doucette, M.B. (2008) *Membertou Heritage Park: community expectations for an Aboriginal cultural heritage centre*, Unpublished MBA research paper, Sydney NS, Cape Breton University.

Douglas, C.H. (2006) Small island states and territories: sustainable development issues and strategies. *Sustainable Development*, 14(2), 75–80.

DPA Consultants (1981) Regional Tourism Development Plan: Cape Breton region. Available from: www.openmine.ca/sites/default/files/4302_bdc.pdf.

Eber, S. (1992) *Beyond the green horizon: a discussion paper on principles for sustainable tourism*, Goldaming, Surrey, Tourism Concern/WWF.

Elliott, J. (1997) *Tourism: politics and public sector management*. New York, Routledge.

Farbotko, C. (2010) 'The global warming clock is ticking so see these places while you can': voyeuristic tourism and model environmental citizens on Tuvalu's disappearing islands. *Singapore Journal of Tropical Geography*, 31(2), 224–238.

Fennell, D. (2003) *Ecotourism*, 2nd edn. London, Routledge.

Golf Digest (2016) The world's 100 greatest golf courses. Available from: www.golfdigest.com/story/worlds-100-greatest-golf-courses-2016-ranking.

Gössling, S., and Wall, G. (2006) Island tourism. In G. Baldacchino (ed.), *A world of islands: an island studies reader*, Luqa, Malta and Charlottetown, Canada, Agenda Academic and Institute of Island Studies, pp. 429–453.

Graci, S. (2007) Accommodating green: examining barriers to sustainable tourism development. Paper presented at TTRA Canada. Available from: www.researchgate.net/publication/228464138_Accommodating_green_Examining_barriers_to_sustainable_tourism_development. Graci, S. (2013) Collaboration and partnership development for sustainable tourism. *Tourism Geographies*, 15(1), 25–42.

Graci, S. (2013) Collaboration and partnership development for sustainable tourism. *Tourism Geographies*, 15(1), 25–42.

Graci, S., and Dodds, R. (2010) *Sustainable tourism in island destinations*. London, Earthscan.

Hampton, M.P. (1998) Backpacker tourism and economic development. *Annals of Tourism Research*, 25(3), 639–660.

Hampton, M.P., and Jeyacheya, J. (2013) *Tourism and inclusive growth in small island developing states*. London, The Commonwealth Secretariat.

Hampton, M.P., and Jeyacheya, J. (2014) *Dive tourism, communities and small islands: lessons from Malaysia and Indonesia*. Hanoi, Vietnam, University of Social Sciences and Humanities (unpublished).

Hampton, M.P., and Jeyacheya, J. (2015) Power, ownership and tourism in small islands: evidence from Indonesia. *World Development*, 70(6), 481–495.

Hashimoto, A. (2004) Tourism and sociocultural development issues. In R. Sharpley and D.J. Telfer (eds.), *Tourism and development: concepts and issues*. Clevedon, Channel View Publications, pp. 202–230.

Hindley, A., and Font, X. (2014) Ethics and influences in tourist perceptions of climate change. *Current Issues in Tourism* (on-line), 1–17.

Hollinshead, K. (2004) Tourism and new sense. In C.M. Hall and H. Tucker (eds.), *Tourism and postcolonialism: contested discourses, identities and representations*, London, Routledge, pp. 25–42.

Icelandic Tourist Board (2009) Your official travel guide to Iceland. Available from: www.visiticeland.com.

Icelandic Tourist Board (2017) Tourism in Iceland in Figures: June 2017. Available from: https://www.ferdamalastofa.is/static/files/ferdamalastofa/Frettamyndir/2017/juli/tourism-in-iceland-2017-9.pdf

King, R. (1993) The geographical fascination of islands. In D.G. Lockhart, D. Drakakis- Smith and J.A. Schembri (eds.), *The development process in small island states*, London, Routledge, pp. 13–37.

Lockhart, D.G. (1994). Island environments: legacies, constraints and development problems. *Journal of Economic and Social Geography*, 84(5), 322–324.

Lockhart, D.G. (1997) Islands and tourism: an overview. In D.G. Lockhart and D. Drakakis-Smith (eds.), *Island tourism: trends and perspectives,* London, Pinter, pp. 3–21.

Löfgren, O. (2002) *On holiday: a history of vacationing*. Berkeley, CA, University of California Press.

Lynch, M.-F., Duinker, P., Sheehan, L., and Chute, J. (2011) The demand for Mi'kmaw cultural tourism: tourist perspectives. *Tourism Management*, 32(5), 977–986.

MacPherson, S., Maher, P.T., Tulk, J.E., Doucette, M.B., and Menge, T. (2016) *Eskasoni cultural journeys: a community-led approach to sustainable tourism development*. TTRA, Canada. Available from: http://scholar works.umass.edu/ttracanada_2016_conference

Mathieson, A. and Wall, G. (1982). Tourism, economic, physical and social impacts. New Zealand: Tourism Research and Development Group. Midland Group of Companies.

McCarten, J. (2016, June 2) Already smitten with Canada's Cabot Links, golf world falls for Cabot Cliffs. *The Globe and Mail*. Available from: www.theglobeandmail.com/feeds/canadian-press/sports/already-smitten-with-canadas-cabot-links-golf-world-falls-for-cabot-cliffs/article30245373/.

McElroy, J.L., and De Albuquerque, K. (1998) Tourism penetration index in small Caribbean islands *Annals of Tourism Research*, 25(1), 145–168.

McElroy, J.L., and Parry, C.E. (2010) The characteristics of small island tourist economies. *Tourism and Hospitality Research*, 10(4), 315–328.

Mowforth, M., and Munt, I. (2008) *Tourism and sustainability: development, globalisation and new tourism in the Third World,* London, Routledge.

Parks Canada (2017) Cape Breton Highlands National Park. Available from: www.pc.gc.ca/en/pn-np/ns/cbreton.

Péron, F. (2004) The contemporary lure of the island. *Journal of Economic and Social Geography*, 95(3), 326–339.

Royle, S.A. (2001) *A geography of islands: small island insularity*. London, Routledge.

Selwyn, P. (1975) *Development policy in small countries*. London, Croom Helm.

Statistics Canada (2016) Census profile: Cape Breton. Available from: www12.statcan.gc.ca/census-recense ment/2016/dp-pd/prof/details/page.cfm?Lang=E&Geo1=CMACA&Code1=225&Geo2=PR&Co de2=12&Data=Count&SearchText=Cape%20Breton&SearchType=Begins&SearchPR=01&B1=All.

Taber, J. (2012, 30 March) Golf course brings hope, and jobs, to Cape Breton community. Available from: www. theglobeandmail.com/news/national/golf-course-brings-hope-and-jobs-to-cape-breton-community/article2387915/.

Tolkach, D., and King, B. (2015) Strengthening community based tourism in a new resource based island nation: why and how? *Tourism Management*, 48(3), 386–398.

Travel and Leisure (2017) The best islands in Canada. Available from: www.travelandleisure.com/worlds-best/islands-in-canada#intro.

Ver Berkmoes, R., Skolnick, A. and Carroll, M. (2009) *Bali and Lombok*, 12th edn. Footscray, Australia, Lonely Planet.

Willmott, L., and Graci, S. (2012) Solid waste management in small island destinations: a case study of Gili Trawangan, Indonesia. *Teoros*, 31(3), 71–76. Available from: https://teoros.revues.org/1974.

11

MIGRATION

John Connell

Introduction

Islands are invariably and increasingly characterised by migration. Their beaches are real and metaphorical borders, crossing points and new destinies. As Braudel suggested, "The commonest ways in which islands entered the life of the outside world was by emigration" (1949/1972, p. 158). Once this concerned exploration, settlement, cultural contestation and carrying capacity, as island populations grew. Canary Islanders went to Cuba, Azores Islanders to Bermuda. More recently, and increasingly in the 21st century, there has been a shift of emphasis towards population decline, as islands often face the loss of vital workers, debates range around local constraints to development within a global political economy and the need for remittances. In contrast, there has been commuting, retirement migration and growing populations in islands closer to mainlands. Broadly, however, the future of island populations has become more challenging.

In many small island states and territories, migration became significant in the 1960s with the 'long boom', rising demand for labour in urban centres and metropolitan states, jet air travel and cheaper fares. Within a decade of that boom, smaller island states and territories, such as Niue and St Kitts, were already losing population and moving towards a 'culture of migration' where migration was normative and central to island life (Connell 2008). Migration became established as an expected and accepted phenomenon, an integral strand in individual, household and national concepts of development and a semi-permanent safety-valve, occasionally for population growth but usually for weak local and national economies (Connell and King 1999). By the 1960s, even on remote atolls such as Pukapuka and Manihiki (Cook Islands), over 90 per cent of adult populations had been away from the island at some point. Mobility was established as an island characteristic.

In the present century, migration has taken new forms in the wake of global economic shifts. It has become more skill-selective, with the 'brain-drain' increasingly exemplifying migration, whether of nurses or footballers, and women are more obviously a key component. Environmental changes have influenced migration and the spectre of climate change hangs over the destiny of many island populations, especially on low-lying coral atolls. Migrants from island states have gone to new destinations. Whereas early migration tended to follow colonial and postcolonial ties as well as linguistic and geographical proximity ties, it has now become gradually more global. Remittances have played an even more important role as island economies have struggled, and commodification of labour has become more apparent, especially in new pressures from many island states to negotiate migration ties with metropolitan nations. Smaller and more vulnerable islands have become more evidently peripheral and dependent parts of a wider world, with the global shift towards neo-liberalism and free trade. The life courses of island people, present or absent, are increasingly embedded in international ties. Island states, individuals, recruiters and various international agencies have attached new and increased significance to migration, remittance flows, return migration and the role of the diaspora, in contexts where 'conventional' development strategies have achieved limited success.

Seas of islands

Islands were always interconnected. None stood alone for long, especially when hazards struck, and even remote islands like Tikopia (Solomon Islands) were part of a 'world system' for thousands of years (Kirch 1986). Migration between islands occurred for various reasons, including hazards, trade and marriage, especially to counter limited productivity and carrying capacity. Some islands like Manam (PNG) had close relationships with mainlanders to exchange goods and to ensure migration options after, in this case, intermittent volcanic eruptions (Connell and Lutkehaus 2017). Occasionally, whole groups resettled on distant islands, as in Bougainville (Papua New Guinea), when 19th century refugees from local warfare managed to establish themselves on a new island site (Terrell et al. 1972); but such possibilities were rare. In pre-contact times, tensions were attached to all relationships across space and with different cultural groups. Nonetheless they were precedents for future migration as islanders sought to sustain and diversify livelihoods.

Early colonial times witnessed the forcible migration, initially as slaves, of plantation labourers from Africa and India to the Indian Ocean islands of Mauritius, the Seychelles and Réunion, unpopulated in the pre-colonial era. In the Caribbean, similar processes on most large islands resulted in African slaves, the cargo of the Middle Passage, replacing existing indigenous populations. Early contact times in the Pacific brought similar exploitation. Almost half the population of the three Tokelau atolls was forcibly taken to Peru in the 1860s as migrant labour, never to return (Maude 1981). At much the same time, blackbirding saw workers recruited from what are now the Solomon Islands, Vanuatu and the Loyalty Islands, to work on Australian plantations. In large parts of the tropics, forced migration in early colonial times both created and destroyed island populations.

When colonialism brought both a tentative peace and the first notions of modernity, with welcome material acquisitions, islanders' lives and livelihoods began to become entangled with distant places. Young men from Tikopia experienced an 'extreme eagerness to see the world . . . to become possessed of knowledge and property from which they can reap an advantage on their return' (Firth 1936, pp. 18–19). In the Caribbean, various local and regional patterns of migration followed emancipation as slaves sought to establish independent livelihoods (Richardson 1980). By the end of the 19th century, many islanders had travelled to work in central America, some remaining there, and, in the first two decades of the 20th century, thousands migrated to construct the Panama Canal, resulting in the first substantial flow of remittances to home islands (Richardson 1986). Migration was beginning to tentatively link islands to wider worlds, although it was another half century before what was still a trickle of migrants became a stream.

A population context?

Most islands and island states have now gone through the demographic (and epidemiological) transition, though population growth rates remain high in some places. Some, as in the Comoros and Solomon Islands, are over 2 per cent per annum. However, in many island states, growth rates are less than 1 per cent, partly because of high levels of outmigration. Life expectancies have risen over the past quarter of a century, but remain relatively low in Melanesian islands. A critical development issue is that of maintaining, let alone improving, present standards of living in the face of continued population increase, the absence of effective population policies and minimal economic growth. In certain localised contexts, population pressure on resources is perceived as a growing problem. Land is not always freely available, but zealously guarded by its traditional owners, and resource pressures have created tensions. In the Solomon Islands, for example, violent turn-of-the-century ethnic conflicts around Honiara were in some part a consequence of Malaitans leaving their own densely populated island and settling on the land of the local Guadalcanal population who resented the greater competition for land and scarce urban jobs (Moore 2004). Population pressures and their socio-economic consequences remain influential in stimulating migration.

In contrast, numerous islands once used for mining or copra plantations have been abandoned, including phosphate mining islands, from Klein Curaçao and Sombrero in the Caribbean, to Manra and Jarvis in Micronesia, and Clipperton in the eastern Pacific. More dramatically, the entire population of Banaba (Kiribati) was shifted by the British colonial government to Rabi island (Fiji), over 1,000 km away, to make way for phosphate mining, and only a few have returned (Teaiwa 2015, Connell and Tabucanon 2016). In the Tuamotus (French Polynesia), several atolls have been depopulated in the post-war years after their copra plantations became uneconomic, while others have been abandoned or left with 'caretaker populations' as they became unviable for social and economic life. Island populations on atolls like Sonsorol (Palau) are barely hanging on (Walda-Mandel 2016), while, on nearby Merir, one observer recorded that community's final phases: "The island is dying, at least as far as the present generation are concerned. . . . The women are too old to cultivate taro in any quantity, and the men cannot keep the taro groves cleared" (Osborne 1966, p. 49). For similar reasons, the isolated Scottish island of St Kilda was depopulated in 1930 after its population had fallen to 36, much like several islands on other European island fringes (Steel 1988, Royle 1999). Yet, other tiny island populations, notably that on Pitcairn, have survived long after their demise was predicted (Connell 1988, Amoamo 2015). Fetlar (Shetland Islands) holds on with fewer than 50 people (Grydehøj 2008); the Out Skerries have 76 but just one child, while Papa Stour survives with a population of fewer than 20 (when once it had over 350), despite feuds within the 'community' and some immigration after an appeal for residents in the 1970s. Such patterns of migration from islands were part of a steady and continuous process of small, remote islands no longer being able to sustain populations, lacking employment and services, with governments often supporting the process of population concentration. They are the contemporary equivalents of small islands like North Rona and the Shiants (Scotland), already depopulated by the end of the 18th century (Nicolson 2001). For centuries, but with gathering pace, island margins have been contracting, whether for environmental, economic or social reasons.

Strategic islands

Exceptionally, and mainly in earlier times, isolation was advantageous for certain activities, as forced migration took prisoners (and warders and support populations) to many remote islands, that became penal settlements and jails, including Devil's Island (French Guiana), St Helena, Norfolk Island and Isle of Pines (New Caledonia). But migration was ephemeral, and several such islands later experienced outmigration after penal functions changed, becoming too costly or archaic. In this century, again, the islands of Manus (PNG) and Nauru have become places of incarceration of migrant refugees (Loyd and Mountz 2014, Baldacchino 2014b), while others, such as Christmas Island, Lesbos and Lampedusa, in different seas, have become staging posts for migrants and refugees (King 2009, Coddington et al. 2012, Triandafyllidou 2014) with inadvertently growing populations.

Various islands, from Greenland to Ascension, have experienced immigration as they acquired strategic functions for defence and communications: contemporary equivalents of such Mediterranean islands as Malta. In the western Pacific, since the war, Okinawa, Guam and Tinian have acquired substantial military facilities, with accompanying immigration, welcomed for its economic impact, but disliked for its cultural impact (Owen 2010), or at a smaller scale, Kinmen (Taiwan) and several tiny islands in the South China Sea. Other islands like Mururoa (French Polynesia) had brief demographic lives, until their strategic and military value shifted. In contrast, in the 20th century, military exercises of various kinds have depopulated several islands, usually unwillingly, notably the Micronesian islands of Bikini and Enewetak (Marshall Islands) to make way for nuclear testing, and the Chagos Islands, in the Indian Ocean, where the indigenous Ilois population were removed to make way for American military establishments (Vine 2009, Gifford and Dunne 2014, Jeffery 2017). Each offer rare examples where isolated islands become desirable and local people are ignored and displaced, rarely to return.

An economic context

Migration is primarily a response to real and perceived inequalities in socio-economic opportunities, within and between islands and states: a straightforward search for social and economic mobility. Social influences are important, especially in terms of access to education and health services, and are in turn often a function of economic issues. Environmental hazards and degradation may also be catalysts. Expectations have risen over what constitutes a satisfactory standard of living, desirable employment and a suitable mix of accessible services and amenities. Employment crises, growing populations, inflation, static (or even falling) commodity prices and the declining availability of land in some areas, slowly increase the gap between expectation and reality. Agricultural work has lost prestige and the declining participation of the young is ubiquitous. Changes in values, following increased educational opportunities and the expansion of bureaucratic (largely urban) employment usually from the 1970s, the period when many island states became independent, have further oriented migration streams outwards, as local employment opportunities have not kept pace with population growth.

Economic growth has often been disappointing since independence, resulting in external interventions, including structural adjustment programmes, in various countries. In the smaller territories, weak economies have explained some reluctance to pursue independence, to retain a dependent political status that offered more economic and outmigration opportunities (Baldacchino 2014a). Only large islands, such as Mauritius, have experienced the migration of an industrial workforce (Lincoln 2009). Problems have intensified, with the shift from an older reliance on commodities following the liberalisation of trade and the multi-nationalisation of production. This has been especially challenging for countries producing commodities like sugar and bananas, notably the island states of the eastern Caribbean, where effective substitutes have yet to be found, and has led to rising unemployment.

Where islands have been able to benefit from tourism, notably the accessible Caribbean and Mediterranean islands, and others like Zanzibar (Gössling and Schulz 2005), but also some 'offshore' British, Australian, Italian and New Zealand islands, such as the Scilly Islands, Aran islands and Skye, Kangaroo Island and Magnetic Island, Capri and Waiheke, both economies and populations have stabilised or grown, or declined rather less than on more remote islands, as in-migration has occurred. Island states such as the Maldives, Cyprus, the Cook Islands and the Seychelles, territories such as Bermuda, Sint Maarten and Bonaire, and islands such as Langkawi, the Gili islands (Indonesia) and the Yasawas (Fiji) have all attributed most contemporary economic growth, and both inmigration and population stability, to tourism. Niche tourism has stabilised some remote places, such as Pitcairn and Fair Isle (Butler 2016), that might otherwise have experienced depopulation. Other niches, as diverse as gourmet foods and tax havens, have had the same effect on islands as different as Alderney, King Island and the Cayman Islands (Connell 1994, 2014, Khamis 2007): in every case, slowing outmigration and encouraging in-migration.

Most island states, particularly dependent states and remote islands, have benefited very substantially from aid and from remittances. Despite efforts at restructuring, the public sector increasingly dominates formal economic activity almost everywhere, turning several islands, like Niue, Norfolk Island, Christmas Island, St Pierre and Miquelon in the north Atlantic and Mayotte into 'government islands' (Connell 2013), where public service employment enables a prosperity and population stability that would be more difficult elsewhere. Many similar islands, such as Fetlar, Anticosti (Canada), Bornholm (Denmark), Heligoland (Germany) and Rapanui (Chile), have managed some diversification into tourism, enabling population stability or growth.

Archipelagic populations have become increasingly concentrated on more central urbanised islands, accentuating problems of service delivery in remote areas, in turn accounting for further migration and the slow depopulation of smaller, remote islands. That is particularly evident in migration from the outer islands of such Pacific states as Kiribati, Tuvalu, Palau, the Federated States of Micronesia (FSM), the Marshall Islands and French Polynesia, but also of migration within such archipelagos as the Aleutians (Alaska, USA), Åland Islands (Finland) and the Hebrides (Scotland, UK), and from what amounts to the 'outer islands' of Greenland (Connell 2016a). Ubiquitously, these flows illustrate an inexorable path dependency, and have intensified urbanisation. Earning power is increasingly concentrated amongst urban bureaucracies, while the absence of developed state mechanisms (such as progressive taxation, unemployment benefits and pension schemes) for affecting transfers of income, minimises redistribution towards rural areas other than through personal remittances. Growing inequalities, coupled with rising expectations, are the concomitants of increased migration.

Unemployment, poverty and the informal sector have become more visible in island towns, with the existence of significant numbers of unemployed and marginalised youth, a youth bulge that poses economic and social problems. Poverty is also hidden in outer islands, where there is a dearth of opportunity and minimal access to employment, education and health services (Connell 2006). Urban and rural safety-nets, where extended families support the unemployed, have gradually broken down. Inequalities are most evident in urban areas. Low incomes and a lack of support during illness or unemployment give a sense of biding time, waiting for uncertain opportunities, locally or overseas, and sometimes working illegally, and abandoning some 'traditional' obligations, simply to survive. The rise of both urban poverty and the informal sector has, in turn, stimulated emigration.

Underpinned by economic forces, an increasing gap between well-being in islands and in metropolitan states has encouraged migration. Perceptions of the wider world, its values, and its material rewards further underlie the migratory experience and, in an electronic era, knowledge of distant places is rather less fragmentary. In many places, and well illustrated by Niue and Cape Verde (Connell 2008, Carling 2002), the culture of migration was accepted as an appropriate and legitimate means towards economic and social well-being; it was neither rupture nor discontinuity with island experiences, but an integral part of island life. When the Marshall Islands, the FSM and Palau negotiated independence through a Compact of Free Association with the

USA, they built in the right to migration, which has now resulted in steady migration from each state to the USA. Tokelau and the Cook Islands have resisted independence in case it jeopardised migration to New Zealand, as have the Caribbean islands of Guadeloupe and Martinique with France. Larger islands and metropolitan countries exercise a powerful allure, offer a sense of future, and simply validate migration.

Society and environment

Economic issues are a critical but not the only influence on migration, especially where small islands have limited social, educational and medical facilities. Migration for education has provided outlets for the young, and reduced return migration while also sustaining the bureaucracy Jacobs and Overton (2016). Many migrate for their children's educational and employment future, and medical referrals may eventually become migration moves. In very small islands, some sporting activities are impossible, marriage partners are scarce, church congregations sparse, life is unduly repetitive and boredom encourages migration (Connell 2013). Political factors have also been significant influences in various contexts, notably in the migration from the Comoros and Fiji after coups, and in skilled migration from Bougainville and the Solomon Islands during violent crises. In Fiji, migration could be seen as 'a barometer of fear' of further conflict (Narayan and Smyth 2003).

Environmental factors have similarly influenced migration. Hazards are catalysts for mobility that might have happened anyway. Cyclones have resulted in depopulation in several Tuamotu islands. Cyclone Heta, which destroyed almost a quarter of the housing stock on Niue in 2004, even prompted some thoughts about the permanent abandonment of the island, as had Cyclone Ofa two decades earlier (Barker 2000). Volcanic eruptions have temporarily depopulated Tristan da Cunha, Vestmann (Iceland) and even more recently the northern Marianas and Manam (PNG) with always incomplete return migration (Connell and Lutkehaus 2017). Half the population of Montserrat left in the aftermath of the eruption of Mount Soufrière in the mid-1990s (Philpott 1999). Migrants have also left atolls like those of Tuvalu, the Carteret Islands (PNG) and the Reef Islands (Solomon Islands), in anticipation of sea level rise but primarily in response to the economic challenges that atolls experience (Connell 2003, 2016b, Mortreux and Barnett 2009, Birk and Rasmussen 2014, Constable 2017, Speelman et al. 2017). In the Maldives and Kiribati, it is anticipated that migration may be necessary in the future (Stojanov et al. 2017, Allgood and McNamara 2017, Hermann and Kempf 2017) so that environmental influences will complement and intensify economic and social factors (Marino and Lazrus 2015). In various islands in the Solomon Islands and Fiji, localised erosion and environmental degradation has resulted in coastal villagers changing their sites, usually by nearby migration inland and upwards (e.g. Charan 2017). In rare instances, as in the islands of Chesapeake Bay, and the ephemeral char islands of Bangladesh, households have migrated as islands have flooded (Gibbons and Nicholls 2006, Poncelet et al. 2010). Small island states are in no position, financially, geographically or politically, to defend themselves against serious environmental threats, so that islanders may one day become a flow of 'environmental refugees' to metropolitan states, in the face of scarce resettlement opportunities nearby (Connell 2012), raising new questions about identity, sovereignty and citizenship (Farbotko et al. 2016) and of the legal and constitutional status of abandoned homelands.

International migration

International migration from islands has accelerated since the 1960s, and again in the present century, as islanders have sought employment and services in metropolitan states. It had precedents in slavery, blackbirding and West Indian migration to construct the Panama Canal, but the second half of the 20th century brought somewhat more enlightened times, as West Indians were contracted to work for London transport and in the National Health Service (e.g. Western 1992). That engendered the first substantial permanent migration streams, raising issues of island identity, belonging and exclusion, examined in academic studies and through a flurry of poetry and novels from such writers as Kamau Brathwaite, Derek Walcott, Caryl Phillips and Jamaica Kincaid (e.g. Hughes 1999). Cultural theorists, such as Stuart Hall (1990) and Paul Gilroy (1987), instigated and developed critical themes that rather later transferred to other island destinations like New Zealand, with similar accompanying literary approaches (Connell 1995). Others remembered island homes, and waxed nostalgic in musical themes (Connell 1999).

For small islands and island states, including the Cook Islands, Niue, Tokelau, Anguilla and Montserrat, migration has been particularly dramatic since a majority of the indigenous population live overseas, and national populations are declining. Several islands experience situations like that of Cape Verde, where as many as half of all those born there live overseas (Carling 2002), and Niue which has a resident population of about 1,200, but over 5,000 Niue-born live in New Zealand, alongside their dependents. Equally for many islands, like Tikopia and Namoluk, most 'islanders' live elsewhere (Marshall 2004). Extremes of depopulation have meant that Niue has sought immigration from Tuvalu, as its population has declined, while Pitcairn has unavailingly sought to encourage its diaspora to return (Amoamo 2015), just as various Scottish islands have launched bids for new settlers.

Islands where migration was of relatively slight importance at independence have become highly dependent upon it. International migration has tended to follow past or present colonial ties, and related cultural and linguistic characteristics. The former US territories in Micronesia – the now independent states of Palau, FSM and the Marshall Islands, and especially American Samoa – have become similar to other Pacific islands: with a growing interest in migration, a steady outflow, relatively permanent urban communities overseas, remittance flows and declining or static populations. Likewise, colonial ties have taken some Caribbean islanders to France and others to the USA, the UK and the Netherlands. Cape Verdeans have mainly gone to Portugal. But migrants take their chances. Africans and Syrians have moved into Malta, Haitians into the Bahamas, and Cubans in the Cayman Islands. In the Indian Ocean, Mayotte (now a French department and part of the EU) has similarly experienced massive migration from the nearby independent Comoros. Wallisians travel to New Caledonia. In these, often unwelcoming, islands population growth has usually been rapid.

Remittances

Migration decisions are usually shaped within a family context, as migrants leave to meet certain family expectations. Migration is directed both at improving the living standards of those who remain at home and improving the lifestyle and income of the migrants. Consequently, as in Tonga, "families deliberate carefully about which members would be most likely to do well overseas and be reliable in sending remittances" (Gailey 1992, p. 465). Through this process, extended households have become 'transnational corporations of kin' which, in some circumstances, strategically allocate family labour to local and overseas destinations to maximise income opportunities, minimise risk, and benefit from resultant remittance flows (Marcus 1981). To a greater extent than for internal migration, international migration is more evidently an economic phenomenon.

In many countries, international remittances form a significant part of national and household incomes. Hence by the 1980s, small Pacific island states (initially Kiribati, Tokelau, Cook

Islands and Tuvalu) had become conceptualised as MIRAB states, where MIgration, Remittances, Aid and the resultant largely urban Bureaucracy were central to the socio-economic system (Bertram and Watters 1985). The notion of MIRAB is also applicable in rather larger states, such as Cape Verde, Samoa and Tonga, where remittances constitute some of the highest proportions of GNP of any country in the world, and in many similar global contexts, where remittances are the single most important source of national income. While the acronym is disliked, for cultural reasons and because of its implication of a 'handout mentality', it nonetheless suggests the centrality of migration and remittances in islands and island states (Connell and Brown 2005, Brown and Connell 2015, Bertram 2006).

Remittances are particularly important in small states and islands, where conventional economic development is difficult (Connell 2013), and have been crucial for replacing income once generated by now stagnating rural economies. Above all, remittances have increased island resilience by enabling diversification of livelihoods (e.g. Christensen and Gough 2012, Wilson 2013) and been invaluable after hazards (Attz 2009, Le De et al. 2013). Remittances are used for debt repayment, new forms of consumption, housing and some community goals (such as schools and churches), air fares, education (the creation of social capital), and various forms of investment, sometimes in the agricultural sector but more frequently in the service sector. Increased use of remittances for investment purposes, in fishing, agriculture, stores and transport businesses, attests to the shift from consumption to investment (Connell and Conway 2000, Lewis and Kirton 2015, Brown and Connell 2015). This transition benefits economic development, but emphasises intra-village (and island) economic inequalities and may hamper social development. Yet, in 'mature remittance economies' when almost all households are linked into remittance flows, as in Tonga and Cape Verde, inequality then declines (Åkesson 2009, 2013, Brown et al. 2014).

Because of the continued and increasing significance of remittances, the sustainability of remittance-dependent development is particularly important but necessarily uncertain. Though even skilled migrants, such as Tongan and Samoan nurses in Australia, sustain remittance levels at high levels over long time periods (Brown and Connell 2004), with family reunification and greater integration of migrants in host communities, the ability and willingness to remit eventually decline. Links with second-generation migrants become more tenuous as newer generations act as individuals rather than perceive themselves as members of wider transnational social groups, especially where the number of overseas born 'islanders' becomes the majority. As overseas generations lose language and cultural skills, or 'marry out', their sense of belonging declines, identity shifts, and without renewed migration island futures are less certain.

Remittances from national migrants are invariably smaller than those from international migrants but of substantial importance in countries like Solomon Islands and Vanuatu, where international migration is unusual other than for seasonal work (Gibson 2015, Cummings 2016). Small islands, such as Mbuke (PNG), where alternative sources of income are scarce, have thus developed a 'singaut economy', where urban migrants are constantly bombarded with requests (Rasmussen 2015) and most islands have a bidirectional remittance system, as those at home send local goods to remind migrants of their obligations and identity (Petrou and Connell 2017). Almost ubiquitously mobile phones and Facebook connections have kept migrants in touch with kin and stimulated remittance flows.

The economic future of several island states, such as Tonga, Samoa and Cape Verde, hinges somewhat on the continued flow of remittances, and hence on continuity of migration (Carling 2002), a situation made more difficult where migration becomes more selective. The possibility of blocked migration in the future, a situation rarely absent from public debate on many islands, also emphasises potential problems of high rates of natural population increase and population pressures on resources if emigration is substantially reduced.

Selectivity and skilled migration

In recent years, migration opportunities in metropolitan states have tended to decline, and be targeted at skilled migrants, rather than enabling family reunions. Male migration usually preceded female or family migration from most islands, especially where plantation labour was required, but preferences have shifted towards women. Some occupational categories are either male or female dominated. For instance, Fijian women have migrated as nurses, domestic helpers and caregivers; while Fijian men have moved overseas as soldiers, tourism workers, pilots and security guards (Voigt-Graf and Connell 2006). More generally, the demand for service sector workers, in areas from domestic service and caring to tourism, information technology and health, has shifted the balance in favour of women.

Migration is always selective. The more educated have tended to migrate first; and migrants have left many rural areas to take advantage of superior urban educational and employment opportunities. Selectivity is increasingly global. Thus migration flows from the Caribbean and Pacific are increasingly likely to be of skilled migrants from various sectors, including health (Connell 2004, 2009, Lewis 2011) and education (Voigt-Graf 2003, Alexander 2016, Jacobs and Overton 2016). Structural changes within metropolitan states have meant that certain sectors, notably health, are short of skilled workers. Island nurses, usually entering the bottom levels of the 'global health care chain', have migrated much greater distances, as demand intensifies. Even newer movements have become particularly important in the last couple of years, with migration from Fiji to the Middle East, emphasising the manner in which new and highly paid overseas employment opportunities are being firmly grasped, even in an unappealing and violent context, and have quickly turned Fiji into a major recipient of remittances (Connell 2006, Kanemasu and Molnar 2017). Rather differently, island rugby players have gone beyond the 'traditional' destinations of New Zealand and Australia to Japan and the UK. Pacific islanders are present almost everywhere that rugby is played professionally (Grainger 2011, Kanemasu and Molnar 2012, Besnier 2012). Cricket and football players have left Caribbean islands, Mauritius and Réunion, to play in Europe, some to be exploited, some to amass wealth, and some, like Dwight Yorke in Tobago and Jonah Lomu from Tonga, to become national symbols of global success.

The proportion of skilled islanders among all migrants is increasing, as a result of shortages in the receiving countries, some of which – as with the health services of New Zealand, the UK and the USA – have led to both private sector recruitment (Connell and Stilwell 2006) and a global ethical code opposing excessive recruitment (Connell and Buchan 2011). Low remuneration, poor promotion opportunities, limited training and educational opportunities, poor working and living conditions, particularly in remote regions, are push factors for skilled migrants (Connell 2009). The growing shortage of skilled workers has also contributed to increased inter-island migration, such as Fijian teachers migrating to the Marshall Islands, and Fijian tourism workers moving to the Cook Islands.

The brain-drain has become excessive in some small island states. There is a widespread shortage of health workers in Caribbean and Pacific islands, with negative impacts on the quality and availability of health care (Connell 2009, Lewis and Kirton 2015). Doctors are twice as likely to migrate as nurses because wage differentials are greater, and because most nurses are women and men are often the primary decision makers regarding migration (Brown and Connell 2004). Many islands and island states now also experience a shortage of airline pilots, accountants, engineers and IT workers, and where losses can be quantified, reduced productivity has occurred, for example, in the Jamaican construction industry (Connell 2013, Lewis and Kirton 2015). Ironically, many skilled migrants also become part of a 'brain loss' or 'brain-waste' because their qualifications, despite gaining them entry, are unrecognised in the destination country. The combination of changing aspirations and the migration of the more educated young contributes to the brain and skill drain especially from national peripheries and from small states. The shift to skilled migration, despite remittances, has been at some cost to development in islands and island states.

Costs to migration can also occur where children are left behind when migrant parents leave. Where children cannot easily be supported by the remaining kin transnational lives become a 'painful necessity' (Åkesson et al. 2012), where some pain is experienced by those who stay, especially the 'barrel children' of the Caribbean (Jokhan 2008). The migration of men may likewise place unacceptably large demands on women even with large remittances and new autonomy.

Return migration and retirement

Rarely has substantial return migration occurred to most islands and island states, despite the centrality of an ideology of return. Return has been greatest where distances have been less and economic opportunities greater, and consequently least in more remote islands. Similarly, circulation is more common where there is ease of contact. In the island Pacific, limited return migration is at least partly due to the great differences in income levels with the metropolitan periphery, but also to a host of social factors. In other contexts, the return of those with distantly acquired skills has gained them new status. Return migration has often been of unskilled workers and retirees, and may be just as diverse as migration, also in its outcomes, leading some to stay and some to migrate once again (Conway et al. 2005, 2012, Maron and Connell 2008).

The return migration of those with skills has tended to be limited, partly where those skills cannot be practised locally, but more frequently because local salaries are comparatively poor. Skilled migrants nonetheless do return, despite the discrepancy in wages and working conditions, for family reasons, to retire or to establish businesses (Brown and Connell 2004). Teachers more often return because of the ease of finding public sector employment (Iredale et al 2015). Since economic opportunities are limited in many islands, return migrants tend to be absorbed within the service sector, rather than directly in production. Hence on small islands, such as Tubuai (French Polynesia) and in Kiribati, although returnees are anxious to invest, opportunities are so limited that this is not always feasible (Lockwood 1990, Borovnik 2006). Disappointments discourage other potential returnees.

Limited return is also a function of a social context where the children of migrants are educated in the destination, have lost some contact with 'home' societies, perhaps losing critical linguistic and other skills, and perceive few opportunities to use and benefit from skills acquired overseas (Connell 2006). More exceptionally, some have been deported, as in the Marshall Islands, Tonga and especially Jamaica, where between 2006 and 2010 some three-quarters of all returning Jamaicans were deportees posing problems for their home states (e.g. Golash-Boza 2014). Rather differently return migration may be enforced by parents anxious to ensure that their children have not lost touch with home cultures (e.g. Lee 2016) or even that they provide a form of 'homeland security' should migrants ever need to return (Rauchholz 2012). Return migration is constantly deferred (such as 'until children leave school', 'until enough money is saved' or 'until retirement') up to the point where it becomes implausible. There can also be resistance to return migration in islands where people who stayed resent returnees as having 'voted with their feet' to leave and but then return to compete for scarce resources (land, accommodation) and opportunities (well-paying jobs). Limited return is also linked to a gradual shift in the demographic balance; relatives are increasingly likely to be found in destinations and thus there is reduced incentive to return to what is less evidently 'home'.

In contrast, in islands of a particular perceived charm, and sometimes elitism (such as Mustique, St Vincent), of advantageous tax status (such as the Channel Islands and Norfolk Island), and, more generally, where they are not too remote, island populations and economies have been boosted by both the emergence of second homes and retirement migration. This has been especially so in and around southern Europe and its mid-Atlantic extension, from the Canaries and Madeira, to Crete, Corfu and Cyprus, stimulated by relatively cheap property, climate and environment, low crime rates, wholesome diets, a semblance of home amenities, and adequate air connections (Connell and King 1999), but also domestically on accessible yet 'cold water' islands in Sweden, Finland, Scotland and elsewhere (King 2009, Philip et al. 2013). Alongside tourism, second home and retirement migration has also transformed some once-dying islands, through new injections of capital – and thus employment – and rising populations (even if sometimes seasonally). Yet ubiquitously, from Papa Stour and Whalsay in the Shetland Islands (Cohen 1982, Grydehøj 2008) to Grand Manan (Canada) (Marshall 1999), much like return migration, new 'outsiders' may be resented, remain temporarily and pursue lives dissonant and apart from those of indigenous islanders, symbolised by the title of one account of recent settlers to Prince Edward Island, Canada: 'Come visit, but don't overstay' (Baldacchino 2012). Nonetheless, the presence of newcomers has sometimes provided a demographic lifeline which return migration may have failed to do.

A policy context?

Within islands and island states, there is some ambivalence towards migration. No island state has recently sought to discourage international migration, though several have expressed particular concerns, mainly attached to the loss of skilled labour. Yet, for the past quarter of a century, most governments have neither sought to intervene in emigration nor tried to curb the loss of skilled and unskilled workers. Indeed, there has been a perceptible shift towards the encouragement of migration, due to difficulties in national economic development, and some preference for skilled migration, described by Anote Tong, the President of Kiribati, as "migration with dignity", since skilled migrants would be more successful and remit greater sums of money.

Colonial administrations sometimes intervened in population mobility. In the 1930s, the UK resettled people from what were deemed overcrowded islands in the Gilbert Islands (now Kiribati) initially to the Phoenix Islands and subsequently, in the 1950s, to another British colony: the Solomon Islands (Maude 1968). Vaitupu (Tuvalu) islanders were resettled on Kioa island in Fiji at around the same time. Such moves across what are now international borders are no longer regarded as politically feasible, and were blighted by disputes over cultural differences, and conflicts over access to land, employment and services (Connell 2012, Donner 2015). A somewhat similar movement was instigated by New Zealand from the densely populated atolls of Tokelau to New Zealand from the 1970s that was initially envisaged to include the entire Tokelauan population. When the scheme was finally ended, a third of the population had been resettled (Wessen et al. 1992). In each case, migration took islanders from relatively impoverished coral atolls, to larger states and high islands. While the motivation was linked to overcrowding and lack of economic opportunities, such moves have resonance and relevance for future contexts where sea level rise may require further uprooting and resettlement.

Increasingly, several small island states have trained workers for overseas employment, notably in the Marine Training Schools of Tuvalu and Kiribati (Borovnik 2006). Kiribati itself has sought to extend the seafarer programme to cover women, nurses and hospitality workers and benefit from superior remittance flows. It has tentatively enabled limited female employment on cruiseships (Kagan 2016). A long history of migration accounts for workers migrating (with colonial or independent government acquiescence) from Caribbean islands to Britain to work in the transport and health sectors, and to France, the latter encouraged after 1963 by a special Bureau to stimulate such migration (Aldrich and Connell 1992). These schemes initiated a major shift in the demographic balance between many Caribbean islands and metropolitan countries.

Late in the 20th century, Canada opened up temporary agricultural employment to migrants from the Caribbean and Mexico, a model that was later extended to New Zealand and Australia, where guest workers from Pacific islands stayed for up to seven months and returned with reasonable incomes, used much as remittances were, but not without social problems reminiscent of those in the blackbirding schemes a century earlier, notably the creation of a precariat(Hennebry and Preibisch 2012, Hammond and Connell 2009, Bedford et al. 2010, Connell 2010, Cummings 2016, Petrou and Connell 2018). More recently, such schemes have been tentatively extended beyond agriculture, with small numbers in the hospitality and fishing industries, for what are seen as the 'orphan states' of Nauru, Tuvalu and Kiribati. Once concerned over the diverse costs of migration, islanders, from individuals and households to national governments, are increasingly likely to seek out new opportunities.

Conclusion: the outward urge

In this century alone, there has been an increase in overseas migration from islands and island states, and an unmet demand for it, from individuals and governments, who have put increased pressure on metropolitan countries as their national economies stagnate. Migration has become a more conscious household strategy, because of its crucial role in diversifying livelihoods and generating remittances, as island households seek education and employment in careers with scope for migration. Migration has become more complex, and more selective by skill and gender, while globalisation has extended the range of destinations beyond former colonial ties and brought longer migration chains. Island states themselves are increasingly training individuals for migration. The extent of international migration, and growing demands for migration opportunities, is indicative of how superior economic opportunities are being perceived to lie overseas rather than in the islands themselves. For even relatively large island states, such as Cape Verde and Samoa, there are now as many islanders overseas as there are at home. The future – a diasporic future – lies elsewhere.

A slow, centuries-long withdrawal is increasingly evident in the smallest, remotest states and islands, where retaining a viable and satisfying lifestyle has proved particularly challenging and relative deprivation increasingly tangible. Limited resources of guano and copra have gone, or been lost to market competition, and sheep husbandry holds few attractions. Without an effective private sector, and where even tourist markets are absent or erratic at best, partly in the absence of adequate transport links, many islands and island territories have become subsidised government islands, in a global climate where aid fatigue and neo-liberalism now threaten such subsidies and the lifestyles they support.

Islands have experienced a growing dependence on external sources of funding, whether from aid, remittances or investment. The rise of neo-liberalism, the shift to free trade and external pressures for privatisation and restructuring, have put new pressures on islands. Islanders have sought remedies beyond their shores, in some respects as they always did, but are vulnerable to more restrictive migration policies. International migration has deferred and mitigated, but not resolved, issues of poverty; while it constitutes one increasingly less hesitant solution: an expanding and unsatisfied outward urge, a bottom-up globalisation that continues to draw the most remote islands and islanders into new international networks (Connell and Corbett 2016).

Migration primarily occurs out of an economic context of relative poverty (of both incomes and opportunities), accentuated by social factors. In the short run, distinct benefits accrue to individual migrants, their families and sending societies. Migration reduces the level of open and disguised unemployment, and produces remittances; but it has also led to the loss of scarce skilled human resources from the 'modern' sector, hindering development. Skill losses are partly compensated for by remittances. However, remittances are primarily private income flows (other than to churches and occasional village projects) whereas the costs of training skilled migrants are usually borne by the public sector. The shift towards more skilled migration has created a new dimension of inequality. The poor are not easily able to move, whereas the relatively rich (or at least those who have acquired training and marketable skills) are actively courted and recruited, even via 'citizenship by investment' schemes. Rural migrants from island states, such as Vanuatu and Tonga, are being displaced by better-off urban workers anxious for quick, short-term gains, threatening policies partly designed to reduce regional inequality. However, in most islands, people have no wish to abandon their homes, even in anticipation of future environmental challenges. Still, uncertainty from the effects of climate change is likely to make some islands less (and others more) habitable, as the strategic value of Arctic islands changes.

In many contexts, as many islanders are now overseas as at 'home', bearing a nomadic identity that can be construed as a powerful transnationalism, as much as a tragic rootlessness. Notions of home thus change, provoking debates over the extent to which cultures are retained or transformed, and become hybrid or syncretic, to reflect the clash or combination of new residences, roots and routes. Traditions become reconstructed, articulated in ever-changing forms and evident across now deterritorialised nations, islands and cultures. Whilst identities may be hybrid and hyphenated, they may also, as in the case of Samoans in New Zealand, represent successful cosmopolitans 'at ease in multiple worlds, rather than natives of place torn by new and multiple allegiances' (Yi-Fu Tuan, quoted in Western 1992, p. 269). While cultures shift, citizenship changes and new loyalties emerge, ties to home have remained remarkably resilient.

Migrants and their children remain 'migrants' though their identities have changed. New technology has made connectivity both more fashionable and more feasible; the island Pacific, fully a third of the earth's surface, has experienced a new "cartography of compression" (Kempf 1999), where telephones, email and chat pages have turned young Polynesians into 'cyber-Polys' (Morton 1999), and electronic identities have brought new transnational ties (Howard and Rensel 2004). Nostalgia and electronics are united. At one level, nothing remains the same. On Kairiru island (PNG), "villagers see migrants as tendrils reaching out from Kragur [village] to the wider world to draw in new strength, while migrants regard themselves as seeds carried from the village to take root elsewhere" (Smith 2013, p. 163). But, as noted by Tancredi in *The leopard* (set in the nineteenth century on the island of Sicily), for things to stay the same, everything must change (Di Lampedusa 1961). Migrants can now be reminded all too easily of their 'duties', obligations, nationhood and identity. Home nations and islands are powerful unifying symbols for migrants and their children. Symbolically, and practically, through land tenure, islanders rarely abandon island homes, and social media contently emphasise continuity. It is a world and centuries away from the plantation labour migration that once dominated mobility.

References

Åkesson, L. (2009) Remittances and inequality in Cape Verde: the impact of changing family organisation. *Global Networks*, 9(3), 381–398.

Åkesson, L. (2013) The queue outside the embassy: remittances, inequality and restrictive migration regimes. *International Migration*, 51(s1), e1–e12.

Åkesson, L., Carling, J., and Drotbohm, H. (2012) Mobility, moralities and motherhood: navigating the contingencies of cape Verdean lives. *Journal of Ethnic and Migration Studies*, 38(2), 237–260.

Aldrich, R., and Connell, J. (1992) *France's overseas frontier: départements and territoires d'Outre-Mer*. Cambridge, Cambridge University Press.

Alexander, R. (2016) Migration, education and employment: socio-cultural factors in shaping individual decisions and economic outcomes in Orkney and Shetland. *Island Studies Journal*, 11(1), 177–192.

Allgood, L., and McNamara, K. (2017) Climate-induced migration: exploring local perspectives in Kiribati. *Singapore Journal of Tropical Geography* (38), 370-385.

Amoamo, M. (2015) Engaging diasporas for development: a case study of Pitcairn Island, *Australian Geographer*, 46(3), 305–322.

Attz, M. (2009) Natural disasters and remittances: poverty, gender and disaster vulnerability in Caribbean SIDS. In W. Naudé, A. Santos-Paulino and M. McGillivray (eds.), *Vulnerability in developing countries*, Tokyo, UNU Press, pp. 249–261.

Baldacchino, G. (2012) 'Come visit, but don't overstay': critiquing a welcoming society. *International Journal of Culture, Tourism and Hospitality Research*, 6(2), 145–153.

Baldacchino, G. (2014a) *Island enclaves: offshoring strategies, creative governance and subnational island jurisdictions*. Montreal, QC, McGill-Queens University Press.

Baldacchino, G. (2014b) Islands and the offshoring possibilities and strategies of contemporary states: insights on/for the migration phenomenon on Europe's southern flank. *Island Studies Journal*, 9(1), 57–68.

Barker, J. (2000) Hurricanes and socio-economic development on Niue Island. *Asia Pacific Viewpoint*, 41(1), 191–205.

Bedford, R., Bedford, C., and Ho, E. (2010) Engaging with New Zealand's recognised seasonal employer work policy: the case of Tuvalu. *Asian and Pacific Migration Journal*, 19(3), 421–445.

Bertram, G. (2006) The MIRAB model in the twenty-first century. *Asia Pacific Viewpoint*, 47(1), 1–14.

Bertram, G., and Watters, R. (1985) The MIRAB economy in South Pacific microstates. *Pacific Viewpoint*, 26(4), 497–520.

Besnier, N. (2012) The athlete's body and the global condition: Tongan rugby players in Japan. *American Ethnologist*, 39(3), 491–510.

Birk, T., and Rasmussen, K. (2014) Migration from atolls as climate change adaptation: current practices, barriers and options in Solomon Islands. *Natural Resources Forum*, 38(1), 1–13.

Borovnik, M. (2006) Working overseas: seafarers' remittances and their distribution in Kiribati. *Asia Pacific Viewpoint*, 47(1), 151–162.

Braudel, F. (1972) *The Mediterranean and the Mediterranean world in the age of Philip II*. New York, Harper and Row (1St edn. 1949).

Brown, R., and Connell, J. (2004) The migration of doctors and nurses from South Pacific island nations. *Social Science and Medicine*, 58(11), 2193–2210.

Brown, R., and Connell, J. (2015) Remittances and migration. In Connell and R. Brown (eds.), *Migration and remittances*, Cheltenham, Edward Elgar, pp. xiii–lxviii.

Brown, R., Connell, J., and Jimenez-Soto, E. (2014) Migrants' remittances, poverty and social protection in the South Pacific: Fiji and Tonga. *Population Space and Place*, 20(5), 434–454.

Butler, R.W. (2016) Changing politics, economics and relations on the small remote island of Fair Isle. *Island Studies Journal*, 11(2), 687–700.

Carling, J. (2002) *Cape Verde: towards the end of emigration?*, Washington, DC, Migration Policy Institute.

Charan, D., Kaur, M., and Singh, P. (2017) Customary land and climate change induced relocation: a case study of Vunidogoloa village, Vanua Levu, Fiji. In W. Filho (ed.), *Climate change adaptation in Pacific Countries*, Cham, Springer, pp. 19–33.

Christensen, A., and Gough, K. (2012) Island mobilities: spatial and social mobility on Ontong Java, Solomon Islands. *Geografisk Tidsskrift – Danish Journal of Geography*, 112(1), 52–62.

Coddington, K., Catania, R., Loyd, J., Mitchell-Eaton, E., and Mountz, A. (2012) Embodied possibilities, sovereign geographies and island detention. *Shima*, 6(2), 27–48.

Cohen, A. (1982) A sense of time, a sense of place: the meaning of close social association in Whalsay, Scotland. In A. Cohen (ed.), *Belonging, identity and social organisation in British rural cultures*. Manchester, Manchester University Press, pp. 21–49.

Connell, J. (1988) The end ever nigh: contemporary population changes on Pitcairn Island. *GeoJournal*, 16(2), 193–200.

Connell, J. (1994) The Cayman islands: economic growth and immigration in a British colony. *Caribbean Geography*, 5(1), 51–66.

Connell, J. (1995) In Samoan worlds: culture, migration, identity and Albert Wendt. In R. King, J. Connell and P. White (eds.), *Writing across worlds: literature and migration*, London, Routledge, pp. 263–279.

Connell, J. (1999) 'My island home': the politics and poetics of the Torres Strait. In R. King and J. Connell, (eds.), *Small worlds, global lives: islands and migration*, London, Pinter, pp. 195–212.

Connell, J. (2003) Losing ground? Tuvalu, the greenhouse effect and the garbage can. *Asia Pacific Viewpoint*, 44(1), 89–107.

Connell, J. (2004) The migration of skilled health professionals: from the Pacific islands to the world. *Asian and Pacific Migration Journal*, 13(2), 155–177.

Connell, J. (2006) Migration, dependency and inequality in the pacific: old wine in bigger bottles? In S. Firth (ed.), *Globalisation and governance in the Pacific Islands*, Canberra, ANU E-Press, pp. 59–106.

Connell, J. (2008) Niue: embracing a culture of migration. *Journal of Ethnic and Migration Studies*, 34(6), 1021–1040.

Connell, J. (2009) *The global health care chain: from the Pacific to the world*. London, Routledge.

Connell, J. (2010) From blackbirds to guestworkers in the South Pacific: plus ça change. . .? *Economic and Labour Relations Review*, 20(2), 111–122.

Connell, J. (2012) Population resettlement in the Pacific: lessons from a hazardous history? *Australian Geographer*, 43(2), 127–142.

Connell, J. (2013) *Islands at risk? Environments, economies and contemporary change*. Cheltenham, Edward Elgar.

Connell, J. (2014) Alderney: gambling, bitcoin and the art of unorthodoxy. *Island Studies Journal*, 9(1), 69–78.

Connell, J. (2016a) Greenland and the Pacific islands: an improbable conjunction of development strategies. *Island Studies Journal*, 11(2), 465–484.

Connell, J. (2016b) Last days in the Carteret Islands? Climate change, livelihoods and migration on coral atolls. *Asia Pacific Viewpoint*, 57(1), 3–15.

Connell, J., and Brown, R. (2005) *Remittances in the Pacific: an overview*. Manila, The Philippines, Asian Development Bank.

Connell, J., and Buchan, J. (2011) The impossible dream? Codes of practice and the international migration of skilled health workers. *World Medical and Health Policy*, 3(3), 1–17.

Connell, J., and Conway, D. (2000) Migration and remittances in island microstates: a comparative perspective on the South Pacific and the Caribbean. *International Journal of Urban and Regional Research*, 24(1), 52–78.

Connell, J., and Corbett, J. (2016) Deterritorialisation: reconceptualising development in the Pacific islands. *Global Society*, 30(4), 583–604.

Connell, J., and King, R (1999) Island migration in a changing world. In R. King and J. Connell (eds.), *Small worlds, global lives: islands and migration*, London, Pinter, pp. 1–26.

Connell, J., and Lutkehaus, N. (2017) Escaping Zaria's fire? The volcano resettlement problem of Manam island, Papua New Guinea, *Asia Pacific Viewpoint*, 58(1), 14–26.

Connell, J., and Stilwell, B. (2006) Merchants of medical care: recruiting agencies in the global health care chain. In C. Kuptsch (ed.), *Merchants of labour*, Geneva, International Labour Office, pp. 239–253.

Connell, J., and Tabucanon, G. (2016) From Banaba to Rabi: A Pacific model for resettlement? In S. Price and J. Singer (eds.), *Global implications of development, disasters and climate change*, London, Routledge, pp. 91–107.

Constable, A. (2017) Climate change and migration in the Pacific: options for Tuvalu and the Marshall Islands. *Regional Environmental Change*, 17, 1029–1038.

Conway, D., Potter, R., and Phillips, J. (2005) The experience of return: Caribbean return migrants. In R. Potter, D. Conway and J. Phillips (eds.), *The experience of return migration: Caribbean perspectives*, Aldershot, Ashgate, pp. 1–26.

Conway, D., Potter, R., and St Bernard, G. (2012) Diaspora return of transnational migrants to Trinidad and Tobago: the additional contributions of social remittances. *International Development Planning Review*, 34(2), 189–209.

Cummings, M. (2016) Uncertain belongings: relationships, money and returned migrant workers in Port Vila. Vanuatu, *Journal of Pacific Studies*, 36(1), 21–33.

Di Lampedusa, T. (1961) *The leopard*. Translated by A. Colquhoun, London, Collins.

Donner, S. (2015) The legacy of migration in response to climate stress: learning from the Gilbertese resettlement in the Solomon Islands. *Natural Resources Forum*, 39(2), 191–201

Farbotko C., Stratford E., and Lazrus H, (2016) Climate migrants and new identities? The geopolitics of embracing or rejecting mobility. *Social and Cultural Geography*, 17(4), 533–552

Firth, R. (1936) *We, the Tikopia*. London, Allen and Unwin.

Gailey, C.W. (1992) State formation, development and social change in Tonga. In A. Robillard (ed.), *Social change in the Pacific Islands*, London, Kegan Paul International, pp. 322–345.

Gibbons, S., and Nicholls, R. (2006) Island abandonment and sea-level rise: an historical analog from the Chesapeake Bay, USA. *Global Environmental Change*, 16(1), 40–47.

Gibson, J. (2015) Circular migration, remittances and inequality in Vanuatu. *New Zealand Population Review*, 41(1), 153–167.

Gifford, R., and Dunne, R. (2014) A dispossessed people: the depopulation of the Chagos archipelago, 1965–1973. *Population, Space and Place*, 20(1), 37–49.

Gilroy, P. (1987) *There ain't no black in the Union Jack.* London, Hutchinson.

Golash-Boza, T. (2014) Forced transnationalism: transnational coping strategies and gendered stigma among Jamaican deportees. *Global Networks*, 14(1), 63–79.

Gössling, S., and Schulz, U. (2005) Tourism-related migration in Zanzibar, Tanzania. *Tourism Geographies*, 7(1), 43–62.

Grainger, A. (2011) Migrants, mercenaries and overstayers: talent migration in Pacific island rugby. In J. Maguire and M. Falcous (eds.), *Sport and migration*, Abingdon, Routledge, pp. 129–140.

Grydehøj, A. (2008) Nothing but a shepherd and his dog: the social and economic effects of depopulation in Fetlar, Shetland. *Island Studies Journal*, 2(2), 56–72.

Hall, S. (1990) Cultural identity and diaspora. In J. Rutherford (ed.), *Identity: community, culture, difference*, London, Lawrence and Wishart, pp. 222–237.

Hammond, J., and Connell, J. (2009) The new blackbirds? Vanuatu guestworkers in New Zealand. *New Zealand Geographer*, 65(3), 201–210.

Hennebry, J., and Preibisch, K. (2012) A model for managed migration? Re-examining best practices in Canada's seasonal worker program. *International Migration*, 50(S1), e19–e40.

Hermann, E., and Kempf, W. (2017) Climate change and the imagining of migration: emerging discourses on Kiribati's land purchase in Fiji. *The Contemporary Pacific*, 29(2), 231–263.

Howard, A., and Rensel, J. (2004) Rotuman identity in the Electronic Age. In T. van Meijl and J. Miedema (eds.), *Shifting images of identity in the Pacific*, Leiden, The Netherlands, KITLV Press, pp. 219–236.

Hughes, R. (1999) Between the devil and a warm blue sea: islands and the migration experience in the fiction of Jamaica Kincaid. In R. King and J. Connell (eds.), *Small worlds, global lives: islands and migration*, London, Pinter, pp. 195–212.

Iredale, R., Voigt-Graf, C., and Khoo, S. (2015) Trends in international and internal teacher mobility in three Pacific island countries. *International Migration*, 53(1), 97–114.

Jacobs, A., and Overton, J. (2017) Tout le monde a sa place? MIRAB, education and society in Wallis and Futuna. *Islands Studies Journal*, 12(1), 151–168.

Jeffery, L. (2017) 'We don't want to be sent back and forth all the time': ethnographic encounters with displacement, migration and Britain beyond the British Isles. *Sociological Review Monographs*, 65(1), 71–87.

Jokhan, M. (2008) Parental absence as a consequence of migration: reviewing the literature. *Social and Economic Studies*, 57(2), 89–117.

Kagan, S. (2016) 'On the ship you can do anything': the impact of international cruiseship employment for i-Kiribati women. *Journal of Pacific Studies*, 36(1), 35–52.

Kanemasu, Y., and Molnar, G. (2012) Pride of the people: Fijian rugby labour migration and collective identity. *International Review for the Sociology of Sport*, 48(6), 720–735.

Kanemasu, Y., and Molnar, G. (2017) Private military and security labour migration: the case of Fiji. *International Migration*, 55(4), 154–170.

Kempf, W. (1999) Cosmologies, cities and cultural constructions of space: oceanic enlargements of the world. *Pacific Studies*, 22(1), 97–114.

Khamis, S. (2007) Gourmet and green: the branding of King Island. *Shima*, 1(2), 14–29.

King, R. (2009) Geography, islands and migration in an era of global mobility. *Island Studies Journal*, 4(1), 53–84.

Kirch, P. (1986) *Island societies: archaeological approaches to evolution and transformation.* Cambridge, Cambridge University Press.

Le De, L., Gaillard, J., and Friesen, W. (2013) Remittances and disaster: a review. *International Journal of Disaster Risk Reduction*, 4(1), 34–43.

Lee, H. (2016) 'I was forced here': Perceptions of agency in second generation 'return' migration to Tonga. *Journal of Ethnic and Migration Studies*, 42(15), 2564–2579.

Lewis, P. (2011) Training nurses for export: a viable development strategy? *Social and Economic Studies*, 60(2), 67–104.

Lewis, P., and Kirton, C. (2015) Migration and remittances in development: a study of Jamaica. In W. Khonje (ed.), *Migration and development. Perspectives from small states*, London, Commonwealth Secretariat, 186–243.

Lincoln, D. (2009) Labour migration in the global division of labour: migrant workers in Mauritius. *International Migration*, 47(4), 129–156.

Lockwood, V. (1990) Development and return migration in rural French Polynesia. *International Migration Review*, 24(2), 347–371.

Loyd, J., and Mountz, A. (2014) Managing migration, scaling sovereignty on islands. *Island Studies Journal*, 9(1), 23–42.

Marcus, G.E. (1981) Power on the extreme periphery: the perspective of Tongan elites in the modern world system. *Pacific Viewpoint*, 22(1), 48–64.

Marino, E., and Lazrus, H. (2015) Migration or forced displacement? The complex choices of climate change and disaster migrants in Shishmaref, Alaska and Nanumea, Tuvalu. *Human Organization*, 74(4), 341–350.

Maron, N., and Connell, J. (2008) Back to Nukunuku: employment, identity and return migration in Tonga. *Asia Pacific Viewpoint*, 49(2), 168–184.

Marshall, J. (1999) Insiders and outsiders: the role of insularity, migration and modernity on Grand Manan, New Brunswick. In R. King and J. Connell (eds.), *Small worlds, global lives. islands and migration*, London, Pinter, pp. 95–113.

Marshall, M. (2004) *Namoluk: beyond the reef*. Boulder, CO, Westview Press.

Maude, H. (1968) *Of islands and men*. Melbourne, Australia, Oxford University Press,

Maude, H. (1981) *Slavers in paradise*. Canberra, Australia, ANU Press.

Moore, C. (2004) *Happy isles in crisis*. Canberra, Australia, Asia Pacific Press.

Morton, H. (1999) Islanders in space: Tongans online. In R. King and J. Connell (eds.), *Small worlds, global lives: islands and migration*, London, Pinter, pp. 235–253.

Mortreux, C., and Barnett, J. (2009) Climate change, migration and adaptation in Funafuti, Tuvalu. *Global Environmental Change*, 19(1), 105–112.

Narayan, P., and Smyth, R. (2003) The determinants of emigration from Fiji to New Zealand. *International Migration*, 41(5), 33–58.

Nicolson, A. (2001) *Sea Room*. London, Harper Collins.

Osborne, D. (1966) *The archaeology of the Palau Islands*, Bishop Museum Bulletin No. 230, Honolulu, HI, Bishop Museum.

Owen, A. (2010) Guam culture, immigration and the US military build-up. *Asia Pacific Viewpoint*, 51(3), 304–318.

Petrou, K., and Connell, J. (2017) Food, morality and identity: mobility, remittances and the translocal community in Paama, Vanuatu. *Australian Geographer*, 48(2), 14–26.

Peterou, K and Connell, J (2018) 'We Don't Feel Free At All'. Temporay Ni-Vnauatu workers in the Riverina, Australia, *Rural Society*, 27(1), 1–14.

Philip, L., MacLeod, M., and Stockdale, A. (2013) Retirement transition, migration and remote rural communities: evidence from the Isle of Bute. *Scottish Geographical Journal*, 129(2), 122–136.

Philpott, S. (1999) The breath of 'the beast': migration, volcanic disaster, place and identity in Montserrat. In R. King and J. Connell (eds.), *Small worlds, global lives: islands and migration*, London, Pinter, pp. 137–159.

Poncelet, A, Gemenne, F, Martiniello, M., and Bousetta, H. (2010) A country made for disasters: environmental vulnerability and forced migration in Bangladesh. In T. Afifi and J. Jäger (eds.), *Environment, forced migration and social vulnerability*, Heidelberg, Springer-Verlag, pp. 211–222.

Rasmussen, A. (2015) *In the absence of the gift*. New York, Berghahn.

Rauchholz, M. (2012) Discourses on Chuukese customary adoption, migration and the laws of states. *Pacific Studies*, 35(1/2), 119–143.

Richardson, B. (1980) Freedom and migration in the Leeward Caribbean, 1838–48. *Journal of Historical Geography*, 6(4), 391–408.

Richardson, B. (1986) *Panama money in Barbados, 1900–1920*. Knoxville, TN, University of Tennessee Press.

Royle, S.A. (1999) From the periphery of the periphery: historical, cultural and literary perspectives on emigration from the minor islands of Ireland. In R. King and J. Connell (eds.), *Small worlds, global lives: islands and migration*, London, Pinter, pp. 27–54.

Smith, M. (2013) *A faraway, familiar place: an anthropologist returns to Papua New Guinea*. Honolulu, HI, University of Hawai'i Press.

Speelman, L., Nicholls, R., and Dyke, J. (2017) Contemporary migration intentions in the Maldives: the role of environmental and other factors. *Sustainability Science*, 12(3), 433–451.

Steel, T. (1988) *The life and death of St Kilda*. London, Fontana.

Stojanov, R., Duzi, B., Kelman, I., Nemec, D., and Prochazka, D. (2017) Local perceptions of climate change impacts and migration patterns in Male, Maldives. *Geographical Journal*, 183(4), 370–385.

Teaiwa, K. (2015) *Consuming Ocean Island: stories of people and phosphate from Banaba*. Bloomington, IN, Indiana University Press.

Terrell, J., Irwin, G., and Irvin, G. (1972) History and tradition in the Northern Solomons: an analytical study of the Torau migration to southern Bougainville in the 1860s. *Journal of the Polynesian Society*, 81(3), 317–349.

Triandafyllidou, A. (2014) Multi-levelling and externalising migration and asylum: lessons from the southern European islands. *Island Studies Journal*, 9(1), 7–22.

Vine, D. (2009) *Island of shame*. Princeton, NJ, Princeton University Press.

Voigt-Graf, C. (2003) Fijian teachers on the move: causes, implications and policies. *Asia Pacific Viewpoint*, 44(1), 163–174.

Voigt-Graf, C., and Connell, J (2006) Towards autonomy? Gendered migration in Pacific island states. In K. Ferro and M. Wallner (eds.), *Migration happens: effects and opportunities of migration in the South Pacific*, Vienna, Austria, Lit Verlag, pp. 43–62.

Walda-Mandel, S. (2016) *'There is no place like home': migration and cultural identity of the Sonsorolese, Micronesia*. Heidelberg, Universitats Verlag.

Wessen, A., Hooper, A., Huntsman, J., Prior, I., and Salmond, C. (1992) *Migration and health in a small society: the case of Tokelau*. Oxford, Clarendon Press.

Western, J. (1992) *A passage to England: Barbadian Londoners speak of home*. London, UCL Press.

Wilson, K. (2013) Wan laki aelan? Diverse development strategies on Aniwa, Vanuatu. *Asia Pacific Viewpoint*, 54(2), 246–263.

12

HEALTH AND WELLBEING

Robin Kearns and Tara Coleman

Introduction

Islands have long been associated with health. Thomas More's *Utopia*, published just over 500 years ago, described an idealised island-state which promoted its residents' wellbeing through being effectively governed and scrupulously planned. Happiness was an outcome of an equitable, yet controlled, society and the absence of private property. The influence of More's idealised political satire has been enduring and the source of one dominant trope within the western mind-set: that islands can offer utopian alternatives to mainland and mainstream societies. Such ideas contribute to the allure of tourism and suggest that islands can offer happiness and wellbeing, whether idealised or real.

Three key aspects of island life are: isolation (evident in the purposeful journey – usually over water – to reach the separate world of the island); boundedness (referring to the distinct physical outline of islands that allows one to picture the island as a whole); and small size (with the potential to know more closely the places and people within the smaller-scale of the island setting) (Royle 2001). Wellbeing, we contend, emerges within the tension between these three characteristics (isolation, boundedness and intimacy) that shapes the contours of island life. Indeed, at its best island life is characterised by a "powerful sense of community and kinship" (Royle 2007, p. 42), a quality acknowledged as a strong determinant of mental and social wellbeing.

This chapter asks: to what extent can islands be 'spaces of wellbeing' (Fleuret and Atkinson 2007)? Two types of wellbeing are considered here. There is the wellness that implies a movement away from a condition of ill-health or impairment (i.e. 'getting well'). This is a conventional concern of medicine: making and keeping people well, but not necessarily having concern for the subtleties of wellness itself (Kearns and Andrews 2010). Second, outside discourses of medicine, and within the places of everyday life, wellbeing is widely evoked as a condition that is important in and of itself. We take the view that, especially with respect to islands that have been historically hard hit by epidemics and often colonised by quarantine functions, health has been too easily medicalised. While understanding that both the structure and delivery of health care on islands is important (as we consider below), a broader vision of place-based wellbeing allows us to embrace a wide set of influences at work, many of which are 'enabling resources' (Duff 2011) intrinsic to the character of islands themselves.

We begin by reviewing characteristics of organised health care on islands, identifying aspects that are shaped by distance, isolation and limited resources. We follow with a series of sections that review aspects of islands that contribute to experiences of wellbeing, draw visitors to islands, and have been influential in enhancing this allure of islands within Western popular culture. Next, we offer closer consideration of the idea of wellbeing through exploration of the value placed on the surrounding seas ('blue space') by residents of two islands that comprise part of Auckland, New Zealand. On Waiheke, the sea has facilitated wellbeing as it was central to both memory and connectedness to a broader ontological security for many residents (Coleman and Kearns 2015). We then, in a second case study, examine Rotoroa Island where, for almost a century, the Salvation Army ran an alcohol treatment facility (Kearns et al. 2015). Here the island's relative isolation was key to its mix of therapeutic and carceral roles until closure in 2005. Its subsequent re-opening in 2011 as a philanthropically-funded reserve for recreation, remembrance and environmental restoration affirms a broader vision of wellbeing contingent upon the activities of diverse human and non-human actors. We close by reflecting on the guiding question we set ourselves in this chapter: in what ways do islands, given their encirclement by the sea, contribute to the experience of wellbeing for residents as well as visitors?

Health care on island settings

While some might regard islands as offering utopian alternatives to life in 'mainland' societies, it is important at the outset to recognise that, regardless of whether experienced as idealised or banal, people need health care on all but unpopulated or rarely-visited islands. Primary care is the foundational level of medical care, including the activities that occur at the first point of entry to a larger health system that typically includes secondary (hospitals) and tertiary (specialist) levels. The general practitioner (GP) is the gateway to wider health services. More radical primary health care ideas have their roots in a declaration that emerged from the World Health Organization's Alma Ata conference in 1978 and endorse 'grass roots' approaches to health care rather than universally imposed Western models of medicine (Crooks and Andrews 2009). To generalise, many island settings offer health care provision that includes a hybrid of orthodox medically-oriented services and more locally-responsive primary health care that is attuned to the particular island-based determinants of health and wellbeing. This hybridity can be borne of accommodation and improvisation, with strong evidence of self-determination in especially more remote locations. For instance, in Vanuatu, there is a strong reliance on traditional remedies in the more remote islands of this Pacific archipelago, in contrast to the islands that are better connected to Port Vila, the capital city where Western medicine predominates (Bradacs et al. 2011).

Wellbeing can be experienced in, or because of, health care services and facilities. However, although hospitals and doctors clearly have concerns for the wellbeing of their patients, a wider network of care needs to be acknowledged. Such networks comprise unpaid voluntary and informal sectors (Milligan 2005). Providing health care and social services on island settings presents particular challenges including: distance to specialist facilities; costs of provision; and availability of suitably qualified staff. The consequences of inadequately addressing these issues include poor access to health care for island residents as well as visitors. This idea of access in the primary care context involves the degree to which people can reach the health services they need to maintain or improve their wellbeing. A particular challenge on islands that are reliant on seasonal tourism is coping with the ebb and flow of demand upon health care services. However, as Gould and Moon (2000) point out with respect to the UK's Isle of Wight, adjusting the funding of care to address the impact of seasonal visitor numbers would be contentious as similar claims could as easily be made by mainland health boards which include coastal resorts within their jurisdictions.

Most literally, access concerns whether services are present. Consideration of access commonly begins by focussing on distance yet a straight line between a clinic location and an island resident's address will rarely reflect either the journey travelled or the obstacles encountered (Kearns 2018). Rather, accessibility to health care – especially on island settings – is complex and multidimensional. It can include the question of service quality (*is a clinic regarded as good enough?*), availability (*do opening hours suit?*), its affordability (*what is the cost?*), continuity (*is it reliably available on an ongoing basis?*), connectivity (*how long does it take to get there; where to stay while there; and how long to get back?*), and acceptability (*is the care offered acceptable to local users?*) (Hanlon 2009, Penchansky and Thomas 1981). This last aspect – acceptability – is the most nuanced. Indigenous populations, for instance, may well postpone or completely avoid contact with a clinic, even if it is located within their own community, if traditional beliefs held by island populations are demeaned or overlooked. Examples abound, however, where sensitivity to traditional culture is strategically flavouring the character of health services. By way of example, the new mental health service at Atoifi Adventist Hospital, Solomon Islands, is described by Maclaren et al. (2009) as incorporating socio-cultural beliefs, such as spirit possession, into routine practice. Such efforts to create 'cultural safety' (Wepa 2005) can be complemented by another set of concerns: that if island-based health care is too steeped in local tradition and practices, then visitors may feel uncomfortable coming forward as patients, thus damaging the reputation of the island and local efforts to promote tourism.

One way of dealing with what are often great distances between islands, or between an island and a mainland, has been a range of technologies labelled 'telehealth'. These digital technology interventions have dissolved the tyranny of physical space, allowing clinical, educational, administrative and research-related communication such that patients and professionals - who would otherwise have restricted access to health care or specialist advice - have contact across vast distances (Cutchin 2002).Yet, as Hanlon and Kearns (2016) ask, can telehealth take the place of a traditional clinic? The answer is that the future may well be a case of 'both/and'. Studies in mainland contexts have analysed the symbolic significance of a clinic as embedded within the fabric of community life (Kearns 1998). Research has concluded that, like schools and shops, primary care clinics are at least potentially sites at which formal delivery of services is often complemented by expressions of social capital like the maintenance of informal relationships, local knowledge and development of trust (Hanlon and Kearns 2016). Hence clinics on island settings are akin to those in isolated rural communities, where other forms of social infrastructure tend to be sparse. They are invariably relational places whose significance is most acutely felt when threatened with closure (Kearns et al. 2009). Their significance to a community is therefore too deep to be effectively replaced by a technological surrogate.

At a broader level, a key influence changing health services in island settings is urbanisation. Recently, for instance, Mauritius has experienced rapid urbanisation and Kalla (1992) offers an account of the reconfiguration of locations, staffing and opening hours in response to this trend. Elsewhere, isolation and smaller population thresholds present challenges of viability to island health care services. Royle (1995), by way of example, discusses the mid-oceanic islands of St Helena and Ascension in the South Atlantic and the challenges of providing health care services within such isolated sites. Solutions include using itinerant health care providers for some specialities (e.g., dentistry) and the establishment of medevac procedures for removing patients to higher-order treatment locations when necessary. Regardless of the island setting, the small-scale hospital invariably occupies a symbolically and practically prominent place in the local landscape, as is the case on the Chatham Islands (see Figures 12.1 and 12.2).

Figure 12.1   Chatham Islands Hospital, Waitangi settlement.
Source: © Robin Kearns.

Figure 12.2   Road sign: hospital and other important amenities.
Source: © Robin Kearns.

Islands and the contours of wellbeing

In addition to utopia, an idealised notion about islands that has come to us from the 16th century, a number of other tropes link islands and wellbeing in the public imagination. First is the idea of islands as *sanctuaries* – literally sites of refuge from pursuit or persecution (Kearns and Collins 2016). Just as islands have been sites of imprisonment (as was Robben Island, off Cape Town, South Africa, or Alcatraz in San Francisco Bay, USA), so too islands can be places of protection from unhealthy mainland influences. We see this today in the way islands are established as wildlife sanctuaries. With respect to people, protection can be for the benefit of mainland society in the case of quarantine islands (Bashford 1998). To this extent, remoteness has benefits: the greater the distance from a mainland, the more likely that an island can offer safety to its inhabitants through the intervening waterway serving as a protective moat. While the ill-effects of toxic products and unhealthy behaviours can take hold, the small scale of many island settlements can facilitate a collective will for change. The declaration of Niuean villages as 'smokefree' offers a good example (see Figure 12.3).

That being said, an island can offer little protection once this protective moat has been breached. By way of example, the island of Rotuma (a northerly Polynesian outlier within the Fiji group) experienced extreme mortality when measles first arrived in 1911. The cumulative measles-related mortality that year of 12.8 per cent indicates that very limited earlier exposure to only a narrow range of microbes (as well as the genetic homogeneity of isolated island populations) can leave islanders liable to deadly outcomes upon first contact with contagious viruses such as measles (Shanks 2011). Similarly, in Samoa, the 1918–1919 influenza epidemic hit hard, with deaths of up to 30 per cent of the men and 22 per cent of women. Beyond the obvious tragedy of such a death rate, this extraordinarily sudden and high level of mortality resulted in rapid change in political structures and acute labour shortages (Tomkins 1992). The message

Figure 12.3   Smokefree Tuapa village, Niue.

Source: © Robin Kearns.

from this historical epidemiology is that remoteness is replete with ambivalence: while a watery distance can afford protection, an island's isolation can also generate vulnerability to threats such as infectious diseases from over the horizon.

A second idea that relates to wellbeing is the notion of islands as *get-aways* (Kearns and Collins 2016). For tourists – whether day-trippers to nearby islands or long-stayers on more remote locations – the 'getting away from it all' involves an *it* that is work, stress, traffic and various other mundane preoccupations. Invariably, however, get-aways involve returns and so the act of 'escape' and the memories upon return potentially add to an unrealistic romanticising of islands. This contrived sense of islands as exotic locations that offer enhanced wellbeing through visitation has been played upon and amplified by what we might call 'wellbeing tourism': the marketing of retreats and spas on islands that require a journey and involve payment in return for pampering. By way of example, the rocky and isolated island of Niue in the South Pacific is one of the world's smallest nations population-wise with just 1,600 residents in 2016. As part of a broader quest to attract visitors, 'Yoga and Wellness Retreats' are now offered, enticing participants with Facebook pages that feature spectacular landscape images (see Figure 12.4). The lone figure in the advertisement is shown reflected in the water, hinting at the retreat offering not only intimate engagement with place but also personal opportunity for reflection. Other associated images suggest meditative quiet and healthy tropical foods, while the sea features prominently.

The promise implied in the advertisement is one of contentment as well as engagement. This focus is on wellbeing as health-outstretched into other domains of human fulfilment. Such a view of marketed wellbeing allows us to integrate the concerns of health and emotional geographies as a specific place is affectively and effectively experienced. These two domains of

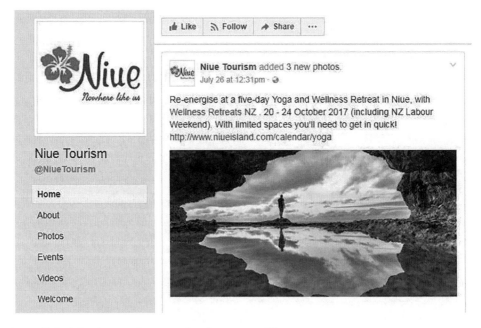

Figure 12.4 Advertisement for yoga and wellness retreat, Niue.

Source: © Niue Tourism. Used with permission.

human experience are interconnected. They are experienced as such especially when one is out of context on an island vacation: *being* well generally involves *feeling* well (Kearns and Andrews 2010). It is, literally, an experienced state of being, incorporating (i.e. embodying) healthiness and happiness. Although the concept of wellbeing remains relatively ill-defined (Fleuret and Atkinson 2007), the field of health geography has built on traditional concerns with the pathological (disease, death), the gastronomical (cuisine, local diet and food security) and sites of care (hospitals, other services) to include this newer curiosity with aspects of wellness such as identity and experience (Kearns and Andrews 2010).

Indigenous cultures on islands present opportunities to see wellbeing in a new but non-commodified light. Hence, the holistic character of wellbeing as defined in Western scholarship and health policy was long preceded in the worldviews of island-based indigenous peoples such as Māori (New Zealand) (Durie 1998). Elsewhere, for instance, in Alaska, USA, the first people of St Lawrence Island have felt a connectedness between body and cosmos, and so tattoo their joints to stop evil entities causing harm (Krutak 1999). To summarise, then, wellbeing implies *being* well, a state that can be sought out in island-based resorts and therapies as well as already-experienced within local traditions and cultural practices.

Wellness-based tourism is further fuelled by ideas of islands as healthy places where longevity and increased health can be experienced. Indeed, residents living in island places such as Okinawa (Japan), Ithaca (Greece) and Dominica are well known for living demonstrably longer and healthier lives (Poulain 2011). Yet, longevity is a complex trait and studies that seek to understand it, such as those investigating the lifespans of the Okinawan population, suggest further knowledge of the interactions between population characteristics, genetic issues and environmental factors are needed (Willcox et al. 2006).

Nonetheless, connecting island places with longevity, health and wellbeing is common and is frequently used to market such island places as differently attractive travel destinations. The concept of 'Blue Zones' also informs these marketing practices. 'Blue Zones' is used to describe the characteristic lifestyles and environments of what are, statistically, the world's longest-lived people, found in five geographic areas: Okinawa (Japan); Sardinia (Italy); Nicoya (Costa Rica); Icaria (Greece) and among the Seventh-day Adventists in Loma Linda, California (Pes et al. 2004). Blue Zones are marketed as tourist destinations where more affluent international tourists may experience and learn behaviours conducive to living a longer and healthier life (e.g. consumption and production of food; traditional cultural practices that support a strong sense of identity; physical activities) (Poulain et al. 2004). For example, in Okinawa, the importance of local food for good health is widely promoted and tourists who visit the area learn about the concept of *ishokudogen*, or 'food as medicine'.

The allure of islands for medical visits

Increasingly, poorer island nations are promoting tourism explicitly aimed at visitors seeking medical care. In the Caribbean region, for instance, medically-focussed 'health tourism' (or, perhaps more appropriately, 'travel') is being seen as a means to diversify the tourism sector and contribute to economic development. This motive can be seen in the case of Barbados where medical tourism policy development is jointly driven by the Ministry of Health, the Ministry of Tourism, and the government corporation Invest Barbados. The central idea is that patients and their supporters (largely from high-cost countries) seek out high quality but lower cost medical care in less developed countries and combine the journey with a period of vacation and/or convalescence. Hence returns to the local economy include expenditures on both medicine and tourism (Johnston et al. 2013).

Regulatory incentives, chiefly surrounding easing international patients' and health professionals' visits to Barbados, are being explored to help support the medical tourism industry (Johnston et al. 2013). Adams et al (2014) argue for cooperation between health officials of countries in the Caribbean region, including regional access to specialised health care as a counter to a trend towards tourism fuelling competition between islands and therefore being a greater strain on political and economic integration. Moreover, at the level of individual island nations such as Barbados, regulations have been adjusted to incentivise travel for both international patients and health professionals to visit Barbados to assist the medical tourism industry (Johnston et al. 2013).

A lurking concern in the face of these developments, however, is health equity: are 'locals' being disadvantaged by the amassing of health care resources in favour of affluent visitors? (Johnston et al. 2015). A further concern is that, more so than even in countries with highly privatised health care, the patients and supporters partaking in medical tourism are purchasing a 'packaged service'. Hence, with this growing commodification of not only health care but also the purportedly therapeutic character of the island landscape, there can be a raised expectation of results and satisfaction. As destinations and enterprises proliferate on islands such as Gran Canaria, Spain or Saaremaa, Estonia – ranging from cosmetic surgery to sauna and spa treatments to dentistry – patient satisfaction becomes an increasingly important outcome to manage (Medina-Muñoz and Medina-Muñoz 2013). In an age of social media and online reviews, the overall experience of medical tourism as well as the surgical outcomes can both be points of vulnerability in the development of corporate strategy.

Connell (2013) points out that medical tourism projects in small Caribbean island jurisdictions such as the Bahamas, Barbados and the Cayman Islands are developed by overseas corporations and are predominantly oriented to a US market. Ethical issues are raised by the fact that business principles rather than health care imperatives dominate development strategies. Benefits to 'locals' are the generation of foreign exchange and the creation of new employment opportunities. According to Connell (2013), medical tourism will be uneven because of an intense competition for patients and therefore has potential to distort national health care systems.

Medina-Muñoz and Medina-Muñoz (2013) prefer to write of 'wellness tourism' and in a sense all travel to holiday destinations could be constructed as such: the quest to be refreshed through the novelty of being elsewhere, often in a more benign climate away from the stresses and strains of everyday life (see the example of the yoga retreat in Niue described earlier). Writing about the pharmaceutical landscape of Mexico's Cozumel Island and its San Miguel's tourism precinct, Hoffman (2017, p. 81) suggests that these features are "enmeshed within discourses that facilitate broader touristic landscape construction as well as expected consumptive practices within the town". Storefront signage in San Miguel appeals to tourists' desires for respite and relaxation, enjoyment and excitement, as well as a belief that Mexico offers consumers an accessible shopping destination at reduced costs (see Figure 12.5). Ultimately, as Hoffman (2017, p. 81) argues, the medical becomes part of the tourist landscape, "a consumer experience which presses to offer tourists that which might not be available at home and/or a memento of their trip abroad".

To summarise, much marketing of island destinations plays up to the imperatives of 'getaway', 'unwind' and 'be pampered'. At times, the invocation of such tropes can lead to a dissonance between expectation and reality and hence there is a need to temper the use of hyperbole in image and description.

Figure 12.5   Pharmaceuticals in the streetscape, Cozumel Island, Mexico.

Source: © Leon Hoffman. Used with permission.

Ambivalence, popular culture and wellbeing

The ebb and flow of visitors, including organised 'medical tourists', can be vital to island economies, but can at times be a source of annoyance. Islands are never closed systems, given their connections to 'mainlands' through flows of people, goods and ideas. Nevertheless, where distances
are sufficient, these flows may be limited and nuanced. The characteristic of insularity often
gives rise to islands being associated with separateness and vulnerability, and island life can be
characterised by a sense of community. This quality was potently observed among the residents
of Ireland's offshore islands whose collective identity was seen to cohere around a positively
perceived isolation and an associated sense of belonging, strong feeling of community, secure
and regular social interactions, and enduring values (Burholt, Sharf and Walsh 2013). However,
as in any community, there are politics of inclusion and exclusion. While islands may enable a
sought-after separation from the mainland, they may also impose a strategic distance from other
opportunities and resources.

Ambivalences can be acute when one considers wellbeing in island-like resort settings. Investigating these settings, Minca (2009) argues that the power relations between visitors and workers results in a 'contact zone' in which ideas about class, gender, sexuality and the body, as well as
health and wellbeing, must be negotiated. In other words, islands are replete with ambivalence
and what on face value may be good for the advancement of wellbeing may, in time or in
excess, have negative impacts on people's experience. By way of example, beaches are intrinsic
to most 'warm water' islands. Wherever beaches are found, they offer potential sites of recreation and relaxation. Yet, as Collins and Kearns (2007) argue, it is all too easy to romanticise such

places when they invariably also incorporate elements of risk and threats to wellbeing (sunburn, drowning, Hepatitis C from contaminated sand).

Notwithstanding these risks, and fuelled by images within tourism and popular culture, islands are coded in the popular imagination as largely happy places. Yet the 'on-the-ground' reality can be abjectly different. A gap between advertising-generated expectations and the actual experience of visitors can be considerable, as research based on Great Barrier Island revealed (Hoffman and Kearns 2013). Moreover, while visitors travel to island resorts in the Pacific and Indian Oceans and the Caribbean for mid-winter holidays, seeking respite from the stresses of employment, 'locals' can experience very different stresses. In a study set in the Solomon Islands, for instance, Blignault et al. (2009) found high rates of drug use, interpersonal violence and teenage pregnancy.

This jarring contrast between visitor and local experience only arises because of the tourist imperative. Why is there such a drive on the part of travellers to seek out particular island destinations? Recent psychotherapeutic research signals how imaging places can be beneficial to mindfulness and how the practice of going to real or imagined 'get-aways' can improve wellbeing (Williams 1998). We can also note links to popular culture. In books including *Robinson Crusoe* and *The Swiss family Robinson* islands are constructed as places replete with treasure, drama and dreams (Connell 2003).

More recently, popular songs commonly involve imaginings of, or metaphorical reference to, islands, something Gastaldo et al. (2004) refer to as "therapeutic landscapes of the mind". Two examples will suffice. Such song lyrics as 'Islands in the Stream' written by the Gibb brothers in 1979 and popularised by Kenny Rogers and Dolly Parton, as well as The Seekers' 1966 hit 'Island of Dreams', suggest a therapeutic soundscape of wellbeing (Andrews et al. 2014): melodies and lyrics which invoke an inner journey to an idealised state of island-related wellbeing. We must end this section, however, noting that islands are not always romanticised in popular music. Rather, emotional projection can see the isolation trope associated with islands played upon in a counter to wellbeing. In a classic example, Paul Simon's 1965 song 'I am a Rock' has the writer saying he has "built walls", and has "no need of friendship", instead seeing himself in his forlorn loneliness as "an island . . . and an island never cries" (also Mezzana et al. 2012).

Case study I: Waiheke

Notwithstanding the potential for isolation to be felt or observed, promoters of islands will invariably seek to present them in the rosiest light. For instance, Waiheke, (land area: 92 km²), 35 minutes by ferry from downtown Auckland, New Zealand, is an island with no traffic lights despite a population of 8,900. It is an island where the sea is frequently visible from peoples' homes. Its stellar rating as one of the world's best ten travel destinations by Lonely Planet (2016) has exponentially increased visitor numbers; its website describes the island as "a utopia of coves, beaches, vineyards, bohemian sensibilities, and above all, fun". While there is a widely-acknowledged therapeutic connection between people and water (Andrews and Kearns 2005, Foley 2011), Conradson (2005) reminds us that therapeutic experiences cannot be presumed but rather are the outcome of peoples' diverse relations with place. Waiheke's permanent population includes a higher proportion of older residents than mainland Auckland. Within this context, Coleman and Kearns (2015) explored relations between the lives of older residents and the sea that surrounds and defines their island home. The study sought to understand the therapeutic properties of the blue spaces surrounding the island. Concern was for how experiences of being aged on Waiheke are formed in relation to the character of the island place itself, including its symbolic and material features. In other words, how does the sea act as a resource to assist independent life and support wellbeing?

Narratives revealed a depth of connection between sea, island living and the wellbeing of older residents. Sam (pseudonyms are used), for instance, said "If I wasn't on this island I wouldn't have even begun to ask questions about my life"; while Michelle saw herself as being "a bit out to sea in more ways than one". Florence explained that pausing to consider the views of blue space from her living room (see Figure 12.6) she felt that "the sea kind of separates me from feeling grief" (Coleman and Kearns 2015). Some residents reported feeling isolation (both welcome and unwelcome), reflecting the way islands can also be sites of exile and influencing experiences of aloneness. For others, their home's vantage point allowed an elevation and safe distance from physical challenges of engaging with the beach and sea. However, it also offered a vicarious opportunity to participate in coastal activities through observing and activating memories of times when more hands-on involvement with the sea was possible. Watching others engage with the sea – even from afar, as Sam described often doing from his dining room – was central to their islander identity; to be denied this intimacy-at-a-distance was felt by older residents as tantamount to letting go of life itself. This reality was noted when two study participants left the island for medical reasons and died soon after.

While it literally reflects light, the sea – as one participant emphasised – can also symbolically reflect one's life back to the self, perhaps re-awakening memories of earlier engagement with the ocean. However, with a subtle but significant shift in valence, the largeness of the ocean can also generate feelings of smallness and insignificance to older island residents and tacitly emphasise their frailty and isolation. The watery expanse between island and mainland can be regarded as a barrier separating the self from others and all that society can offer in terms of assistance and support. Thus, a key characteristic of 'islandness' is illustrated: that it is possible to have both a

Figure 12.6   View from Florence's window, Waiheke Island.

Source: © Tara Coleman.

positive (e.g. calming) and challenging (e.g. isolating) experience of island settings since they are surrounded by the sea. This case study suggests that the valences of therapeutic experience of places change with a person's age and ability to engage with the island itself. For older island residents, the sea can be thus both an omnipresent material dimension of the natural world and a metaphorical resource that is drawn upon to make sense of the changes in the body which, in a sense, is the landscape closest to home.

Case study II: Rotoroa

Belief in the ability of coastal environments to restore mental and physical wellbeing is long-standing, but few studies trace how this understanding has been acted upon in particular places, or how such places have changed over time. On Rotoroa Island, 82 ha in size and 32km from downtown Auckland, the Salvation Army ran a facility for the treatment of alcoholism from 1911. Kearns et al. (2015) outline the story of its admission of almost 12,000 men and women, voluntarily and by court order, over its near-century of operation until its closure in 2005 with the wider move away from isolated institutional forms of treatment. The authors follow Völker and Kistemann's (2011) call to understand water not merely as a common feature of places with therapeutic intentions, but also as a potentially salutogenic element in its own right. In the context of Rotoroa, seawater was not only what – to be banal – defined the island itself and 'contained' detainees. It was also experienced as an activity space (for swimming, fishing), and through participation in these sea-based activities, had social and symbolic dimensions. From this vantage point, the health-enhancing qualities of blue space are a relational outcome,

Figure 12.7   Remnant of Salvation Army buildings, Rotoroa Island.

Source: © Robin Kearns.

emerging from the intersection of human agency (setting up the facility), social relations (organised routines), and built and natural environments (the chapel and dormitories with sea views as well as the beaches themselves) (Conradson 2005) (see Figure 12.7). It is a combination of this set of relational resources that led many detainees to seek a return once back on the mainland (see Figure 12.8).

Kearns et al. (2015) examine Rotoroa Island's associations with health and wellbeing through drawing on Duff's (2011) concept of 'enabling places'. This notion focuses on the diverse actor-networks that generate resources in place, which in turn enable some places to promote health more than others. In developing this idea, Duff (2011) identifies three categories of enabling resources: social (centred on intimacy, trust and reciprocity), affective (centred on feelings, disposition and action-potential) and material (centred on access to goods, services and the physical environment). In this case study, the authors considered the enabling resources created and experienced at Rotoroa by patients, residents and staff during its time as a site of detention/ treatment, and more recently by recreational visitors who come to experience the island's heritage, its rare wildlife and its art installations. This focus on 'enabling' resources helpfully emphasises that the restorative benefits of place are not necessarily a function of the setting itself but are rather continually made, and re-made, through the interactions and practices of a diverse set of human and non-human actors. The story of Rotoroa is one involving an island being reconstituted as a place for a broader vision of wellbeing. While founded in a time of the medicalisation of alcoholism, the island has outlived the ideological acceptability of confined treatment. Now

Figure 12.8   Signage commemorating Salvation Army's alcohol addiction centres' detainees Rotoroa Island, New Zealand.

Source: © Robin Kearns.

the island has been re-worked and re-imagined as a place where members of the public can enjoy new expressions of recreational wellbeing.

Conclusion

Islands and the waters that surround and define them are mutually interdependent, with water self-evidently being the defining topographic character of any island. Our chapter has explored a question that logically flows from this interdependence: how islands, given their encirclement by the sea, contribute to the experience of wellbeing. Understanding the structure of health care on islands is important and both informal care and organised health services may be shaped by distance, isolation and absence of economies of scale. Meanwhile, critical consideration of place-based wellbeing brings into view a wider set of 'enabling resources' (Duff 2011) intrinsic to the character of islands themselves and extending into more nuanced and holistic approaches to health and wellbeing.

Our two brief case studies sketching island wellbeing showed, first, the potency of the sea as a daily distraction for older Waiheke Island residents and the way it can be a means to cope with day-to-day challenges such as grief. Second, in the case of Rotoroa, we indicated that healthy 'islandness' is created and sustained through enabling encounters, networks and associations involving a gestalt between an island and its institutional context.

A final point is the danger in over-attributing salutogenic effects to islands themselves. Doing so risks, perhaps, the granting of too much agency to the island itself. Rather, just as visitors and 'locals' experience the same island differently, so too do different permanent residents. One might experience claustrophobia, preferring to escape from stifling intimacy; or else choosing a delicate balance, living for a time away on the mainland. Others might instead relish the isolation and closely connected community and wild landscape. In other words, and as our examples earlier indicated, therapeutic experiences of islands and their encircling blue spaces vary according to the relational dynamic between self and place. Experiences of wellbeing also change as the places themselves change under urbanising developments.

Island places are often seen as spaces of health and wellbeing through their construction as sanctuaries, getaways and places where it is possible to cultivate contentment as one 'takes in' (i.e. embodies) the qualities of the setting (e.g. a view of blue space) and performs situated 'healthy activities' (e.g. reflecting on one's life). While tourism and popular imagery suggest that island settings can be sources of happiness and increased health, islands can be both ordinary and exceptional settings in terms of human wellbeing. For instance, in their study of children's wellbeing in an island setting, Pivik (2012) found that children reported many of the same needs and desires as mainland counterparts. Four broad categories emerged as important for these island children and youth regardless of age: a sense of safety; the positive influence of the natural environment; a close-knit community; and available resources, programmes and services. Arguably this is not an unexpected set of needs and is also reflective of the wellbeing needs of children in cities (Ergler et al. 2017) and other population centres.

The diversity of islands needs to be acknowledged, along with the difficulty – and risks – in attempting to generalise. Yet, we must generalise because the alternative would simply be an archipelago of unconnected case data. We can observe, for instance, that improved transport links have made many small islands more accessible to tourists and, for island residents, offer enhanced access to mainland health care facilities. Such connectivity can make the difference between life and death, and impact on the decision as to whether to move to or continue living on the island, or else pack up and leave. Yet, at the very time the health needs of island residents are potentially being better addressed, we need to recognise that small islands are becoming more vulnerable to climate change with potential impacts on public health such as inundation of villages and increased exposure to parasitic diseases (Hernández-Delgado 2015).

The continuity within island studies is the globally vast and diverse numbers of islands themselves. With health and health care granted an expansive orientation so as to accommodate the diverse axes of human experience and feeling well, we have the opportunity to observe and analyse a rich suite of relationships anchored in the links between island places and human wellbeing.

References

Adams, K., Snyder, J., Crooks, V., and Hoffman, L. (2014) Medical tourism in the Caribbean: a call for cooperation. *WIMJ Open*, 1(1), 70–73.

Andrews, G.J., and Kearns, R.A. (2005) Everyday health histories and the making of place: the case of an English coastal town. *Social Science & Medicine*, 60(12), 2697–2713.

Andrews, G J., Kingsbury, P., and Kearns, R.A. (eds.) (2014) *Soundscapes of wellbeing in popular music,* Farnham, Ashgate.

Bashford, A. (1998) Quarantine and the imagining of the Australian nation. *Health*, 2(4), 387–402.

Blignault, I., Bunde-Birouste, A., Ritchie, J., Silove, D., and Zwi, A.B. (2009) Community perceptions of mental health needs: a qualitative study in the Solomon Islands. *International Journal of Mental Health Systems*, 3(1), pp. 3–6.

Bradacs, G., Heilmann, J., and Weckerle, C. (2011) Medicinal plant use in Vanuatu: a comparative ethnobotanical study of three islands. *Journal of Ethnopharmacology*, 137(1), 434–448.

Burholt, V., Sharf, T., and Walsh, K. (2013) Imagery and imaginary of islander identity: older people and migration in Irish small-island communities. *Journal of Rural Studies*, 31(1), 1–12.

Coleman, T.M., and Kearns, R.A. (2015) The role of blue spaces in experiencing place, aging and wellbeing: insights from Waiheke Island, New Zealand. *Health & Place*, 35(2), 206–217.

Collins, D., and Kearns, R. (2007) Ambiguous landscapes: sun, risk and recreation on New Zealand beaches. In A. Williams (ed.), *Therapeutic landscapes*, Ashgate, London, pp. 15–31.

Conradson, D. (2005) Landscape, care and the relational self: therapeutic encounters in rural England. *Health & Place*, 11(4), 337–348.

Connell, J. (2003). Island dreaming: the contemplation of Polynesian paradise. *Journal of Historical Geography*, 29(4), 554–581.

Connell, J. (2013) Medical tourism in the Caribbean islands: a cure for economies in crisis? *Island Studies Journal*, 8(1) 117–133.

Crooks, V., and Andrews, G. (2009) Thinking geographically about primary health care. In V. Crooks and G. Andrews (eds.), *Primary health care: people, practice, place*, Aldershot, Ashgate, pp. 1–20.

Cutchin, M.P. (2002) Virtual medical geographies: conceptualizing telemedicine and regionalisation. *Progress in Human Geography*, 26(1), 19–39.

Duff, C. (2011) Networks, resources and agencies: on the character and production of enabling places. *Health & Place*, 17(1), 149–156.

Durie, M. (1998) *Whaiora: Maori health development*, 2nd edn. Auckland, Oxford University Press.

Ergler, C., Kearns, R., and Witten, K. (eds.) (2017) *Children's health and wellbeing in urban environments*. London, Routledge.

Fleuret, S., and Atkinson, S. (2007) Wellbeing, health and geography: a critical review and research agenda. *New Zealand Geographer*, 63(2), 106–118.

Foley, R. (2011) Performing health in place: the holy well as a therapeutic assemblage. *Health and Place*, 17(2), 470–479.

Gastaldo, D., Andrews, G., and Khanlou, N. (2004) Therapeutic landscapes of the mind: theorising some intersections between health geography, health promotion and immigration studies. *Critical Public Health*, 14(2), 157–176.

Gould, M.I., and Moon, G. (2000) Problems of providing health care in British island communities. *Social Science & Medicine*, 50(7), 1081–1090.

Hanlon, N. (2009) Access and utilization reconsidered: towards a broader understanding of the spatial ordering of primary health care. In V. Crooks and G. Andrews (eds.), *Primary health care: people, practice, place*, Ashgate, Aldershot, pp. 43–56.

Hanlon, N., and Kearns, R. (2016) Health in rural places. In *International handbook of rural studies*, London, Routledge, pp. 62–70.

Hernández-Delgado, E.A. (2015) The emerging threats of climate change on tropical coastal ecosystem services, public health, local economies and livelihood sustainability of small islands: cumulative impacts and synergies. *Marine Pollution Bulletin*, 101(1), 5–28.

Hoffman, L. (2017) Pharmaceuticals and tourist spaces: encountering the medicinal in Cozumel's linguistic landscape. *ACME: An International Journal for Critical Geographies*, 16(1), 59–88.

Hoffman, L., and Kearns, R. (2013) A necessary glamorisation? Resident perspectives on promotional literature and images on Great Barrier Island, New Zealand. In J. Lester and C. Scarles (eds.), *Mediating the tourist experience: from brochures to virtual encounters,* Surrey, Ashgate, pp. 57–74.

Johnston, R., Crooks, V., Snyder, J., Fraser, H., Labonté, R., and Adams, K. (2013) An overview of Barbados' medical tourism industry: v 2.0. *SFU Medical Tourism Research Group*. Available from: www.sfu.ca/medicaltourism/An%20Overview%20of%20Barbados'%20Medical%20Tourism%20Industry%20-%20Version%202.0.pdf.

Johnston, R., Adams, K., Bishop, L., Crooks, V.A., and Snyder, J. (2015) 'Best care on home ground' versus 'elitist healthcare': concerns and competing expectations for medical tourism development in Barbados. *International Journal for Equity in Health*, 14(1), 1–12.

Kalla, A.C. (1992) From labour lines to urban living: health transition in Mauritius. *GeoJournal*, 26(1), 69–73.

Kearns, R.A. (1998) Going it alone: Community resistance to health reforms in Hokianga, New Zealand. In R.A. Kearns and W.M. Gesler (eds.), *Putting health into place: landscape, identity and well-being*, Syracuse, NY, Syracuse University Press, pp. 226–247.

Kearns, R.A. (2018, in press) The place of primary care clinics. In V. Crooks, G. Andrews and J. Pearce (eds.), *International handbook of health geography*, London, Routledge.

Kearns, R., and Collins, D. (2016) Aotearoa's archipelago: re-imagining New Zealand's island geographies. *New Zealand Geographer*, 72(3), 165–168.

Kearns, R., and Neuwelt, P. (2009) Within and beyond clinics: Primary health care and community participation. In V. Crooks and G. Andrews (eds.), *Primary health care: people, practice, place*, Ashgate, Aldershot, pp. 203–20.

Kearns, R.A., and Andrews, G. (2010) Wellbeing. In S. Smith, R. Pain, S. Marston and J. Jones III (eds.), *Handbook of social geographies*, London, Sage, pp. 309–328.

Kearns, R.A., Collins, D., and Conradson, D. (2015) A healthy island blue space: from space of detention to site of sanctuary. *Health & Place*, 30(1), 107–115.

Krutak, L. (1999) Saint Lawrence Island joint-tattooing: spiritual/medicinal functions and intercontinental possibilities. *Etudes/Inuit/Studies*, 23(1–2), 229–252.

Lonely Planet (2016) Waiheke island. Available from: www.lonelyplanet.com/travel-tips-and-articles/best-in-travel-2016-top-10-regions/40625c8c-8a11-5710-a052-1479d276e5c7.

Maclaren, D., Asugeni, J., Asugeni, R., and Kekeubata, E. (2009) Incorporating sociocultural beliefs in mental health services in Kwaio, Solomon Islands. *Australasian Psychiatry*, 17(1), S125 – S127.

Medina-Muñoz, D.R., and Medina-Muñoz, R.D. (2013) Critical issues in health and wellness tourism: an exploratory study of visitors to wellness centres on Gran Canaria. *Current Issues in Tourism*, 16(5), 415–435.

Mezzana, D., Lorenz, A., and Kelman, I. (2012) Islands and islandness in rock music lyrics. *Island Studies Journal*, 7(1), 67–96.

Milligan, C. (2005) From home to 'home': situating emotions within the caregiving experience. *Environment and Planning A*, 37(12), 2105–2120.

Minca, C. (2009) The island: work, tourism and the biopolitical. *Tourist Studies*, 9(2), 88–108.

Penchansky, R., and Thomas, J. (1981) The concept of access: definition and relationship to consumer satisfaction. *Medical Care*, 19(1), 127–140.

Pivik, J.R. (2012) Living on a rural island: children identify assets, problems and solutions for health and well-being. *Children Youth and Environments*, 22(2), 25–46.

Poulain, M. (2011) Exceptional longevity in Okinawa: a plea for in-depth validation. *Demographic Research*, 25(7), 245–284.

Poulain, M., Pes, G.M., Grasland, C., Carru, C., Ferucci, L., Baggio, G., Franceschi, C., and Deiana, L. (2004) Identification of a geographic area characterized by extreme longevity in Sardinia island: the AKEA study. *Experimental Gerontology*, 39(9), 1423–1429.

Royle, S.A. (1995) Health in small island communities: the UK's South Atlantic colonies. *Health & Place*, 1(4), 257–264.

Royle, S.A. (2001) *A geography of islands: small island insularity.* London, Routledge.

Royle, S.A. (2007) Island definitions and typologies. In G. Baldacchino (ed.), *A world of islands: an island studies reader*, Luqa, Malta and Charlottetown, Canada, Agenda Academic and Institute of Island Studies, pp. 33–56.

Shanks, D.G., Lee, S.-E., Howard, A., and Brundage, J.F. (2011) Extreme mortality after first introduction of measles virus to the polynesian island of Rotuma, 1911. *American Journal of Epidemiology*, 173(10), 1211–1222.

Tomkins, S.M. (1992) The influenza epidemic of 1918–19 in Western Samoa. *Journal of Pacific History*, 27(2) 181–197.

Völker, S., and Kistemann, T. (2011) The impact of blue space on human health and well-being – Salutogenetic health effects of inland surface waters: a review. *International Journal of Hygiene and Environmental Health*, 214(6) 449–460.

Wepa, D. (ed.) (2005) *Cultural safety in Aotearoa, New Zealand.* Auckland, Pearson.

Willcox, D., Willcox, B., Hsueh, W., and Suzuki, M. (2006) Genetic determinants of exceptional longevity: insights from the Okinawa centenarian study. *Age*, 28(4), 313–332.

Williams, A. (1998) Therapeutic landscapes in holistic medicine. *Social Science & Medicine*, 46(9), 1193–1203.

13

LITERATURE AND THE LITERARY GAZE

Elizabeth McMahon and Bénédicte André

Introduction

Literary studies is an integral component of island studies. At the most basic level, oral and written texts provide records of experience, the habits of daily life and the relationship of individuals and communities to the major forces of nature, culture and history, as these become part of cultural understanding. In this way also, literature operates as a kind of archaeological artefact that can be animated through reading or listening. US novelist Marilynne Robinson recently argued that literature grants access to "the humanity of the long dead". She says: "People don't wonder why people look for DNA in dinosaur bones but they wonder why you would find a poem written in Middle English. For me, it is the discovery of a living voice" (Robinson 2016). Oral and written literatures can bridge time, reach back through and beyond colonial modernity so that we can hear voices of the past in the living present. They can trace continuums into the present and, conversely, lead us to wonder at the ruptures and estrangements of community or individuals over time. Alongside these attentive encounters with the acts of the literary imagination is the involuntary wiring of language in the human subject that is performed by every culture, so that linguistic structures and metaphorics become indivisible from material reality and lived experience as they shape their perception. As scholars in the geohumanities have alerted us, there is no unmediated access to space, its inhabitation, traversal or apprehension. Inspired by Henri Lefebvre's *La production de l'espace* (1974), Edward Soja (1996) contends that space is simultaneously lived, perceived and conceived, reminding us that spatiality is always culturally mediated. Like all human experiences, it is channelled through the perceptive apparatus of language systems and the rhetorical Unconscious, bringing material and imaginary domains into a relational dynamic of relentless co-constitution.

As lived experiences, islands are no different from Soja's notion of spatiality, but they do stand out from other spatial forms because of the place they hold in our collective psyche. As real spaces constantly remoulded by our imaginary (Fougère 1995, p. 303; our translation), they are particularly marked by the mutuality of material and imaginary spaces, which are in play in all topographies and topologies. Furthermore, islands have come to represent the creative imperative in and of themselves. This does not mean that the island becomes abstracted into pure fantasy when it enters forms of creative expression such as literature. On the contrary, the link to the real prohibits that illusion. So, for instance, in Shakespeare's *The tempest* (1611), Prospero is exiled on an island which he turns into a dominion for his own ends. This island/dominion/stage is, variously, a microcosm of the world; the theatrical stage, which is the playwright's world; and a representation of an island of the New World in the Caribbean about which Shakespeare had read. The tensions between reality and illusion, art and history that the play enacts sustain their effect though the indivisibility of these various operations, which we, the audience, try at certain points in the action to prise apart and at others points to bring together; the island, as a liminal space *par excellence* (André 2016), exacerbates such tensions. The indivisibility of the island's reality from its imaginative charge is integral to the web of illusions – the 'magic' – that Prospero weaves. Conversely, this entanglement also alerts us to the powerful fantasies that fuel historical and material realities, especially the way that the island has long operated in the Western imaginary as a *topos* of ownership and absolute power. This condensed set of volatile meanings explains why *The tempest* is also a productive site of postcolonial critique, as in Aimé Césaire's *Une tempête* (1969), wherein Prospero embodies a coloniser attempting to take away the island from islanders Caliban and Ariel, who become, respectively, allegorical figures of Malcolm X and Martin Luther King in Césaire's work. Written at the heart of the Black Civil Rights Movement, Césaire's play is a call for freedom by means of psychological decolonisation through a black consciousness, which he famously termed *Négritude*.

Along with unpacking political discourses of domination, agency and reappropriation, Island literary studies also explores the unique ways that island topographies and topologies have structured literary forms, as evidenced in narratives from the classical era (Homer's *Odyssey*) the early-modern period (Garci Rodríguez de Montalvo's *Amadis de Gaula* (1508); Miguel de Cervantes' *Don quixote* (1605); Thomas More's *Utopia* (1516)); and, later, Daniel Defoe's *Robinson Crusoe* (1719) and Jonathan Swift's *Gulliver's travels* (1726). As we will show, in these cases the island becomes the site on and from which new literary forms are imagined, invented and developed.

Since the time of the Columbian discoveries, island literature has become inseparable from historical, political and geographical globalisation, though as Rod Edmond and Vanessa Smith note, at the time of their writing, "islands had tended to slip the net of postcolonial theorising" (2003, p. 5). Modelling ways of reading island literature in the context of modern world-making, the following discussion will examine three of the dominant binaries by which islands have been understood and bifurcated: reality and fantasy, utopia and dystopia, isolation and connection.

Our intention is twofold. First, this chapter will demonstrate the inevitable contagion between these three sets of binaries, which readily collapse if we dislocate the imperial eye. Second, it will set out some of the key issues for literary studies in the context of island studies where 'the island' is uniquely poised between real and imaginary domains.

We use this approach in the full knowledge that binary structures are inherently problematic – from the gender system to the 'us and them' that configures racism – and are no less thorny in relation to islands. However, as the organising structure of Western language, hence conscious and unconscious thought, binaries have a particular significance in literary studies. The need to identify and undo their unconscious hold on our understanding requires continual attention to their operations. In literary studies, this specific form of unravelling is termed 'deconstruction'; a reading practice that traces the interconnected binaries of a text to the point where they inevitably collapse, where the very logic of the binary undoes itself from the inside (Derrida 1981, pp. 61–156). We are, of course, tired of binaries in relation to islands – indeed in relation to everything – but we cannot move 'beyond' them, for this 'beyond' is an imaginary place without language or culture; for us, it does not exist.

The range of literature we discuss is also limited by our own literary heritages: French and English. We have attempted to extend that range through translated texts but these are far outweighed by our own areas of expertise. What we have attempted is a cross-cultural dialogue to open a wider conversation about and between the world's literary islands.

Real and phantasmatic islands

The inseparability of material and imaginary experiences of islands in the Western tradition can be traced back at least as far as the episodic structure of Homer's *Odyssey* (c. 800 BC) in which each island Odysseus visits is the site of a self-contained adventure or experience. In this way, the archipelagic geographies of the Aegean and Ionian islands explicitly structure literary form, as each island is like a chapter division along the archipelagic thread of narrative. Conversely, literary forms such as the Irish Immram, (most famously the voyages of St Brendan (c. 484 – c. 577), in which heroes undertake journeys to the islands off the Atlantic coast of Ireland in search of the Otherworld, directly informed Columbus's navigation of the Atlantic (Dathorne 1994, p. 95; Flint 1992, 86).

Islands have also played a fundamental role in the construction of modern literary forms. In *Le livre des îles* (2002), French literary historian and Renaissance specialist, Frank Lestringant, shows the ways in which maps, which pre-date fiction in many instances, also conditioned the emergence, organisation and reading of literature (2002, p. 34). The narrative process of the *Isolario* or *Insulaire* (an atlas exclusively dedicated to islands, whose prototype dates back to Cristoforo Buondelmonti's 1420 *Liber insularum*) is textualised in what he calls "Insulaire-récit" or archipelagic narrative. Quoting literary scholar Réal Ouellet, Lestringant demonstrates how the art and function of cartography evolved through time: whilst ancient maps were 'relational', in that they evinced relationships between events, people and places by means of added para-textual comments in prose or verse, their modern counterparts sought to categorise, standardise and quantify these same entities (Lestringant 2002, p. 188). In other words, maps used to tell complex stories the same way novels do, stories that even served political interests at times, especially during European explorations. The *relation* between space and text is at the heart of Lestringant's definition of the 'Insulaire-récit'. A perfect illustration of this relationship lies with François Rabelais' *Fourth* and *Fifth* books of his pentalogy originally published in 1552 and 1564 respectively, a seminal literary work whose influence can be traced notably to Swift's *Gulliver's travels* (Eddy 1922), and whose first two books, *Gargantua* and *Pantagruel*, although not explicitly dealing with islands, are perceived as the first forms of modern novels. Drawing on Lestringant, Simone Pinet has shown how the dialogue between the cartographic form of the *isolario* and chivalric Iberian texts of the medieval period shaped both the modern novel and the atlas, and traces a line to Cervantes' *Don Quixote* (1605, 1615), considered to be the first modern novel (Pinet 2003, p. 173; 2011, pp. 141 ff). Indeed, island narratives are at the heart of many modern forms of fiction. More's island narrative *Utopia* (1516) is, in Fredric Jameson's words, "one of those rare texts that inaugurated a whole new genre of fiction" (Jameson 1977, p. 4). Margaret Cavendish's *The blazing world* (1666) is a key forerunner of science fiction and of the more influential *Gulliver's travels* by Swift (1726, 1735). Coincident with Swift is Defoe's *Robinson Crusoe* (1719), which is the first realist novel in English.

Due to their versatile and hybrid (Lestringant 2002) and even paradoxical (Girault-Fruet 2010) nature, islands have emerged as an enduring locus of satire to critique institutions, customs, movements and ideas of the historical moment. For this reason and because of the inherent play with distance underpinning satire, islands can be read, particularly in postcolonial narratives, as what Michel Foucault calls *heterotopias* or spaces of otherness, which he introduces as "counter-sites . . . of effectively enacted utopia in which the real sites, all the other real sites that can be found within the culture, are simultaneously represented, contested, and inverted" (Foucault

1986, p. 24). Given their relation to social, political and geographical spaces that not only constitute them but which they confront, heterotopias are the spatial manifestation of ambivalence, in turn offering marginal characters the possibility to experience a sense of place and belonging. Heterotopias become the postcolonial subjects' *home*, who, far from being subjected to the confines of marginality, reclaim the space they occupy, be it the island itself or its heterotopic representations like the hospital, the school canteen or the local fair (André 2016, p. 77). In terms of critique and satire, whilst Rabelaisian islands display various forms of a mythical paradise, every one of them unmasks an illusionary utopia (Duval 1998, p. 21). In both Cavendish and Swift, fantastical island worlds provide a site on which to satirise the new experimental (empirical) philosophy, such as that promoted by the London's Royal Society (of the sciences) (Anstey and Vanzo 2012, pp. 399–418). Pierre de Marivaux's 1725 play, *L'Île des esclaves*, critiques the corrupting influence of power. As a rewriting of *Robinson Crusoe* at the time of an emerging moral geography (Lestringant 2002, Chapter 10), the play reveals a growing interest in more 'realistic' literatures, in simplicity and humility, refuting Rabelaisian vulgarity and grotesque exaggerations on the one hand, and the pedantry of the *Précieux* on the other.

Defoe's novel in particular directs us to the historical and political contexts in which these new literary forms proliferated including the role of the real and fantastical in the processes of globalisation. For Defoe's hero is (almost certainly) based on the real-life castaway, Alexander Selkirk, who spent four solitary years on Más a Tierra island, off modern day Chile; Selkirk published an account of this experience in 1712 (Souhami 2002, pp. 163–180). Selkirk had been a sailor with various buccaneering expeditions in the newly accessible terrains of the Pacific and the Americas. Hence, we can claim that the fictional island that provides the site for a new form of writing (realism) is based on a real island that is integrally bound up with historical ('real') events. This is a much larger claim than saying that the content of the novel is based on real events or real places, which is self-evident. It is claiming that new thought and new ways of being in a new world are negotiated at and through the crossroads of a real and an imaginary island. It is the quality of this island as simultaneously real and historically significant, as well as a site of a particular imaginary experimentation, that provides the space on which crucial developments in modern philosophical, economic and political thought, creative form and human subjectivity are mapped (McMahon 2016, pp. 19 ff, 87 ff).

Moreover, it is the island, not the continent or even the archipelago, that is the charged *topos* of this negotiation in Western modernity including literature. The particular intensity and concentration of meaning invested in the literary island from this period marks a key turn in modern thought at the nexus of reality and fantasy that has been played out with devastating effects for many islands and islanders themselves. Gilles Deleuze's essay on 'Desert islands', opens by viewing the modern 'geographic' (actually Darwinian) understanding of two kinds of islands: oceanic and continental, as confirmation of what the human imagination *already* knew about islands: namely, that they may figure a new beginning (the oceanic island) or alienation (the continental island that has broken from the main) (Deleuze 2004, p. 1). These two conditions may also be fused or in contradiction within the human mind. For these profound reasons, islands can be beautiful and productive topologies "to think with" (Gillis 2004) in the re-creation of self and world. Furthermore, just as the circle is the grapheme of belonging, and islands are imagined as bounded circular shapes (Baldacchino 2005), the island not only figures home place but the place where self and space are perfectly aligned; the space where the self is contained, sheltered, held. However, for contemporary Haitian writers Jean-Claude Fignolé, Frankétienne, and René Philoctète, "seeking to narrow the divide between the written and the lived" (Glover 2010, p. xii), the island becomes the fertile ground of a spiralic aesthetic, subverting both the generalised representations of islands as totalising units, and localised authoritarian regimes such as the Duvalier father–son dictatorship (1957–1986) type. Spiralism, as both practice and world vision, refuses definitions – the movement never produced a manifesto – and prefers an ever-questioning Socratic approach to uncover its Haitian realities. In the first comprehensive study dedicated to Fignolé, Frankétienne, Philoctète, and their overlooked, if not forgotten movement, Kaiama Glover explains the importance of the spiral as follows:

> A delicate balance of centripetal and centrifugal forces – of opposing pressure to at once collapse inward and release outward – the spiral effectively allegorises the tension between the insular and the global at work in their fiction. It offers a path via which the three authors have been able to universalise their creative perspective without literally or figuratively abandoning the particular space of their island.
>
> *(Glover 2010, p. xiii)*

When this profound ontology of space and identity became instrumentalised in the related processes of European colonisation, Enlightenment discovery and industrialisation, islands presented themselves to the Imperial eye as 'natural' colonies (Caribbean), natural laboratories (Galápagos, Australia), natural prisons (Van Diemen's Land/ Tasmania), natural lazarets (Isla de Flores, Moloka'i) and natural estates or plantations (Barbados) (Grove 1995). Literature is one of the key sites where this maze of projections, both conscious and unconscious, is encrypted and traversed. Hence its examination allows us, in very particular ways, to unravel colonial desire and trace its operations in history.

The shifts and overlaps between these imaginaries and histories can be clarified by textual comparison. The classic novel of the Caribbean renaissance, *Crick crack money*, published by Trinidadian writer Merle Hodge in 1970, is structured around impossible dualities and binaries. At the level of family, the young protagonist Tee is torn between two aunts of different social standing and values, both of whom have care of her. This duality is embedded in the broader racialised bifurcation of the island, and Tee's inability to reconcile these impossible differences at either family or social levels leads her to formulate her future in terms of being either in or out of the island. These profound divisions, experienced as binary oppositions, are illuminated further when viewed as the corollary of a colonial imaginary (ideology) founded on the island's self-alignment, its imagined homogeneity, an illusion in conflict with the lived reality of its inhabitants. A similar pattern is evident in V. S. Naipaul's account of growing up in Trinidad, which he gave in his Nobel Prize address titled 'Two worlds' (2001). Here, again, the island experience is figured in terms of irreconcilable and recursive dualisms from the family to the broader society and the (im)possibility of imaginative creativity while remaining within the island's borders. He writes:

> My grandmother's house in Chaguanas was in two parts . . . So as a child I had this sense of two worlds, the world outside that tall corrugated-iron gate, and the world at home – or, at any rate, the world of my grandmother's house.
>
> *(Naipaul 2001)*

The world "outside" is described as culturally diverse but access to this diversity is inhibited by the over-riding (island) dualism of inside and outside, either/or, enclosed or free.

Francophone island writing rehearses similar dichotomies. In his examination of travel narratives published between 1598 and 1750, a genre in which the island is a destination of choice, Arlette Girault-Fruet shows the fine balance between the supply and demand of *credible* texts, her notion of credibility being based on what readers were *expecting* the texts to reveal: cyclones,

life rafts, dodo birds, coconuts, and other products of idealisation. Through compiling a repos
itory of island *topoï* in the Mascarene archipelago, Girault-Fruet shows how writers of this
particular genre sought to create both ideal travel narratives – even if it meant adding extra
stereotypes to comply with readers' expectations of exotic tokens of 'elsewhere' – and ideal
landscapes through an exacerbated exoticisation whose purpose was to elicit the admiration of
readers of the time, and increase publication sales. This powerful imprint on the imaginary led
to (at least) two outcomes: either a dichotomous rewriting of the colonised island, or its erasure
from the collective imaginary. In both cases, the island is always defined in the negative: it is that
which it is not. This trend would find one of its most controversial forms in representations of
the Caribbean towards the end of the nineteenth century in the works of Doudouist poets such
as the Martinician 'Prince of Poets', Daniel Thaly (1879–1950). As Ernest Pépin observes, while
Doudouist writers celebrated "blue seas, golden sands, humming birds, luxuriant vegetation,
and the physical grace of the Creole *doudou*" (term of endearment for a woman in the French-
speaking Caribbean) "guaranteed to inspire a cheap wonder based on the illusion of an innocent
paradise", Doudouism was in fact a "crudely staged sham" concealing "reality behind a mask
that serves a poetics of deterritorialisation". In other words, Doudouism as a literary movement
epitomised "the bankruptcy of a space that has become fetishised and dehumanised, emptied of
all meaningful human value" (Pépin 2003, pp. 2–3). By refusing to engage with non-idealised,
unexpected, and disruptive representations of island reality, especially before the 1946 law proclaiming the Departmentalisation of Guadeloupe, French Guyana, Martinique and Réunion,
Doudouist literature both deprived islands of their respective geographical and historical particularities, as well as presented and perpetuated a 'suitable' and refined version of colonial
reality. As a way to refuse this form of *islandedness* (McCusker and Soares 2011), omitting the
word 'island' became a mode of resistance for many postcolonial writers. Pitting themselves
against literary forms fraught with such essentialist and distorted views, Martinican, Reunionese
and Tahitian contemporary writers like Patrick Chamoiseau, Axel Gauvin and Chantal Spitz
explore the problematic attachment to the island-home in their work.

Rejecting the Doudouist model in which the island was deployed to conjure the stereotypes
highlighted above, Patrick Chamoiseau's literary and theoretical texts rarely mention the word
île (island), preferring the term *pays* (country). As such, he "emphasizes the strangeness of the
word [island] . . . oblivious to the horror of slavery" and argues that "the sense of insularity is
imposed from without, a by-product of the colonial gaze" (McCusker 2011, pp. 48–49).

Similar to Chamoiseau's works, the term 'island' is almost absent in Gauvin's five novels. *Train
Fou* (2000) particularly, demonstrates how for the outsider (the continent, the *métropole*), the
island must correspond to certain criteria – exoticism, luxuriance, tropicality, sensuality, immu-
tability and isolation – which only accentuates geographical distance (mainland–island) to create
a reciprocal distancing (island–mainland). Yet this process, or *mise à distance*, goes further: as part
of the dominant discourse on islands, these criteria become internalised by local rhetoric, which
in turn projects them reflexively onto their own environment. Silenced, ignored or set apart for
different reasons, the island remains the geographical reification of desired conquests, appropria-
tions and exploitations, which has respectively lead to uprooting, assimilation and violence.

In a 2012 interview given to the television station France 3, Chantal Spitz, author of the
acclaimed *L'Île des rêves écrasés* (1991), the first novel written by an Indigenous Tahitian writer,
deplores the enduring influence of Louis-Antoine de Bougainville's account of Tahiti. According to Spitz, since Bougainville, Tahitians no longer exist as human beings because they have
been replaced with myths, notably that of the noble savage and the lascivious *vahine*. *Cartes
Postales* (2015), her first short story collection, addresses these misrepresentations by focusing on
the life of undesirables as a way to compensate for the absence of marginal voices. Spitz's style

is as brutal and confronting as her title is misleading, forcing her readers to reflect on their own misconceptions of island life, far from all European fetishisation.

Island as fetish object

The particular qualities of the island gaze – what, in literary studies, is also termed narrative perspective or focalisation – are integral to literature's negotiation between fantasy and reality, utopia and dystopia. Gillian Beer notes the centripetal movement of the colonial gaze of British colonist Stanford Raffles in his *History of Java* (1817 p. clvi) as his eye moves from the abstract conception of "country" to telescope in on the "a sea", "an island", "a mountain", "a fort", "palace" and finally "a jewel" at the centre. Each of the objects in this list, from 'an island' onward, is interchangeable, for each is imagined as a contained unit of space or an object that is graspable or obtainable. However, each is also a step along a diminishing line of scale to the precious centre, to an object that the gaze can encompass, fathom and possess. Louise Pratt's compelling analysis of the coloniser as the 'monarch of all I survey' suggests that it is a commonplace of postcolonial theory for the colonial gaze, embedded as it is in literary accounts, to serve as a mechanism of possession that peers out to the distance of the horizon taking in all land along this vista (Pratt 1992, p. 166). The gaze upon the island shares elements of this scopic drive, for it imagines the island, even when it is relatively proximate, as somehow always on the horizon. For the island's real or imagined diminutive size creates the illusion of a perennial distance – and distance is the space of desire, of the condition of suspension, of never arriving at the destination – hence the island's allure. However, this suspension is simultaneous with the contradictory apprehension of the island as a *topos* of obtainable materiality, as we saw with Raffle's jewel.

Many authors manipulate the conventions of this perspective in their project of challenging this island gaze. Chloe Hooper's *The Tall Man* is a study of past and present life on Palm Island, off the coast of northeastern Australia, which became a prison for mainland Aboriginal people in the early 20th century and continues to battle the historical and ongoing effects of this racialised colonialism. The description of the author's journey to the island first claims an omniscient aerial perspective, only to overturn it. The island is giganticised and engulfs the author/viewer in a scene of multiple "unfolding[s]":

> Travelling to Palm Island had been like a sequence in a dream: the pale green so luminous and the plane flying so close I could almost see the life in it – dugongs, giant turtles, whales. All around were moored small pristine islands. Then on the horizon like a dark green wave, came a larger island. As the plane turned to land the wilderness unfolded. Mountains of forest met the palm-lined shore, which met mangrove swamps, the coral reef. Then the dream shifted.
>
> *(Hooper 2009, p. 8)*

The tall man's (2009) "dream" unfolds into history including the recent death in custody of a local man and the ensuing trial of the relevant police officer.

Literature rehearses the continuous interplay of these two contradictory elements of the colonial island gaze: the illusion that the island can be grasped whole, and the illusion that the island is always in the distance, which is the space of desire. The island's capacity to figure both unending desire *and* the very object of desire, can result in the travesty of an endless, insatiable quest for island possession, as witnessed in Europe's expansion across the oceans since Columbus. Alternatively, in utopian fiction, the interplay of these contradictory elements produces stasis, so that the travellers or castaways who find themselves on an island utopia are initially rendered immobile, as in *Gulliver's Travels*, where Gulliver wakes to find he has been literally pinned down with ropes by the islanders of Lilliput. This immobility relates to the island's object status, its imagined perfection or jewel-like quality. In the utopian scenario the island is unchanging because it is imagined as existing outside time. So, too, utopian fiction from More onward characteristically includes long passages of description in which the traveller and the island inhabitants swap accounts of the ways their respective worlds operate. These descriptions fend off the progression of the narrative that is brought about by events, for events would compromise the island's condition of static perfection by the introduction of time – hence decay – into the text.

Louis Marin provides one of the great insights into utopian fiction that pertains to island fiction and the island as a figure of the imagination. Marin identifies the key error in reading utopian texts as the reader's misrecognition of the island as a *concept*, rather than as a *figure*:

> Utopia is a discourse, but not a discourse of the concept. It is a discourse of the figure: a particular figurative mode of discourse. It is fiction, fable-construction, "anthropomorphised" narratives, concrete descriptions, exotic, novel, and pictorial representation, these are all of its nature. It is one of the regions of discourse centred on the imaginary and no matter how forceful or precise, how correct or coherent, are its theses, utopia will never become a concept. It will always stay wrapped in fiction and fable-making.
>
> *(1984, p. 8)*

That is to say, the operations of utopia pertain to the imagination and futurity, a work in progress, even while the boundedness – the very objectness – of utopian space (an island, or a planet in science fiction) leads the viewer or reader into misrecognising this discourse as an object, a product of thought complete unto itself. This misreading is compounded in utopian fiction by the convention of the philosophical dialogue staged between utopian inhabitants and the visitor, as we see in Aldous Huxley's *Island* (1962). In this novel, cynical journalist Will Farnaby is shipwrecked on the fictional island of Pala in the Pacific. Pala is a utopian island founded on Hindu/Buddhist/Taoist thought and practice. Farnaby undertakes an education in Palaian thought during his one-month stay on the island, which alters his own thinking and being. However, Farnaby enters into this process with a corrupted purpose for, in addition to a genuine interest, he also wants to negotiate an oil deal with Pala, from which he would make enormous sums of money. He is ultimately instrumental in the island's demise. He is the agent who takes the island into time, into the endgame of industrial modernity and decay.

Even this truncated account of the novel's narrative conveys the central importance accorded to the interrogation of concepts, including modes of living, human relationships, the determining force of capital. However, these concepts, variously considered and dismissed and modified throughout the narrative, do not constitute the utopia. Rather, utopian discourse is the 'no place' where these profound questions are endlessly considered and successive alternatives dismissed. Each idea is presented only for its flaws to be exposed. Ultimately, in *Island*, the ethos and practice of Pala is accepted as a form of Truth, but that realisation is simultaneous with destruction. Utopia hence remains unrealised; its time is always futural, yet to come.

These operations, especially the insights into misreading the discourse as a concept (an object) rather than a literary figure (a process) and the unremitting procedure of considering and dismissing concepts, are greatly enabling for dismantling objectifying, fetishistic perspectives of islands. They focus attention on the fictionality of the alignment between a bounded geographical space, temporal stasis and the human subject, as the island is commonly imagined. Hence this recognition also identifies dynamism: this is the history and futurity at the heart of what is commonly misrecognised as a static object or jewel, the precious but isolated anachronistic island.

Moreover – and this is the paradoxical twist at the core of the utopia's imaginative figurations – the Utopian reading process ultimately leads us to history and material reality. Jameson observes that the footnotes in More's *Utopia*, which refer the reader to the links between the fiction and its historical context, are integral to utopian operations in that they engage the reader in a process of measuring the fantastical speculations of the fiction against history. This calibration is not incidental but embedded in the structure of the text, which calls on the reader to entertain a succession of speculative social structures, only to dismiss each in turn – both statement and counter-statement – as internally flawed and contradictory. The relentless process of consideration then refusal of the possibilities presented obliges readers to move between the fiction and their own historical context in the formation of judgment. In this way, the utopian text's speculation on other worlds actually leads readers back into their own time, place and culture.

Following Marx's theorisation of the fetish as the object that hides the labour that created it, work is one of the principal ways islands are fetishised and defetishised in literature: from the apathy of the lotus-eating islanders in the *Odyssey* to the island resort of contemporary tourism. Accordingly, the focus on labour is also one of the significant ways that islands are defetishised, as is clear from the opening scenes of many island texts. From its title and prologue, *Crown jewel* (1952) by the Trinidadian/Australian writer Ralph de Boissière, dismisses the island fetish-jewel of Trinidad by foregrounding the labour politics of the island colony in the 1930s. The novel opens in a scene of mid-work:

> They gripped the spokes of the two-wheeled timber cart and threw their bodies forward.
>   Heave! . . . And again! . . . And again!
>   The flooring boards projected five feet front and rear. Grasping the boards with his massive hands, the helmsman struggled with an angry desperate look to balance the cart while steering it through the traffic according to the wheelman's shouted orders.
>
> *(de Boissière 1981, p. 4)*

Similarly, Trinidadian/Canadian writer Dionne Brand opens her novel of the Grenadian revolution, *In another place, not here* (1986), by foregrounding hard rural labour:

> Grace is grace. Yes. And it take it, quiet, quiet, like thiefing sugar. From the word she speak to me and the sweat running down she in the sun, one afternoon as I look up saying to myself, how many days these poor feet of mine can take this field, these blades of cane like razor, this sun like a coal pot. Long as you have to eat, girl.
>
> *(Brand 1996, p. 3)*

The opening of the short story 'Boat Girl' (1995) by Cook Islands writer Florence Syme-Buchanan immediately redresses the image of languorous island girl integral to the fantasy of the Pacific islands. The story opens with 17-year-old Rima keeping watch for the arrival of the goods ship in port when she sexually services the sailors: "Boats mean business for this island girl and her clientele's harsh crudeness is accepted without complaint" (p. 53).

A formalist perspective on a range of French texts complements the reading of utopian fiction as a figure that overturns the imagined stasis and objectified condition of islands by exposing the interplay between imaginary and materiality. In terms of figures, social geographer Anne Meistersheim discusses the following in her *Les figures de l'île* (2001): the island microcosm, the island in archipelago, the united island, the island paradise, the island labyrinth, the island of masks, the island conservatory, the island laboratory and the island system. As both the imagined and material aspects are tightly intertwined, it is worth considering the question from a different angle: for instance, that of the genre. To do so, it is necessary to distinguish utopia as a programme (that of an ideal state) from utopia as a narrative, a conjectural literary form which deploys motifs of reversibility (Atallah 2016). To illustrate the former, and inspired by Michel Racault's Chapter III of *Mémoire du grand océan* (2007 pp. 81–102), let us turn to two 17th-century indianoceanic utopias: *La République de l'Île d'Eden* and *La République du Libertalia*. In 1689, Henri Duquesne publishes *L'Île d'Eden: recueil de quelques mémoires servant d'instruction pour l'établissement de l'isle d'Eden* in Holland where many persecuted Huguenots fled following the revocation of the Edict of Nantes in 1685, taking away their right to safely practise their religion. In his *Recueil*, Duquesne details his plan to start a Protestant colony away from persecution on an Indian Ocean island: l'île Bourbon (today's Réunion) which, according to Etienne de Flacourt's paradisiacal description, was also "the healthiest island in the world" (de Flacourt, quoted in Dumas-Champion 2008, p. 23). This "utopie-programme" as coined by Racault – "a didactic programme to be realised at a displaced point in time" (Meding 1998, p. 88) – was taken on by French explorer François Leguat, who, while he was in hiding in Holland, decides to participate in Duquesne's vision, and sets sail for the Mascarenhas Archipelago in the southwest of the Indian Ocean. In his 1708's *Voyage et Aventures de François Leguat et de Ses Compagnons en Deux Îles Désertes des Indes Orientales*, he shares the tale of how he unsuccessfully endeavoured to establish an ideal society on the desert island of Rodrigues after failing to settle on Bourbon Island. Half-way between a programme (drastically different to that of Duquesne) and a narrative (the story of how the eight men survived for two years before leaving onboard a makeshift boat for the Dutch colony of what is now Mauritius), Leguat's adventures were criticised for lacking credibility at a time where publications were frequently edited to match readers' desire and expectations of elsewhere. Following suit, the *République de l'Île d'Eden* influenced another utopia known under the name of "République du Libertalia". Said to have been set up in the north of Madagascar, it was introduced in the second volume of *A general history of the pyrates, from their first rise and settlement in the island of Providence, to the present time* (1724). Authored by the mysterious Captain Charles Johnson, this section of the book shares the story of two philosopher pirates, Olivier Misson (the captain) and Father Caraccioli (an unfrocked priest) who fled Europe to set up a libertarian pirate haven in the bay of today's Diego Suarez/Antsiranana, the port city in northern Madagascar. Misson and Caraciolli's accounts might have in fact been written by no other than Daniel Defoe who is believed to have found inspiration in Leguat's account. Coincidentally, or not, Johnson/Defoe's character Olivier Misson shares his last name with Maximilien Misson, who contributed to the publishing of Leguat's adventures.

Once again, island narratives have the propensity to blur the lines between genres and subgenres, between fact and fiction. Suggesting both a place found nowhere (u-topia) or a blissful locale (eu-topia), utopias constantly play on inverted images: self and other, positive and negative, simulacrum and reality (Fougère 2004 p. 17). As presented in narrative form, utopias follow a circular three-step sequence of events detailed by Racault: first, a departure (or rupture with the 'real') followed by a voyage (marked with preparatory incidents); second, an entrance into utopia (boundary crossing, symbolic death, new birth) followed with a detailed description (encounters, dialogues); and last, an exit from utopia, followed by a return voyage and an arrival in the 'real' world (that which was left upon departure) (Cooper 2015, p. 49). While *Paul et Virginie* (1788) is usually read as an exotic pastoral novel, Racault suggests that it can be envisaged as a utopia in so far as the novel follows the steps highlighted above. More interestingly however, Bernardin de Saint-Pierre's indianoceanic writings "illustrate the shifting ties uniting utopian projects striving to materialise, travel narratives, fictional elaborations of social models, and reverie of another world" (Racault 2003, p. 375; our translation).

According to Jacqueline Dutton, "utopia is in fact predicated on desire: the desire for change, the desire for the Other, the desire for that which cannot be achieved in the current order of the society" (2002, p. 20). However, utopia as a narrative should not be confused with the Robinsonade, a genre based on Defoe's model that pits man, nature and God against one another. Robinson Crusoe can be seen as an epitome of "imperial manhood" (Weaver-Hightower 2007, p. 65), while two of his French epigones appear to challenge the traditional genre of the Robinsonade. Published in 1921, Jean Giraudoux's *Suzanne et le Pacifique* transgresses Defoe's model. After her shipwreck, Suzanne, who refuses to domesticate the islands she explores, never gives in to a providential interpretation of her fate (Gauvin 1999, pp. 84–85). At once contemplative and undaunted, she distances herself from her predecessor, Defoe's hero, "this Puritan who was loaded down with reason", proclaiming: "He was perhaps the one man – so superstitious and such a busy-body did I find him – that I would not have liked to meet on an island" (Giraudoux 1975, pp. 225–226). Correspondingly, according to Mario Tomé there are two types of Robinsons: the civilising hero (focused on exploring, building and ordering) or the mystic, for whom the discovery of "another island" will lead to an "inner transformation" (Tomé 1997), as is the case of Michel Tournier's *Vendredi, ou les limbes du Pacifique* (1967). Translated as *Friday, or, the other island* two years later, this award-winning rewriting of Crusoe's adventures introduces new perspectives on the triangular relationship connecting Robinson, the island of Speranza (*Hope* in Italian) and Friday. Whereas traditional Robinsonades textualise the tensions arising from binary oppositions of alienation and domination, Tournier's dialectics generate a revelation. In the pivotal sixth of 12 chapters, Speranza becomes "*l'autre île*" [*the other island*] (p. 105, original italics), that Robinson learns to embrace and accept. If Tournier's island remains stereotypically feminised (e.g. Le Juez and Springer 2015), oscillating between the figure of the mother (and her womb-like cave), and that of the wife (whom he impregnates), Speranza is the primary catalyst of Robinson's final metamorphosis: without 'her' (island is a feminine noun in French), he learns that he cannot experience ultimate otherness, the last phase of his ontological journey, sublimated through his encounter with Friday. Furthermore, in the title's English translation, liminality gives way to otherness as the original eponymous 'limbes' [*limbo*] (*Vendredi ou les Limbes du Pacifique*) are replaced with an 'other' (*Friday, or, the other island*). As a lived experience, the liminal space of the island allows for a critical consciousness, or conscientisation of otherness, a restructuring of Robinson's understanding of his place in his island world.

This dual recognition of the island as a living process and of the necessity of otherness provides a starting point for considering the perceived connection between islands and isolation in the Western imaginary. This connection is iconically represented by *The Tempest* and *Robinson Crusoe* and by John Donne's axiom "No man is an island" which, as Gillian Beer argues, also invokes its opposite to suggest that in fact man *is* an island (Beer 1997, p. 43). These various forms of isolation, banishment and alienation invoke the continental island of Darwin and Deleuze, discussed above, as much as the oceanic island of new beginnings. If we pan outwards from the isolated, gendered subject of the Robinsonade and other castaway fictions, we can recognise how these conceptions also adhere to whole island cultures. We become attuned also to the various refusals of this construction rehearsed in island literatures as well as paradigms of difference, specifically of island interconnection.

Archipelagic interconnection, diasporic interconnection

Western thought has long imagined island connectedness, as is clear in the ancient term 'archipelago' to designate the Aegean, which was renamed Adalar Denizi – sea of islands – in the Ottoman period (Constantakopoulou 2007, p. 1). However, colonialism's possessive drive, including its fetishisation of islands as bounded objects, has informed dominant global perceptions of islands as disconnected topographies. In literary studies, the most thoroughgoing riposte to the perception of islands and islanders as isolated has come from late twentieth-century writers from the Caribbean (Walcott, Brathwaite, Glissant) and the Pacific (Hau'ofa), and these alternatives serve to alert us to the many more understandings in operation in island cultures themselves.

Through powerful metaphors, island critical writing has more often than not interrogated the finitude of its own geographical boundaries to offer creative conceptual perspectives on questions of identity, mobility, and globalisation. Thanks to the translation and dissemination of their work, the voices of poets, thinkers and activists from such island regions as the Mediterranean, the Caribbean, the Pacific and the Indian Ocean loudly resonate in a network of shifting centres and decentred perspectives. From the seminal works of Édouard Glissant (Martinique) and Epeli Hau'ofa (Tonga/Oceania), to the lesser known but nonetheless relevant reflections of Carpanin Marimoutou and Françoise Vergès (Réunion), and Issa Asgarally (Mauritius), islands undeniably emerge as a privileged vantage point to challenge and deconstruct old (colonial) hegemonies.

The suggestive power of the archipelago is central to the thought of Martinican poet, philosopher and theorist Edouard Glissant, and more specifically to his poetics of relation. In Glissant's work, the archipelago does not invoke fixated notions of identity and territory, but rather calls for a vision of the world marked by errantry, openness and shifts. Drawing from chaos theory, and echoing the works of Deleuze, Guattari and Derrida, the archipelagic thought, whose main characteristic is what Glissant calls the trembling [*le tremblement*], acts as an operatory scheme allowing us to rethink relation and encounter with places and people. Set in opposition to notions of the absolute, this "non-systematic system of relationships . . . allows us to intuit the unforeseeability of the world" (Glissant 1997, p. 248). Paradoxically, or unsurprisingly, the island of Martinique, along with its landscapes and flora, has shaped Glissant's "deterritorialised writings" (McCusker 2011, p. 57) allowing him to question the relationship between the individual and the collective, but also to envisage possibilities of anchorage within alienating spaces in constant negotiation:

> Dans le panorama actuel du monde, une grande question est celle-ci: comment être soi-même sans se fermer à l'autre et comment s'ouvrir à l'autre sans se perdre soi-même. [*In the current grand scheme of things, a critical question to ask is this: How to be oneself without closing oneself to others; and how to open up to others without losing oneself.*]
>
> (Glissant 1995, p. 20; our translation)

Envisaging the world as an archipelago is not making use of a denaturing metaphor; it is a poetics rooted in an inspired island imaginary, a complex vision of our connectedness where all beings can claim their right to *opacité*. Recalling Victor Segalen's definition of exoticism whereby this notion "is not the perfect comprehension of something outside oneself which one would contain in one self, but the acute and immediate perception of an eternal incomprehensibility" (Segalen 1955, p. 44), Glissant's opacity recognises the existence of cultural traits particular to a group of people that would be quite incomprehensible to others (Mbom 1999, p. 248). In other words, the finite space of the archipelago world we all inhabit is characterised by the incommensurable enmeshing of relation and opacity, not in a dualistic way but as a trembling entanglement, in all its unforeseeability, vulnerability and *diversalité*.

Drawing on Glissant's work on creolisation – what he calls "the encounter, the interference, the clash, the harmonies and disharmonies between cultures in the accomplished totality of the earth-world" – Réunionese scholars Carpanin Marimoutou and Françoise Vergès resort to the metaphor of the mooring to interrogate the possibility of an Indianoceanness. As the authors of *Amarres: Créolisation India-Océanes* explain,

> We want to suggest an Indianoceanness that comprises both anchorages and moorings. We highlight the metaphor of anchorage because it helps us think about exile and displacement, movement and flux, without forgetting about the territory we have left. . . . The island remembers its continents. We see a to-and-fro movement, a hither and thither, between continents and the island, between the island and the world of islands.
>
> *(Vergès and Marimoutou 2003, p. 15)*

Whilst Marimoutou and Vergès envisage creolisation as "a methodology of living together", their perspective contradicts the underpinnings of Glissant's *surgissement* and *imprévisibilité*, the aim of a methodology being to predict a result. Nonetheless, *Amarres* adds to Glissant's reflection by reframing the implications of such a process through a regional starting point.

Published two years after *Amarres* and prefaced by Nobel laureate J.M.G. Le Clézio, Issa Asgarally's *L'Interculturel ou la guerre* (2005) proposes to focus on diversity in unity (interculturality) as opposed to unity in diversity (multiculturalism). Denouncing the walls Mauritian multiculturalism has erected throughout the so-called rainbow nation (Mauritius' nickname), Asgarally calls instead for a shift in perspective, placing our human unity before our diversity. Highlighted by the anthropologist Thomas Eriksen (2007), the observation that Mauritian identity is marked by tensions that occur between creolisation and multiculturalism does not, however, necessarily suggest two ways of life separately operating in a vacuum. Rejecting any form of totalising perspectives in her study of Mauritian literary ethnotopographies, Srilata Ravi (2007) draws on the imagery of the island geography to demonstrate how various reading levels (individual, communal, colonial/national) intersect within the literary space of the novel to interrogate prevailing ethnocultural and geopolitical determinisms. Her detailed analysis of several novels written in French, from the canonical *Paul et Virginie* (Bernardin de Saint-Pierre 1788), to the confronting *Moi, l'interdite* (Ananda Devi 2000), reveals what happens precisely at the intersection of various *ethniles*, or ethnocultural spaces within the island geography of Mauritius.

Also interested in the spatial representation of postcolonial island identity, Raylene Ramsay has published a number of studies devoted to the literary works of Pacific women, particularly that of Kanak writer and activist Déwé Görödé. Preferring the notion of hybridity to that of creolisation, Ramsay brings to the fore the centrality of place, in imaginaries constructing "the particular power and knowledge that attach to Kanak forms of postcoloniality" (2014, p. 11). While she acknowledges the importance of Hau'ofa's seminal envisioning of Oceania as a sea of islands, "a wholistic perspective in which things are seen in the totality of their relationships" (Hau'ofa 2008, p. 31), Ramsay remains critical of the usefulness of voyaging imaginaries in the historical context of land dispossession. There are undeniable ties, or moorings, between these various envisaging of island (postcolonial) identities, nonetheless underlining similarities remain: the centrality of the influence of the island space on processes of identity construction.

Conclusion

The unique bond between literature and islands plunges the reader of island literature into a productive, generative space between material, historical and imaginary worlds to reveal some of the interactions between these domains of experience. In these entanglements we can identify the political charge of the island daydreams of Western fantasy, the intense identification of identity with space, the real and imaginary processes of creative renewal that islands enable. It is necessary and revelatory to prise apart this compacted matrix of desires and anxieties from their lived effects for island peoples and island habitats more broadly. Such a project does not oppose the literary imagination with the real life but re-discovers their mutual constitution, even within island studies *per se* (Fletcher 2011). After all, literature is a primary site where this re-negotiation of psyche, space, history and materiality is re-formed, re-figured and lived.

References

André, B. (2016) *Iléité: perspectives littéraires sur le vécu insulaire*. Paris, Les Editions Pétra.

Anstey, P., and Vanzo, A. (2012) The origins of early modern experimental philosophy. *Intellectual History Review*, 22(4), 399–418.

Asgarally, I. (2005) *L'Interculturel ou la guerre*. Port-Louis, Mauritius, self-published.

Atallah, M. (2016) Utopie et dystopie. *Fabula*. Available from www.fabula.org/atelier.php?Utopie_et_dystopie.

Baldacchino, G. (2005) Islands: objects of representation. *Geografiska Annaler*, 87B(4), 247–251.

Beer, G. (1997) The making of a cliché: no man is an island. *European Journal of English Studies*, 1(1), 33–47.

Bernardin de Saint-Pierre, J.-H. (1788) 1984 *Paul et Virginie*. Paris, Gallimard.

Brand, D. (1996) *In another place, not here*. New York, Grove.

Brathwaite, E.K. (1974) *Contradictory omens: cultural diversity and integration in the Caribbean*. Mona, Jamaica, Savacou Publications.

Cavendish, M. (1994) *The blazing world and other writings*. Edited by K. Lilley. London, Penguin.

Cervantes, M. de. (1605, 1615) 2003 *Don Quixote*. Trans. Edith Grossman New York: HarperCollins.

Césaire, A. (1969) *Une tempête*. Paris, Seuil.

Constantakopoulou, C. (2007) *The dance of the islands: insularity, networks, the Athenian empire and the Aegean world*. Oxford, Oxford University Press.

Cooper, B.T. (2015) Utopie et esclavage au théâtre français pendant la première moitié du XIXe siècle: l'exemple du Marché de Saint-Pierre d'Antier et Decomberousse (1839). *Tropics*, 2, 47–60. Retrieved at http://tropics.univ-reunion.fr/accueil/numero-2/.

Dathorne, O.R. (1994) *Imagining the world: mythical belief versus reality in global encounters*. Westport, CT, Greenwood Publishing.

De Boissière, R. (1952) 1981 *Crown jewel*. London, Picador.

de Marivaux, P. (1725/2004) *L'île aux esclaves*. Paris, Gallimard.

Defoe, D. (1719) 2003 *Robinson Crusoe*. London, Penguin.

Deleuze, G. (2004) *Desert islands and other texts (1953–1974)*. Translated by Michael Taorima. Los Angeles CA, Semiotext(e) Foreign Agents Series.

Derrida, J. (1981) *Dissemination*. Translated by B. Johnson. Chicago, IL: University of Chicago Press.

Devi, A. (2000) *Moi, l'interdite*. Paris, Dapper.

Dumas-Champion, F. (2008) *Le mariage des cultures à la Réunion*. Paris, Karthala.

Duquesne, H. (1689/1989) *L'Île d'Eden: recueil de quelques mémoires servant d'instruction pour l'établissement de l'isle d'Eden*. Sainte-Clothilde, Association Region Sud: Terres créoles.

Dutton, J. (2002) Feeding utopian desires: Examples of the Cockaigne legacy in French literary utopias. *Nottingham French Studies*, 41(2), 20–36.

Duval, E.M. (1998) *The design of Rabelais's quart livre de Pantagruel*. Geneva, Librairie Droz.

Eddy, W.A. (1922) A source for Gulliver's Travels, *Modern Languages Notes*, 37(7), 416–418.

Edmond, R., and Smith, V. (2003) Introduction. In R. Edmond and V. Smith (eds.), *Islands in history and representation*, London, Routledge, pp. 1–18.

Eriksen, T.H. (2007) Creolisation in anthropological theory and in Mauritius. In C. Stewart (ed.), *Creolisation: history, ethnography, theory*. Walnut Creek CA, Left Coast Press, pp. 153–177.

Fletcher, L. (2011) ' . . . some distance to go': a critical survey of Island Studies. *New Literatures Review*, 47–48, 17–41.

Flint, V. (1992) *The imaginative landscape of Christopher Columbus*. Princeton, NJ, Princeton University Press.

Foucault, M., and Miskowiec, J. (1967/1986) Of other spaces. *Diacritics*, 16(1), 22–27.

Fougère, E. (1995) *Le Voyage et l'ancrage: représentation de l'espace insulaire à l'âge classique*. Paris, L'Harmattan.

Fougère, E. (2004) *Escales en littérature insulaire: Îles et balises*. Paris, L'Harmattan.

Gauvin, L. (1999) La bibliothèque des Robinsons. *Etudes Françaises*, 35(1), 79–93.

Gauvin, A. (2000) *Train fou*. Paris, Seuil.

Gillis, J.R. (2004) *Islands of the mind: how the human imagination created the Atlantic world*. New York, Palgrave Macmillan.

Giraudoux, J. (1921/2015) *Suzanne et le Pacifique*. Paris: Ligaran.

Giraudoux, J. (1975) *Suzanne and the Pacific*. New York: H. Fertig. Translated by B.R. Redman

Girault-Fruet, A. (2010) *Les voyageurs d'îles: sur la route des Indes aux XVIIe et XVIIIe siècles*. Paris, Classiques Garnier.

Glissant, E. (1995/1996) *Introduction à la poétique du Divers*. Paris, Gallimard.

Glissant, E. (1997) *Le traité du tout-monde*. Paris, Gallimard.

Glover, K.L. (2010) *Haiti unbounds: a spiralist challenge to the postcolonial canon*. Liverpool, Liverpool University Press.

Grove, R.H. (1995) *Green imperialism: colonial expansion, tropical island Edens and the origins of environmentalism, 1600–1860*. Cambridge, Cambridge University Press.

Hau'ofa, E. (2008) *We are the ocean: selected works*. Honolulu, HI, University of Hawai'i Press.

Hodge, M. (1970) *Crick crack monkey*. London, Andre Deutsch.

Homer (800 BC) *The Odyssey*: Massachusetts Institute of Technology Internet Classics Archive. Available from http://classics.mit.edu/Homer/odyssey.html.

Hooper, C. (2009) *The tall man: death and life of Palm Island*. Melbourne, Penguin.

Huxley, A. (1962/2005) *Island*. London, Vintage.

Jameson, F. (1977) Of islands and trenches: neutralisation and the production of utopian discourse. *Diacritics*, 7(2), 2–21.

Le Juez, B., and Springer, O. (2015) *Shipwreck and island motifs in literature and the arts*. New York, Rodopi.

Johnson, C. (1724). *A General History of the Pyrates, from their First Rise and Settlement in the Island of Providence, to the Present Time*. London, Printed for, and sold by T. Warner. Retrieved at https://archive.org/details/generalhistoryof00defo

Lefebvre, H. (1974) *La production de l'espace*. Paris, Anthropos.

Leguat, F., and Duquesne, H. (1708) 1995 *Voyage et aventures de François Leguat et de ses compagnons en deux îles désertes des Indes orientales*. Paris, Editions de Paris.

Lestringant, F. (2002) *Le livre des îles: atlas et récits insulaires de la Genèse à Jules Verne*. Geneva, Librairie Droz.

Marin, L. (1993) Frontiers of utopia: past and present. *Critical Inquiry*, 19(3), 197–220.

Mbom, C. (1999) Edouard Glissant, de l'opacité à la relation. In J. Chevrier (ed.), *Poétiques d'Edouard Glissant*, Paris, Presses Universitaires de Paris-Sorbonne, pp. 246–254.

McCusker, M. (2011) Writing against the tide? Patrick Chamoiseau's i(s)land imaginary. In M. McCucker and A. Soares (eds.), *Islanded identities: constructions of postcolonial cultural insularity*, New York, Rodopi, pp. 41–61.

McCusker, M., and Soares, A. (2011) *Islanded identities: constructions of postcolonial cultural insularity*. New York, Rodopi.

McMahon, E.M. (2016) *Islands, identity and the literary imagination*. New York, Anthem Press.

Meding, T.A. (1998) Diana's domain: the displaced centre of the feminine utopia in Honoré Ufré's L'Astrée. In A. Stroup (ed.), *Utopia: 16th and 17th Centuries*. Charlottesville, VA, Rookwood Press, pp. 84–114.

Meistersheim, A. (2001) *Figures de l'île*. Ajaccio, Corsica, DCL.

More, T. (1516/1969) *Utopia*. Harmondsworth, Penguin.

Naipul, V.S. (2001) 'Two worlds'. Nobel lecture. *Nobel Academy*. Available from: www.nobelprize.org/nobel_prizes/literature/laureates/2001/naipaul-lecture-e.html.

Pépin, E. (2003) The place of space in the novels of the Créolité movement. In M. Gallagher (ed.), *Ici-là: place and displacement in Caribbean writing in French*, New York, Rodopi, pp. 1–23.

Pinet, S. (2003) On the subject of fiction: islands and the emergence of the novel. *Diacritics*, 33(3–4), 173–187.

Pinet, S. (2011) *Archipelagoes: insular fictions from chivalric romance to the novel*. Minneapolis, MN, University of Minnesota Press.

Poulle, S., and Desvaux, L. (2012, 3 March) Interview of Chantal Spitz. In *Thalassa, en Polynésie*. [Television broadcast]. Paris, France, France 3.

Pratt, M.L. (1992) *Imperial eyes: travel writing and transculturation*. London, Routledge.

Rabelais, F. (1995) *Oeuvres complètes*. Edited by G. Demerson. Paris, Seuil.

Racault, J.M. (2003) *Nulle part et ses environs: voyage aux confins de l'utopie littéraire classique (1657–1802)*. Paris, Presses de l'Université Paris-Sorbonne.

Racault, J.M. (2007) *Mémoires du Grand Océan: des relations de voyages aux littératures francophones de l'océan Indien*. Paris, Presses de l'Université Paris-Sorbonne.

Ramsay, R. (2014) Kanak imaginaries: a sense of place in the work of Déwé Görödé. *Imaginations*, 5(1). Available from: http://dx.doi.org/10.17742/IMAGE.periph.5-1.2.

Ravi, S. (2007) *Rainbow colours: literary ethno-topographies of Mauritius*. Lanham, MD, Lexington Books.

Robinson, M. (2016) Marylinne Robinson on fear. *Heart and soul*. London, BBC World Service. Available from: www.bbc.co.uk/programmes/p03nbzwg.

Segalen, V. (1955) *Essai sur l'exotisme*. Paris, Mercure de France.

Shakespeare, W. (c. 1611) (1999) *The tempest*. Edited by V. Mason Vaughan and A.T. Vaughan). London, Arden, Bloomsbury.

Soja, E. (1996) *Thirdspace: journeys to Los Angeles and other real-and-imagined places*. Oxford, Basil Blackwell.

Souhami, D. (2002) *Selkirk's island: the original Robinson Crusoe*. London, Orion.

Spitz, C. (1991/2003) *L'Île des rêves écrasés*. Papeete, Tahiti, Au Vent des îles.

Spitz, C. (2015) *Cartes postales*. Papeete, Tahiti, Au Vent des îles.

Swift, J. (1726/2004) *Gulliver's travels*. Syracuse, NY, Barnes & Noble.

Syme-Buchanan, F. (1995) Boat girl. In A. Wendt (ed.) *Nuanua: Pacific writing in English since 1980*. Honolulu, HI, University of Hawai'i Press, pp. 53–56.

Tome, M. (1997) Odyssees et robinsonades: l'aventure insulaire. In D. Reig (ed.), *Ile des merveilles: mirage, miroir, mythe*, Paris: L'Harmattan, pp. 265-278.

Tournier, M. (1967/1983) *Vendredi, ou, les limbes du Pacifique*. Paris: Gallimard.

Vergès, F., and Marimoutou, C. (2003) 2005 *Amarres: créolisation india-océanes*. Paris, L'Harmattan. English translation available from: http://epress.lib.uts.edu.au/journals/index.php/portal/article/view/2568/2884.

Walcott, D. (1992) 'The Antilles: fragments of epic memory'. Nobel Lecture. *Nobel Academy*. Available from: www.nobelprize.org/nobel_prizes/literature/laureates/1992/walcott-lecture.html.

Weaver-Hightower, R. (2007) *Empire islands: castaways, cannibals and fantasies of conquest*. Minneapolis, MN, University of Minnesota Press.

14

CITIES AND URBANISATION

Adam Grydehøj and Ramanathan Swaminathan

Introduction: islands and cities

In recent decades, the interdisciplinary field of island studies has undergone considerable theoretical sophistication. Yet until very recently, its burgeoning lines of inquiry have nevertheless been peculiarly narrow, focusing largely on remote, peripheral, or isolated island communities – or at the very least on interpreting island communities in terms of their remoteness, peripherality, or isolation. Indeed, there at times emerges a circular element to enquiries into islandness: Because islands are regarded as remote by definition, research into the characteristics of islandness often identifies factors related to remoteness as core characteristics of islands.

It is necessary to challenge the all-too-easy association between islandness and remoteness or peripherality: although some islands may be geographically remote, it would be a mistake to try to define islandness exclusively in terms of remoteness. Indeed, it is helpful if we take a step back and consider a very basic, non-theorised definition of an island. According to the *Encyclopædia Britannica* (Baldacchino, 2015), an island is quite simply "any area of land smaller than a continent and entirely surrounded by water". This and similar definitions are so straightforward as to prompt little debate within island studies. Furthermore, it is generally felt within the field that the simple presence of a bridge or other kind of fixed link may alter the dynamics of an island community but does not in itself stop a place from being an island. Otherwise, such indisputably insular places as Prince Edward Island, Canada; Isle of Skye, Scotland; and Lofoten, Norway would need to be excluded from the list. Barrowclough (2010) argues that connections – both physical and immaterial – between islands and mainlands are key (rather than antithetical) to islandness.

So far, so peripheral. But, what of Hong Kong Island (China; see Figure 14.1), Manhattan Island (USA), Mombasa (Kenya) and Stockholm (Sweden): are not these islands that are cities; or, equally, cities that are islands? Such places lie outside island studies' comfort zone for two reasons. First, it is only with great difficulty that they can be described as remote, peripheral, or isolated. Second, they are urban by all definitions, with the transmissive and absorptive capacities of urbanism on full display. Such places nevertheless include 'island' (or an equivalent term in a different language) in their toponyms, so that it might seem an act of stubbornness to deny them island status.

Figure 14.1 City island, island city and subnational jurisdiction: Hong Kong and a view of its Victoria Harbour.

Source: © 2017 Adam Grydehøj.

Yet, many major cities have their cores on pieces of land surrounded by water or had their cores on pieces of land surrounded by water but have since merged with the mainland through land reclamation. Amsterdam (Netherlands), Paris (France) and Tokyo (Japan) all rose out of wetland environments but did so through highly divergent political and economic processes and ultimately developed very different urban morphologies as a result. Cities such as Busan (South Korea) and Mumbai (India) developed due to their favourable port geographies; while Tenochtitlan (the precursor to Mexico City, Mexico) was constructed on an entirely artificial lake island. In our experience, anecdotal though it may be, there remains a wariness within the field to regard such places as islands; it remains commonplace to ignore such places when listing or describing islands in a given region. As we shall discuss below, however, these cities' historical (and in some cases present-day) island geographies have had a significant impact on their development.

If it is not bridges and fixed links that render such places non-insular, then what is it? It cannot be high population size for, after all, places such as Jamaica, New Guinea and Sicily are happily embraced by island studies (e.g. Baldacchino 2015, Brooks 2016, Allen 2017), despite having populations larger than those of many cities. Nor can it be these island cities' high population densities, for even on relatively 'remote' and 'peripheral' islands and archipelagos, high population densities are quite common, so that places such as Malé (Maldives), Barbados and Bermuda feature uncommonly high numbers of residents per square kilometre.

This suggests that it is not simply a matter of densely populated near-shore small islands being denied insular status. The study of islands has traditionally struggled to examine even its traditional research sites in a manner that is sensitive to their urban aspects. It is thus, for instance, that Malta, which is both one of the birthplaces of modern island studies *and* an exceptionally urbanised small state, has been examined within island studies primarily as a periphery or border zone rather than as an urban environment (e.g. Bernardie-Tahir and Schmoll 2014) or as a city-state. Similarly, while the wider scholarship often places considerable emphasis on major population centres when the focus is on a place that happens to be an island – for instance, St John's in Newfoundland (Canada) and Las Palmas in the Canary Islands (Spain) – island studies itself has tended to either elide the entire territory into a single unit of remote islandness or to re-focus on peripheralities within the island or archipelago. (It is perhaps telling that Pacific island studies – with strong roots in development studies rather than in human geography – represents something of an exception to this tendency, with significant emphasis on island urbanisation *per se*; e.g. Connell and Lea 2002, Jones 2007.) The problem seems to lie in an unwillingness to simultaneously analyse a place in terms of its islandness and its urbanity.

This unwillingness has its roots in island studies' development as a distinct research field from the 1990s onwards, with attention consistently being placed on countering stereotypes and clichés regarding islands: the island as isolated, as paradise, as backward, as timeless and so on. Because such stereotypes frequently meet the needs and desires of mainland, continental and metropolitan publics, scholars who focus on islands *per se* often implicitly or explicitly seek to consider islands "on their own terms" (McCall 1994).

This goal, though not without its problems (Baldacchino 2008), is well-intentioned. Yet, it may also be partially self-defeating. For although it has become commonplace in the literature to assert that islands are part of global and regional processes and that remoteness need not imply marginality, this very line of argumentation serves to emphasise the essential association between islandness and distance from centres of power. There can be good reason for seeking to critically claim a voice and representational dignity for disadvantaged or unfairly characterised remote island communities. But care must be taken not to confuse the effects of island status with those of marginalisation more generally. To do so is to place islanders within a peripherality trap; for, while we may conceivably succeed in helping to advance the agendas of colonised or exploited populations or even simply of communities outside the economic and social mainstream, our aim cannot be to render islands less insular, to divest islanders of their island characteristics – whatever these may be. Indeed,

> emphasis on "the islanded few" does not merely protest against how peripheral island communities are denied centrality; it also denies island cities true island status. Ironically, island studies has failed precisely in presenting island communities as centres: Island cities are regarded as not truly insular, and peripherality becomes a defining feature of islands. The terms of the debate are skewed against understanding islands on their own terms. In such a discourse, we cannot hope to place island communities at the centre; we can at best succeed in presenting islands as self-centred.
>
> *(Grydehøj et al. 2015, p. 4)*

To insist upon only viewing islands through the lens of remoteness and peripherality is to over-look the very real urban dynamics affecting many – if not all – island communities.

Over the past few years, there has been growing recognition of the extent to which tradi-tional island studies has overlooked urban issues. An *Island cities and urban archipelagos* research network was set up in 2014, producing both a conference series of the same name and an open-access journal, *Urban Island Studies*, and serving to introduce an urban perspective to island stud-ies. Urban island studies has since been embraced not only by a great many researchers in the fields of urban studies, planning and architecture, who have appreciated the contribution that an island perspective can make on their own research areas, but also by many scholars in island studies proper. Nevertheless, urban island studies has remained a distinct subfield within island studies and, despite the overlap in researchers and research sites, it can be difficult to discern much influence from urban island studies on wider island studies thinking. This is not to criticise those who have contributed to the formation of urban island studies; it is just to highlight that more work must be done to counter the peripheralising tendencies of island studies as a whole.

Thus, this chapter reviews the 'state of the art' of urban island studies, making the case not just for island cities being important objects of study, but also for island cities being important for and to island studies.

Defining island cities and urban island studies

Broadly speaking, island cities can be broken down into two primary categories: (1) major population centres of larger island geographies; and (2) densely urbanised small islands and archipelagos.

The first of these primary categories can be broken down into two distinct subcategories:

(a)   Major population centres of large islands, such as Nuuk (Greenland) (see Figure 14.2), Antananarivo (Madagascar), Hobart (Tasmania, Australia) and Malabo (Bioko, Equatorial Guinea); and
(b)   Major population centres of archipelagos, such as Port Vila (Efate Island, Vanuatu), Kirkwall (Orkney, Scotland, UK), Honolulu (Hawai'i, USA), Yuzhno-Kurilsk (Kuriles, Russia).

The second of the primary categories includes four distinct subcategories:

(a)   Cities that are contiguous (or nearly contiguous) with one or more small islands (such as Banjul, Miami Beach, Portsmouth, Singapore);
(b)   Cities that are significantly or substantially located on one or more densely urbanised small islands (such as Florianópolis, Guangzhou, Lagos, Montreal, Nantes);
(c)   Small islands within cities that are largely located on the mainland (such as the islands of Copenhagen, Hamburg, Manila, Paris, Shanghai, Toronto); and
(d)   Small islands or archipelagos that are not typically regarded as a single urban zone but that are nonetheless densely urbanised as a whole (such as Bahrain, Bermuda, Jersey, Malta, Penang Island).

These are not, of course, precise categories and subcategories. There is, moreover, considerable overlap between them. For instance, Makassar is the major population centre of the large island of Sulawesi, which is itself a relatively peripheral part of the Indonesian archipelago – for many islands, designation as a single island or part of an archipelago is a matter of scale and perspective. Similarly, a place like Abu Dhabi is simultaneously a city that is substantially located on multiple small islands and an island with a distinct urban centre that provides certain urban functions for the remainder of the territory.

An urban island studies perspective could be used to analyse Abu Dhabi in such terms as: (1) the densifying effect of island spatiality; (2) the use of island nature areas to provide ecological and leisure functions to the wider city (in the case of Saadiyat Island); and (3) island city–mainland city relationships. As such, we argue that urban island studies is less the study of particular places that we can designate as 'island cities' than it is a method of applying urban and island approaches to the study of places.

Figure 14.2 Blocks of flats in Nuuk, Greenland: the major population centre of the world's largest island.
Source: © 2017 Adam Grydehøj.

Island population centres

With the above in mind, it is possible to set forth some motivations for taking an urban approach to islands. Although island studies has remained largely stuck in somewhat essentialised centre–periphery and island–island (archipelagic) analyses, which serve to reinforce the importance of remoteness and peripherality for islands, recent developments in urban studies have expanded the field outward from the classically conceived city and have facilitated the wider deployment of an urban approach.

Traditional island studies has not given much attention to island population centres as such. Yet, elsewhere in the social sciences, urban perspectives are increasingly being applied not just to large cities but to population centres in general. An island population centre's density or absolute size (the latter measured in such terms as resident population, economic heft or land area) is less important than its relative impact on, or fulfilment of, urban functions for its hinterlands. For instance, Reykjavik (Iceland) and Saint-Denis (Réunion) are relatively large cities on larger – and relatively remote – islands; yet, their function within their wider island contexts is in many respects similar to that of Stanley (Falkland Islands, UK) and Futami Port (Ogasawara, Japan), which are scarcely more than villages in the grand scheme of things yet provide vital urban func-tions for their wider island hinterlands. Whether large or small, such population centres provide government, retail and other services to their wider island or archipelago communities as well as provide a home to many of the islanders themselves. It might be imagined that advances in communication and transport technology have lessened the significance of such population centres, but in fact such developments have merely changed the manner in which these population centres are significant.

By serving as information and communication technology and transport nodes, connecting rural island communities with metropolitan centres on distant mainlands, these 'island cities' grow in importance in parallel to globalisation. They are the gateways through which the world and the island interact (Pons et al. 2014) as well as through which different parts of a single island or archipelago interact (Yue et al. 2017, Johnson 2016). The greater the complexity of, and resources and skills required for, embedding new technologies into island communities, the more vital the role of the urban hub; and the more dependent the island or archipelago becomes on global networks of skills, knowledge and resources (Al Khalifa 2016).

Ultimately, such cities, towns and villages play a major role in the lives of most islanders; yet, so enduring is the association of islands with remoteness and peripherality that urban settings are typically secondary to scenes of wild or idyllic nature in the mainland imagination of islands. Paradoxically, the vision of natural island paradises encourages diverse forms of tourism urbanisation that can have profound effects on island landscapes and communities (e.g. Ou and Ma 2017, Maguigad et al. 2015, Brooks 2016).

In recent years, however, urban studies has witnessed a movement toward an even broader interpretation of 'the urban'. Brenner (2013) has revived Lefebvre's (1970) theory of 'complete urbanisation' to posit a state of 'planetary urbanisation' in which all regions of the world are drawn into urban processes, as "centres of agglomeration and their operational landscapes are woven together in mutually transformative ways" (Brenner 2014, pp. 17–18). For Brenner (2014, p. 21):

> Traditional notions of the hinterland or the rural [cannot explain] the processes of extended urbanization through which formerly marginalized or remote spaces are being enclosed, operationalized, designed and planned to support the continued agglomeration of capital, labour and infrastructure.

That is, understanding even remote, sparsely populated islands requires an approach that is sensitive to urban processes. In contrast, efforts to conceptually isolate and de-urbanise island communities reinforce a kind of peripherality that may poorly reflect island realities today.

There is also an emerging school of thought that argues that the imagination of the urban must make a decisive break from any counter-imagination of the rural. The argument is that the hyphenated bond between urban and rural, almost like a binary scale, which has defined the urban and the rural in relational terms, no longer holds true. It has been decoupled, as it were, by a suite of digital technologies. These construct narrative bridges that transmit, absorb, transform and mutate existing spaces, obliterating conventional differences between urban and rural, village and city, island and mainland, thrusting upon us a pressing need to create a new politics of space and place:

> There isn't one "single" politics of digitally mediated spaces. There is, however, without a doubt, a socio-technical connectedness between the multiple politics of the hybridized techno-spaces that creates a real world spatiality and territoriality of social singularity. There is also not a single meta-narrative of these spaces; yet the micro-narrative of each one of these spaces mimics the nodal architecture of network systems, contributing to and drawing from other narrative nodes in a nearly instantaneous manner. The logic of the digital has permeated the consciousness of daily common sense, existing, as it were, without any material foundations. Ironically, this logic, while fragmenting the

material and non-material means and modes of social production, the sociality of life, is based on a singular imperative of order and control, which ranges from the genetic manipulation of ecology and ecosystems to reconfiguration of physical and imagined spaces.

(Swaminathan 2014b, p. 101)

Island cities are often powerful nodes of continuous transmission and reception, aided in no small measure by material manifestations of inherent geographical advantages and non-material articulations of information flows and network centricity. Daily urban processes continuously create and re-create spaces (Soja 1989), complicating simple definitions of islandness. In this cauldron of "social production of space" (Lefebvre 1991), in which different urbanisms structure multiple social realities, from daily life processes to global flows (Hayden 1995, Soja 2000), space acquires characteristics of "place and non-place" (Augé 2008), creating islands within islands, of insular mindscapes within a common landscape (Swaminathan 2014a). These can exist, as it were, as separate bodies of a single soul, displaying all the characteristics of isolation, remoteness and peripherality. The logic of the urban presupposes the existence of a landscape attuned to "information flows underpinned by a network" (Castells 1996), in which each hub is interconnected to the other by a variety of material and non-material bridges. The urban logic also injects a landscape, typically a geographical entity, with subjectivities that transform it into a series of 'scapes': ideoscapes, mediascapes, financescapes, technoscapes and ethnoscapes (Appadurai 1990).

Ultimately, it has become imperative to consider islandness through the lens of the digital processes that anchor almost all forms of urbanism, making some of the theorisations surrounding remoteness, peripherality, isolation and insularity seem redundant and obsolete. Such digital processes – simultaneously hyper-urban, global and local – are becoming new frameworks for intuiting sense and meaning, drawing new contours in the urban landscape and creating 'digital bubbles' that are islandic in nature, character and spirit (Swaminathan 2015a, 2014b).

From this perspective, islands are geographical entities, but their geographies consist of more than just stone, sand and sea. Both 'central' and 'peripheral' islands are integrated into material and intangible global processes related to the planetisation of urbanism itself.

Densely populated small islands

We have seen that traditional sites for island studies research can benefit handsomely from an urban perspective. And yet, however, once one begins adopting such an urban perspective, it becomes clear that 'the urban' is not simply something that occurs on islands and mainlands or even that urbanisation is of a special type of significance in island communities. It also becomes clear that islandness has an impact on how 'the urban' develops. Although islands are often presented as the antithesis to the city (Grydehøj 2014), ideal visions of the city have long been associated with island qualities (Pigou-Dennis and Grydehøj 2014).

Indeed, there is an exceptionally strong association between small islands and major cities. Of the world's ten mostpopulous urban agglomerations (Brinkhoff 2014), four (Guangzhou, Jakarta, Manila and Tokyo) are centred on former estuary islands, one (Mexico City) emerged on an artificial island in a lake and one (Mumbai) began on a volcanic archipelago. Eight of the world's ten busiest ports by volume are located on islands: Busan, Guangzhou, Hong Kong, Ningbo-Zhoushan, Rotterdam, Shanghai, Singapore and Tianjin. The largest and most densely populated cities in sub-Saharan Africa (Lagos) and the USA (New York City) are based on small islands; South America's largest city (São Paulo) is a historical offshoot of the island city of São Vicente; and numerous European capitals (Amsterdam, Copenhagen, London, Paris and Stockholm) were established on small islands. What is the reason for the striking propensity for major regional, national and world cities to be located on small islands?

Grydehøj (2015a) identifies three primary benefits that small island spatiality lends to efforts to establish trading posts and/or centres of political power: territorial, defence and transport. Territorial benefits involve the political use of the clear spatial demarcations associated with small islands to lay claim to a particular piece of territory and, ultimately, to project power out from this territory. This was the case with numerous major European cities that arose from the Classical period through to the Early Modern period as well as in sub-Saharan Africa and colonial America: political actors established control over small, easily manageable and circumscribed spaces, then used these as power bases to assert control over wider areas. Interestingly, such territorial benefits are culturally conditioned: in China, where conceptions of statehood developed at a much earlier period and where the centre of regional power was located deep in the continental interior, small island spatiality first began offering territorial benefits in the context of interactions between local and Western powers, with small islands being exploited as overseas power centres by the Netherlands, Portugal, the UK and Germany, and with the Chinese state designating island trading posts in an attempt to curb the spread of colonial power.

Territorial benefits involve internal control and are closely related to defence benefits, which involve protection from external forces. The defensive attributes of small island spatiality have been exploited in numerous geographical and historical contexts, including Tokyo's Edo Castle, Copenhagen's Slotsholm, St Petersburg's Zayachy Island (see Figure 14.3), Belize City's Fort George and Tyre in Ancient Lebanon. In such cases, the fortification of a very small island allowed a local or colonial power to establish a well-protected trading post or residence, from

Figure 14.3   Fortress of St Peter and St Paul, Zayachy Island, Saint Petersburg, Russia.

Source: Wikipedia, https://en.wikipedia.org/wiki/Wikipedia:Featured_picture_candidates/Aerial_view_of_Peter_and_Paul_Fortress_(Saint_Petersburg)#/media/File:RUS-2016-Aerial-SPB-Peter_and_Paul_Fortress_02.jpg.

which power was projected out into the surrounding area. Commercial, governmental and residential functions would congregate in the vicinity of this islanded centre of power and, as the usefulness of island spatiality for defensive purposes declined, the city would often be extended out from or toward the island as a result of land reclamation.

Given the comprehensive nature of their land–water interface, small islands also provide benefits for the waterborne transport of goods and people. Although no longer the fastest means of transport, ships remain the most preferred and cheapest means of transporting large quantities of goods over great distances today (Urry 2014). Remote island communities – the traditional objects of island studies – are characterised by a lack of immediate hinterlands (Baldacchino 2006); yet, small islands located close to the mainland have long been important nodes for trade. In particular, delta and river mouth islands have historically and continue to serve as easily administrable junctions for maritime and terrestrial supply routes. Perhaps surprisingly, despite the massive changes in shipping technology that have occurred over the past centuries, small islands remain favoured sites for major port operations worldwide. Over the past century, comprehensive water access has also made small islands valuable as sites for coastal residential development (Hayward and Fleury 2016).

It is common to differentiate between artificial and natural islands, so that we today treat 'reclaimed' islands – such as those of the Port of Los Angeles, Port of Yokohama, and Port of Rotterdam – differently from the 'natural' islands of historic port cities such as New York City, Singapore and Lübeck. However, the distinction between 'artificial' and 'natural' islands is unclear. Most small islands that have been subject to lengthy human habitation have experienced significant landscape change and coastal engineering, with the result that even 'natural' islands are likely to possess more-or-less artificial port geographies (Grydehøj 2015b). In addition, even completely human-made islands tend to be 'naturalised' over time, ceasing to be regarded as 'artificial' by those who live, visit, or work there (such as Venice, Miami Beach and Christianshavn). By their very nature, in the absence of human efforts to sculpt and control the landscape, river and delta islands are changeable and shifting landforms. The fact that such delta islands are the site of so many major ports presents a challenge to the imagination of 'natural' port geographies. Island cities of this kind are in fact sites of constant and exceptionally intensive human–environment and land–water interaction (Hayward 2015).

Many densely populated island cities have combined two or all three of these primary benefits to island spatiality. The island cities along southeast Africa's Swahili Coast (such as Lamu, Zanzibar, Mombasa and Kilwa), the Western colonial outposts in 20th-century China (Macau, Hong Kong, Qingdao, Shamian in Guangzhou and Gulangyu in Xiamen) and the fortified small islands found in so many regions of the world (Island of Mozambique, Hiroshima, Strasbourg, Kunta Kinteh and the sea forts of Maharashtra) all possess or have possessed complex mixes of functions and spatial benefits. Grydehøj (2015a, p. 433) posits that the various benefits of small island spatiality combine with land scarcity to encourage urban densification, which in turn leads to a form of "circular causation" (Fujita et al. 2001, p. 4) on account of urban agglomeration benefits.

The above discussion of densely urbanised small islands has focused on why cities are especially likely to develop on small islands; yet, the emerging field of urban island studies has also devoted attention to how small island spatiality affects life in cities. Swaminathan (2014a, 2015b) has explored how urban island imaginaries coincide with geographically facilitated hub, node and port functions in Mumbai, while Su (2017) has shown how shifting island city geographies have affected livelihoods and identifications among the Hui minority nationality in Guangzhou.

It has similarly been argued that the archipelagic enclave spatiality of the hyper-dense city of Macau has fostered not only an unusual pattern of urban expansion (Sheng et al. 2017) but also a unique form of political development (Li 2016). Steyn (2015) has, for his part, analysed Mombasa's architectural history and future in terms of island spatial benefits; while Casagrande (2016) has honed in on the demographically disastrous combination of island city density and heritagisation that has played out in Venice.

Conclusion

Contemplating the production of islandness with regard to the multi-islanded Chinese city of Zhuhai, Hong (2017) argues that:

> In the absence of a holistic and dynamic understanding of the spatiotemporal, socio-geographic coproduction of an island city as both focus and locus, we risk what I wish to call *island aphasia*, symptomised in a series of *fixations* on mainland, particular islands, particular interpretations of the island/mainland dialectic, particular functions purported for islands, or particular social groups' imaginaries and deployment of islands.

This serves as a warning against essentialising narratives of all kinds and as a call for a more nuanced island–city relationality. It is tempting to react against island studies' traditional lack of interest in the urban by emphasising connections between the island and the city. Yet, it is vital to remain conscious of similar connections between the mainland and the island as well as the mainland and the city. It is necessary to highlight the problems that arise from excluding the urban from island analyses; but attentiveness to urban processes must not result in a focus on "mechanical spatial arrangement" (Hong 2017): this would be tantamount to a geographical determinism shorn of its socio-temporal context.

Among the tasks for urban island studies researchers going forward is to find a means of integrating an urban perspective into mainstream island studies; to explore how the island and the city interact across their multitude of forms.

References

Al Khalifa, F. (2016) Achieving urban sustainability in Bahrain: university education, skilled labour and dependence on expatriates. *Urban Island Studies*, 2(1), 95–120.

Allen, M.G. (2017) Islands, extraction and violence: mining and the politics of scale in island Melanesia. *Political Geography*, 57(1), 81–90.

Appadurai, A. (1990) Disjuncture and difference in the global cultural economy. *Theory, Culture & Society*, 7(2), 295–310.

Augé, M. (2008) *Non-places: an introduction to supermodernity*. Translated by J. Howe. New York, Verso.

Baldacchino, G. (2006) Managing the hinterland beyond: two ideal-type strategies of economic development for small island territories *Asia Pacific Viewpoint*, 47(1), 45–60.

Baldacchino, G. (2008) Studying islands: on whose terms? Some epistemological and methodological challenges to the pursuit of island studies. *Island Studies Journal*, 3(1), 37–56.

Baldacchino, G. (2015) Lingering colonial outlier yet miniature continent: notes from the Sicilian archipelago. *Shima: The International Journal of Research into Island Cultures*, 9(2), 89–102.

Barrowclough, D.A. (2010) Expanding the horizons of island archaeology. *Shima: The International Journal of Research into Island Cultures*, 4(1), 27–46.

Bernardie-Tahir, N., and Schmoll, C. (2014) Opening up the island: A 'counter-islandness' approach to migration in Malta. *Island Studies Journal*, 9(1), 43–56.

Brenner, N. (2013) Theses on urbanisation. *Public Culture*, 25(1_69), 85–114.

Brenner, N. (2014) *Implosions/explosions*. Berlin,Jovis.

Brinkhoff, T. (2014) *Major agglomerations of the world*. Available from: www.citypopulation.de/world/ Agglomerations.html.

Brooks, S. (2016) Informal settlements in Jamaica's tourism space: urban spatial development in a small island developing state. *Urban Island Studies*, 2(1), 72–94.

Casagrande, M. (2016) Heritage, tourism and demography in the island city of Venice: depopulation and heritagisation. *Urban Island Studies*, 2(1), 121–141.

Castells, M. (1996) *The rise of the network society*. Oxford, Blackwell.

Connell, J., and Lea, J. (2002) *Urbanisation in the island Pacific: towards sustainable development*. New York, Routledge.

Fujita, M., Krugman, P.R., and Venables, A. (2001) *The spatial economy: cities, regions and international trade*. Boston, MA, MIT Press.

Grydehøj, A. (2015a) Island city formation and urban island studies. *Area*, 47(4), 429–435.

Grydehøj, A. (2015b) Making ground, losing space: land reclamation and urban public space in island cities. *Urban Island Studies*, 1(1), 96–117.

Grydehøj, A., Pinya, X.B., Cooke, G., Doratlı, N., Elewa, A., Kelman, I., Pugh, J., Schick, L., and Swaminathan, R. (2015) Returning from the horizon: introducing urban island studies. *Urban Island Studies*, 1(1), 1–19.

Grydehøj, A. (2014) Constructing a centre on the periphery: urbanization and urban design in the island city of Nuuk, Greenland. Island Studies Journal, 9(2), 205-222.Hayden, D. (1995) *The power of place: urban landscapes as public history*. Cambridge, MA, MIT Press.

Hayward, P. (2015) The aquapelago and the estuarine city: reflections on Manhattan. *Urban Island Studies*, 1(1), 81–95.

Hayward, P., and Fleury, C. (2016) Absolute waterfrontage: road networked artificial islands and finger island canal estates on Australia's Gold Coast. *Urban Island Studies*, 2(1), 25–49.

Hong, G. (2017) Locating Zhuhai between land and sea: a relational production of Zhuhai, China as an island city. *Island Studies Journal*, 12(2), 7–24.

Johnson, H. (2016) Encountering urbanisation on Jersey: development, sustainability and spatiality in a small island setting. *Urban Island Studies*, 2(1), 50–71.

Jones, P. (2007) Placing urban management and development on the Pacific islands. *Australian Planner*, 44(1), 13–15.

Lefebvre, H. (1970) *La révolution urbaine*. Paris, Gallimard.

Lefebvre, H. (1991) *The production of space*. Translated by D. Nicholson-Smith. Oxford, Blackwell.

Li, S. (2016) The transformation of island city politics: the case of Macau. *Island Studies Journal*, 11(2), 521–536.

Maguigad, V., King, D., and Cottrell, A. (2015) Political ecology, island tourism planning and climate change adaptation on Boracay Island. *Urban Island Studies*, 1, 152–179.

McCall, G. (1994) Nissology: a proposal for consideration. *Journal of the Pacific Society*, 63–64(17), 93–106.

Ou, Z., and Ma, G. (2017) Marginalisation of the Dan fishing community and relocation of Sanya fishing port, Hainan island, China. *Island Studies Journal*, 12(2), 143–158.

Pigou-Dennis, E., and Grydehøj, A. (2014) Accidental and ideal island cities: islanding processes and urban design in Belize City and the urban archipelagos of Europe. *Island Studies Journal*, 9(2), 259–276.

Pons, A., Rullan, O., and Murray, I. (2014) Tourism capitalism and island urbanization: tourist accommodation diffusion in the Balearics, 1936–2010. *Island Studies Journal*, 9(2), 239–258.

Sheng, N., Tang, U.W., and Grydehøj, A. (2017) Urban morphology and urban fragmentation in Macau, China: island city development in the Pearl River Delta megacity region. *Island Studies Journal*, 12(2), 199–212.

Soja, E. (1989) *Postmodern geographies: the reassertion of space in critical social theory*. London, Verso.

Soja, E. (2000) *Postmetropolis: critical studies of cities and regions*. Oxford, Blackwell.

Steyn, G. (2015) The impacts of islandness on the urbanism and architecture of Mombasa. *Urban Island Studies*, 1(1), 55–80.

Su, P. (2017) The floating community of Muslims in the island city of Guangzhou, China. *Island Studies Journal*, 12(2), 83–96.

Swaminathan, R. (2014a) The epistemology of a sea view: mindscapes of space, power and value in Mumbai. *Island Studies Journal*, 9(2), 277–292.

Swaminathan, R. (2014b) The politics of technoscapes: algorithms of social inclusion and exclusion in a global city. *Journal of International and Global Studies*, 6(1), 90–105.

Swaminathan, R. (2015a) The emergent artificial intelligence of green spaces: digital gaze and urban ecology in Asia. *Asiascape: Digital Asia*, 2(3), 238–278.

Swaminathan, R. (2015b) Ports and digital ports: the narrative construction and social imaginaries of the island city of Mumbai. *Urban Island Studies*, 1(1), 35–54.

Urry, J. (2014) *Offshoring*. London, Polity.

Yue, W., Qiu, S., Zhang, H., and Qi, J. (2017) Migratory patterns and population redistribution in China's Zhoushan archipelago in the context of rapid urbanisation. *Island Studies Journal*, 12(2), 45–60.

15

RIVERS AND ESTUARIES

Mitul Baruah and Jenia Mukherjee

Introduction

Islands have occupied humankind's literary imagination for a long time. However, *island studies* as a more coherent scholarly field is relatively new. Grydehøj (2017) distinguishes the early studies that were mostly about individual islands and the manifestation of their insularity from the studies that have come out since the 1980s, the latter being more general studies of islands and island societies *per se*. There has since been a burgeoning of resources in island studies, including the formation of the International Small Island Studies Association(ISISA), and journals such as *Island Studies Journal*, *Shima* and *Urban Island Studies*, among others (Grydehøj 2017).

Despite such a growing interest and scholarship in island studies, river islands have been largely ignored. Until recently, river islands have also been neglected by fluvial geomorphologists, since the Eurocentric nature of the discipline has ensured that tropical rivers remained outside the purview of its mainstream discussions. Lahiri-Dutt (2014a, p. 22) aptly argues:"[river islands] exist in the vocabulary neither of those who study rivers, nor those who study islands, and have largely remained beyond the mainstream discussions on nature/culture." Referring to the lack of attention to islands in general, Gillis (2014, p. 157) talks about a "blue hole" in environmental history. We would argue, however, that the "brown hole" in environmental scholarship is bigger than the "blue hole", that is to refer to the even bigger gap in scholarship on river and estuarine islands. So, when Baldacchino (2006b, 2008) highlights that most of the world's islands are in the temperate and sub-arctic zones of the northern hemisphere and *not* in the tropics, despite its factual veracity, and despite understanding the specific context in which he is making this point, we fear that such a statement will provide much fodder to an already Eurocentric literature in island studies as well as fluvial geomorphology. Focusing on Majuli, the largest river island in the world, situated in the Brahmaputra River, this chapter is an attempt to foreground river islands in island studies scholarship.

We argue that while riverine and estuarine islands are generally characterised by 'islandness' (Baldacchino 2006a, p. 9), they are distinct in certain ways. These islands are far more ephemeral than islands in the ocean due to their specific fluvial-geomorphological processes (cf. Lahiri-Dutt 2014a). Key to the formation and re-formation of these geographies is the role of the specific sediment dynamics of riverscapes. And, like sediments in rivers, these geographies are not *emplaced* for long: they are transient, travelling constantly with the flow of the river, always re-shaping, and sometimes disappearing entirely (Mukherjee 2011). This study brings forth the question of sedimentation, thereby enhancing the concept of the hydrosocial cycle and island studies scholarship. Furthermore, despite being highly susceptible to various natural disasters, these islands lack the attention that their island cousins in the open sea now tend to secure, the latter being salient in contemporary debates around sea-level rise due to climate change and the vulnerability of island geographies. Our study, therefore, attempts to highlight the distinct features and vulnerability faced by riverine and estuarine islands and the need for island studies to pay attention to these marginalised and hazardous geographies.

Situating tropical river islands in island studies: theoretical frameworks and directions

The neglect of tropical river islands in island studies is rooted in the modernist Eurocentric hydrological knowledge that had instituted a sharp separation between 'land' and 'water'. Indeed, pre-modern conceptions of rivers seem to oscillate between a primary attention to water – for instance, the ancient Greek word for river, *potamos*, means 'water that flows' – and the recognition of the importance of sediment; immense use of river silt is attested in Mesopotamian and Egyptian civilisations as key soil fertiliser, hence livelihood supplier.

The European environmental imagination considered soil–liquid hybrids (in the form of marshes, swamps, fens) as uninhabitable and treacherous places and crafted innumerable drainage, reclamation and embankment campaigns to deploy pumps, dredging devices, locks, and sluices to transform these precarious waterscapes into firm and durable landscapes between 16th and 17th centuries (D'Souza 2009). The European modernist interventions effected 'separations' between land and water. As D'Souza (2009, pp. 3–4) comments:

> land exorcised of water was transformed into property, to be then elaborated as socio-economic-legal objects . . . Flowing waters telescoped into contained channels, on the other hand, were revealed principally as engineering visions . . . rationalised chiefly as communication, transport and movement rather than as a site for production.

The Western hydraulic knowledge and rationality, loaded with new economic, legal, and quantitative calculations was imposed on the tropics. The distinct nature of tropical deltaic rivers, and the lack of colonial knowledge about them, led to the implementation of a series of aggressive interventions causing severe socio-ecological disruptions. The colonial legacy continued in both academic and policy circles making scholars and policy-makers blind towards multiple and multi-layered realities surrounding human engagements with muddy terrains in tropical–deltaic–estuarine geographies.

Recent scholarship had emerged to critically address the land/water binary or separation by looking into sediments as physical and social sites of interactions. Environmental history in south Asia, more specifically water history, had reflected on the impact of colonial hydrological knowledge and interventions on local societies and environment. There is a rich gamut of literature focusing on eastern India (D'Souza 2002, 2003, 2009, Klingensmith 2007, Mishra 2008, Singh 2011) that portrays how the colonial land/water intervention converted the region from a flood dependent agrarian regime to a flood vulnerable scape. A recent, critical literature has emerged in anthropology and geography dedicated to *muddy terrains*, or those places where sediments, rivers and societies intersect (Lahiri-Dutt 2014a, Krause 2017). Lahiri-Dutt (2014a, p. 1) argues for the need of "[reconsidering] one of the foundational binaries [of geographies], that of land and water". She revisits the 'wet theory' conceived by anthropologists like Appadurai and Breckenridge (2009) in order to bring "more fluidity in speaking of hybrid environments" (Lahiri-Dutt 2014a, p. 2), noting that most geographical metaphors are related to land only. As an instance of not excluding complexities or ambiguities, she further invites critical geographers "not to give up mud and silt in favour of either land or water" in their explorations of hybridity (Lahiri-Dutt 2014a, p. 8), drawing empirical insights from deltaic Bengal. Krause (2017) proposes an 'amphibious anthropology' to adequately account for lives in deltas. This approach advances Lahiri-Dutt's work on hybridity by encompassing concepts of wetness (recognising the spectrum of realities between dry and wet, and their local importance), volatility (instability and fluidity of humans and non-humans' interactions) and rhythms (analysis of clashing and/or corresponding ecological and social interrelated rhythms) (Krause 2017).

The emerging scholarship on hydrosociality, within a political ecology of water studies, deals with how water and society make and remake each other over space and time (Linton and Budds 2014). While early hydrosocial literature mainly dealt with political and social injustices around water services in urban landscapes (Bakker 2002, 2003, Swyngedouw 2004,2009), where water is piped water, the recent thrust has been on rivers, specifically issues surrounding dam construction or controversies (Swyngedouw 2007,Hommes et al. 2016), irrigation maintenance and conflicts (Boelens 2014,Mollinga 2014) and river basin governance (Bouleau 2014,Bourblanc and Blanchon 2014,Budds 2009,Budds and Hinojosa 2012,Fernandez 2014). What happens when hydrosociality encounters sediments? Focusing on the riverine islands (*chars*) of the Lower Gangetic Basin, Mukherjee and Lafaye de Micheaux (2016) postulate the hydro (sediment) social cycle concept by advancing a hydrosocial analysis with an enhanced consideration of the sediment component of river materiality. The hydro (sediment) social cycle implies the manifold ways in which humans and non-humans mesh in relations that are simultaneously social and hydrological, in the sense of involving or impinging on the circulation, distribution and quality of water, as well as on the materials that water gathers: muddy sediments (Mukherjee 2018). Mukherjee and Lafaye de Micheaux (2016) explore how human intervention – such as the construction of the Farakka Barrageon the River Ganges in its lower stretch – have altered the deposition pattern of its alluvial sediments, transforming these into hazardscapes and changing perceptions towards riverine islands among multiple social actors across temporal trajectories. The continuous emergence, submergence, re-emergence and re-submergence of *chars*during the post-Farakka period have influenced social processes of settlement, displacement, resettlement and re-displacement among *choruas*(people inhabiting *chars*). These *chars* were also looked into as revenue assets during the pre-Farakka period: this is evident in the introduction of the Bengal Alluvion and Diluvion Act (BADA) of 1825; the key to establishing land rights in the court of law was the payment of rent, even on diluviated land. Massive colonial survey operations were also initiated to produce cadastral maps for revenue survey lists. The postcolonial state considered running *chars* as *sikasti* (water) to legitimise its strategy of not providing infrastructural provisions in these fragile spaces, in turn affecting the livelihood of *choruas* who keep on migrating from one island to the other (Mukherjee and Lafaye de Micheaux 2016).

This scholarship and frameworks provide pathways and directions towards tropical river island studies and provokes empirical research that should further inform and complicate it. The present chapter, focusing on Majuli, is an attempt towards this direction.

Majuli River Island, Brahmaputra Valley, Assam

The Brahmaputra is one of the most gigantic river systems in the world, with a 580,000 km² basin, spreading across China, India, Bangladesh and Bhutan. Originating in the great glacier mass of Chema-Yung-Dung in the Kailash range in southern Tibet, the river traverses a distance of 2,880 km through Tibet, India and Bangladesh, receiving as many as 58 major tributaries in its journey, finally emptying into the Bay of Bengal (Goswami 1985, 2008). It is one of the most sediment-charged large river systems in the world, second only to the Yellow river in China, and it is the fourth largest river in the world in terms of average discharge in its mouth (ibid.). In Assam, the Brahmaputra covers the entire length of the state while its dozens of tributaries connect almost the entire width of the state with the river. Because of its massive scale and the significant role that it plays in the lives and livelihoods of the Assamese people, the Brahmaputra is popularly known as the *MahabahuBrahmaputra* (the Mighty Brahmaputra) in Assam.

Majuli river island is the largest river island in the world (India Today 2016). It is situated in the Brahmaputra Valley in Assam, India, close to the Himalayan foothills. It is bounded by the main channel of the Brahmaputra on the south and the Luit and the Kherkatia *sutis* (*suti* means branch) on the northwest and the northeast respectively (see Figure 15.1). The island is predominantly rural, with a total of 243 small and large villages, inhabited by 167,000 people belonging to a mix of different castes and tribal groups (Government of India 2011). Majuli is not a single island; instead, it consists of a cluster of islands, locally called *chars* or *chaporis*. There are about three dozen of such *chaporis* in Majuli, all of which are lived-in geographies, centering ona large contiguous landmass, which is interestingly referred to as the *mainland* by the *chapori*-dwellers. This 'mainland' is relatively much larger than the *chaporis*, houses all the government offices and business centres, and has better infrastructure compared to the *chaporis*. In this chapter, I will be also referring to this contiguous portion of the island as the mainland as a contrast to the rest of the smaller, more ephemeral islands: the *chaporis*. Majuli is home to numerous wetlands, a diverse range of rural livelihoods, and the *sattras*:the religious and cultural institutions that emerged in Assam as part of a 15th-century neo-Vaishnavite reform movement. However, all this is now under serious threat due to the twin processes of flooding and riverbank erosion. It is to this question of flooding and erosion, and the processes of reproduction of the Majuli hazardscape, that we now turn our focus.

Flooding and riverbank erosion have defined the Brahmaputra valley landscape in general and Majuli river island in particular over the long haul of history. During the monsoon season, the Brahmaputra and its numerous tributaries swell up and inundate the valley in multiple waves within a year. The Brahmaputra is also a highly-braided river, characterised by the processes of channel migrations and resultant riverbank erosion all along its course (Goswami 1985, Sarma and Phukan 2004). The high seismicity of the Brahmaputra valley further adds to the volatility of the local environments. The 1897 and 1950 earthquakes, for instance, which had a magnitude of 8.1 and 8.7 on the Richter scale respectively, reconfigured the Brahmaputra river system significantly, leading to new patterns of flooding and erosion in the valley (Saikia 2013). The 1950 earthquake has also had particularly drastic impacts on the courses and configuration of the Brahmaputra and its several tributaries (and as reported by various old people who lived through the event): the riverbed of the Brahmaputra was drastically raised up by a few metres in some places; new river channels were formed; some old channels were completely diverted; and silt load in the river dramatically increased (Goswami 2008). As a result, the riverbank erosion processes in the valley has accelerated, Majuli being a case in point. Between 1901 and 2001, the island has shrunk from a landmass of 1,255 km²to 422 km², with most of this erosion occurring in the past five decades (Sarma and Phukan 2004). This has resulted in large-scale displacement of families on the island and breakdown of their traditional livelihoods. At the same time, flooding on the island has been consistent as well. Household surveys conducted by Baruah (one of the authors) among 110 households in three villages in Majuli, located along the riverbanks, suggest that between 1962 and 2012, 13 catastrophic floods have devastated the entire island, while relatively moderate floods have affected the low-lying villages on an annual basis. Overall,

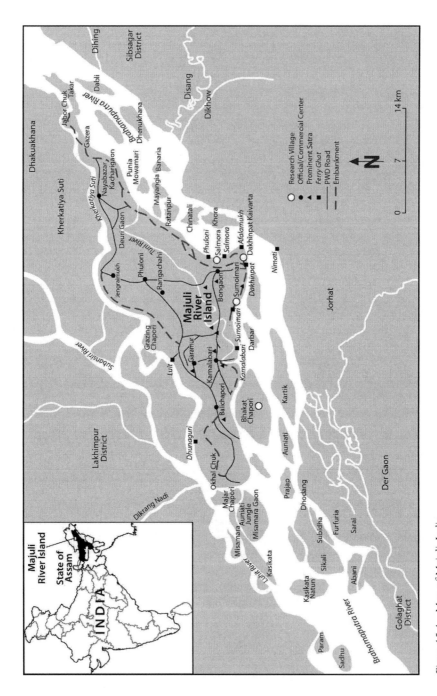

Figure 15.1   Map of Majuli, India.

© Mitul Baruah and Joe Stoll, Syracuse University Cartography Lab, USA.

the combined processes of flooding and erosion have rendered close to 10,000 families on the island homeless over the past half century (Circle Office, Kamalabari 2013).

Flooding however is an annual occurrence in Assam and is integral to the natural history of the Brahmaputra valley (Saikia 2013, p. 10). Traditionally, the Assamese society lived intelligently with annual flooding as well as the processes of mild riverbank erosion. Borrowing Spivak's (1994, p. 48) apt description of the Bangladeshi peasants, it can be said that the rural communities in the Brahmaputra valley "learned to manage [the annual floods], welcome them, and build a life-style with respect for them" (also Lahiri-Dutt and Samanta 2013). Indeed, to paraphrase Linton and Budds (2014, p. 176), as part of the hydrological processes in the valley, annual floods have found their place within the *hydrosocial cycle*, thereby producing 'rhythms' against which the Assamese society has organised its economic and cultural activities. Floodwaters enhanced agricultural productivity in the valley through the deposition of fertile silt (*polosh*); the rivers and wetlands were supplied with fresh fish stock, thus supporting local, fishery-based livelihoods; and the flood season also helped support trading and transportation activities. Flood dependency for the Assamese peasantry was, thus, an "accepted wisdom" for generations (Saikia 2013, p. 6). It was only when massive infrastructures, ostensibly aimed at 'flood control', began re-shaping the hydraulic processes in the Brahmaputra valley, and the reconfiguration of the power dynamics in Brahmaputra valley hydraulic regime therewith, that flood events became dramatic and disastrous. The society that was "living in the rhythm of water" (Spivak 1994, p. 55) suddenly started to witness unprecedented events, and the "hydrosocial cycle" was forever broken (Linton and Budds 2014).

Now, flooding and erosion have acquired a particularly devastating character in Majuli and *constant displacement* has become a part of life for thousands of families living along the riverbanks. For instance, Salmora village on the south of Majuli, where one of these authors conducted fieldwork, was located 2–3 km away from its current location just 50 years ago. But that old area is now in the middle of the main channel of the Brahmaputra, as riverbank erosion took away much of Salmora's original land mass, and the village has been on a constant process of moving inward. Between 1962 and 2012, some 500 families from this village out-migrated to different parts of Assam. Those who continued to stay in the village have been constantly relocating, often without any assistance from the government. And the case of Salmora is not unique; most of the villages located near the river in Majuli are facing similar crises of constant displacement, loss of land and livelihoods, and a steady flow of outmigration of families.

Infrastructures and hazardscapes

In Assam, flood control measures such as embankments go back to the medieval era when the Ahom rulers built such infrastructures by deploying *corvée labour* (that is, a form of unpaid labour instituted by the Ahom state on its subjects) to expand wet rice cultivation in the Brahmaputra floodplains (Guha 1967). However, the scale of these interventions was small then. It was mainly in the 20th century that the hydrological regime in the Brahmaputra valley has gone through unprecedented changes due to massive techno-engineering interventions: first under the colonial regime and later accelerated by the postcolonial state (Saikia 2013). In the postcolonial era, the Assam Embankment and Drainage Act of 1953 was instrumental in allowing rapid expansion of flood control measures in the state. Within half a century of independence, the Assamese state built a large network of about 5,000 kilometres of embankments along the Brahmaputra and its tributaries (Government of India 2013). In addition, a range of other infrastructures such as boulder spurs, geo-textile bags and mattresses, RCC porcupines and screens, and small and large dams have been put in place as flood and erosion control measures.

In Majuli, embankments and other cognate infrastructures, as mentioned above, largely pro-liferated during the postcolonial period, especially in the last few decades. Since the late 1980s, more than 100 km of embankment have been constructed on this small island (Govt of Assam 2013). These infrastructures were built by the state's Water Resources Department, previously known as the Embankment and Drainage Department. In recent years, the Brahmaputra Board, a federal agency, has been tasked with controlling riverbank erosion in Majuli and, to this end, the agency has added a new set of infrastructures on the island. However, *because* of these infrastructures, flooding and erosion have worsened in Majuli. The infrastructures have recon-figured the waterscape of Majuli with devastating socio-ecological implications. Processes that were defining characteristics of the island for generations (for instance, monsoonal flooding and sedimentation of the river and wetlands in and around the Majuli terrain) have now turned disastrous, rupturing the 'rhythm' in the hydrosocial metabolism, thereby turning Majuli into a "hazardscape" (Mustafa 2005).

The embankments have aggravated the flooding and erosion crises in Majuli in several ways. First and foremost, the embankments have divided the island into two separate zones: *inside* and *outside*. The 'inside' zone refers to the 'mainland' portion of the island, which is to say the large contiguous – more developed – portion of the island, surrounded by embankments. The 'out-side' zone, on the other hand, refers to all the areas, including the *chaporis*, which fall outside the embankments, thus being fully exposed to the rivers. By dividing the island into such separate zones, the embankments have produced a distinct spatiality of vulnerability within the island, thereby further deepening the existing societal inequities.

Although the embankments have created a sense of safety in the 'inside' zone, it is a false sense of safety. In reality, neither of the two zones is safe from flooding and erosion. The frequency of flooding in the 'inside' zone may have reduced to some extent due to the embankments, but flood events in this zone have now become much more disastrous. Because the embankments have confined the course of the river, the water level in the river now rises much higher during the monsoon season than it did before, thereby resulting in frequent embankment breaching (Colten 2005). Unlike the regular monsoonal flooding, embankment breaching causes cata-strophic flooding since it is unpredictable and sudden. The agrarian livelihoods in the 'inside' zone have also become much more vulnerable now. Earlier, floodwater used to revitalise the soil on the island by depositing *polosh* (alluvium) on an annual basis. But, ever since the embank-ments came up, this process of natural revitalisation of the soil has come to an abrupt end, leading to declining agricultural productivity. Hence, farmers on the island now resort to heavy doses of chemical fertilisers and pesticides in their fields, which is both economically and eco-logically unviable in the longer run. Mrigen Kutum, a villager in Majuli, explained the impacts of the embankments thus:

> We had our best times here before the *mathauris*[embankments] were built. There was no scarcity of food or fish. But after the embankments were built, there is neither water nor fish flowing into our *beels* and *pukhuris* [the wetlands]. The embankments have also destroyed our agriculture since the lands do not get sufficient and timely water any more. . . . Moreover, with the breaching of embankments, floods started to wreak havoc, which was not the case earlier. Before, the water used to come in gracefully, spread evenly across the island, and dry up sooner.
> *(Interview with Mrigen Kutum by one of the authors, 27 February 2013)*

Kutum's statement deftly captures the multifarious nature of the damages that the embank-ments have caused to rural livelihoods on the island across spectrums. On the other hand, for the 'outside' zone, the embankments have produced a condition of *permanent flooding*. Since floodwater cannot easily enter the mainland anymore due to embankments, the 'outside' zone experiences flooding more regularly and for longer durations. As the water level in the Brahmaputra rises with the onset of the monsoon season, these areas are submerged, and the water does not recede from there for months. Sandwiched as they are between the river and the embankment, the families living in the 'outside' zone own little agricultural land. And when that little plot of land, too, remains flooded for a good part of the year, these families are pushed further to the margins.

The embankments are, however, not the only infrastructures contributing to the crises in Majuli. As mentioned earlier, over the years, government agencies have built a range of other infrastructures in Majuli (see Figure 15.2 and Figure 15.3 for a sample of such infrastructures), ostensibly for flood and erosion control. But far from preventing disasters, some of these infrastructures have instead worsened the situation on the island. Finding their natural courses obstructed by the above hydraulic infrastructures in and around Majuli, the Brahmaputra and the Subansiri now constantly shift their courses, thereby aggravating the processes of flooding and erosion. As for the boulder spurs, some people in Majuli referred to these infrastructures as *mrityubaan* (weapon of death), indicating the potential danger that these structures pose to the nearby villages in case any of these were to be toppled. Indeed, instances of toppling of spurs and washing away of neighbouring habitations have become common in Majuli in recent years.

Large dams are probably the most catastrophic of all hydraulic infrastructures to have come up on the Brahmaputra river. Efforts towards damming the Brahmaputra go back to the early 20th

Figure 15.2   RCC porcupine system.

(Photo: © Mitul Baruah, Majuli, 2013).

Figure 15.3 Boulder spur and geo-bags.
(Photo: © Mitul Baruah, Majuli, 2013).

century when the colonial administration surveyed the hydroelectricity potential of the river for industrial purposes and proposed a number of dams in the valley (Saikia 2013). However, these projects could not take off then due to the global political and economic uncertainties of the time (Saikia 2013). In the postcolonial era, however, keeping in line with the overall countrywide push towards big dams (D'Souza 2008), in Assam, too, dams have gained renewed salience. This obsession for dams continues to date. In recent years, for instance, there has been a special impetus for dams in the northeastern region of the country, as the Indian state has declared this region as the "future powerhouse" of the country (Vagholikar and Das 2010). Thus, in a "great leap forward in hydropower" (Baruah 2012, p. 44), the Indian state has proposed as many as 168 large dams in the state of Arunachal Pradesh with an estimated capacity of 63,328 MW of electricity (Vagholikar and Das 2010). The largest among these, the Lower Subansiri dam with 2,000 MW capacity, where the work is currently ongoing, is in fact within 50 kilometre upstream of Majuli river island.

By now, there are ample studies globally on the devastating impacts of large dams (see, for example, Bandyopadhyay 1995, Baviskar 1995, D'Souza 2006, McCully 2001, Mehta 2001, Sneddon and Fox 2008). Indeed, studies abound on the long-term devastating impacts of large hydraulic infrastructures in general (Colten 2005, D'Souza 2006, 2008, Swyngedouw 1999, White 1995, Worster 1985). The experience in Majuli, as discussed above, offers further testimony to infrastructure-induced disasters. Yet, much like Akhter's (2015) description of the "infrastructure nation" (also see Parenti 2015), the Indian state's interventions in Assam with respect to flood and erosion control pivot around massive hydraulic infrastructures. In the case of big dams on the Brahmaputra, a Technical Experts' Committee instituted by the Government of India has strongly recommended against, since the seismo-tectonic volatility of the local environment makes such infrastructures 'unsafe' (*Down to Earth* 2013). Besides, there are popular resistances in the valley against these gigantic hydraulic projects. But the government seems determined to build these "temples of modern India" (Roy 1999) even as the local inhabitants in the Brahmaputra valley live under constant threat of being wiped out of their land. In fact, the current government in Assam has proposed a US$6.2 billion project to dredge the entire length of the Brahmaputra in Assam as a measure to eradicate the flood and erosion crisis (Sharma 2017). This project has already faced severe criticisms from various corners for its ecological and economic unviability (Akhtar 2017); and yet, 'high modernist' projects of such nature, to borrow Scott's (1998) term, tend to gain salience in government policy parlance in India.

In Majuli, flooding and riverbank erosion have rendered thousands of families homeless over the years, and there has been a continuous wave of out-migration of such families to places outside the island. However, on the basis of data obtained from the Circle (Revenue) Office at Kamalabari, Majuli, in 2013, it appears that the role of the state in rehabilitating these families has been inadequate, with only a handful of them receiving land outside Majuli so far. Similarly, the government has not provided viable livelihood alternatives to communities who are increasingly dispossessed of their traditional livelihoods due to the loss of natural resources. The economy of the potter community (*Kumar*) in Salmora village, for instance, has dwindled significantly as the clayey soil along the riverbanks is fast disappearing, thereby forcing many potter families into wage-labour for their livelihoods. Yet there is no support available from the state and non-state agencies to address the growing livelihood crises on the island. In a nutshell, then, the hazardscape in Majuli is re-produced by the state and the political economic processes, inscribing it, nevertheless, on the body of a fragile, highly vulnerable island ecosystem.

Re-imagining river and estuarine islands: what we learn from these communities

So far in this chapter, our focus has been on the disastrous nature of river and estuarine islands, and the specific risks and vulnerabilities faced by the population in these geographies. We have shown how, in Majuli, flood and riverbank erosion have caused irreparable socio-ecological damage over the years, and how the state is crucial to the processes of re-production of hazard-scape in these geographies. As mentioned earlier, the Brahmaputra is a heavily sediment-laden river; sediments make the islands on the river what they are. However, the relentless techno-engineering interventions on the river over the past century have had significant impacts on the sediment dynamics in the Brahmaputra. Now, the intensity and patterns of sedimentation in the course of the river have changed, and so have the island geographies borne out of the processes of sedimentation. The erosion and re-formation of these landmasses, their susceptibil-ity to flooding and other such disasters, and the stability of their natural resources base have now been altered. The ephemeral nature of these geographies has both accelerated and attained a much more erratic pattern now. It is important to note, though, that the communities living on these islands are not mere victims of disasters; instead, despite the slate of disasters and the lack of adequate government support, these communities have shown enormous perseverance and ingenuity to adjust to their hostile environments.

In Majuli, traditionally, stilt houses have been identified with the Misings, a tribal community. Because the Mising population has historically settled near rivers, stilt houses have helped them avoid flooding and, in due course, these have become part of their culture. In recent years, how-ever, stilt houses have gained acceptance among other communities as well. In Salmora village, for instance, all households have shifted to stilt housing (see Figure 15.4) in the past two to three decades as a measure to withstand regular flooding of their homes. During fieldwork in Majuli,

Figure 15.4 Stilt house in Salmora, Majuli
(Photo: © Mitul Baruah, 2013).

one of these authors also came across people belonging to different communities that were contemplating stilt houses as a flood-prevention measure. People in Majuli have also developed strategies of flood preparedness. During the monsoon season, communities on the island make use of their social networks to spread news about the status of embankments, any impending breaching of these infrastructures, and the level of water in the rivers to ensure necessary preparations in case of a sudden flood. Such informal systems have evolved over the years, especially in the absence of adequate government measures for flood forecast and preparedness.

Similar ingenuity is noticed in livelihoods. The Kaivarta community at Dakhinpat is a case in point. The Kaivartas have been traditionally a fishing community. However, fishing as a primary source of livelihood has become increasingly unviable in Majuli as the water resources on the island have been severely depleted (Baruah 2016). For the Kaivarta population in Dakhinpat, the fishing crisis reached its critical point in the late1980s when a fish epidemic hit the entire island, forcing many families in this village to depend on wage-labour within a short span of time. It was at that point that a few fishers in Dakhinpat decided to experiment with goat trading as an alternative source of livelihood. They chose goat trading because as fish traders previously, they were somewhat also acquainted with goat trading. While visiting the fish and meat market in Jorhat (the nearest big town outside the island) regularly to sell fish, these fish traders in Dakhinpat had built some connections with meat contractors and butchers in the market, and had also realised the high demand for goat meat in the Jorhat market. Jorhat is a district headquarters, and is also one of the large cities in Assam with diverse economic activities, including tourism and a large tea industry. The demand for meat is thus always high in the Jorhat market. The Kaivarta men from Dakhinpat grasped the business opportunity well. Today, goat trading is the dominant source of livelihoods in this village. Indeed, Dakhinpat Kaivarta village is popularly called the *Sagoli Beparir Gaon* (the village of goat traders) in Majuli.

Goat trading for the petty traders is a backbreaking profession, however. Since these traders are not goat-herders themselves, they have to cycle 10–12hours a day to far-flung villages within and outside Majuli to procure goats. And they do that three days a week, round the year, regardless of the weather. These goats are then loaded onto a ferry and then a onto truck to sell in the Jorhat market three days a week. This is a tough profession and the income is meagre. Still, these erstwhile fishers in Dakhinpat have been able to sustain their families in this way. The traders in the village have also formed a committee that looks after various day-to-day issues facing the business. Furthermore, this committee has set up a community fund, based on periodic contribution from all the traders, which is utilised towards supporting cultural events in the village as well as to lend money to needy families within the village.

Initiatives like the above are not the most desirable solutions to the growing livelihood crises facing Majuli. In fact, such local level initiatives may render rural families much more precarious, since they are not geared towards stabilising the local resource base, nor are they backed up by adequate support from the state and non-state agencies. Nonetheless, the flood- and erosion-ravaged Kaivarta families in Dakhinpat are praiseworthy for their ingenuity, perseverance, and success in establishing an alternative form of livelihood on their own. The Kaivarta families in Dakhinpat are not alone, though; indeed, many other communities on the island have also shown similar ingenuity and hard work to find alternative livelihoods. These islanders possess courage, perseverance and creativity: qualities that have allowed them to survive hostile environments with very little external support. Gillis (2013, p. 13) refers to islanders as "jacks-of-all trades" due to their unique adaptability to some of the most hostile environments. Hay (2013) points out that because islanders live in a transition zone that is ever-changing and challenging, they are exposed to "edge effects" – edges being always "fraught with tension" – and it is this exposure to edge effect that make islanders exceptionally adaptive. The case of Majuli, as discussed above, testifies this phenomenon. So, we ask whether these communities can be made effective partners of environmental governance processes and, if so, how? How would that change the nature of the crises currently facing these islands? Can we re-imagine the fate of river and estuarine islands through meaningful inclusion of the island communities in the governance of their islands, their own land, and their own water?

Conclusion

The inclusion of tropical river islands in island studies is an academic and political imperative. It is significant in both river research and island studies, enabling the rethinking of several categories and waves of thought that formulate mainstream perspectives and practices. While one might investigate rivers just in terms of the water they transport, they also erode, carry and deposit sediments. Human technology, livelihoods, social relations and power equations are just as imbricated with river sediment processes as they are with simply water. Though river sediments are less present in discourses and governance paradigms, they cannot be removed from river-society realities. These interrelations are nowhere more striking than in meanders, deltas and estuaries, where the boundaries between land and water are not fixed, neither over seasonal nor over longer-term cycles. Lack of nuanced understandings and explorations of these dynamic, overlapping and heterogeneous realities in these ephemeral geographies have led the state apparatus to assume the guise of a disruptive and interventionist regime. The hybrid and fluid categories of the *chars* run counter to state attempts at legibility, infused with the 'land/water' divide frame of thought (Scott 1998).

Yet, in spite of being vulnerable, the *chars/chaporis*remain attractive to the marginalised communities inhabiting these islands due to the availability of a wide range of rich ecosystem services. The way of life, knowledge and capabilities of *choruas* (people inhabiting*chars*) are forged by and tuned to these ephemeral island spaces. *Choruas* have perceptions, cultural practices and livelihood capabilities within these hybrid lands that could provide insights to document coping mechanisms in the climate change perspective. Though it is important not to romanticise coping measures, yet it is important to assert that the existing local knowledge is neither valued, nor scaled up. The hydro (sediment) society approach across historical scales provides a critical alternative paradigm as to how the liminal spaces of hybrid water/lands are reframed "not [only as] lines of separation but zones of interaction . . . transformation, transgression and possibility" (Howitt 2001, p. 240). Tropical river islands provokes critical interrogation of the modernist Eurocentric watertight divide of water and lands that had robbed the rivers of their histories and extracted them from their social contexts of human experience (Lahiri-Dutt 2014a) Passively ignorant or deliberately neglectful, missing river islands/sediments means becoming blind to the vast numbers of people who live by these realities.

The connection between the global discourse of the Anthropocene, borne of an acute awareness of anthropogenic climate change in the form of melting of glaciers and sea-level rise, etc. and the vulnerability of sea islands had received significant attention in contemporary island literature. Tropical river island research complicates this global discourse, mainly dominated by natural science, by bringing in power equations and cultural dynamics prevalent within local contexts. The political–ecological lens of hazardscape draw our attention to complex array of global, national and local variables that make way to the constant reconfiguration of these ephemeral geographies. Exploration and understanding of local environments and local lived realities expose both challenges and potentials in these constantly 'moving' terrains.

References

Akhtar, M. (2017, 29 August) Dredging the Brahmaputra. *The Third Pole*. Retrieved from: https://www.thethirdpole.net/2017/08/29/dredging-the-brahmaputra/

Akhter, M. (2015) Infrastructure nation: state space, hegemony and hydraulic regionalism in Pakistan. *Antipode*, 47(4), 849–870.

Appadurai, A., and Breckenridge C.A. (2009) Foreword. In A. Mathur and D. Da Cunha (eds.), *Soak: Mumbai in an estuary*, New Delhi, Rupa Publications, pp. 1–3.

Baldacchino, G. (2006a) Islands, island studies, island studies journal. *Island Studies Journal*, 1(1), 3–18.

Baldacchino, G. (2006b) Editorial introduction. In *Extreme tourism: lessons from the world's cold water islands*, Oxford, Elsevier, pp. 3–14.

Baldacchino, G. (ed.) (2007) *A world of islands: an island studies reader*. Charlottetown, Canada and Luqa, Malta, Institute of Island Studies and Agenda Academic.

Baldacchino, G. (2008) Studying islands: on whose terms? Some epistemological and methodological challenges to the pursuit of island studies. *Island Studies Journal*, 3(1), 37–56.

Bakker, K. (2002) From state to market? Water mercantilización in Spain. *Environment and Planning A*, 34(5), 767–790.

Bakker, K. (2003) Archipelagos and networks: urbanisation and water privatisation in the South. *The Geographical Journal*, 169(4), 328–341.

Bandyopadhyay, J. (1995) Sustainability of big dams in Himalayas. *Economic and Political Weekly*, 30(38), 2367–2370.

Baruah, S. (2012) Whose river is it anyway? Political economy of hydropower in the eastern Himalayas. *Economic and Political Weekly*, 47(29), 41–52.

Baruah, M. (2016) *Suffering for land: environmental hazards and popular struggles in the Brahmaputra Valley (Assam), India*. Doctoral dissertation. New York, Syracuse University.

Baviskar, A. (1995) *In the belly of the river: tribal conflicts over development in the Narmada Valley*. New Delhi: Oxford University Press.

Boelens, R. (2014) Cultural politics and the hydrosocial cycle: water, power and identity in the Andean highlands. *Geoforum*, 57, 234–247.

Bouleau, G. (2014) The co-production of science and waterscapes: the case of the Seine and the Rhône Rivers, France. *Geoforum*, 57, 248–257.

Bourblanc, M., and Blanchon, D. (2014) The challenges of rescaling South African water resources management: catchment management agencies and interbasin transfers. *Journal of Hydrology*, 519, 2381–2391.

Budds, J. (2009) Contested H$_2$O: Science, policy and politics in water resources management in Chile. *Geoforum*, 40(3), 418–430.

Budds, J., and Hinojosa, L. (2012) Restructuring and rescaling water governance in mining contexts: the co-production of waterscapes in Peru. *Water Alternatives*, 5(1), 119–137.

Circle Office, Kamalabari (2013) Data obtained from the Circle Office, Kamalabari, Majuli, through personal meetings with officials.

Colten, C.E. (2005) *An unnatural metropolis: wresting New Orleans from nature.* Baton Rouge, LA, Louisiana State University Press.

Down to Earth. (2013, 12 March) *Subansiri dam unsafe: Experts committee.* New Delhi.

D'Souza, R. (2002) Colonialism, capitalism and nature: debating the origins of Mahanadi Delta's hydraulic crisis (1803–1928). *Economic and Political Weekly*, 37(13), 1261–1272.

D'Souza, R. (2003) Damming the Mahanadi river: the emergence of multi-purpose river valley development in India (1943–46). *The Indian Economic and Social History Review*, 40(1), 81–105.

D'Souza, R. (2006) *Drowned and damned: colonial capitalism and flood control in Eastern India.* New York, Oxford University Press.

D'Souza, R. (2008) Framing India's hydraulic crisis: the politics of the modern large dam. *Monthly Review*, 60(3). Available from: https://monthlyreview.org/2008/07/01/framing-indias-hydraulic-crisis-the-politics-of-the-modern-large-dam/.

D'Souza, R. (2009) River as resource and land to own: the great hydraulic transition in Eastern India. In *Asian environments shaping the world: Conceptions of nature and environmental practices*, Singapore, National University of Singapore, pp. 1–25.

Fernandez, S. (2014) Much ado about minimum flows: unpacking indicators to reveal water politics. *Geoforum*, 57, 258–271.

Gillis, J.R. (2014) Not continents in miniature: islands as ecotones. *Island Studies Journal*, 9(1), 155–166.

Goswami, D.C. (1985) Brahmaputra River, Assam, India: physiography, basin denudation, and channel aggradation. *Water Resources Research*, 21(7), 959–978.

Goswami, D.C. (2008) Managing the wealth and woes of the river Brahmaputra. *Ishani*, 2(4).

Government of Assam (2013) Internal document obtained from the Water Reosurces of Department, Majuli Division. Ministry of Water Reosurces, Government of Assam.

Government of India (2011) Census of India 2011. Office of the Registrar General Census Commissioner, India. Indian Census Bureau. Retrived from: https://data.gov.in/catalog/villagetown-wise-primary-census-abstract-2011-assam

Government of India (2013) Data obtained from the Office of the Fishery Department, Government of Assam, Garamur, Majuli.

Grydehøj, A. (2017) A future of island studies. *Island Studies Journal*, 12(1), 3–16.

Guha, A. (1967) Ahom migration: its impact on the rice economy of medieval Assam. *Arthavijnana*, 9. Pune.

Hay, P. (2013) What the sea portends: a reconsideration of contested island tropes. *Island Studies Journal*, 8(2), 209–232.

Hommes, L., Boelens, R., and Maat, H. (2016) Contested hydrosocial territories and disputed water governance: struggles and competing claims over the Ilisu Dam development in southeastern Turkey. *Geoforum*, 71(1), 9–20.

Howitt, R. (2001) Frontiers, borders and edges: liminal challenges to the hegemony of exclusion. *Australian Geographical Studies*, 39(2), 233–245.

Hunter, W.W. (1872) *Orissa, or the vicissitudes of an Indian province under native and British rule*, Vol. 2. London, Smith, Elder and Co.

India Today (2016, 3 September) Majuli is declared the largest river island in the world by Guinness World Records: 10 facts about it. Available from: http://indiatoday.intoday.in/education/story/majuli-largest-river-island/1/755914.html.

Klingensmith, D. (2007) *'One valley and a thousand': dams, nationalism and development.* New York, Oxford University Press.

Krause, F. (2017) Towards an amphibious anthropology of delta life. *Human Ecology*, 45(3), 403–408.

Lahiri-Dutt, K. (2014a) Chars. *Shima*, 8(2), 22–38.

Lahiri-Dutt, K. (2014b) Beyond the water-land binary in geography: water/lands of Bengal. Re-visioning hybridity. *ACME: An International E-Journal for Critical Geographies*, 13(3), 505–529.

Lahiri-Dutt, K., and Samanta, G. (2013) *Dancing with the river: people and life on the chars of South Asia*. New Haven, CT, Yale University Press.

Linton, J., and Budds, J. (2014) The hydrosocial cycle: defining and mobilising a relational–dialectical approach to water. *Geoforum*, 57(1), 170–180.

McCully, P. (2001) *Silenced rivers: the ecology and politics of large dams*. London, Zed Books.

Mehta, L. (2001) The manufacture of popular perceptions of scarcity: dams and water-related narratives in Gujarat, India. *World Development*, 29(12), 2025–2041.

Mishra, D.K. (2008) *Trapped between the devil and the deep waters: story of Bihar's Kosi river*, New Delhi, People's Science Institute and SANDRP.

Mollinga, P.P. (2014) Canal irrigation and the hydrosocial cycle. *Geoforum*, 57(1), 192–204.

Mukherjee, J. (2011) *No voice, no choice: riverine changes and human vulnerability in the 'chars' of Malda and Murshidabad*. Ocassional Paper 28, Kolkata, India, Institute of Development Studies.

Mukherjee, J. (2018) From hydrology to hydrosociality: historiography of waters in India. In J.L. Caradona (eds.), *Routledge handbook of the history of sustainability*, London: Routledge.

Mukherjee, J., and Lafaye de Micheaux, F. (2016) Intervened river, transformed muddyscapes: exploring clashing perceptions across state-society interface in the 'chars' of Lower Gangetic Bengal, India. Paper presented at American Sociological Association Conference. Durham, July.

Mustafa, D. (2005) The production of an urban hazardscape in Pakistan: Modernity, vulnerability, and the range of choice. *Annals of the Association of American Geographers*, 95(3), 566–586.

Parenti, C. (2015) The environment making state: territory, nature and value. *Antipode*, 47(4), 829–848.

Roy, A. (1999, 24 May) The greater common good. *Outlook Magazine*. Available from: www.outlookindia.com/magazine/story/the-greater-common-good/207509.

Saikia, A. (2013) *Ecology, floods and the political economy of hydropower: The Brahmaputra river in the 20th century*. NMML Occasional Paper. New Delhi, Nehru Memorial Museum and Library.

Sarma, J.N., and Phukan, M.K. (2004) Origin and some geomorphological changes of Majuli Island of the Brahmaputra River in Assam, India. *Geomorphology*, 60(1), 1–19.

Sharma, S. N. (2017). Dredging of Brahmaputra will begin this winter: Assam, CM Sarbananda Sonowal. Economic Times, July 16, 2017. Retrieved from: https://economictimes.indiatimes.com/opinion/interviews/dredging-of-brahmaputra-will-begin-this-winter-assam-cm-sarbananda-sonowal/articleshow/59611855.cms

Scott, J. (1998) *Seeing like a state: how certain schemes to improve the human condition have failed*, New Haven CT, Yale University Press.

Singh, P. (2011) Flood control for North Bihar: an environmental history from the 'ground-level', 1850–1954. In D. Kumar, V. Damodaran and R. D'Souza (eds.), *The British empire and the natural world: environmental encounters in South Asia*, Oxford, Oxford University Press, pp. 160–178.

Sneddon, C., and Fox, C. (2008) Struggles over dams as struggles for justice: the world commission on dams (WCD) and anti-dam campaigns in Thailand and Mozambique. *Society and Natural Resources*, 21(7), 625–640.

Spivak, G.C. (1994) Responsibility. *Boundary 2*, 21(3), 19–64.

Swyngedouw, E. (1999) Modernity and hybridity: nature, regeneracionismo and the production of the Spanish waterscape, 1890–1930. *Annals of the Association of American Geographers*, 89(3), 443–465.

Swyngedouw, E. (2004) *Social power and the urbanization of water: flows of power*. Oxford, Oxford University Press.

Swyngedouw, E. (2007) Technonatural revolutions: the scalar politics of Franco's hydro-social dream for Spain, 1939–1975. *Transactions of the Institute of British Geographers* (New Series), 32(1), 9–28.

Swyngedouw, E. (2009) The political economy and political ecology of the hydro-social cycle. *Journal of Contemporary Water Research & Education*, 142(1), 56–60.

Vagholikar, N., and Das, P.J. (2010) *Damming Northeast India*. Pune, India, Kalpavriksh publication.

White, R. (1995) *The organic machine: the remaking of the Columbia River*. New York, Hill and Wang Press.

Worster, D. (1985) *Rivers of empire*. New York, Pantheon.

16

SOCIETY AND COMMUNITY

Godfrey Baldacchino and Wouter Veenendaal

Introduction

In the academic literature, island societies are frequently presented as friendly and easy-going environments in which citizens live together in harmony. According to Dag Anckar (2002, p. 386) "remote and small units are likely to promote feelings of fellowship and a sense of community. When people live at a distance from the outside world, they share a feeling that they are, so to speak, alone in the world." Similar notions are expressed by scholars who aim to find explanations for the well-established statistical relationship between islandness and democracy (Anckar 2008, Clague et al. 2001, Congdon Fors 2014, Hadenius 1992, Srebrnik 2004). Yet, the evidence forthcoming from small islands around the globe suggests that island societies are often characterised by longstanding rivalries, profound antagonism, and deep-seated conflicts among their populations (Austin 2000, Baldacchino 2012, Richards 1982). These conflicts seldom give rise to armed conflict or violence; indeed, most small island units maintain formally democratic political institutions. All in all, it would be fair to state that, in practice, island societies are not always as friendly as the literature may suggest, especially for those members of society who do not conform to dominant norms and practices (Baldacchino 2012). In their case, the pressures to leave and settle elsewhere would be immense.

Most island jurisdictions have small territories and small populations, as a result of which it may be hard to separate the (causal) effects of smallness and islandness (Anckar 2006). In terms of their societal characteristics, being small and enisled is hypothesised to produce the relative isolation, remoteness and social cohesiveness that are commonly deemed to stimulate democratic development and consensual politics. Since most islands are small and most small jurisdictions are islands, there is considerable overlap between the two sets of polities in practice, which further hampers opportunities to differentiate and unpack between the condition of being small and that of being an island. Moreover, if a purely geographical definition of being an island is exchanged for a more psychological or perceptual one, referring to a state of disconnection and detachment from the mainland (however defined) (Baldacchino 2006a; McCall 1994), then any society that is peripheral or isolated from the core might be viewed as 'insular'. In this case, insularity can also stem from geographical conditions like mountains, rivers or deserts, or even from socially constructed boundaries such as ghettos, enclaves or other borders between countries; although we must also acknowledge that there is nothing quite similar to the sea in how it acts as transportation highway, border, climate moderator and source of protein (Hay 2013). Precisely because smallness and islandness are so intertwined both in theory and in practice, and without falling into the trap of definitional empiricism, this chapter focuses explicitly on the quintessential social, economic and political features of small island communities.

A social 'ecology' of smallness: the MITE syndrome

Building on the existing academic literature as well as the experience of small islands around the world, in this chapter we discuss and outline some of the core social characteristics of small, often island communities. We organise this discussion around four central themes which, we argue, to a greater or lesser extent, play a role in island societies around the world: *monopoly*, *intimacy*, *totality* and *emigration* (or collectively, the 'MITE' syndrome). This 'ecology of smallness' was first proposed in Baldacchino (1997) and deployed with respect to Tonga by Puniani Austin (2002). We do not aim to provide an all-encompassing model of small island societies, and we also do not claim that each of the four factors we highlight necessarily plays a role in every island community. Rather, our goal is to propose a tentative but plausible and sound conceptual and analytic framework, on the basis of which social dynamics in island societies can be understood. The same framework is also a leitmotif which allows for comparisons between various island jurisdictions in different corners of the world. Our four core features of island societies can furthermore be seen as root causes or drivers of various historical, economic and political developments and dynamics, but we will not discuss these other effects in detail in this chapter.

We offer here a sequential discussion of each of the core attributes of island societies that we have proposed. We introduce each of our four concepts on the basis of a brief outline of the relevant academic literature, and we subsequently refer to examples from various islands around the world to demonstrate how these dynamics play out in practice. Finally, we pay attention to a long-recognised social phenomenon that appears to have become increasingly salient in small and island communities of late: the tension between 'native' and 'settler' communities (Houbert 1985, Clark 2007) Such tensions can be increasingly witnessed in continental societies that are affected by sudden and intensive migration; but we argue that island societies, by virtue of their greater social intimacy and interconnectedness, and the manner in which their small size exacerbates social dynamics, can in many ways be regarded as archetypal cases for these larger mainland communities. We conclude by discussing some of the broader implications of the social dynamics of island societies, and provide a number of avenues for future research.

Monopoly

> Island communities can, in some situations, experience higher prices because of a lack of competition arising from classic local monopoly conditions. The small market may simply be unable to sustain a rigorously competitive distribution system.
>
> *(Armstrong et al. 1993, p. 317)*

Small islands are often characterised by what they lack: their resources, in terms of natural properties and human capital stock, tend to be scarce or limited. By definition, communities with small populations only have a small pool from which to draw and recruit candidates for influential positions (Bray 1991, Sutton 2007). The result of these restrictions is *monopolisation*, in the sense that power and influence tends to be concentrated in the hands of single or few individuals, institutions or organisations. In the economic sphere, such monopolies and oligopolies translate into the sectoral dominance of single companies and businesses, producing a lack of competition, higher prices and diminished economic efficiency. In the societal and political arena, monopoly entails that a few skilled individuals can obtain supremely powerful positions, producing a natural tendency towards domination, power concentration and oligarchy (Dahl and Tufte 1973, Gerring and Zarecki 2011, Richards 1982). These monopolies have an unfavourable impact on diversity: (serious) alternatives to the owners of economic, political or social power are either weak or lacking. As Richards argues (1982, p. 157), the end result is "a tendency towards convergence of the elite structures in the economic, political and social fields".

From a democratic perspective, the lack of diversity and alternatives creates problems for political contestation. Small communities come along with higher levels of attitudinal homogeneity (Anckar 1999), as a result of which ideologies play a much more limited role in the political arena. Political competition thus often boils down to personal differences, which entails that voters are generally not presented with a choice between substantially different political visions (Veenendaal 2014). This absence of ideological diversity thwarts the formation of a powerful political opposition, because opposition parties and politicians are often unable to formulate serious political alternatives to the (all-powerful) government. In addition, power concentration leads to a disrupted balance between different political institutions, in the sense that the political executive of small islands commonly dominates other institutions and branches of government. While small island jurisdictions are significantly more likely to maintain a democratic political system than larger and continental polities, the presence of a formally democratic political framework may disguise a political reality that is marked by the concentration of power in the hands of single individuals (Peters 1992).

Examples of monopolisation in small islands abound, and particularly so when we deal with small island jurisdictions where the local is national and the national is local. National airlines, national universities, national banks, national internet service providers, national ferry companies, and national water and electricity companies tend to produce a convergence between political and business elites. These and other sectors provide such important economic services that they tend to be dominated or occupied by just one provider – a state or private monopoly – for better or for worse. (The alternative may be to have none at all, which is not so palatable from a practical, political and nation-building perspective.) In practice, this translates into economic inefficiency and diminished competition, as well as strong political and societal dependence on these single providers. In addition, due to the convergence of political and business elites, and the fact that many monopolies are held by the government or other political actors, the private sector of small island territories is often relatively weak or underdeveloped. As we highlight in the section on 'totality' below, the fact that the state is by far the largest employer in any small island jurisdiction also increases the dependency of citizens on the government and politicians in office, which heightens citizens' personal stakes in politics. In Cape Verde, for example, "state institutions, and more particularly the ruling elite, have a sprawling presence all over the economic and social fabric, which promotes the emergence of strong partisan networks" (Alves Furtado 2014, p. 442).

In addition to power concentration, monopoly also entails that those who are not part of the elite are effectively powerless, able only (at best) to oppose. To a much greater extent than larger mainland units, small island societies can be divided into strict 'in' and 'out' groups, or 'haves and have nots'. In the political sphere, this means that opposition groups or parties have virtually no political function, except for contrasting the government and hoping for an electoral victory in the future (Richards 1982). In society at large, smallness and social intimacy commonly produce a lack of political anonymity, in the sense that politicians know with a large degree of certainty which citizens support them and which ones do not (Veenendaal 2013a). This may produce strong inequality between citizens, in the sense that supporters of the ruling parties have access to the political elite, state resources and jobs in public office, while supporters of the opposition are excluded or even victimised. In the Caribbean, the winner-take-all political environment means that "at any time almost half of the population of any given Caribbean society is marginalised and alienated from participation in the development of their society" (Ryan 1999). In the African island nation of São Tomé and Príncipe, "[i]t has been a recurrent feature in the country that relatives expect and demand favours and protection from power and office holders on the basis of their kinship ties" (Seibert 1999, p. 388). The provision of such favours to family and friends means that some citizens are privileged over others, creating strong inequalities. This practice can even be seen as legitimate, as long as power oscillates and alternates; such that those currently in the political wilderness can realistically hope to get back into the limelight, and vice versa.

Intimacy

> Within the enclosed and entangled world of island society, no one was given the freedom of being a stranger.
>
> *(John Enright, Fire knife dancing (2013))*

A second feature of island societies is intimacy, which refers to a social environment characterised by pervasive personal connections and overlapping role relationships (Benedict 1967, Lowenthal 1987, Ott 2000). Whether or not they like each other, inhabitants of small states generally know each other very well, and are compelled to deal with each other on multiple occasions and while fulfilling different societal roles (Lowenthal 1987, pp. 38–39). The ubiquity of such "multiple role relationships" (Ott 2000) entails that private and professional relationships tend to become blurred and intertwined; this results in professional interactions which often become personalised, and vice versa, with personal relationships that may transition imperceptibly to professional arrangements: trends that are often described as clientelism, patronage or nepotism (May and Tupouniua 1980, p. 424). Citizens of small societies have detailed knowledge about each other's personal lives, and "learn to get along, like it or not, with one another, knowing that they are likely to renew and reinforce relationships with the same persons in a variety of contexts over a whole lifespan" (Baldacchino 1997, p. 77). In the political domain, this intimacy means that personal relationships have a great influence on public affairs, since "the political decisions are left squarely with those who have known each other since childhood" (Wood 1967, p. 34).

Intimacy and the ensuing social cohesiveness have been deemed to create a social environment that is friendlier, more harmonious and consensus-oriented (Anckar 2002, Srebrnik 2004). According to Dag Anckar, small island societies are inclined to be charged with "a spirit of fellowship and community" that is associated with "feelings of tolerance and understanding" (2002, pp. 386–387). Yet, while social intimacy may indeed produce enduring friendships and bonds of affection, it can also create profound antagonism, hostility and longstanding feuds. Because individuals who do not like each other also have to constantly interact and engage, however, this antagonism will rarely be openly displayed or expressed. Out of this context develops a complex situation of "managed intimacy" (Lowenthal 1972, Bray 1991), in which citizens are forced to minimise or mitigate open conflict. "The [social-psychological] environment breeds dissimulation, a guardedness that one can never be completely divested of without fear of negative consequences" (Spiteri 2016, p. 300). These dynamics have all the colour and drama of game theory settings (such as prisoner dilemmas); their end result could well be a *prima facie* "concerted social harmony" (Sutton 2007, p. 204), and the prevalence of strong pressures to conform to "dominant cultural codes" (Baldacchino 2012). In short, while social intimacy may be linked to a semblance of harmony and consensus, this may in fact be a carefully managed façade that obscures deep-seated (but subtly handled) personal conflicts.

As a consequence of the strong pressures to avoid disagreements, individuals from small island societies who openly voice their opposition or dissent run the risk of social exclusion and ostracism. This feature obviously has negative repercussions for the development of a vigorous political opposition, since criticism of the government is generally not taken well. Yet, while conflicts between citizens or groups tend to be smothered or shunned, if and when they occur, they are likely to be personal, more explosive, and more likely to polarise every part of society (Dahl and Tufte 1973, pp. 92–93; Srebrnik 2004, pp. 334–338). As a result of the closeness between citizens and politicians and the resulting politicisation of society, small island jurisdictions can be remarkably polarised along party or other social or ethnic lines. For decades, Malta has experienced intense divisions between supporters of the Nationalist Party and the Labour Party; and in Seychelles the rift between the Marxist Seychelles People's Progressive Front and the liberal Seychelles Democratic Party worried British colonial administrators so much that they would only consent to independence if the two parties agreed to form a coalition government. In Jamaica, an entrenched two-party political system can be accompanied by outbursts of armed violence between rival factions (Clarke 2006). Especially in islands with deep social divisions or ethnic, religious and linguistic pluralism, the risk of vicious conflict looms large. As Austin (2000, p. 61) highlights, "when small island communities are mixed . . . resentment is both mutual and strong". On culturally heterogeneous islands like Cyprus, Fiji, Ireland, Solomon Islands, Sri Lanka, Timor, Trinidad and Tobago, and Vanuatu, violent conflicts have occurred, and social tensions between groups occasionally still resurface (e.g. Baldacchino 2014, Srebrnik 2008). Particularly in archipelagic units that bind multiple small islands into a single jurisdiction, island-based nationalism and separatist sentiments may cause upheaval and harbour a risk of state breakdown (Baldacchino 2005, Hepburn and Baldacchino 2013).

And yet, despite strong political and societal polarisation, island societies are often remarkably stable, and are mostly characterised by a large degree of "pragmatic conservatism" (Sutton 2007, pp. 204–205), in the sense that momentous political and social changes rarely occur. Small islands rarely experience severe political crises, armed conflict, or other forms of political violence. And while relations between two individual politicians may be hostile – think Gaston Flosse and Oscar Temaru in French Polynesia; or James Mancham and Albert René in the Seychelles – at one point, when the circumstances change (for example due to the emergence of a common enemy) it is quite possible that political cooperation between former adversaries can be established. When the 'founding father' of politics in the Marshall Islands – Amata Kabua – died in office in 1996, initially an uneasy alliance was formed between his two designated successors (his son Imata Kabua and high chief Litokwa Tomeing). The two soon ran into conflict, and Tomeing formed an alliance with opposition leader Kessai Note, which succeeded in unseating Imata Kabua's government. Subsequently, a conflict between Note and Tomeing ensued, after which Tomeing was elected President with the support of his former ally and adversary Imata Kabua. Examples of such shifting alliances between individual politicians in small island societies abound; while they often produce instability in the short term, in the long run they do not undermine (and perhaps even contribute to) political stability.

Totality

> In some countries, the opposition might not be very powerful, but at least it has
> influence, it has a voice. When the government is as powerful as it is here . . . then
> whatever the opposition says can be totally ignored.
>
> *(Respondent, St Kitts and Nevis, in Veenendaal (2013b, p. 168))*

The remarkably strong sense of a tight community on small islands is often accompanied by
an equally strong presence of the state. Ubiquitous and omnipresent, the ramifications of the
state in small island jurisdictions are extensive and are reminiscent of totalitarian regimes where
'Big Brother' is watching you: "In a small island . . . everybody depends on the government for
something, however small, so most are reluctant to offend it" (Lewis 1965, p. 10; also Connell
1988, p. 5). While most small islands are formally democratic, the tendency toward totality means
that their informal dynamics are often characterised by a remarkable degree of authoritarianism
(Erk and Veenendaal 2014). On the Caribbean island of Antigua, for example, where the Antigua
Labour Party (ALP) had been in power between 1976 and 2004:

> [t]he ALP's domination of the broadcast media, its inordinate use of public resources
> to influence the vote, and its vast, unaccountable spending made for a highly uneven
> playing field and confirmed that rule in Antigua and Barbuda is still based more on
> power and the abuse of authority than on law.
>
> *(Payne 1999, p. 1)*

The relative lack of a vibrant private sector economy in many small islands means that many
islanders will depend on the state, directly or indirectly, for employment or contracts. Bertram
and Watters postulated the MIRAB model to explain how small island systems survive on a frugal diet of remittances and aid transfers (e.g. Bertram 2006); in doing so, they also acknowledged
the pivotal role of bi-lateral or multi-lateral state aid in securing those vital finances which, in
turn, makes it possible for the state to employ (and pay) its public servants. Indeed, the MIRAB
acronym – **MI**gration to fuel **R**emittances; **A**id to support the state **B**ureaucracy – has been
tweaked by Howard Brookfield (quoted in Bertram and Watters 1985, p. 497) to read MIRAGE:
in which case, the letters 'GE' stand blatantly for government employment. The clutches of the
state are so expansive that even civil society, where it exists on small islands, may often organise
itself primarily to lobby and seek resources from the state.

From studies of the Caribbean, Sutton (2006, pp. 13–15) characterises the performance of
the public sector in small island states as manifesting four particular traits: (1) an 'exaggerated
personalism', referring to the strong influences of government representatives who tend to
use their power to control or influence appointments and contracts; (2) 'limited resources' that
push most citizens (and including civil servants) to perform multiple roles and responsibilities,
even though these may not fit so neatly with their specialisation and they were not adequately
trained to perform them; (3) an 'inadequate service delivery' due to a general inability of reaching economies of scale; and (4) a 'relatively high degree of dependence on foreign management
consultants', typically sourced from the former colonial power, who may promote and apply
practices that do not necessarily suit the small island state context.

In the Anglophone Caribbean, political leaders remain in office for a comparatively long
time, and are often able to dominate the entire political arena. As Sutton (1999, p. 73) highlights,
"short of defeat at a general election, the prime minister is invincible". The relative weakness

of Parliament, the political opposition, the media, and other institutions that are supposed to function as a check on executive power entails that Caribbean governments are omnipotent; and, given the acute personalisation of politics, this often translates into an accumulation of vast powers by single individuals. Leaders such as Vere Bird of Antigua and Barbuda, Lynden Pindling of the Bahamas, Eric Gairy of Grenada, John Compton of St Lucia, and Claude Wathey of St Maarten were not only known for their remarkably long tenures in office (sometimes over 30 years), but also for their authoritarian ruling style. In similar fashion, South Pacific island societies are dominated by 'big men' and traditional leaders who sometimes possess tremendous personal powers (McLeod 2007). Traditional leaders commonly combine their chiefly titles with owning a business and running for elected office, leading to a convergence of traditional, economic and political power. The upshot of such developments is a lack of power-sharing as well as the absence of economic and political pluralism, which may ultimately undermine effective governance.

This intensity of the state's presence in island life is exacerbated by personality politics. Political contests are of course accompanied by media broadcasts, televised meetings and debates, and now also social media, as in other larger jurisdictions. However, in small societies, the voter and the voted are known, and they will typically make it a point to connect at a physical and face-to-face level. This voter–politician relationship is facilitated by the relatively smaller number of votes required to elect a small island representative to office. This occurs because Parliaments, even in small jurisdictions, will always have a minimum number of members and so loom disproportionately large.

Just by way of example, there are 577 seats in the French National Assembly (Parliament), with 46 million eligible voters at the latest general election: this means that each parliamentarian represents around 80,000 eligible voters. In contrast, for the Dominica House of Assembly, there were 72,500 voters registered for the latest general election. With 21 seats in the House, this translates as just 3,450 eligible votes per parliamentarian. Thus, even though Dominica can field only 1/634th of the voters that France has, its legislature is an impressive 1/27th the size of that of France.

Such low numbers and ratios make personal relationships inevitable, and both the voter and the candidate will go out of their way to make themselves known to each other. Indeed, small island societies can rightly claim to manifest some of the world's highest voter turnouts. In Malta and Iceland, voter turnout is typically 90% or more (even though voting is not compulsory). Such behaviour is not necessarily equivalent to strong democratic values: in such small societies, it may not be a good idea to be seen, and therefore known, *not* to vote (Hirczy 1995).

Another justification for the extraordinary role of the state in small island societies has to do with critical mass. This suggests that any society, and especially a jurisdiction, will require a set number of roles typically performed by the state: a chief electoral officer, a commissioner of police, a chief justice, an accountant general, a postmaster general . . . whether we are talking of India (the world's largest democracy by population) or Tuvalu (the world's smallest sovereign island state), these roles tend to exist. If unable to spread out these roles to as many people, small societies can, and do, combine some of these roles within the job description of the same person. Thus: "Not only are there fewer roles in a small scale society, but because of the smallness of the total social field, many roles are played by relatively few individuals" (Benedict 1967, p. 26)

This practice of compression, alas, leads to its own set of concerns, since overlapping roles can lead to situations of role conflict. In many Caribbean and Pacific islands for instance, politicians are also active as radio and television producers (and vice versa), and therefore have a large influence on both the political agenda and the agenda of the news media (Puppis 2009). The same individuals are brought into contact over and over again in various activities, where these persons are notionally playing different roles. In *gemeinschaft* – often understood to mean pre-industrial – type communities, with limited mobility outside the group, such interactions may last a lifetime (Brint 2001). In small island social systems, ascribed criteria trump achieved criteria, even in notionally meritocratic contexts. Primary and secondary school mates reconnect in higher education, in the workplace, in other social, political and religious activities. Friendships last decades and can encourage subtle forms of preference, discrimination and favouritism; meanwhile, enmities too will last a lifetime and can trigger serious rivalries and antipathetic relations, reminiscent of mafia-esque affairs. Thus, a small island citizen will grow up in a dense network of family, friends and "friends of friends" (Boissevain 1974) leading to a potentially claustrophobic "straightjacket of community surveillance" (Weale 1992, p. 9). S/he will eventually construct and sustain his/her own network, while also, often unwittingly, building an anti-network of those with whom s/he does not sympathise (Baldacchino 1997).

Thus, (social) intimacy, (economic) monopoly and (political) totality combine to create a very distinct community fabric which one must learn to negotiate and possibly manage in order to survive and lead a decent livelihood. Small island life consists of regular bouts of managing intimacy, of suffering monopolies while also exploiting them for one's own benefit, and of living cheek by jowl with politicians. Should this cocktail of island life be found to be too stifling and unbearable, the solution may not lie in transferring to the next village, town or city. Hence, the fourth and final leg of our small island syndrome: exile (or emigration).

Emigration

> I know I can't live away from Samoa for too long. I need a sense of roots, of home – a place where you live and die. I would die as a writer, without roots; but when I go home, I am always reminded that I'm an outsider, palagified.
>
> *(Albert Wendt, in Beston and Beston (1977, p. 153))*

We will not review the important yet ambivalent topic of island migration here – that is handled, more appropriately, in a separate and stand-alone chapter (see Connell, this volume). Suffice it to argue here that migration becomes almost the inescapable option for those who cannot survive or tolerate the often toxic MIT combination expounded above. Young people in many small island societies feel a strong urge to leave their families and seek adventure and fortune elsewhere; often, it is the unmistakable allure of better paying and more stable jobs, of higher and better quality education and of spending time with other (and already uprooted) family members which serve as amongst the pivotal 'pull' factors. However, the clannish nature of island life, the difficulty of striking out on your own shorn of the baggage and 'social capital' that comes from family and tribe, or the difficulties involved in starting a new life (after, say, a period of imprisonment, drug rehabilitation, educational failure or a business fiasco) may be simply too much to bear, especially for the young and adventurous. These legacies of belonging become the subtle but nagging 'push' factors that are conducive to migration, initially to the next village, town, city or island; progressively, and if the circumstances and finances permit, to a different country altogether.

Reactions to the MITE syndrome

These dynamics may appear self-evident, both to those who live in small island societies, other small and tight-knit communities, as well as to those who know them well enough to understand 'what is really going on' beneath the veneer of respectability, market forces and legal-rationality. Yet, this is often not the case in practice. Other than emigration, which we have already acknowledged, there are three general types of response when confronted with the power and ubiquity of the small-scale syndrome (Baldacchino 1997).

The first is *denial*. It is not just mainland observers, freelance consultants and dispassionate academics but many islanders themselves who may find it hard to believe and admit that monopoly, intimacy and totality, and their multiple interconnections, have such a grip on island life. To them, any such admission may be tantamount to accepting that their island is somewhat a 'second class' jurisdiction, not abiding by the canons of legality, objectivity and impartiality that they are meant to aspire and adhere to.

The second is *correction*. This is a more common response, whereby individuals do accept that such 'tendencies' exist, but that they can be somehow weeded out, if suitable 'therapy' is applied. Training, following 'best practice', the adherence to 'standard operating proecdures' and other precautionary measures would 'protect' islanders from having to succumb to the notorious, demeaning and corrupt practices of patronage and clientelism.

A third response is *control*. This *may* be common but will not be easily acknowledged. It involves understanding the ingrained dispositions of the small-scale system and seeking to manipulate these in one's favour. Deftly exploiting political connections and deploying friendship networks, become lynchpins of a supreme survival strategy and the route to satisfying ambitions.

These responses are not necessarily mutually exclusive. A judge or magistrate, say, will uphold the rule of law during court sittings; but may not think twice about informing a medical specialist working in the national state hospital – with whom he had gone to school many years back – that his partner is heading into hospital for an operation . . . you know, just for some special attention, care and overall peace of mind.

Native versus settler tensions

In addition to emigration, and as a result of globalisation, small islands in different world regions are also increasingly affected by immigration movements. Of course, some small islands – most notably those in the Caribbean, Mediterranean and South China Seas, proximate to continents – have historically always experienced a high influx of migrants, whether they were European navigators and colonists, slaves from Africa, indentured labours from the Indian subcontinent, or guest workers from other neighbouring islands and territories. In many ways, such islands can now be seen as prototypical migration societies, and some of them have evolved today as remarkably diverse multicultural societies as a consequence of these (voluntary or involuntary) migration flows. Trinidad in the Caribbean, Mauritius in the Indian Ocean, Singapore in South East Asia and Fiji in the Pacific are examples of multicultural islands that were formed as a consequence of such large-scale migration movements: these four cases refer to an India diaspora instigated by the post-slavery labour policies of the British Empire (Srebrnik 1999, Sriskandarajah 2005). And while ethnic, religious and linguistic diversity is commonly regarded as a feature that undermines the likelihood of democratic development and political stability, many of these islands have secured and held on to remarkably stable and democratic regimes since the attainment of independence (Bräutigam 1997, Meighoo 2008).

Despite this achievement, it is obvious that immigration can also generate conflicts and tensions, both historically and in contemporary times. Yet while this is probably true for any society affected by immigration, such conflicts and tensions are arguably even more likely to emerge in small island societies. This is partially due to sheer numerical factors: in small societies, newly arriving migrants are more easily noticed; their relative presence is more significant, and the effects of their presence are more directly felt. In addition, immigration is likely to contribute to, and reinforce, powerful and pre-existing sensations of island nationalism, which can be observed in (small) island territories around the world (Baldacchino 2012). As Doumenge (1985) argued, the geographical boundaries posed by water, enhanced by jurisdictional and administrative borders, intensify feelings of community and shared identity among island populations, and thereby also contribute to a feeling of 'us versus them', a distinct 'otherness' vis-à-vis the inhabitants of other islands and political units. From Bougainville to New Caledonia, and from Åland to Nevis, such forms of island nationalism and ethnocentrism fuel secessionist and separatist tendencies and threaten a political fission or fragmentation that looms large in various multi-island and island–mainland units around the globe. The disposition is also known as the 'Tuvalu effect', given the successful disengagement of some 12,000 Tuvaluans from a larger number of some 100,000 I-Kiribati in the late 1970s, leading to the Gilbert and Ellice Islands securing independence from Britain as two separate sovereign states: Tuvalu and Kiribati (MacIntyre 2012)

In the context of migration, island nationalism often results in the development of powerful 'in-group' and 'out-group' identities. In small islands, many social dynamics and interactions revolve around the questions of who 'belongs' and who does not; who is a real islander (from here) and who is an outsider (a 'come from away'). Such identities are often quite flexible, in the first place because second generation migrants may already be part of the 'belongers' (especially if a new group or wave of migrants has arrived in the meantime), and in the second place because people may interchangeably be classified in the 'in' or 'out' group, depending on the circumstances. In the Dutch Caribbean island of Curaçao, an informal distinction is often made between those who call themselves *Yu di Kòrsou* (or YDK – 'child of Curaçao' in Papiamentu) and those who cannot lay claim to this status. However, because the group of YDKs has been increasing and diversifying, the new label of *Bon yu di Kòrsou* ('good child of Curaçao') was recently invented to further distinguish between insiders and outsiders.

The insider status often comes with important (even if imagined) prerogatives, because 'belongers' commonly affirm their rights regarding employment (especially in the public sector), housing, and a variety of public goods. Many immigrants are attracted by potential employment opportunities; but when outsiders are perceived to take away these jobs from 'belongers', social tensions can easily flare up. For example, when pro-democracy protesters took to the streets of Nuku'alofa, Tonga, in November 2006, several of the larger Chinese shops, as well as those owned by ethnic Indians, were targeted for looting and burning (The Guardian 2006).

Incidentally, a similar dynamic is often experienced by 'local' students who, after pursuing their education abroad, return to their home-island to find a job in which they can put their expertise to good practical use. Here, they may find themselves excluded from prestigious positions due to their newly acquired outsider status (Lee 2016); or otherwise advantaged by their stint overseas which may give them a vital edge against others who stayed at home. Intriguingly, many small island jurisdictions have become both emigrant and immigrant societies, with sizable overseas diasporas as well as growing migrant populations at home. For example, while more than half of all Cape Verdeans now live outside their homeland, on the Cape Verde islands "[i]mmigration is becoming a contentious issue and workers from West Africa (primarily Guinea Bissau) enter the country and work for lower rates than Cape Verdeans" (Baker 2006, p. 503). The societal tensions sparked by both emigration and immigration are also reflected in sharp discussions about voting rights; in many small island territories, the extension of universal suffrage to both overseas citizens and newly arriving immigrants is a highly charged affair.

As a consequence of their greater economic openness and vulnerability to fluctuations in the global economy, small islands are more likely to experience economic instability, and – as the recent examples of Cyprus, Iceland and Nauru demonstrate – high economic growth rates can suddenly turn into sharp economic decline. The labour force may be swelled with non-locals in times of economic prosperity, risking an invasion that could easily see locals evicted from affordable property and quality employment (Maurer 1993). Meanwhile, the high unemployment rates that may follow in a period of economic contraction can lead to rapidly rising tensions between in- and outsider groups (McElroy and De Albuquerque 1988).

Another flashpoint for 'settler versus native' animosities on small islands is typically the provision of housing. Island geographies come with finite land areas, and the availability of suitable and attractive land is further constrained by the inevitability of reconciling industrial, leisure, infrastructural, commercial and military interests, particularly along highly coveted coastal areas. This combination typically leads to very expensive beach and coastal properties, which many locals may find are totally beyond their financial reach. A gentrification of such properties becomes hard to avoid, as these are progressively sold to well-heeled foreigners (e.g. Clark et al. 2007, Clark and Kjellberg, this volume). Various small island jurisdictions have sought to avoid this using various policy tools: Prince Edward Island, Canada, charges non-residents double the property tax payable by 'Islanders' (Financial Post 2015); Jersey and Guernsey, crown colonies of Britain, each have two separate housing markets, with property ring-fenced for foreign purchase costing much more than those earmarked for domestic sale (Renouf and Du Feu 2012, Sebire and David 2012). Bermuda has a fixed number of houses available for sale to foreigners: since this number is capped, and foreign workers must be resident on the island, housing availability effectively controls the influx of skilled foreign labour (Low Tax 2017).

Conclusion

As we have highlighted throughout this chapter, small island territories around the world experience a variety of strong informal dynamics that have a profound effect on the practical workings of island economics, societies and politics. Due to the mostly informal nature of these dynamics, they are often invisible to non-island outsiders, whether they are academics, tourists or entrepreneurs. As a result, small islands that on the surface appear to be paradisiacal places with amicable societies might in fact be plagued by pervasive divisions, enduring tensions and longstanding conflicts. Hence, in order to better grasp and analyse the characteristics of island life, an examination of its rich informal practices, and particularly of the 'ecology of smallness', is indispensable. In this chapter, we have provided a tentative conceptual and analytical framework on the basis of which these dynamics of island societies can be better understood.

Therefore, we contend that small island societies in different world regions share a number of characteristics; yet, we plainly concede that there are also many differences and idiosyncrasies that are worth highlighting. In our view, much can be gained by paying more attention to identifying these differences and examining their effects. For example, despite demographic and geographic similarities, there has long been a somewhat artificial divide between scholars of small sovereign and small non-sovereign island jurisdictions (SNIJs), with little comparisons being made between these two sets of cases (for exceptions, see Baldacchino 2006b). In addition, little systemic analysis has been conducted on the difference between single-island and multi-island (i.e. archipelagic) units, and the effects of these different constitutional arrangements on island societies, economics and politics (Baldacchino 2013).

Reflecting our discussion of rising tensions between 'native' and 'settler' communities and the societal effects of immigration and emigration, we contend that future studies should pay more attention to the social and societal effects of migration, and its political repercussions, in small island contexts. In culturally contested islands like Bonaire in the Caribbean, Greenland in the Arctic, and Guam and New Caledonia in the Pacific, migration produces increasing tensions between native or 'insider' groups and newly arriving immigrants, who often carry different norms, values and worldviews. While many Western societies are affected by the arrival of immigrants with lower levels of education and wealth than the autochthonous populations, in many small islands an inverse migration trend can be observed. Here, many immigrants (generally coming from western Europe or North America) are wealthier and more educated than the local populations. Because of their greater capacities and resources, such groups tend to rapidly form a new societal elite, and lay claim to the most scenic locations, the best houses, and the best paid jobs on the island. The result is that the pre-existing or 'native' population is outperformed and outcompeted, creating rising antagonism vis-à-vis the new immigrant cadres. In non-sovereign territories around the world, such migratory trends, and the ensuing social tensions, increasingly instigate and drive political status debates, indicating that these tensions also have political ramifications.

References

Alves Furtado, C. (2014) Social movements in Cabo Verde: processes, trends and vicissitudes. In N. Samba Sylla (ed.), *Liberalism and its discontents: social movements in West Africa*, Dakar, Senegal, Rosa Luxemburg Foundation, pp. 419–461.

Anckar, D. (1999) Homogeneity and smallness: Dahl and Tufte revisited. *Scandinavian Political Studies*, 22(1), 29–44.

Anckar, D. (2002) Why are small island states democracies? *The Round Table: Commonwealth Journal of International Affairs*, 91(365), 375–390.

Anckar, D. (2006) Islandness or smallness? A comparative look at political institutions in small island states. *Island Studies Journal*, 1(1), 43–54.

Anckar, C. (2008) Size, islandness and democracy: a global comparison. *International Political Science Review*, 29(4), 433–459.

Armstrong, H., Johnes, G., Johnes, J., and MacBean, A. (1993) The role of transport costs as a determinant of price level differentials between the Isle of Man and the United Kingdom, 1989. *World Development*, 21(2), 311–318.

Austin, D. (2000) Contested islands. *The Round Table: Commonwealth Journal of International Affairs*, 89(353), 59–63.

Baker, B. (2006) Cape Verde: the most democratic nation in Africa? *Journal of Modern African Studies*, 44(4), 493–511.

Baldacchino, G. (1997) *Global tourism and informal labour relations: the small scale syndrome at work*. London, Mansell.

Baldacchino, G. (2005) The contribution of 'social capital' to economic growth: lessons from island jurisdictions. *The Round Table: Commonwealth Journal of International Affairs*, 94(378), 31–46.

Baldacchino, G. (2006a) Islands, island studies, Island Studies Journal. *Island Studies Journal*, 1(1), 3–18.

Baldacchino, G. (2006b) Innovative development strategies from non-sovereign island jurisdictions? A global review of economic policy and governance practices. *World Development*, 34(5), 852–867.

Baldacchino, G. (2012) Islands and despots. *Commonwealth and Comparative Politics*, 50(1), 103–120.

Baldacchino, G. (ed.) (2013) *Archipelago tourism: policies and practices*. London, Routledge.

Baldacchino, G. (ed) (2014) *The political economy of divided islands: unified geographies, multiple polities*. New York, Springer.

Benedict, B. (1967) Sociological aspects of smallness. In B. Benedict (ed.), *Problems of smaller territories*, London, University of London and Athlone Press, pp. 45–55.

Bertram, G. (2006) Introduction: the MIRAB model in the twenty-first century. *Asia Pacific Viewpoint*, 47(1), 1–13.

Bertram, G., and Watters, R. (1985) The MIRAB economy in South Pacific microstates. *Pacific Viewpoint*, 26(3), 497–519.

Beston, J., and Beston, R. (1977) An interview with Albert Wendt. *World Literature written in English*, 16, 152–159.

Boissevain, J. (1974) *Friends of friends: networks, manipulators and coalitions*. New York, St Martin's Press.

Bräutigam, D. (1997) Institutions, economic reform and democratic consolidation in Mauritius. *Comparative Politics*, 30(1), 45–62.

Bray, M. (1991) *Making small practical: the organisation and management of ministries of education in small states.* London, Commonwealth Secretariat.

Brint, S. (2001) Gemeinschaft revisited: a critique and reconstruction of the community concept. *Sociological Theory*, 19(1), 1–23.

Clague, C., Gleason, S., and Knack, S. (2001) Determinants of lasting democracy in poor countries: culture, development and institutions. *The Annals of the American Academy of Political and Social Science*, 573(1), 16–41.

Clark, E., Johnson, K., Lundholm, E., and Malmberg, G. (2007) Island gentrification and space wars. In G. Baldacchino (ed.), *A world of islands: an island studies reader*, Charlottetown, Canada and Luqa, Malta, Institute of Island Studies and Agenda Academic, pp. 483–512.

Clarke, C. (2006) Politics, violence and drugs in Kingston, Jamaica. *Bulletin of Latin American Research*, 25(3), 420–444.

Congdon Fors, H. (2014) Do island states have better institutions? *Journal of Comparative Economics*, 42(1), 34–60.

Connell, J. (1988) *Sovereignty and survival: island microstates in the Third World* (Monograph No. 3. Sydney, Department of Geography, University of Sydney.

Dahl, R.A., and Tufte, E.R. (1973) *Size and democracy*. Stanford, CA, Stanford University Press.

Doumenge, F. (1985). The viability of small tropical islands. In E. Dommen and P. Hein (eds.), *states, microstates, and islands,* Dover, NH, Croom Helm, pp. 70–118.

Enright, J. (2013) *Fire knife dancing*. Newburyport, MA, Thomas and Mercer.

Erk, J., and Veenendaal, W.P. (2014) Is small really beautiful? The microstate mistake. *Journal of Democracy*, 25(3) 135–148.

Gerring, J., and Zarecki, D (2011). Size and democracy revisited. *DISC Working Papers*. Budapest, Hungary, Central European University.

The Guardian (2006, 17 November) State of emergency after Tonga riots. Available from: www.theguardian.com/world/2006/nov/17/1.

Hadenius, A. (1992) *Democracy and development*. Cambridge, Cambridge University Press.

Hay, P. (2013) What the sea portends: a reconsideration of contested island tropes. *Island Studies Journal*, 8(2), 209–232.

Hepburn, E., and Baldacchino, G. (eds.) (2013) *Independence movements in subnational island jurisdictions.* London, Routledge.

Hirczy, W. (1995) Explaining near-universal turnout: the case of Malta. *European Journal of Political Research*, 27(2), 255–272.

Houbert, J. (1985) Settlers and natives in decolonisation: the case of New Caledonia. *The Round Table: Commonwealth Journal of International Affairs*, 74(295), 217–229.

Lee, H. (2016) 'I was forced here': perceptions of agency in second generation 'return' migration to Tonga. *Journal of Ethnic and Migration Studies*, 42(15), 2573–2588.

Lewis, W.A. (1965) *The agony of the eight*. Bridgetown, Barbados, Advocate Press.

Low Tax (2017) Bermuda: business environment, residence and property. Available from: www.lowtax.net/information/bermuda/bermuda-residence-and-property.html.

Lowenthal, D. (1972) *West Indian societies*. London, Oxford University Press.

Lowenthal, D. (1987) Social features. In C. Clarke and A. Payne (eds.), *Politics, security and development in small states*, London, Allen and Unwin, pp. 26–49.

MacIntyre, W.D. (2012) The partition of the Gilbert and Ellice Islands. *Island Studies Journal*, 7(1), 135–146.

Marr, G. (2015) Prince Edward Island, the one place in Canada where foreign property buyers must check in. *Financial Post*, 15 May. Available from: http://business.financialpost.com/personal-finance/mortgages-real-estate/prince-edward-island-the-one-place-in-canada-where-foreign-property-buyers-must-check-in.

Maurer, B. (1993) 'Belonging,' citizenship and flexible specialization in a Caribbean tax haven (British Virgin Islands). *PoLAR: Political and Legal Anthropology Review*, 16(3), 9–18.

May, R., and Tupouniua, S. (1980) The politics of small island states. In R.T. Shand (ed.), *The island states of the Pacific and Indian Oceans: anatomy of development*, Canberra, Australia, ANU Press, pp. 419–437.

McCall, G. (1994) Nissology: the study of islands. *Journal of the Pacific Society*, 17(2–3), 1–14.

McElroy, J.L., and De Albuquerque, K. (1988) Migration transition in small Northern and Eastern Caribbean states. *International Migration Review*, 22(3), 30–58.

McLeod, A. (2007) *State, society and governance in Melanesia*. Discussion Paper 2008/6. Leadership models in the Pacific. Available from: http://ssgm.bellschool.anu.edu.au/sites/default/files/publications/attach ments/2015-12/08_06_0.pdf.

Meighoo, K. (2008) Ethnic mobilisation vs. ethnic politics: understanding ethnicity in Trinidad and Tobago politics. *Commonwealth & Comparative Politics*, 46(1), 101–127.

Ott, D. (2000) *Small is democratic: an examination of state size and democratic development*. New York, Garland.

Payne, D. (1999) The failings of governance in Antigua and Barbuda: the elections of 1999, Washington, DC, Center for Strategic and International Studies.

Peters, D.C. (1992) *The democratic system in the Eastern Caribbean*. New York, Greenwood Press.

Puniani Austin, S. (2002) Issues in human resource management for small states: a Tonga perspective. In G. Baldacchino and C.J. Farrugia (eds.), *Educational planning and management in small states: concepts and experiences*, London, Commonwealth Secretariat, pp. 24–38.

Puppis, M. (2009) Media regulation in small states. *International Communication Gazette*, 71(1–2), 7–17.

Renouf, J.T., and Du Feu, T. (2012) Jersey. In G. Baldacchino (ed.), *Extreme heritage management: the practices and policies of densely populated islands*, New York, Berghahn Books, pp. 95–115.

Richards, J. (1982) Politics in small independent communities: conflict or consensus? *Journal of Commonwealth & Comparative Politics*, 20(2), 155–171.

Ryan, S.D. (1999) *Winner takes all: the Westminster experience in the Caribbean*. St Augustine, Trinidad and Tobago, UWI Press.

Sebire, H., and David, C. (2012) Guernsey. In G. Baldacchino (ed.), *Extreme heritage management: the practices and policies of densely populated islands*, New York, Berghahn Books, pp. 75–95.

Seibert, G. (1999) *Comrades, clients and cousins: colonialism, socialism and democratisation in São Tomé and Príncipe*. Ridderkerk, Ridderprint.

Spiteri, S. (2016) Developing a national quality culture for further and higher education in a micro-state: the case of Malta. *Malta Review of Educational Research*, 10(2), 297–313.

Srebrnik, H.F. (1999) Ethnicity and the development of a 'middleman' economy on Mauritius: the diaspora factor. *The Round Table: Commonwealth Journal of International Affairs*, 88(350), 297–311.

Srebrnik, H.F. (2004) Small island nations and democratic values. *World Development*, 32(2), 329–341.

Srebrnik, H.F. (2008) Indo-Fijians: marooned without land and power in a South Pacific archipelago. In P. Raghuram, A. Kumar Sahoo, B. Maharaj and D. Sangha (eds.), *Tracing an Indian diaspora: contexts, memories, representations*, New York, Sage, pp. 75–95.

Sriskandarajah, D. (2005) Development, inequality and ethnic accommodation: clues from Malaysia, Mauritius and Trinidad and Tobago. *Oxford Development Studies*, 33(1), 63–79.

Sutton, P.K. (1999) Democracy in the Commonwealth Caribbean. *Democratization*, 6(1), 67–86.

Sutton, P.K. (2006) *Modernising the state: public sector reform in the Commonwealth Caribbean*. Kingston, Jamaica, Ian Randle Publishers.

Sutton, P.K. (2007) Democracy and good governance in small states. In E. Kisanga and S.J. Danchie (eds.), *Commonwealth small states: issues and prospects*, London, Commonwealth Secretariat, pp. 201–217.

Veenendaal, W.P. (2013a) Political representation in microstates: St Kitts and Nevis, Seychelles and Palau. *Comparative Politics*, 45(4), 437–456.

Veenendaal, W.P. (2013b) *Politics and democracy in microstates: a comparative analysis of the effects of size on contestation and inclusiveness*. Published PhD thesis, Leiden, The Netherlands, University of Leiden.

Veenendaal, W.P. (2014) *Politics and democracy in microstates*. London, Routledge.

Weale, D. (1992) *Them times*. Charlottetown, Canada, Institute of Island Studies, University of Prince Edward Island.

Wood, D.P.J. (1967) The smaller territories: some political considerations. In B. Benedict (ed.), *Problems of smaller territories*, London, University of London and Athlone Press, pp. 23–34.

17

RESILIENCE AND SUSTAINABILITY

Ilan Kelman and James E. Randall

Introduction: how two words stand out

Two words are now habitually used to represent positive goals for which humanity should aspire, from local to planetary scales: sustainability (e.g. Meadows et al. 1972) and resilience (e.g. Alexander 2013). In ensuing debates, these words have been (re)interpreted, (re)defined, promoted, criticised and interrogated to the point they could mean virtually anything to anyone, depending on how they are (mis)applied in any specific context.

No matter how severe the critiques, both words are now fully part of discussions across fields and sectors, spawning multiple books and various academic journals, including two titled *Resilience* and *Sustainability*. Often lost are meanings on the ground, not just of the terms, but also of the processes. How do peoples, places, and communities understand and apply resilience and sustainability to their lives and livelihoods?

Islands are locations where much has been made of different concepts of and approaches to resilience and sustainability, yielding significant diversity and divergences. This chapter synthesises and critiques research on resilience and sustainability from an island perspective, exploring island futures beyond this flavour-of-the-decade labelling.

This discussion is needed to fill in significant gaps and to set firm directions for island studies. Thus, for example, the island studies reader (Baldacchino 2007) represented a seminal piece of scholarship covering wide-ranging island issues and bringing together many of the leading island studies researchers of the time. Despite indirect references throughout the volume to sustainability and resilience, it did not explicitly address these foundational topics.

Since the publication of that volume, these concepts have been made to assume even greater theoretical and empirical salience for island governments, societies and scholarly debates at local, national and international scales.

Defining and critiquing resilience

The term 'resilience' has a long pedigree across multiple disciplines. Alexander (2013) provides a rich analysis of the evolution and various applications of the term, from the Roman author and naturalist Pliny the Elder to its current use in surgery, organisational theory, psychology and human ecology. His paper describes common elements of resilience as "rebounding, adapting, overcoming and maintaining integrity" (Alexander 2013, p. 2710). In another review of 'resilience', Bhamra et al. (2011) note that the concept has become more applicable because of growing social, technological and environmental interconnectedness, a feature often associated with globalisation. In other words, there are now more external events and features that can affect what might have previously been (assumed to be) relatively closed systems at the local level. Although the tropes of islands as pure, closed systems or isolated laboratories, worthy of investigation for these reasons, has long been challenged (Baldacchino 2007, Greenhough 2006), human influences on islands and island societies (systems) are increasingly being witnessed in external social and environmental realms, such as immigration laws, long-lived pollutants and anthropogenic climate change.

From an ecological perspective, Holling (1973) applied the concept of resilience to ecological systems and he is often credited with having founded resilience ecology, leading to the Resilience Alliance which has used its own island focus through meetings (Parker and Hackett 2012) while failing to engage with island studies. Yet, resilience had appeared in ecological work long before (e.g. Blum 1968, Person 1960) even while Holling (1973) significantly advanced its theoretical constructs for aspects of ecology. The end result from ecology was a textbook definition (Townsend et al. 2003) with parallels from engineering (Fiering 1982), examining two nuanced meanings of resilience: (1) how quickly a system might return to stability after being disturbed; and (2) the extent to which a system can be disturbed without breaking down. The definitions are linked through (1) the time dimension, namely the length of time required after a disturbance before a definite decision is made about the system's current state compared to the pre-disturbance state and (2) the assumption that systems are stable and have a fixed trajectory.

These links are overly assumptive, working well as part of mathematical modelling when systems are defined as a set of equations. The assumptions break down in reality, because equations can never capture all real-world behaviours, especially for islanders whose experience demonstrates the instability of and recurrent changes within living systems (Hau'ofa 1994, 1998). Ecology-based resilience researchers today interpret this reality in terms of 'social-ecological systems', 'coupled human and natural systems', and related phrases. These notions have a much earlier baseline in studies within the field of 'human ecology' which often researched island communities (e.g. Catala 1957).

Such threads of resilience and the continual interconnectedness and evolution of systems have more recent explorations. Island societies are often touted as being especially resilient due to higher levels of social bonding and kinship than mainland locations (Baldacchino 2011, Randall et al. 2014). These conclusions speak more to the social science definition of 'resilience' from psychology which, to a large degree, focuses on how individuals and families deal with failure, tragedy, adversity and stress. This approach has been applied especially to disasters in the island context of New Zealand (Paton and Johnston 2001).

Island lessons in the context of resilience reveal particular limitations in the ecological definition, in terms of taking 'resilience' to mean that a system has a specific state which it should retain or to which it should return or bounce back after a disturbance. Island societies thrive on openness and change, recognising the volatility of the environment and using it for living and livelihoods (Hau'ofa 1994, 1998) while having numerous migrant-based societies and extensive diasporas (Randall et al. 2014, King and Connell 1999). Embracing change makes island communities able to continue island life; that is, change makes them resilient. For example, a steady stream of island migrants to other centres, often on the mainland, supports home island life through reducing demand on local resources, lowering pressure on local job creation and returning remittances (Connell and Brown 1995). Meanwhile, with many livelihoods emerging from their coastal and oceanic environments, islanders are used to dealing with change at multiple time scales, from waves to migrating animals (Hau'ofa 1994, 1998).

This attitude does not mean that all island change is seen as being positive or supporting resilience. An analysis of islanders in three coastal fishing villages in the Solomon Islands by Schwarz et al. (2011) found that they were most fearful of the consequences of climate change (external) and the deterioration of social values (internal). Caution is needed in interpreting these results, because the study's framing might have influenced its outcome. As island studies has long queried: when communities are labelled as being 'remote', exactly what are they remote from and how does the 'remote' label influence their feelings of dependency? Meanwhile, disaster research (e.g. Hewitt 1983) has long aimed to eschew the language of 'shocks', instead recognising that disasters are inherently foreseeable (e.g. Glantz 2003) because they are caused by the everyday, inherent condition of vulnerability rather than by hazards which sometimes can manifest rapidly. Calculated correlations may be questioned when causal explanations do not factor in established science.

The UK Overseas Territory of Montserrat in the Caribbean further illustrates islander attitudes to change. Since 1995, most of the island's infrastructure has been destroyed in a continuing volcanic eruption. Over half the resident population at the time out-migrated. Many have ended up being disappointed in the morals which their children learn in school in London (Shotte 2006), showing that not all change in attitudes is welcome. Meanwhile, some Montserratians remaining behind opposed the UK government's aims for the territory to decriminalise homosexual acts between consenting adults (Skelton 2000) despite this ban violating fundamental human rights. Since not all change is supported, islanders might sometimes assume that resilience is achieved through the lack of change.

Resilience has been misappropriated in other ways. Neo-liberalism has used the concept to explain that people must learn to help themselves, doing so in place, because neither government support nor migration can be guaranteed (Pugh 2014). Why should coastal island communities such as Shishmaref (Alaska, US) and Funafuti (Tuvalu) be concerned about sea-level rise if they can learn to be resilient by bouncing back to their state from before they were forced to move? Resilience becomes an excuse to do nothing for communities in need, instead letting them work out difficulties on their own, even if these difficulties were caused mainly by external forces, as are most climate change impacts on island locations.

Furthermore, concepts of resilience have little meaning in many island cultures, such as in the Arctic and the Pacific. The word does not exist in their languages and is difficult to translate culturally. 'Social-ecological systems', 'coupled human and natural systems', dealing with adversity, and managing change are what many indigenous islanders have always known within their knowledge and how they have always lived: it is island life.

Defining and critiquing sustainability

'Sustainability' effectively means something continued or upheld; that is, literally, to be sustained. In the context of life and livelihoods, the word takes on a deeper meaning with a seminal definition (followed by its critiques) from WCED (1987): "Humanity has the ability to make development sustainable to ensure that it meets the needs of the present without compromising the ability of future generations to meet their own needs" (Chapter I.3.27). In other words, sustainability refers to sustaining humanity into perpetuity on Earth. Many definitions and debates have since entered science, policy and practice (Glavič and Lukman 2007).

The most internationally visible policy applications of sustainability and sustainable development emerged largely from the United Nations, such as the Conference on the Human Environment held in June 1972 in Stockholm, and in later decades taken over by climate change discourses. Meadows et al. (1972), Brandt (1980) and WCED (1987) were pivotal points, but were based on much earlier work. In parallel, science on the changing climate was rapidly expanding, leading from MIT (1970, 1971) to the first assessment report of the Intergovernmental Panel on Climate Change published in 1990. The UN Earth Summit in Rio de Janeiro followed in 1992 and founded three UN conventions and secretariats: The Convention on Biological Diversity, the Convention to Combat Desertification and the Framework Convention on Climate Change (UNFCCC).

Island governments embraced these international sustainability fora. They led the world through the Small States Conference on Sea Level Rise from 14 to 18 November 1989 which produced the *Malé Declaration on global warming and sea level rise* (1989) and the 1990 founding of the Alliance of Small Island States (AOSIS). AOSIS continues to promote island interests, especially at the UN. The Small Island Developing States (SIDS) category arose, also at the UN, based on an assumption of common sustainability threats and opportunities across their widely varying human, physical and economic geographies. Using the sustainability ethos, SIDS held the 1994 meeting for the Barbados Programme of Action (BPoA) for the Sustainable Development of Small Island Developing States (UN 1994), then continued through decadal meetings and declarations in Mauritius (UN 2005) and Samoa (UN 2014).

SIDS played a further prominent role in such other international fora as the World Summit on Sustainable Development (WSSD) in Johannesburg, South Africa in 2002 through to the 2012 omnibus United Nations Conference on Sustainable Development in Rio de Janeiro, Brazil, which covered Stockholm+40, WCED+25, Rio+20, and WSSD+10. In parallel, the eight Millennium Development Goals from 2000 to 2015 and the 17 Sustainable Development Goals from 2015 to 2030 have helped to shape the world's high-level sustainability agenda.

SIDS exerted their influence prominently with respect to climate change through the UNFCCC negotiations (Betzold 2010, Betzold et al. 2012). The main results from UNFCCC reflect significant island interests, but the Kyoto Protocol from 1997 is seen as largely ineffective, the Copenhagen Accord from 2009 had few substantive measures, and the Paris Agreement from 2015 is principally voluntary. The UNFCCC meetings are notable for AOSIS pleading for climate change agreements to protect their countries' and peoples' existence; often invoking sustainability discourse, with powerful appeals such as "1.5 to stay alive". 1.5°C refers to the maximum suggested rise in global annual mean temperature, compared to the pre-industrial revolution era, which should be permitted in order to avoid major damage to low-lying SIDS.

Yet, the science on climate change impacts on low-lying islands (Kench et al. 2015, Rankey 2011) does not suggest that disappearance or submergence is inevitable; but nor does it deny the prospect. Nonetheless, AOSIS' role in climate change negotiations turned the public's gaze to the lack of island environmental sustainability, especially in the face of climate change. This organisational capacity emerged on the basis of assumed threats to the physical environment, translating into presumed threats to island cultures and peoples. Island sustainability, the argument implied, was set to become impossible under climate change. This attention and discourse may have proved useful for increasing interest in reducing greenhouse gases and increasing their sinks (the definition of climate change 'mitigation') as well as for generating commitments (not always fulfilled) for addressing climate change impacts (the definition of climate change 'adaptation'). It has also reinforced a stereotype that SIDS and other island communities are only capable of achieving sustainability with the assistance of external bodies.

Generating dramatic angles of menaces to island sustainability produced publicity. In 2002, a court case was threatened by the Prime Minister of Tuvalu against the US and Australian governments due to their leaders at the time declining to ratify the Kyoto Protocol. The case did not proceed because Tuvalu's Prime Minister lost an election. From 2008, Mohamed Nasheed, President of Maldives, represented the international face for the alleged plight of SIDS under climate change. As one of a slew of movies of varying scientific quality covering islands and climate change, he related his story through the 2011 Hollywood-produced documentary *The island president*, repeating and reinforcing many island icons. In fact, in October 2009, Nasheed had held an underwater cabinet meeting to iconise drowning islands, generating broad media attention. Nasheed was deposed in 2012.

These approaches to portraying climate change impacts on low-lying islands lend themselves to a (re)conceptualisation of islands as closed systems or living laboratories, which are self-sustaining and demonstrate how sustainability can ideally be emulated and achieved; until outside forces swamp (literally and figuratively) their sustainable paradise. Yet these tropes are not overly realistic when applied to most islands. The ecosystems, livelihoods and cultures of many island communities have been strongly linked to each other and to mainland areas, and hence to more-than-local influences for millennia.

Iconisation includes delineation, such as drawing boundaries around islands and assuming isolation within those lines. Within their coasts, islands provide us with "a world of stable boundaries around, and absolute spaces of, knowledge" (Farbotko 2010, p. 53). Farbotko (2010) uses Tuvalu, which remains in the public consciousness along with Maldives (see Figure 17.1), the Marshall Islands and Kiribati as symbols for climate change's consequences, to analyse the view that Tuvalu's sustainability can be understood only by understanding climate change impacts.

Baldacchino and Kelman (2014) critique this morphing of island sustainability into a focus on climate change impacts on islands. Their concern is that this conflation of sustainable development with climate change has marginalised the many other causes of unsustainable practices, internal and external. In some cases, these may be linked to climate change; for example, when food supplies and people's health are adversely affected by freshwater deterioration in quality and quantity due to sea-level rise. In other cases, a failure to achieve a more sustainable future has more to do with governance choices, such as corruption or maintaining longstanding social and political inequalities, than it has to do with any physical environmental changes on the islands, whether from nature, human activity, or (as with contemporary climate change) both.

These concerns mirror Mentz's (2012) underpinning critique of the conceptualisation of sustainability being a 'kind of fantasy pastoral' (p. 586). This view extends from the criticism that

Figure 17.1 Maldives, with a highest natural point above sea level of 2.4 metres.

Source: Photo by Ilan Kelman.

sustainability is equivalent to a sort of stasis – an image which seems to have emerged for islands fighting against transformation due to climate change, rather than change being a fundamental feature of all systems with little orderliness about it – which was the pre-climate change understanding and acceptance of many island cultures.

The non-sustaining embracing-change aspect of sustainability in no way justifies climate change impacts on islanders which might well be devastating and which are already the main factor forcing the relocation of island communities in places such as Alaska and Papua New Guinea. It does imply that sustainability might be about accepting change while aiming to minimise adverse consequences from change. It should also dispel the myth that islands are somehow more sustainable just because their government ministers and top civil servants speak of sustainability and sustainable development while tabling sustainability plans and leading initiatives to address climate change

Resilience and sustainability interactions

With all the challenges to and differences in the definitions and understandings of 'sustainability' and 'resilience' discussed above, how do the terms and processes interact on, in, and for island communities? The answers and examples demonstrate the concurrent existence of contrasting approaches.

Researchers are increasingly approaching resilience and sustainability questions through combining different knowledge forms (Gaillard 2007, Mercer et al. 2007). These discussions segue into aiming to explore and express who sets the island resilience and sustainability agenda(s), especially through defining the terms and seeking connections amongst them (e.g. Baldacchino 2008). Melding perspectives and knowledges illustrates how social movements with strong culturally destructive elements, such as indigenous islanders converting to Christianity which undermined traditional approaches to resilience and sustainability, have later been found to support some useful elements – for example, environmental conservation – when combined with traditional practices, so losing neither the traditional nor the contemporary (Jacka 2010).

Integrating traditional and non-traditional approaches has shown numerous other successes for island resilience and sustainability, such as in the aftermath of specific disasters. When Cyclone Pam swept through Vanuatu in March 2015, a combination of traditional knowledge and information technology is likely to have contributed to a lower-than-expected death toll (Jean Mitchell, personal communication 2016). Island communities that had not yet incorporated more recent building practices rebuilt quicker than those which had. Similarly, although local food supplies were decimated, once new crops were harvested, local villages were better able to sustain themselves without external assistance.

This self-sufficiency demonstrates local resilience and sustainability working together so that people can assist themselves in addressing their own vulnerabilities. In Fiji, Johnston (2015) described how communities which received substantial post-disaster aid were less able to deal with cyclones than communities which had received much less aid. Lewis (2009) identified similar situations across many other Pacific island communities for a variety of environmental hazards.

Yet, island communities committed to tourism, migration and trade, can hardly be self-sufficient and support their own resilience and sustainability. The communities of Tikopia and Anuta, Solomon Islands, survived Cyclone Zoë in December 2002 through their traditional knowledge; but much of their infrastructure was destroyed and outside assistance was helpful for reconstruction (Treadaway 2007). People from Simeulue, Indonesia, knew to move away from coastlines after the earthquake of 26 December 2004, so they survived the subsequent tsunami (Gaillard et al. 2008), but aid helped them to recover from the ensuing devastation.

Consequently, sustainability and resilience can support each other beyond the local level. Lauer et al. (2013) researched the response to the 2 April 2007 tsunami on Simbo, Solomon Islands, suggesting that both local approaches and globalisation have increased the communities' general resilience; while however reducing their resilience to such natural hazards as may occur less frequently. In doing so, the study neglects the many communities such as Simeulue that have maintained resilience to low-frequency hazards, suggesting perhaps some missing elements not examined for Simbo. It also raises the question regarding the meaning of 'general resilience', particularly given sustainability discourses. If a community can deal with environmental hazards for a few centuries and thrives in its environment while doing so, but is then destroyed by a low-frequency event, is it appropriate to describe the community as being generally resilient since it is clearly not sustainable?

Globalisation has further brought rapid dissemination of disaster news followed by the availability and rapid transfer of remittances: this condition has been highly effective for post-disaster recovery in Samoa (Le De et al. 2015, see Figure 17.2). No claim is made that remittances substitute for local knowledge, because survivors must exist to receive any remittances. Conversely, knowing that remittances will be available to rebuild livelihoods can encourage evacuation and survival, rather than attempts to stay behind to defend property which led to numerous fatalities in the Philippines during Typhoon Haiyan (Yolanda) in 2013 (Aaron Opdyke, personal communication 2017).

Yet, globalisation has weakened political institutions at many scales and across many islands, leading to livelihood reliance on external factors such as tourism, currency exchange rates and international trade agreements. Irrespective of assumptions and expectations of external post-disaster aid, vulnerabilities to disasters can increase in island communities through increased connectedness. Kaloumaira (2002) and Lightfoot (1999) discuss drought-related impacts of El Niño on sugar cane in Fiji. Sugar cane (see Figure 17.3) is a cash crop which has diverted

Figure 17.2 Overseas remittances often help Samoans repair storm–damaged structures.

Source: Photo by Ilan Kelman.

Figure 17.3 The fields of Fiji.

Source: Photo by Ilan Kelman.

land away from producing more locally based livelihoods, not dependent on external markets. Kaloumaira (2002) and Lightfoot (1999) thus promote resilience and sustainability approaches for Fijian droughts involving agricultural diversification, avoiding a culture of aid dependency, and improving local water management.

To a large degree, this discussion and its examples demonstrate that the core of resilience and sustainability express the same fundamental ethos from island studies. The key point is not islander survival despite adversity, but instead islanders thriving in the communities and environments in which they live. The definitional discussion of the terms, and the exploration of their interactions through an island lens, indicate that interactions between resilience and sustainability produce significant overlaps, which however, are not necessarily harmonious. The next section focuses on island studies contributing to resilience and sustainability by examining resilience–sustainability tensions and synergies.

Island studies and resilience–sustainability tensions

One core element of many 'resilience' definitions is the implication of remaining unchanged or sustained, through being able to withstand or recover from external impositions: what, in ecology, is termed a 'disturbance'. In considering modern economies and globalisation, the paradigm of 'development as growth' tends to dominate and to be assumed to be desirable. Consequently, resilience implies 'sustainable development' or sustained economic growth.

Meanwhile, one core element of many 'sustainability' definitions is managing without self-destruction through living within the resources available into perpetuity. Considering the Earth, continual economic or population growth is not possible without self-destruction, because eventually the rate of resource consumption must outstrip the energy demands required for resource renewal (Bartlett 2004).

An inherent contradiction emerges between resilience protecting growth and sustainability seeking to end it. Traditional island economies were not always premised on growth, instead aiming to live within the available resources and often using emigration; and at times even induced abortion and infanticide (Ulijaszek 2006) as a means of stabilising population levels: further challenging the 'closed system' idea, but also indicating that growth is acceptable through its export and offloading. The shift towards the growth paradigm indicates that traditional, sustainable ways of living were not necessarily resilient because they were supplanted by a less sustainable approach which has remained remarkably robust despite numerous challenges, including financial crashes, leadership changes and wars.

In fact, resilience's meaning of returning to the older state as quickly as possible contrasts with sustainability seeking change to a newer and better state. Migration, for example, has long been debated as being either an adjustment to changing conditions or as a failure to adjust to changing conditions, with island communities at the forefront of examples. The ultimate conclusion is that migration represents both circumstances – and those in between – depending on the specific case study.

Many examples exist of volcanic eruptions forcing the temporary abandonment of entire islands, such as the 1973 eruption on Vestmannaeyjar, Iceland (Sowan 1985) and the 1961 eruption on Tristan da Cunha in the South Atlantic (Baker et al. 1964). It is hard to be sustainable in-place when a volcano erupts alongside one's community. Meanwhile, changes to the ocean and climate potentially led to island community abandonment around the Pacific in the 14th century (Clark and Reepmeyer 2012, Nunn et al. 2007). Such an action might have been a resilience strategy, the ultimate outcome of failure to be resilient, or a combination of the two.

Much of this history and discussion is forgotten or bypassed in current rhetoric associated with islander mobility related to climate change, including using phrases such as 'environmental refugees' and 'climate (change) refugees'. These images are often linked to an icon of permanent island abandonment, with the land sinking or slipping below the rising waters: a view which is inaccurate. This fiction might even have emerged from the trope of islands as miniature worlds or scaled-down versions of Planet Earth. In a putatively closed system, the possibility for external migration does not exist, so the system must be resilient. Even in the island utopia and tourism paradise stereotypes, the community should be sustainable and happy, so the need to migrate would not even arise.

Taken from a longer historical perspective, mobility has long been an integral element of island life (King and Connell 1999, Randall et al. 2014, Wertheim 1959), often supporting island life through reduced local resource pressure and remittance inflows. Another criticism of assumptions of involuntary, undesired migration under environmental change is that it implies that islanders have no choice or control in the matter. It constructs mobility as being forced upon them with decisions being out of the hands of islanders. Finally, it neglects non-mobility or 'the right to stay', both forced and voluntary, as being as important as mobility.

People with the fewest choices and resources for moving tend to be the poorest and most marginalised (Felli and Castree 2012); they appear to be resilient because they are unchanging. Their situation is hardly sustainable in not having enough resources to be able to make choices. They are only resilient in the sense that they still survive in the face of severe adversity, but it is not a thriving or welcome resilience – and this situation might not last for long given high mortality rates *in situ*. The two terms 'sustainability' and 'resilience' thus become difficult to interpret, perhaps even taking on unintended connotations while demonstrating how certain statements ring hollow.

The prevalence of confusing and often contradictory buzzwords pertaining to resilience and sustainability, especially with regards to climate change – such as 100 per cent renewable, zero-carbon, carbon neutral, adaptive capacity, transformation, and eco-island – is unsatisfactory from island perspectives and for islander needs, with definitions and connotations typically emerging from mainland perspectives (see Robertson, this volume). One challenge in deconstructing resilience and sustainability on and for islands is the incredible diversity of island contexts. Although the public stereotype of islands may be influenced by pristine white sand beaches, warm climates, ubiquitous palm trees, and the backdrop of the azure blue sea, these may be the exception to the norm, and are sometimes even built artificially, in many islands. In seeking tourism income, this picture becomes an island icon, not entirely false, but not necessarily depicting the everyday resilience and sustainability challenges and opportunities for island peoples and communities.

Island communities themselves willingly adopt the branding, such as their attempts to define and market themselves as being green, eco-friendly, and hence sustainable while being resilient in the face of globalisation challenges. This marketing can be accepted uncritically by media and researchers as being at the forefront of achieving 'sustainability', such as through energy self-sufficiency (Mitra 2006, Chen et al. 2007).

Through the UN, several countries have reiterated their pledge at the annual climate change meetings to become carbon neutral by 2020, including Maldives and Tuvalu. Generating 100 per cent renewable energy is highlighted, but the incorporation of aviation and shipping emissions into the calculations is usually sidestepped. Many renewable energy technologies, such as wind turbines and photovoltaics, are not manufactured on the islands, so products and servicing would require transport by air or ship, adding to energy costs.

Several European islands have been presented as exemplars for sustainability and resilience through their public and largely unaudited quest to achieve carbon neutrality. These include El Hierro, Canary Islands, Spain (Iglesias and Carballo 2011); Porto Santo, Madeira, Portugal (Duić and da Graça Carvalho 2004); Samsø, Denmark (Jørgensen et al. 2007); and Flores, Azores, Portugal (Silva and Ferrao 2009). The latter two have branded themselves as versions of 'Green Islands', at least partly to fill a tourism niche. Their other intended or unintended goal is to foster an international consultancy industry explaining how they achieved their sustainability goal and how other jurisdictions might emulate their achievements. This despite the fact that they have not really achieved the goal even while providing positive and creative examples of reduced energy consumption and other sustainability endeavours (e.g. Baldacchino and Kelman 2014, Grydehøj and Kelman 2017).

Despite the critique (e.g. Baldacchino and Kelman 2014, Grydehøj and Kelman 2017), many of the opportunities are real and legitimate or, at the very least, constitute "geographies of hope" (Turner 2007). They support sustainability and add dimensions of resilience to island life and livelihoods, such as being able to restore electricity supplies more quickly after a disaster due to the local, decentralised nature of the networks. As Baldacchino (2007) describes, it helps to move the framing of sustainable island futures from what these island societies do *not* have to what they are capable of achieving, even if the communities are not fully there. Islands might not be miniature worlds, but they assist in providing microcosms which help to better understand the wider world of resilience–sustainability tensions, not just through the microcosm itself, but also through the connections beyond (e.g., Selvon 1955).

Island studies demonstrating resilience–sustainability synergies

Resilience and sustainability narratives converge on ideas of supporting self-help and self-sufficiency, covering numerous sectors including energy, food, water, building materials, disaster risk reduction and dealing with change. Island communities and islander interests have found such approaches to be particularly apposite, since large distances must often be traversed to reach other locations. Ostensible remoteness, though, does not necessarily mean a lack of connectedness, either today with telecommunications systems or earlier when long voyages were part of island life.

Hau'ofa (1994, 1998) offers his 'sea of islands' perspective about connections amongst Pacific peoples. For millennia, islanders in the Pacific have interacted through complex networks of trade, intermarriage, violence and support (Gough et al. 2010, Ward 1989). Island connections and separations complement each other to avoid isolated and distinct populations and without creating dependency on external assistance. The insularity which undermines resilience is avoided while sustainability is promoted through exchange and mutual support.

DeLoughrey (2007) echoes this ethos of island 'Routes and Roots' (her book's title) in that the water surrounding the land has been a medium to connect, rather than isolate, island communities. *Roots* is a metaphor for place attachment, or *topophilia* (Tuan 1974) and a link between identity and territory, while *routes* suggests mobility, exchange and knowledge gained from exposure to a mixture of experiences and values (Gustafson 2001). Baldacchino (2008) speaks of routes in a slightly different manner, suggesting that islanders have an urge to escape and observe one's island from a 'glocal' perspective. The land has served to reinforce cultural traditions and values for thriving more than surviving through centuries of change. Resilience and sustainability dovetail, by using what the land, water, and in-between zones offer to provide and maintain island life and livelihoods.

This synergistic relationship between resilience and sustainability has been recast recently as an opportunity for the interplay of the so-called blue (or oceans) economy and the green economy (Smith-Godfrey 2016, UNEP 2012, Visbeck et al. 2014). The former represents the sustainable use of resources derived from the marine environments surrounding islands and coastal areas, including fisheries, aquaculture, tourism, as well as offshore and seabed resource extraction such as oil, gas and minerals. The latter is a more terrestrial-based balance between economic activities, the physical environment and community well-being.

Turner et al. (2003) suggest that the edges of marine and the terrestrial environments have commonalities in being zones of incredible ecological and cultural richness and diversity. Indigenous peoples are drawn to these biodiversity zones and, in so doing, build a strong cultural baseline that enhances flexibility, resilience and hence sustainability. Although the examples they provide are limited to freshwater continental shorelines, the same cultural–ecological synergy exists along the saltwater coasts of islands. For example, Bridges and McClatchey (2009) show that Marshall Islanders living on Rongelap atoll have been forced to develop a rich cultural understanding and management of food and water resources at the shorelines of atolls which they characterise as being vulnerable to extreme environmental conditions, including drought, but which actually represent typical environmental variabilities. Their words repeat a long-accepted island and island studies truism that "atoll life forces recognition of the boundedness of small ecosystems" (p. 143).

What next?

Who decides island goals; who is affected by efforts to achieve these goals; and who gains and loses from achieving or not achieving them? With respect to financing or accruing benefits from the process of aiming for island goals, some financing comes from international aid efforts, but much comes from the islanders themselves, through allocating government revenues (including international aid) and remittances.

For instance, sustainability might be assumed to be reachable by building sea walls (see Figure 17.4), especially against sea-level rise. These sea walls could then reflect wave energy

Figure 17.4   Seawalls, such as this one from the British Virgin Islands, can do more harm than good for reducing property damage.

Source: Photo by Ilan Kelman.

inhibiting sediment release and further eroding islands. The next storm could cause significant damage leading to an increased need for international aid and remittances. Conversely, without engineered solutions for low-lying atolls, how many people could stay in their island communities and how many would have to move due to sea-level rise?

No matter how and how much resilience and sustainability are sought, major environmental and cultural changes will be foisted on islanders in the coming years and decades. These changes may be comparable to the island changes experienced due to the arrival of Christianity, the motor car, air travel and the internet. Laracy (1983) reports how some Tuvaluans saw the Second World War as being the best time of their lives, because the American occupation brought more cigarettes, soap and kerosene than they had ever seen. The widespread availability of Spam and other cheap foods is seen as a boon by many Pacific islanders, because getting fat is now cheap and easy, rather than the continual hunger felt when having to produce all of one's own food. The induced double burden of malnutrition and obesity in many Pacific island communities (Haddad et al. 2014) may be far more sustainable in the modern era than was a constant food supply in earlier times.

Overall, much regarding island resilience and sustainability depends on definitions of the words; and perhaps even on the definition of 'island'. Thus, one agenda item is to understand better how different island cultures and island contexts define and apply 'resilience' and 'sustainability'. In which island languages do the words and concepts exist and not exist? Where the terms exist, how are they applied and not applied? How might islanders wish to see them being applied or avoided? How might the consequences and effectiveness of island resilience and sustainability initiatives be measured, monitored and evaluated?

To pursue such questions, one challenge to overcome is the lack of island-specific data. Many future projections and analyses are based on a limited number of island observation stations and case studies. Many are too coarse spatially to consider numerous inhabited islands, because the islands are much smaller than the grid size. Increased local and downscaled social and environmental modelling for islands would be useful; but this could never be the only or most important consideration, especially since islands are not simply synecdoches, miniature continents that can be scaled down (Nunn 2004). Instead, and just like the definitions of 'resilience' and 'sustainability' – and 'islandness' – "What next?" should be answered by island peoples, in consultation with others through exchange and mutual support.

References

Alexander, D.E. (2013) Resilience and disaster risk reduction: an etymological journey. *Natural Hazards and Earth System Sciences*, 13(11), 2707–2716.

Baker, P.E., Gass, I.G., Harris, P.G., and Le Maitre, R.W. (1964) The volcanological report of the Royal Society expedition to Tristan da Cunha, 1962. *Philosophical Transactions of the Royal Society of London. Series A, Mathematical and Physical Sciences*, 256(1075), 439–575.

Baldacchino, G. (ed.) (2007) *A world of islands: an island studies reader*. Luqa, Malta and Charlottetown, Canada, Agenda Academic and Institute of Island Studies, University of Prince Edward Island.

Baldacchino, G. (2008) Studying islands: on whose terms? Some epistemological and methodological challenges to the pursuit of island studies. *Island Studies Journal*, 3(1), 37–56.

Baldacchino, G. (2011) Prince Edward Island settlement strategy: a summary. In J. Biles, M. Burstein, J. Frideres, E. Tolley and R. Vineberg (eds.), *Integration and inclusion of newcomers and minorities across Canada*, Montreal and Kingston, Canada, McGill-Queen's University Press, pp. 373–379.

Baldacchino, G., and Kelman, I. (2014) Critiquing the pursuit of island sustainability: blue and green, with hardly a colour in between. *Shima: The International Journal of Research into Island Cultures*, 8(2), 1–21.

Bartlett, A.A. (2004) *The essential exponential*. Lincoln, NE, University of Nebraska Press.

Betzold, C. (2010) 'Borrowing' power to influence international negotiations: AOSIS in the climate change regime, 1990–1997. *Politics*, 30(3), 131–148.

Betzold, C., Castro, P., and Weiler, F. (2012) AOSIS in the UNFCCC negotiations: from unity to fragmentation? *Climate Policy*, 12(5), 131–148.

Bhamra, R., Dani, S., and Burnard, K. (2011) Resilience: the concept, a literature review and future directions. *International Journal of Production Research*, 49(18), 5375–5393.

Blum, J.L. (1968) Salt marsh spartinas and associated algae. *Ecological Monographs*, 38 (3), 199–221.

Brandt, W. (1980) *North – South: a programme for survival*. Report of the Independent Commission on International Development Issues. Boston, MA, MIT Press.

Bridges, K.W., and McClatchey, W.C. (2009) Living on the margin: ethno-ecological insights from Marshall islanders at Rongelap atoll. *Global Environmental Change*, 19(2), 140–146.

Catala, R.L.A. (1957) Report on the Gilbert islands: some aspects of human ecology. *Atoll Research Bulletin*, No. 59. Washington, DC, Pacific Science Board.

Chen, F., Duic, N., Alves, L.M., and da Graca Carvalho, M. (2007) Renewislands: renewable energy solutions for islands. *Renewable and Sustainable Energy Reviews*, 11(8), 1888–1902.

Clark, G., and Reepmeyer, C. (2012) Last millennium climate change in the occupation and abandonment of Palau's Rock islands. *Archaeology in Oceania*, 47(1), 29–38.

Connell, J., and Brown, R.P. (1995). Migration and remittances in the South Pacific: towards new perspectives. *Asian and Pacific Migration Journal*, 4(1), 1–33.

DeLoughrey, E.M. (2007) *Routes and roots: navigating Caribbean and Pacific island literatures*. Honolulu, HI, University of Hawai'i Press.

Duić, N., and da Graça Carvalho, M. (2004) Increasing renewable energy sources in island energy supply: case study Porto Santo. *Renewable and Sustainable Energy Reviews*, 8(4), 383–399.

Farbotko, C. (2010) Wishful sinking: disappearing islands, climate refugees and cosmopolitan experimentation. *Asia Pacific Viewpoint*, 51(1), 47–60.

Felli, R., and Castree, N. (2012) Neoliberalising adaptation to environmental change: foresight or foreclosure? *Environment and Planning A*, 44(1), 1–4.

Fiering, M.B. (1982) Alternative indices of resilience. *Water Resources Research*, 18(1), 33–39.

Gaillard, J.C. (2007) Resilience of traditional societies in facing natural hazards. *Disaster Prevention and Management*, 16(4), 522–544.

Gaillard, J.C., Clavé, E., Vibert, O., Azhari, D., Denain, J.-C., Efendi, Y., Grancher, D., Liamzon, C.C., Sari, D.S.R., and Setiawan, R. (2008) Ethnic groups' response to the 26 December 2004 earthquake and tsunami in Aceh, Indonesia. *Natural Hazards*, 47(1), 17–38.

Glantz, M.H. (2003) *Climate affairs: a primer*. Covelo, CA, Island Press.

Glavič, P., and Lukman, R. (2007) Review of sustainability terms and their definitions. *Journal of Cleaner Production*, 15(18), 1875–1885.

Gough, K., Bayliss-Smith, T., Connell, J., and Mertz, O. (2010) Small island sustainability in the Pacific: introduction to the special issue. *Singapore Journal of Tropical Geography*, 31(1), 1–9.

Greenhough, B. (2006) Tales of an island-laboratory: defining the field in geography and science studies. *Transactions of the Institute of British Geographers*, 31(2), 224–237.

Grydehøj, A., and Kelman, I. (2017) The eco-island trap: climate change mitigation and conspicuous sustainability. *Area*, 49(1), 106–113.

Gustafson, P. (2001). Roots and routes: exploring the relationship between place attachment and mobility. *Environment and Behaviour*, 33(5), 667–686.

Haddad, L., Cameron, L., and Barnett, I. (2014) The double burden of malnutrition in SE Asia and the Pacific: priorities, policies and politics. *Health Policy and Planning*, 30(9), 1193–1206.

Hau'ofa, E. (1994) Our sea of islands. *The Contemporary Pacific*, 6(1), 147–161.

Hau'ofa, E. (1998) The ocean in us. *The Contemporary Pacific*, 10(2), 392–410.

Hewitt, K. (ed.) (1983) *Interpretations of calamity*. London, Allen & Unwin.

Holling C.S. (1973) Resilience and stability of ecological systems. *Annual Review of Ecology and Systematics*, 4, 1–23.

Iglesias, G., and Carballo, R. (2011) Wave resource in El Hierro: an island towards energy self-sufficiency. *Renewable Energy*, 36(2), 689–698.

Jacka, J.K. (2010) The spirits of conservation: ecology, Christianity and resource management in Highlands Papua New Guinea. *Journal for the Study of Religion, Nature and Culture*, 4(1), 24–47.

Johnston, I. (2015) Disaster management and climate change adaptation: a remote island perspective. *Disaster Prevention and Management*, 23(2), 123–137.

Jørgensen, P.J., Hermansen, S., Johnsen, A., Nielsen, J.P., Jantzen, J., and Lundén, M. (2007) *Samsø: a renewable energy island. 10 years of development and evaluation*, Samsø, Denmark, PlanEnergie/Samso Energy Academy.

Kaloumaira, A. (2002) *Reducing the impacts of environmental emergencies through early warning and preparedness. The case of El Niño Southern Oscillation (ENSO): the Fiji Case Study*, SOPAC Technical Report 344, Suva, Fiji, SOPAC (South Pacific Applied Geoscience Commission).

Kench, P.S., Thompson, D., Ford, M.R., Ogawa, H., and McLean, R.F. (2015) Coral islands defy sea-level rise over the past century: records from a central Pacific atoll. *Geology*, 43(6), 515–518.

King, R., and Connell, J. (eds.) (1999) *Small worlds, global lives: islands and migration*. London, Pinter.

Laracy, H. (ed.) (1983) *Tuvalu: a history (with seventeen Tuvaluan contributors)*. Funafuti, Institute of Pacific Studies, University of the South Pacific and the Ministry of Social Services, Government of Tuvalu.

Lauer, M., Albert, S., Aswani, S., Halpern, B.S., Campanella, L., and La Rose, D. (2013) Globalisation, Pacific islands, and the paradox of resilience. *Global Environmental Change*, 23(1), 40–50.

Le De, L., Gaillard, J.C., Friesen, W., and Smith, F.M. (2015) Remittances in the face of disasters: a case study of rural Samoa. *Environment, Development and Sustainability*, 17(3), 653–672.

Lewis, J. (2009) An island characteristic: derivative vulnerabilities to indigenous and exogenous hazards. *Shima: The International Journal of Research into Island Cultures*, 3(1), 3–15.

Lightfoot, C. (1999) *Regional El Niño social and economic drought impact assessment and mitigation study*. Suva, Fiji, SOPAC (South Pacific Applied Geoscience Commission).

Malé Declaration on global warming and sea level rise (1989) Malé, the Maldives: small states conference on sea level rise 14–18 November 1989. Available from: www.islandvulnerability.org/slr1989/declaration.pdf.

Meadows, D.H., Meadows, D.L., Randers, J., and Behrens III, W.W. (1972) *The limits to growth*. New York, Universe Books.

Mentz, S. (2012) After sustainability. *The Proceedings of the Modern Language Association of America*, 127(3), 586–592.

Mercer, J., Dominey-Howes, D., Kelman, I., and Lloyd, K. (2007) The potential for combining indigenous and western knowledge in reducing vulnerability to environmental hazards in small island developing states. *Environmental Hazards*, 7(4), 245–256.

MIT (1970) *Man's impact on the global environment: report of the study of critical environmental problems (SCEP)*. Cambridge, MA, MIT Press.

MIT (1971) *Inadvertent climate modification: report of the study of man's impact on climate (SMIC)*. Cambridge, MA, MIT Press.

Mitra, I. (2006) A renewable island life: electricity from renewables on small islands. *Refocus*, 7(6), 38–41.

Nunn, P.D. (2004) Through a mist on the ocean: human understanding of island environments. *Tijdschrift voor Economische en Sociale Geografie*, 95(3), 311–325.

Nunn, P.D., Hunter-Anderson, R., Carson, M.T., Thomas, F., Ulm, S., and Rowland, M.J. (2007) Times of plenty, times of less: last-millennium societal disruption in the Pacific basin. *Human Ecology*, 35(4), 385–401.

Parker, J.N., and Hackett, E.J. (2012) Hot spots and hot moments in scientific collaborations and social movements. *American Sociological Review*, 77(1), 21–44.

Paton, D., and Johnston, D.J. (2001) Disasters and communities: vulnerability, preparedness and resilience. *Disaster Prevention and Management*, 10(4), 270–277.

Person, O.P. (1960) A mechanical model for the study of population dynamics. *Ecology*, 41(3), 494–508.

Pugh, J. (2014) Resilience, complexity and post-liberalism. *Area*, 46(3), 313–319.

Randall, J.E., Kitchen, P., Muhajarine, N., Newbold, B., Williams, A., and Wilson, K. (2014) Immigrants, islandness and perceptions of quality-of-life on Prince Edward Island, Canada. *Island Studies Journal*, 9(2), 343–362.

Rankey, E.C. (2011) Nature and stability of atoll island shorelines: Gilbert Island chain, Kiribati, equatorial Pacific. *Sedimentology*, 58(7), 1831–1859.

Schwarz, A.-M., Béné, C., Bennett, G., Boso, D., Hilly, Z., Paul, C., Posala, R., Sibiti, S., and Andrew, N. (2011) Vulnerability and resilience of remote rural communities to shocks and global changes: empirical analysis from Solomon Islands. *Global Environmental Change*, 21(3), 1128–1140.

Selvon, S. (1955) *An island is a world*. London, Allan Wingate.

Shotte, G. (2006) Identity, ethnicity and school experiences: relocated Montserratian students in British schools. *Refuge*, 23(1), 23–39.

Silva, C. A., and Ferrão, P. (2009) A systems modeling approach to project management: the green islands project example. In *Second International Symposium on Engineering Systems* (pp. 1–12). Cambridge, MA, MIT.

Skelton, T. (2000) Political uncertainties and natural disasters: Montserratian identity and colonial status. *Interventions*, 2(1), 103–117.

Smith-Godfrey, S., (2016) Defining the blue economy. *Maritime Affairs: Journal of the National Maritime Foundation of India*, 12(1), 58–64.

Sowan, P.W. (1985) Living with volcanoes: turning potential disaster to good account in Iceland. *Geography*, 70(1), 67–69.

Townsend, C.R., Begon, M., and Harper, J.L. (2003) *Essentials of ecology*. Oxford, Blackwell.

Treadaway, J. (2007) *Dancing, dying, crawling, crying: stories of continuity and change in the Polynesian community of Tikopia*. Suva, Fiji, IPS Publications, University of the South Pacific.

Tuan, Y.-F. (1974) *Topophilia: a study of environmental perception, attitudes and values*, New York, Columbia University Press.

Turner, C. (2007) *A geography of hope: a tour of the world we need*. Toronto, ON, Random House.

Turner, N.J., Davidson-Hunt, I.J., and O'Flaherty, M. (2003) Living on the edge: ecological and cultural edges as sources of diversity for social–ecological resilience. *Human Ecology*, 31(3), 439–461.

Ulijaszek, S.J. (ed.) (2006) *Population, reproduction and fertility in Melanesia*. New York, Berghahn Books.

UN (1994) *Report of the global conference on the sustainable development of small island developing states*. Document A/CONF.167/9 (October, 1994) from Global Conference on the Sustainable Development of Small Island Developing States. Bridgetown, Barbados, United Nations.

UN (2005, 13 January) *Draft Mauritius strategy for the further implementation of the programme of action for the sustainable development of small island developing states*. Document A/CONF.207/CRP.7. Port Louis, Mauritius, United Nations.

UN (2014) *Draft outcome document of the 3rd international conference on small island developing states. Apia, 1–4 September 2014*. Apia, Samoa, United Nations.

UNEP (2012) *Green economy in a blue world: synthesis report*. New York, UNEP. Available from: www.unep.org/pdf/green_economy_blue.pdf.

Visbeck, M., Kronfeld-Goharani, U., Neumann, B., Rickels, W., Schmidt, J., Van Doorn, E., Matz-Lück, N., Ott, K., and Quaas, M.F. (2014) Securing blue wealth: the need for a special sustainable development goal for the ocean and coasts. *Marine Policy*, 48(1), 184–191.

Ward, R.G. (1989) Earth's empty quarter? The Pacific islands in a Pacific century. *Geographical Journal*, 155(2), 235–246.

WCED (World Commission on Environment and Development) (1987) *Our common future*. Oxford, Oxford University Press.

Wertheim, W.F. (1959) Sociological aspects of inter-island migration in Indonesia. *Population Studies: A Journal of Demography*, 12(3), 184–201.

18

BRANDS AND BRANDING

Godfrey Baldacchino and Susie Khamis

Introduction

Eyjafjallajökull is an Icelandic mountain glacier with an ice cap that covers the caldera of a volcano. The stratovolcano mountain stands 1,651 metres (5,417 feet) at its highest point, and has a crater 3–4 km (1.9–2.5 miles) in diameter, open to its north. The volcano has erupted relatively frequently since the last glacial period, and most recently in 2010. That spring, a series of eruptions that saw meltwater seeping into the volcanic vent caused powerful explosions that threw fine volcanic ash several kilometres into the atmosphere. The result was a region-wide air travel disruption in northwest Europe for six days from 15 to 21 April 2010 and, again, in May 2010. Airspace over many parts of Europe had to be closed due to the smoke and ash particles and associated electrical storms. At one stage, 1.2 million passengers were affected, with 100,000 flights grounded across Europe. Eyjafjallajökull became the most famous volcano in the world; and Iceland became the world's most reviled nation. "I hate Iceland" twice blurted a frustrated traveller at Edinburgh airport, Scotland, his travel plans dashed, on 16 April 2010: in those three words, he captured the sentiment of many. His outburst was caught by SkyNews and went viral on social media (Sky News 2010).

The episode, brief but spectacular, proved a seminal one for Iceland's global image. Personnel tasked with boosting the nation's tourism turned what was seemingly a disaster into an opportunity. 'Promote Iceland', the main body responsible for supporting Iceland's good image and reputation (through tourism, trade and investment) launched 'Inspired by Iceland', the nation's first integrated tourism campaign. A sophisticated media strategy was designed, and it reconfigured the huge (albeit negative) publicity that had been given to this marginal island in the North Atlantic into a timely opportunity (Pálsdóttir 2016). Campaign imagery capitalised on the link with volcanic eruptions, and hinged on promoting a feisty island state whose dynamic geology and natural beauty – a 'land of ice and fire' – were waiting to be discovered. For a country that straddles the North Atlantic, a modest 3–5 hour flight from most of western Europe and the eastern seaboard of North America (Benediktsson et al. 2011), it was a bold move; and it worked. Tourist visitor arrivals shot up from just under 500,000 in 2009 to 2.1 million in 2016 (Icelandic Tourist Board 2017). The Icelandic brand was not just saved; it was successfully catapulted to unprecedented popularity (Ólafsdóttir et al. 2016). In short, the episode produced a signature story for the island – a narrative that was intriguing, authentic, engaging and strategic (Aaker and Aaker 2016) – around which its place brand could cohere, and from which myriad other associations would be generated and leveraged for commercial gain. It was, in other words, an effective example of place and island branding.

The growth of place branding

A strong brand represents more than a cute picture or a smart sound bite. It is a long-term reputational asset (Bell 2016): a logo or product identifier that, once encountered, ideally triggers a positive response from consumers. As such, branding is neither organic nor haphazard. Brand strategists meticulously craft a brand's identity over time in order to inspire other individuals to believe in it enough to spend their money in assured and specific ways. A brand is imbued with personality and charisma; the means by which the 'essence' of something, or somewhere, is relayed to consumers (Govers and Go 2016).

Recent years have seen growing interest in the concept and practice of place branding: the marketing of places as brands with a view to engage various stakeholders on multiple fronts. This in turn sits within the burgeoning general interest in branding, as well as the increasing sophistication of marketing and organisational development, in an age driven by the pursuit of "intangible investment": not just branding, but also design, research and software (Haskel and Westlake 2017).

Put simply, all firms pursue a loyal (and thus captive) consumer base. They often spend large amounts of money on research and advertising in order to develop such loyalty, always with a keen eye on the competition. For this, marketing campaigns appeal to the reliability, safety, quality, durability and/or exceptionality of the product or service for sale (amongst other qualities); as well as to its freshness, vitality and relevance in changing times. Amidst shifting consumer habits, dynamic fiscal conditions and erratic fashion cycles, advertising speaks to the relevance and suitability of commercially marketed goods or services. In this situation, branding prioritises those associations that ostensibly signpost some distinction or differentiation, marking out the brand's superiority in an otherwise competitive and cacophonous field (Wood 2000). The more crowded or ruthless the marketplace, the higher the premium on branding, whereby differences are appreciated by consumers at a connotative level, if not through direct experience.

This need to project a reliable, memorable and endearing brand has extended well beyond 'everyday' goods and services. Politicians, religions and celebrities are similarly branded, as are places. Within the context of advanced globalisation, trade liberalisation and the growing accessibility of international travel (Yousaf et al. 2017), nations, cities and regions now vie for a lucrative share of global attention, spend, reputation and goodwill (Anholt 2007). Thus, places 'speak' and 'behave' through multiple contact points with consumers and stakeholders. Since 2008, a Nation Brands Index (NBI) seeks to capture perceptions of place brands, by canvassing six areas that ultimately contour how consumers perceive a place: exports, governance, culture, people, tourism, and immigration and investment. Such a survey nudges governments to identify gaps between a nation's strengths and capabilities, and its brand image: where consumers under-appreciate a place's diversity or potential, a brand strategy should focus on correcting such misconceptions (AGNBI 2017).

Like all branded commodities, place brands are designed for competitive edge: to stand out in a global marketplace replete with consumer options. The brand cues a suite of associations to register levels of salience, relevance and likeability. Therein lies brand equity: the extent to which consumers see value in the brand. For Aaker (1992), this value rests on five key assets: brand loyalty, brand awareness, perceived brand quality, associations that stem from perceived quality and other proprietary assets (such as patents or trademarks). As such, brand equity pivots on consumers' experience of the brand, which will either confirm or contradict whatever promises underpinned the brand's marketing. Therefore, whilst strategists create brand identity, what Vela (2013, p. 255) terms "an *a priori* perception", it is consumers that form the brand *image* (Kavaratzis and Hatch 2013). For this reason, a three-part approach to successful place branding is proposed: (1) a strategy that reflects diverse stakeholders' interests and is both feasible and inspiring; executed with (2) substance, that is, activities, laws, reforms or investments that serve the strategy; and are (3) communicated with symbols that 'cut through' with conviction and charisma (Anholt 2008). Striking symmetry between identity and image thus compels the brand to have some basis in 'reality': if consumers fail to experience the brand in ways consonant with how strategists have shaped it, the brand's image will suffer and its ever-mercurial equity will be dented. At best then, a brand with high equity resonates with consumers to such an extent that their engagement with it becomes akin to a relationship (Keller 2016, p. 5): emotive and enduring, and both valuable and valued (Florek and Kavaratzis 2014, p. 103).

Given the crucial difference between brand identity (as defined by advertisers) and brand image (as determined by consumers), it follows that the concept of place branding – a relatively recent addition to brand literature and research – is highly contentious. Grafting the logic and language of a marketing tool designed for products (sneakers, soap and so on) onto a place is messy, complicated and divisive, given: the diversity of stakeholders invested in a place (literally and figuratively); the likelihood that places are both perceived and experienced differently depending on numerous variables and vantage points; and that places change – often without warning and sometimes in ways that could not have been predicted or planned for – which in turn produces varying reactions from both inhabitants and observers (Gartner 2014). Put simply, and unlike sneakers or soap, the brand of a place is not readily contained (Khamis 2012).

The contemporary salience of place branding has coincided with the growth of social media; this has highlighted its inherent tensions, ironies and possibilities. On the one hand, social media offer awesome possibilities to corporations as well as governments to infiltrate multiple online platforms with strategic marketing, communicating messages to so many, so cheaply, so instantaneously. On the other hand, the inevitable plurality of voices and representations of place challenge any attempt to force a particular image; barring a deliberate monitoring of Internet traffic and content, as practised in some countries. Put another way, social media defy uniform or linear articulations of anything. Since consistency is integral to any brand campaign, which serves the integrity and cohesion of the brand story, there is obvious potential for any marketing message to veer off-course once alternative (and even combative) voices join the digital conversation.

This chapter surveys the complexities and contours of place branding through an island lens. At the very least, the 'separateness' of an island, its obvious demarcation, plays nicely to the aim of all place branding initiatives: to project a distinct and desirable identity from which various dividends flow; from tourism and investment to exports and diplomacy. Moreover, and given the extent to which a brand image is largely formed in the minds of consumers, islands have long enjoyed and endured key associations and connotations, from idyllic notions of paradise and bliss to less attractive assumptions of backwardness and insularity.

The island: a powerful draw

The word 'island' is one of the most heavily romanced in the English language. Along with the forest, the seashore and the valley, the island is a natural environment that has figured prominently in humanity's dreams of the ideal world (Tuan 1990, p. 24). It is no wonder that a TV series like *Lost* (2004–2010), which recycled all these tropes, was a worldwide success. The association of islands with mystery, fantasy, redemption, utopia and refuge is a long-lasting one that continues to be exploited by global media: consider the TV series *Fantasy island* (1977–1984) and such movies as *Blue lagoon* (1980), *The beach* (2000) and *Shutter island* (2010). To these, one could add *The Martian* (2015): the movie rendition of an exo-planetary Robinson Crusoe. The shipwreck motif, and the ensuing dialectic between islanded and island, runs through literature in such works as the *Odyssey* (Homer), *Swiss family Robinson* (Wyss 1812), *The mysterious island* (Verne 1874) and *Lord of the flies* (Golding 1954) (see introduction, this volume). The global tourism industry is rife with competing 'island paradise' destinations: one of the most extravagant of these, in the Bahamas, had its name changed to Paradise Island from the less enticing Hog Island, no doubt intended as a marketing stunt (Albury 2004). While there are exceptions – Malta, Manhattan, Singapore – most islands suggest themselves as (ideally) empty spaces, waiting and wanting to be possessed and explored. As notes DH Lawrence (1926, p. 1):

> There was a man who loved islands. He was born on one, but it didn't suit him, as there were too many other people on it, besides himself. He wanted an island all of his own: not necessarily to be alone on it, but to make it a world of his own.

Islands enjoy a lingering charm, allure and fascination: qualities that are well suited to tempt visitors eager for salvation, reinvigoration or escape. 'The Island' is so thoroughly seeped in emotional geography that it is perhaps impossible to determine where island dreams stop and island realities start. Such is the inexorable coupling of geographical materiality and metaphorical allusion.

Islands as prototype brand exercises

Islands have been 'branded' long before the concept found its way into management schools and contemporary marketing discourse. As early as the 10th century AD, Eric the Red, an early settler on a large and remote island, is reported in old Icelandic sagas to have named that new territory *Greenland*: a marketing move to attract other settlers there. The strategy worked: some 500 men and women returned to the island with him, albeit many perishing *en route*, to set up Greenland's first permanent Viking settlement (Attwooll 2013). Five hundred years ago, Genoese navigator Giovanni Caboto claimed that one could harvest cod from the rich waters off the island of Newfoundland, now Canada, simply by lowering a basket into the sea (Kurlansky 1997, p. 28): hyperbole that served as evidence of the fertility of the 'new found land', with Caboto keen to secure further commissions. Such and similar islands have served as prototypes, targets for some of the earliest systematic attempts at branding: advancing and romancing a meaningful and desirable difference in a world crowded by competitive categories (Martin 1989).

There are at least five interlocking reasons as to why islands have become so emphatically branded. First, there is an enduring Western tradition which has held islands in high regard, assigning them a key role in the economic, political, religious and social life of the Mediterranean and then Atlantic worlds, given the way that myth, icon and narrative of and from islands have functioned for mainland cultures (Gillis 2003). Second, and starting at around the European 'Age of Discovery', is the construction of islands as outposts of aberrant exoticism, peopled by innocent and exuberant 'South Sea maidens', with natives captured so tantalisingly in the paintings of Frenchman Paul Gauguin and the photographs of the American Captain Francis Rickman Barton (Sturma 2002). Third is the island as background for the enactment of a (usually) male and heroic tribute to colonialism and settler endurance (Hymer 1973): the subject of Robinson Crusoe-style stories that extend up to the present in the likes of the Robert Zemeckis film *Cast away* (2000). Fourth is the development of the notion of going on vacation as a regular activity by the world's burgeoning middle classes: whether for relaxation, adventure or self-discovery (Löfgren 2002); islands project themselves as ideal destinations for such catharsis, with the air or sea travel involved obliging necessary ruptures with *terra firma* and the ennui that it stands for. Fifth is the realisation by many developing island states and territories that they can 'sell' their freely available 'sea, sun and sand' to these visitors, appealing to their constructed modern need for travel, and thus carve out for themselves an easy route to development. Crucial here is that one-fifth of the world's sovereign states are islands or archipelagos, and all small island states except one (Iceland) are to be found in tropical or temperate regions. Other attractive, physical and psychological characteristics can be added: jurisdictional specificity, cultural difference, 'getting away from it all', and the possibility of claiming an understanding of the totality of the island locale as trophy (Baum 1997).

The sum total of these phenomena is that islands are now, often unwittingly, the objects of what may be the most lavish, global and consistent branding exercise in human history. They are presented as locales of desire, platforms of paradise, habitual sites of fascination, emotional offloading or religious pilgrimage. The metaphoric deployment of (the tropical) island, with the associated attributes of small physical size and warm water, is a central and compelling metaphor within Western discourse. Weaver (2017, p. 11) refers to a "virtuous periphery syndrome", the product of both necessity and factor endowment, which "positions small islands as sites of impressive resilience and innovation capable of providing peak experiences that support robust tourism sectors within contexts of balanced autonomy and cultural distinctiveness" (ibid.).

Added to these tropes of late, in a context of growing global awareness of climate change, is the representation of islands as "geographies of hope" (Turner 2010): places that strive to become – and demonstrably so – beacons of renewable energy and sustainable living. Samsø (Denmark), El Hierro (Canaries, Spain) and other islands lead the way in advancing 'blue' and 'green' agendas for development – (see Robertson, this volume) – even though the costs of such projects per capita tend towards an inefficient use of human and material resources and financial outlays (Grydehøj and Kelman 2017). In any case, the optics are good, and the publicity is welcome: this is how many small islands make the news now, by showcasing their own efforts to combat, mitigate or adapt to the effects of the climatic changes that would ultimately seek to destroy them, particularly via sea level rise.

Challenges

With these advantages of positioning within this global effort for the reputation management of place, however, come a series of challenges. The most salient include: (1) how to secure visibility; (2) how to influence an island's construction of its overall repute; and (3) the absence of congruence between 'the island' as a self-defined and stand-alone brand; and the various product or service brands that emerge from, or are associated with, that same island more generally.

Being noticed

First, there is the challenge of being noticed. There are hundreds of thousands of islands: many are not populated, are owned privately and/or are deliberately cast as nature reserves, inviting little to no attention from the curious or paparazzi. For these and their human and/or non-human inhabitants, invisibility is a blessing. Not so for the Galápagos, though, where even a hefty entrance fee to the national park does not dissuade tourists from coming in droves (Larson 2002).

For the rest, many offer what appear to be similar experiences or products – the generic trinity of 'sun, sand and sea' – and so these islands must somehow stand out. The fate of an island that is 'off the radar' is to remain so. Some islands must bear the humility and embarrassment of being excluded from maps; or, at best, suffer the indignity of being an out-of-scale inset tucked in a corner. For example, the far-flung French overseas departments of Réunion, Martinique, Guadeloupe and Mayotte (all islands) plus French Guyana are integral to the French state. Yet, they appear only as insets along the edges of maps of continental France. This situation can get worse, however: various maps of 'France' completely exclude these territories, only portraying the (much closer) island of Corsica (if at all) (Baldacchino 2013, p. 8).

Other islands seem permanently eclipsed by more visible, often larger or better-known competitors. Judging simply by size of population, one is more likely to know of the Dominican Republic than Dominica. To make matters worse, the citizens of both countries are called Dominicans (when spoken, the stress falls on a different syllable; but this nuance is lost in writing). Much smaller Dominica tries to capitalise on the resultant confusion, declaring itself "the incredible Caribbean island you never thought to visit" (Bignell 2017). Also, there is Barbuda, part of the archipelago state of Antigua and Barbuda, and not to be confused with either Bermuda or Barbados, as Wikipedia (2017) warns. Then there are the many times that Grenada, the Caribbean island state, has been confused with Granada, the UNESCO World Heritage city in Spain, 6,500 km away. In 2014, an American couple flew by mistake to Grenada in the Caribbean and not Granada in Spain; they lost their lawsuit against British Airways for the ticket mix up (Nye 2014).

In some cases, smaller islands suffer from double insularity and are sidelined by the marketing drive of a larger neighbouring island. For example, and in relation to Mauritius, there is the island of Rodrigues, its sub-national jurisdiction (Wergin 2012). Or, in the case of Grenada, there are neighbouring Carriacou and Petit Martinique (Baldacchino, forthcoming). Even smaller are the islands of Gozo and Comino, next to Malta. Here, being a jurisdiction is a net advantage. It secures dedicated administrative machinery (with, for instance, an island-specific tourism or investment promotion agency) that has a vested (and funded) interest in promoting the island as a destination for tourism, investment, adventure, immigration or retirement. Otherwise, there is some scope to benefit from being 'out of sight, out of mind': focusing on a narrow, niche concept of supply, where exclusivity becomes the draw and knowledge of the privileged place is disseminated mainly by word of mouth. The five-star island of St Barthélemy in the French Caribbean has achieved this (Serra and Theng 2015).

Forms of representation

The second challenge is the ability to autonomously fashion the island's reputation. The smaller, poorer or less populated the island is, and the more bereft of the "resourcefulness of jurisdiction" (Baldacchino and Milne 2000), then the more likely is it that its web, textual and literary content, its very representation, is dictated, penned or determined exogenously. All too often, the *subject* matter – the island as well as the islanders – becomes *object* matter: a 'looked at' reference group (Edmond and Smith 2003, Baldacchino 2008). Thus unfolds historically a summative objectification of, first, the island/s *per se*, and then the islanders, singly or collectively, by those that gaze upon them, connecting deliberately or subconsciously the persons and their setting to such wider and broader tropes as domestication, fertility and reproduction, pastoral livelihoods, scientific curiosa, resilience and 'noble savages' in need of civilising missionary work. Such representations were addressed invariably to distant, metropolitan interests, fashions and tastes. In the course of this carto/pictographic and narratological exhibition, the island(er)s find themselves ritually "aesthesicised, sanitised and anaesthetised" (Connell 2003, p. 568), problematised spatially and temporally, lurching from paradise to prison: from candidates deserving preservation (as the last and relictual upholders of a vanished Western pastoral idyll) to deserving translocation and displacement (and thus being saved from their tragic isolation and primitive existence).

The proponents of such a ritualised gaze were mainly explorers, missionaries and traders in the past. Today, these are more likely to be airline companies, hotel chains, media firms and corporate interests. This seems the case when the name of any (small) island is fed into an online search engine; it is highly unlikely that the resultant find (textual and visual) was produced by and for its actual inhabitants. That said, the social media explosion has liquidated any attempts at impression management via the determination of island representations, whether by insiders or outsiders. Until a few years ago, island tourism agencies would have had a strong say in crafting the limited sets of colour postcards for sale to tourists. Today, millions of photos are posted online daily, creating a feast of visual representations for all to witness. What islands stand for now is the outcome of a complex, anarchic and dynamic process that is channelled and populated by the algorithms of Internet search engines.

Island brands and 'the island' as a brand

The third challenge is this: any populated island inadvertently rides on parallel reputational messages. These include those pertaining to the island itself; and those that arise from the particular products and services that the (same) island has to offer.

Niche manufacturing is emerging as a viable industry for many small islands. Lacking economies of scale, small islands are simply unable to compete at cost with similar goods that are more cheaply produced elsewhere (Briguglio 1998). Still, the NISSOS Project (2002–2004) identified 140 firms in five European island regions – Åland, Iceland, Malta, Saaremaa (in Estonia) and the Scottish isles of Skye and Shetland (in the UK) – that were locally owned yet export-oriented manufacturing firms, had fewer than 50 employees each, and used some form of home-grown or adapted technology (Baldacchino and Vella Bonnici 2006, NISSOS Project 2017) (see Figure 18.1). When island products are smartly branded in alignment with the island with which they are associated, they *can* stand out and out-compete cheaper brands because of the exclusive and deliberate connection to place (Punnett and Morrison 2006). In some exceptional cases, particularly in relation to products that do not source local raw material (besides entrepreneurial acumen), an association with the small island context is *not* sought because it may discredit, belittle or unduly exoticise the product.

We therefore end up with the two faces of island brands: the island as a brand unto itself, as well as the brands of island products or services associated with that same island.

Here, brand convergence and complementarity help to convey congruent and mutually reinforcing signals: say, spicy pepper sauce is a good fit with the vibrant party culture of the Bahamas; and the pirate history of rum resonates nicely with 'fun and frolic' in Jamaica (Pounder 2010). There is often intent in the juxtaposition of the two island brand types presented above in the representation and marketing of Bahamian conch, Chios mastic, Corsican brocciu, Fair Isle knitted sweaters, Faroese lamb, Gozo lace, Guernsey cows, Islay whisky, Kinmen knives, Shetland ponies, Texel sheep and Trinidad hot sauce. Their producers seek to evoke a strong appeal to the appreciation of the product as a distinctly *local* speciality (Baldacchino and Baldacchino 2012, Butler 2015). Indeed, it is not just the name of the island but also often the visual image of the island, or its map outline, that also feature prominently in the products' promotion.

But the risk of a non- or misalignment of the two brand strategies is there; motives can clash and messages can jar. Prince Edward Island (PEI), Canada's smallest (and only fully enisled) province, takes pride in calling itself 'the gentle island'. Its most famous (albeit fictitious) personality

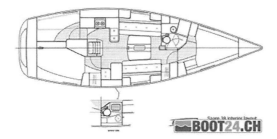

Figure 18.1   Scandinavian Design Blinds by Snickarboden from Åland, Lysi from Iceland, Mdina Glass from Malta, Saare Paat from Saaremaa and Fair Isle Sweaters from Fair Isle, Shetland, Scotland: all local manufacturing firms on small islands sourcing local raw material and local talent, but serving export markets.

is the rambunctious 'Anne of Green Gables' (Montgomery 1908). 'Anne' may be good for drawing tourists – especially from Japan – to PEI; but its associated message of bucolic farming and/or sedate scenery does not easily equate with a place that is nurturing cutting-edge aerospace and video game industries (Baldacchino 2010a).

Advertising template

Typically, the strategic branding of island products proceeds according to a simple protocol. Overt reference to the basic local ingredients is made – such as clean water, stone, wood, wool, sand or natural produce – preferably items that lend themselves to a deep historicity and, ideally, with a connection to a specific and interesting human episode. Marketing will often spotlight local production procedures, tools, technologies, rituals and tips. That is the basis of a place-specific story, invariably overlaid with poignant language: authentic, wholesome, frugal, rustic, genuine, time-honoured, 'handed down from generation to generation'. In this way, the brand speaks to a rich, exclusive and revered tradition (Baldacchino 2015).

And yet, within the predictability of this technique, lie the seeds of its own undoing. Large corporate interests recognise a threat – as well as an opportunity – when they see one, and they can adopt smart strategic positioning accordingly. As such, the reaction to a successful local product could include a leveraged buyout, or a price war that forces the local producer out of business, or the production of what are presented as 'local' products – such as beer – that are actually part of a repertoire of brands owned by bigger firms (Baldacchino 2010b). The larger, global corporate businesses cannot be discounted from venturing into niche markets. They have the ability to develop specialised designer products that seek to appeal to specific market niches, posing a real threat to local fare (which the global firms can imitate to a fault). Indeed, many so-called 'local products' are produced elsewhere; others, while produced locally, may be owned by off-island companies, and so are less local than they appear. For instance, consider 'Fiji Water', a leading brand of bottled water in the USA and a significant export of the Fiji Islands. While containing water extracted from Fiji, it is owned by an American couple (Connell 2006, Kaplan 2007). Its dominant market presence has given the company "profound developmental influence" and this has brought both costs and benefits at a local socio-environmental scale (Jones et al. 2017).

As an expression of the dangers of excessive protection against market forces, there are also risks of monopoly or oligopoly status achieved by *local* producers, breeding inefficiencies and eroding or eliminating competitive pressures. These result in a reduced choice of goods that are usually of lower quality, as well as more expensive, to local consumers. Producing local products to satisfy loyal local and/or tourist demand *can* exploit consumers, insofar as buyers come across as an easy, sympathetic, protectionist-friendly and perhaps even a not-too-demanding captive market. Price and product competition is less prevalent in small and remote markets, where conditions of 'natural monopoly' may prevail, and where bigger players may not be interested in staking a presence. A single local producer, supplier or processor may therefore quickly come to dominate the market, and effectively set prices. Local cartels – whether in production, in processing, in merchandise trade or in combinations of these – may well explain why local products or services in small and remote rural and island communities are more expensive than on the mainland or in the city. Transport and delivery costs and diseconomies of scale are not always to blame, though they too can be impacted by a situation of imperfect competition (Armstrong et al. 1993, Gyimóthy and Mykletun 2009, p. 265).

Next, what may be marketed as a quaint, rustic, family product could very well be produced industrially and on a larger scale. In other words, commodification is not just the consequence of large and external corporate interests driven by profit margins; it could be locally self-inflicted. A brand that begins as a traditional, family and hand-crafted affair may become so successful and generate such demand that its operation gets rescaled beyond recognition: for example, local kitchen equipment is replaced with stainless steel components and bulk industrial processes are introduced (Welz and Andilios 2004). In the interests of cutting costs, traditional ingredients that had been sourced locally may start being imported from elsewhere, and even replaced by trendier fare: bread replaces breadfruit; aquavit replaces sour milk, and so on. Hygiene, occupational safety and animal welfare regulations impose other obligations. Private profit seeking and conventional employment relations typically supersede any communal collaborative element and voluntary family labour. Of course, past homely idylls about the making of such a product are likely to persist in its branding and marketing strategies. The main way to beat the urge to grow is to stick stubbornly to tradition, hoping that its allure would allow the product to develop a sense of exclusivity and its own price-insensitive niche (Baldacchino 2015).

Conclusion

Globalisation has paradoxically led to a renaissance of locality: places and products have discovered that having and flaunting indigenous and native credentials helps them to stand out from the (mass market and price-sensitive) competition and legitimate their higher prices with discerning buyers. The alignment of product and place thus becomes a lynchpin of branding strategies. The explicit association of products with the islands where they are made is a strategic decision that rides on this new-found fascination with the local and the different, deploying popularly agreeable narratives of prestige, purity and ecological responsibility, while also tapping into the contemporary allure of island spaces. Here is an economic opportunity for small islands that can help them escape the strictures of aid, remittance or tourism dependency (Connell 2013). In this task, islands are both agents and victims of the designs, practices and consequences of place branding.

References

Aaker, D.A. (1992) The value of brand equity. *Journal of Business Strategy*, 13(L4), 27–32.

Aaker, D., and Aaker, J.L. (2016) What are your signature stories? *California Management Review*, 58(3), 49–65.

AGNBI (2017) *Anholt GfK Nation Brands Index*. Available from: http://nation-brands.gfk.com/.

Albury, P. (2004) *The Paradise Island story*, 2nd edn (rev. A. Lawlor and J. Lawlor). Oxford, Palgrave Macmillan.

Anholt, S. (2007) *Competitive identity: the new brand management for nations, cities and regions*. Basingstoke, Palgrave Macmillan.

Anholt, S. (2008) Place branding: is it marketing, or isn't it? *Place Branding and Public Diplomacy*, 4(1), 1–6.

Armstrong, H.W., Johnes, G., Johnes, J., and MacBean, A. (1993) The role of transport costs as a determinant of price level differentials between the Isle of Man and the United Kingdom. *World Development*, 21(2), 311–318.

Attwooll, J. (2013, April 15) Greenland: Eric the Red's white and green land. *The Telegraph*. Available from: www.telegraph.co.uk/travel/destinations/europe/greenland/articles/Greenland-Eric-the-Reds-white-and-green-land/.

Baldacchino, A., and Baldacchino, G. (eds.) (2012) *A taste of islands: 60 recipes and stories from our world of islands*. Charlottetown, Canada: Island Studies Press.

Baldacchino, G. (2008) Studying islands: on whose terms? Some epistemological and methodological challenges to the pursuit of island studies. *Island Studies Journal*, 3(1), 37–56.

Baldacchino, G. (2010a) Island brands and 'the Island' as a brand: insights from immigrant entrepreneurs on Prince Edward Island. *International Journal of Entrepreneurship and Small Business*, 9(4), 378–393.

Baldacchino, G. (2010b) Islands and beers: toasting a discriminatory approach to small island manufacturing. *Asia Pacific Viewpoint*, 51(1), 61–72.

Baldacchino, G. (ed.) (2013) *Archipelago tourism: policies and practices*. London, Routledge.

Baldacchino, G. (2015) Feeding the rural tourism strategy? Food and notions of place and identity. *Scandinavian Journal of Hospitality and Tourism*, 15(1–2), 223–238.

Baldacchino, G. (2018, forthcoming) 'Together, but not together, together': the politics of identity in island archipelagos. In Y. Martinez San Miguel and M. Stephens (eds.), *Archipelagic thinking: towards new comparative methodologies and disciplinary formations.*

Baldacchino, G., and Milne, D.A. (eds.) (2000) *Lessons from the political economy of small islands: the resourcefulness of jurisdiction.* Basingstoke, Palgrave Macmillan.

Baldacchino, G., and Vella Bonnici, J. (2006) *Real stories of small business success: insights from five European island regions.* Msida, Malta, Malta Enterprise with the support of the Leonardo da Vinci Programme of the European Commission.

Baum, T.G. (1997) The fascination of islands: a tourist perspective. In D.G. Lockhart and D.W. Drakakis-Smith (eds.), *Island tourism: trends and prospects*, London, Pinter, pp. 21–35.

Bell, F. (2016) Looking beyond place branding: the emergence of place reputation. *Journal of Place Management and Development*, 9(3), 247–254.

Benediktsson, K., Lund, K.A., and Huijbens, E. (2011) Inspired by eruptions? Eyjafjallajökull and Icelandic tourism. *Mobilities*, 6(1), 77–84.

Bignell, P. (2017, 31 January) The incredible Caribbean island you never thought to visit. *The Telegraph (UK)*. Available from: www.telegraph.co.uk/travel/destinations/caribbean/dominica/articles/dominica-nature-retreat-unspoilt-caribbean/.

Briguglio, L. (1998) Small country size and returns to scale in manufacturing. *World Development*, 26(3), 507–515.

Butler, R.W. (2015) Knitting and more from Fair Isle, Scotland: small island tradition and micro-entrepreneurship. In G. Baldacchino (ed.), *Entrepreneurship in small island states and territories*, London, Routledge, pp. 83–96.

Connell, J. (2003) Island dreaming: the contemplation of Polynesian paradise. *Journal of Historical Geography*, 39(4), 554–581.

Connell, J. (2006) The taste of paradise: selling Fiji and Fiji Water. *Asia Pacific Viewpoint*, 47(3), 342–350.

Connell, J. (2013) *Islands at risk? Environments, economies and contemporary change.* Cheltenham, Edward Elgar.

Edmond, R., and Smith, V. (eds.) (2003) *Islands in history and representation.* London, Routledge.

Florek, M., and Kavaratzis, M. (2014) From brand equity to place brand equity and from there to the place brand. *Place Branding and Public Diplomacy*, 10(2), 103–107.

Gartner, W.C. (2014) Brand equity in a tourism destination. *Place Branding and Public Diplomacy*, 10(2), 108–116.

Gillis, J.R. (2003) *Islands of the mind: how the human imagination created the Atlantic world.* New York, Palgrave Macmillan.

Golding, W. (1954/1983) *Lord of the flies.* Harmondsworth, Penguin.

Govers, R. and Go, F. (2016) *Place branding: glocal, virtual and physical identities, constructed, imagined and experienced.* New York, Springer.

Grydehøj, A., and Kelman, I. (2017) The eco-island trap: climate change mitigation and conspicuous sustainability. *Area*, 49(1), 106–113.

Gyimóthy, S., and Mykletun, R.J. (2009) Scary food: commodifying culinary heritage as meal adventures in tourism. *Journal of Vacation Marketing*, 15(3), 259–273.

Haskel, J., and Westlake, S. (2017) *Capitalism without capital: the rise of the intangible economy.* Princeton, NJ, Princeton University Press.

Hymer, S. (1971) Robinson Crusoe and the secret of primitive accumulation. *Monthly Review*, 63(4), 11–36. Available from: https://monthlyreview.org/2011/09/01/robinson-crusoe-and-the-secret-of-primitive-accumulation/.

Icelandic Tourist Board (2017) Tourism in Iceland in figures. Available from: www.ferdamalastofa.is/en/research-and-statistics/tourism-in-iceland-in-figures.

Jones, C., Murray, W.E., and Overton, J. (2017) Fiji Water, water everywhere: global brands and democratic and social injustice. *Asia Pacific Viewpoint*, 58(1), 112–123

Kaplan, M. (2007) Fijian water in Fiji and New York: local politics and a global commodity. *Cultural Anthropology*, 22(4), 685–706.

Kavaratzis, M., and Hatch, M.J. (2013) The dynamics of place brands: an identity-based approach to place branding theory. *Marketing Theory*, 13(1), 68–86.

Keller, K.L. (2016) Reflections on customer-based brand equity: perspectives, progress and priorities. *AMS Review*, 6, 1–16.

Khamis, S. (2012) Brand Australia: half-truths for a hard sell. *Journal of Australian Studies*, 36(1), 49–63.

Kurlansky, M. (1997) *Cod: a biography of the fish that changed the world.* Toronto, ON, Vintage Canada.

Larson, E.J. (2002) *Evolution's workshop: God and science on the Galápagos islands.* New York, Basic Books.

Lawrence, D.H. (1926) *The man who loved islands.* Available from: http://riverbendnelligen.com/the manwholovedislands.html.

Löfgren, O. (2002) *On holiday: a history of vacationing.* Berkeley, CA, University of California Press.

Martin, D.N. (1989) *Romancing the brand: the power of advertising and how to use it.* New York, American Management Association.

Montgomery, L.M. (1908) *Anne of green gables.* New York, Grosset and Dunlap.

NISSOS Project (2017) *Small business from small islands: the NISSOS Project.* Available from: http://projects. upei.ca/iis/islands-and-small-businesses/.

Nye, J. (2014, 27 August) Couple flown to Grenada in the Caribbean and not Granada in Spain lose US$34,000 lawsuit against British Airways for ticket mix up. *Daily Mail (UK).* Available from: www. dailymail.co.uk/news/article-2735359/Couple-flown-Grenada-Caribbean-not-Granada-Spain-lose-34-000-lawsuit-against-British-Airways-ticket-mix-up.html#ixzz4qK8VEd26.

Ólafsdóttir, R., Sæþórsdóttir, A.D., and Runnström, M. (2016) Purism scale approach for wilderness mapping in Iceland. In S.J. Carver and S. Fritz (eds.), *Mapping wilderness: concepts, techniques and applications,* Dordrecht, The Netherlands, Springer, pp. 157–176.

Pálsdóttir, I.H. (2016) Promoting Iceland: the shift from nature to people's power. *Place Branding and Public Diplomacy,* 12(2–3), 210–217.

Pounder, P. (2010) Branding: a Caribbean perspective on rum manufacturing competitiveness. *International Journal of Entrepreneurship and Small Business,* 9(4), 394–406.

Punnett, B.J., and Morrison, A. (2006) Niche markets and small Caribbean producers: a match made in heaven? *Journal of Small Business & Entrepreneurship,* 19(4), 41–353.

Serra, K.O., and Theng, S. (2015) Entrepreneurial traits, niche strategy and international expansion of a small firm on a small island territory. In G. Baldacchino (ed.), *Entrepreneurship in small island states and territories,* London, Routledge, pp. 135–153.

Sky News (2010, 16 April) Available from: www.youtube.com/watch?v=15joCwPYYk8.

Sturma, M. (2002) *South Sea maidens: Western fantasy and sexual politics in the South Pacific.* New York, Praeger.

Tuan, Y.F. (1990) *Topophilia: a study of environmental perceptions, attitudes and values.* New York, Columbia University Press.

Turner, C. (2010) *The geography of hope: a tour of the world we need.* Toronto, ON, Vintage Canada.

Vela, J. de San Eugenio (2013) The relationship between place branding and environmental communication: the symbolic management of places through the use of brands. *Place Branding and Public Diplomacy,* 9(4), 254–263.

Verne, J. (1874/1988) *The mysterious island.* New York, Simon and Schuster.

Weaver, D.B. (2017) Core-periphery relationships and the sustainability paradox of small island tourism. *Tourism Recreation Research,* 42(1), 11–21.

Welz, G., and Andilios, N. (2004) Modern methods for producing the traditional: the case of making halloumi cheese in Cyprus. In P. Lysaght and C. Burckhardt-Seebass (eds.), *Changing tastes: food culture and the processes of industrialisation,* Proceedings of the 14th Conference of the International Commission for Ethnological Food Research, Basel and Vevey, Switzerland, pp. 217–230.

Wergin, C. (2012) Trumping the ethnic card: how tourism entrepreneurs on Rodrigues tackled the 2008 financial crisis. *Island Studies Journal,* 7(1), 119–134.

Wikipedia (2017) Barbuda. Available from: https://en.wikipedia.org/wiki/Barbuda.

Wood, L. (2000) Brands and brand equity: definition and management. *Management Decision,* 38(9), 662–669.

Wyss, J. (1812/1991) *The Swiss family Robinson.* Oxford, Oxford Paperbacks.

Yousaf, A., Amin, I., and Gupta, A. (2017) Conceptualising tourist based brand-equity pyramid: an application of Keller brand pyramid model to destinations. *Tourism and Hospitality Management,* 23(1), 119–137.

Television series

Fantasy Island (1977–1984) Producer: A. Spelling. American Broadcasting Corporation. Available from: www.tv.com/shows/fantasy-island/.

Lost (2004–2010) TV series. 121 episodes. Producers: J.J. Abrams and D. Lindelof. American Broadcasting Corporation. Available from: www.tv.com/shows/lost/watch/.

Filmography

The Beach (2000) Director: Danny Boyle. Official Trailer available from: www.youtube.com/watch?v=MqDoxUxpobg.

Blue Lagoon (1980) Producer: Randal Kleiser. Official Trailer available from: www.youtube.com/watch?v=EFBj3sP7x-4.

Cast away (2000) Director: Robert Zemeckis. Official Trailer available from: www.youtube.com/watch?v=PJvosb4UCLs.

The Martian (2015) Director: Ridley Scott. Official Trailer available from: www.youtube.com/watch?v=lQqhfq87FgY.

Shutter Island (2010) Director: Martin Scorsese. Official Trailer available from: www.youtube.com/watch?v=5iaYLCiq5RM.

19

COMMONING AND ALTERNATIVE DEVELOPMENT

Eric Clark and Siri M. Kjellberg

Introduction

Research and policy analyses on island development commonly assume perspectives emphasising characteristics that distinguish island development from development in continental contexts, while drawing upon development theories advanced and honed in continental contexts (Clark 2009). Consequently, island development is widely understood in terms of policy choices conducive to either gentrification (growth, regeneration, reinvestment) or decline (decay, degeneration, disinvestment). We argue that this approach builds on a myth that poses a false choice propagated by and supportive of powerful interests, and suggest there are alternatives to this false choice that involve neither dystopian decline nor the normalised utopia of externally driven investor-oriented development. In so doing, we draw on two bodies of literature: gentrification research that casts light on uneven development and the false choice assumption; and research on commons and commoning that expounds alternatives to the 'regenerate (read gentrify) or decline' imperative. Gentrification research relevant to critiquing the false choice between regeneration and decline is first briefly presented, followed by an overview of island gentrification. We then consider commons and commoning as alternatives to this false choice. We finally narrow in on a small set of examples, drawn from island settings, where practices are geared to navigate development between gentrification and decline, highlighting the environmental and social advantages of such developmental pathways.

Gentrification or decline: a false choice

The gentrification of an area is characterised by both a marked upward shift in occupancy in terms of class and socio-economic position, and associated reinvestment in the built environment. Originally coined to designate processes of inner-city residential change (Glass 1964), half a century of theoretical debate harnessed to now voluminous empirical analyses has revealed a vastly broader scope of contexts in which structurally similar processes have been taking place. Gentrification accordingly came to be seen as a much more general process. We began to understand that it is the underlying mechanisms and associated structural relations that are central to identifying and delineating the process, not particular features in specific contexts (Clark 2005). Contrary to early formulations, gentrification does not occur only in inner cities, it does not manifest itself only through renovation, it is not only market-driven, it is not limited to residential spaces, and it is not even limited to specific social classes, regardless of etymology. Consequently, a number of corresponding qualifiers have flourished: rural gentrification (Phillips 1993, 2005), new-build gentrification (Davidson and Lees 2005), state-led gentrification (Cameron 2003, Slater 2004), commercial gentrification (Kloosterman and van der Leun 1999, Bridge and Dowling 2001), supergentrification (Lees 2003), and yes, island gentrification (Clark et al. 2007), to name a few.

Gentrification has become "a global urban strategy" amid the rush for global urban competitiveness whereby place politics is reduced to attracting capital investment, based on "the mobilisation of urban real-estate markets as vehicles of capital accumulation" (Smith 2002, pp. 437, 446; cf. Harvey 1989). Given the scope of the process – well beyond inner-city, working-class residential space – gentrification might be more adequately understood as a generic form of accumulation by dispossession, driven by the power of landed developer interests (Harvey 2003a, 2006, 2010). The commodification of space through the imposition of real-estate markets on the web of life opens up space for the flow of capital onto 'underutilised' land, facilitating 'highest and best' land uses to supplant present uses (Blomley 2002).

Power over space

The greater the inequalities in a society, the more marked and conflict-ridden gentrification becomes. A host of other relations concerning property rights and political regulations of modern capitalist space economies influence where and how the process plays out; but essentially it involves the re- or dis-placement of current residents/land-users by relatively more powerful and resourceful residents/land-users. It comes down to power over space, and whose visions of a place actually take place in that space (Hägerstrand 1986). And, as Bauman (1998, p. 86) points out: "[t]hose 'low down' happen time and again to be thrown out from the site they would rather stay in."

Modernity, says David Harvey (2003b, p. 1), is "always about 'creative destruction', be it of the gentle and democratic, or the revolutionary, traumatic, and authoritarian kind". Gentrification is a common manifestation of modernity, a mode of creative destruction that in gentle and democratic form easily passes unnoticed. Displacement is a key issue behind critical perspectives on gentrification (Atkinson 2000). It would be easy to link replacement with gentle and democratic forms of creative destruction, and displacement with revolutionary, traumatic and authoritarian forms. Some would associate *laissez-faire* with gentle and democratic, however revolutionary, traumatic and authoritarian the ball and chain of market transactions can be. But: these divisions are seldom clear cut. Were there no market pressures behind the eviction of 1.5 million residents in Beijing to make way for the 2008 Olympic Games (COHRE 2009)? Did democracy make any difference in the scale of evictions in Atlanta, Georgia, USA (1996) and Athens, Greece (2004), for their own hosting of the mega-games (COHRE 2007)? Is it gentle replacement (rather than traumatic displacement) when residents of gentrified off-shore economies, unable to afford to stay at home, 'choose' to emigrate?

Gentrification is seldom referred to by the rent-seeking proponents of redevelopment and reinvestment in places considered to be 'underutilised'. Instead redevelopment is presented in terms of 'revitalisation', 'regeneration' and 'improvement'. For all practical purposes, gentrification has become essentially synonymous with development and growth. Any rearguard mention of gentrification as a consequence of redevelopment is in response to struggles against the threat of displacement. Gentrification is then commonly defended with the ideology of trickle-down economics and the now threadbare metaphor of a 'tide that lifts all boats', with callous disregard of research that consistently shows these to be fallacious arguments. Like development and growth, gentrification is cast either as entirely beneficial, or at worst as a mixed blessing with largely positive outcomes.

In ways similar to many urban neighbourhoods, island communities increasingly face "the 'choices' of either continued disinvestment and decline . . . or reinvestment that results in their displacement" (DeFilippis 2004, p. 89). But, as Slater (2006, p. 753, emphasis in original) argues, "*either* unliveable disinvestment and decay *or* reinvestment and displacement is actually a *false choice*" foisted upon communities whose homes and livelihoods are considered by both (local) state and market to be 'lower and worse' land uses than what free-roaming finance capital deems to correspond to the potential yield of the land. To paraphrase Slater (2009, p. 297, emphasis in original):

> Gentrification is treated as *the only conceivable remedy* for pathological '[island] dereliction and decay'. Those in the path of [island] transformation are presented with a false choice: they can either have decay or gentrification. There is no alternative.

But, there *are* alternatives: many communities, including urban and island communities, have sought out development paths that refuse to accept the underlying assumptions of the simplistic dichotomy of false choices.

Of particular interest, both for island societies and for gentrification research, is to bring greater clarity to issues surrounding displacement and the relation between gentrification and local economies and cultures. Empirical studies addressing impacts of gentrification indicate considerable detrimental effects and social costs, suggesting a "moving around of social problems rather than a net gain" (Atkinson 2002, p. 21). These detrimental effects are above all due to gentrification generating displacement, topocide (death of place; *topos* place, *cidium* killing; Porteous 1988), domicide (death of home; *domus* house, *cidium* killing; Porteous and Smith 2001, Shao 2013), and demise of local economies and cultures. The 'choice' between gentrification and decline is often false; not only because there are alternatives, but because gentrification *entails* decline for the displaced, however much claims are made about revitalisation, regeneration and improvement. It is against this background that alternatives have been sought.

Island gentrification

Many islands around the world have experienced, are experiencing, or are on the verge of experiencing, gentrification processes, some gentler, others more traumatic: 'paradises' 'discovered', El Dorados occupied, visions developed, potentials realised. Peaks Island, Maine, USA: "a working class island feeling the grip of gentrification" (Bouchard 2004). The Maine coast: "All along the coast, land values are skyrocketing due to second-home ownership and gentrification" (Snyder 2006). New England: "The process of gentrification and coastal transformation is accelerating in New England as it is in most coastal areas of the US" (Hall-Arber et al. 2001). Staten Island, New York: "our neighbourhoods are being gentrified. Our people displaced" (Fighting the Gentrification of the North Shore of Staten Island 2017). Smith Island, Maryland: "working waterfront, threatened by gentrification" (Horton 2005). St Simons Island, Georgia: "Developers are sweeping through this island. . . . They've got a plan in place, and it does not include us" (Jonsson 2002). Hilton Head and other islands in Beaufort County, Georgia: "Now they are involved in an ongoing conflict wherein their land is quickly being taken away by wealthy elderly people and multi-million-dollar resort companies" (Yagley et al. 2005). Salt Spring Island, British Columbia, Canada: "In the future, the people who contribute to the uniqueness of Salt Spring . . . won't be able to afford to live here" (Shilling 2004). Puerto Rico: "I'm sure it can turn into a place that the land can be out of reach to the locals" (Román 2003), while "[s]peculators from New York and San Juan are buying, buying, buying" (Dreifus 2004). Tenerife, Spain: development pressure from the tourist industry has generated gentrification of whole districts involving evictions of over 5,000 residents and over 200 businesses (Garcia Herrera and Smith 2005). Styrsö, Sweden: "It is getting more and more difficult for people with low-incomes who work in health care or in the public sector to be able to afford housing on the islands. They are forced to move to the mainland and commute" (Clark et al. 2007, p. 502). Bruny Island, Tasmania, Australia: long-term residents are "concerned that their children will not be able to afford to buy property on the island" (Jackson 2006, p. 207). Waiheke Island, New Zealand: "due to gentrification . . . many long-standing Waiheke families are simply no longer able to afford to own or rent a house on Waiheke" (New Zealand Herald 2017).

The list could easily be multiplied. In a world of burgeoning tourism, rampant property development and booming markets for island havens (tax and finance) and retreats (from hyper-connectivity), island gentrification research and policy analyses will not suffer from lack of raw material for fieldwork. In spite of decades of research, "a geography of gentrification is only in its infancy" (Slater 2004, p. 1191). Perhaps a case could be made that studies of island gentrification can contribute to theoretical development on pivotal issues – including assumptions of and alternatives to false choice – in much the same way as island studies in other fields provide fertile ground for theoretical refinement (Baldacchino 2004, 2006a).

Alternatives: commons and commoning

> My great novel is now completed . . . A novel focused on ideas (no folksiness), enacted by the sea. . . . I myself find it grandiose . . . in an aristocratic, modern spirit, it asserts the incontestable right of the Stronger (i.e. the shrewder), to oppress and treat the Underdog as dung for his own good, he doesn't understand.
> *(August Strindberg, letter to Ola Hansson, 13 June 1890, about I havsbandet*
> *(By the open sea) (Strindberg 2005 [1890]), translated by Sven Heilo)*

At the core of geography (not the discipline, but real places) are "struggles for power over the entry of entities and events into space and time" (Hägerstrand 1986, p. 43). Gentrification is one kind of social geographic outcome of these 'space wars' (Lund Hansen 2003). But the outcome is neither necessary nor natural. There are degrees of action space in which resistance and the creative construction of alternatives can make a difference (DeFilippis 2004). That such resistance faces strong forces is underscored in Strindberg's archipelagic novel, accurately characteris-ing the modern spirit of creative destruction as driven by a belief in the incontestable right of the strong to oppress the 'small people'. There are, for instance, cases of neighbourhood rehabili-tation without dislocation or gentrification (Carmon and Hill 1988), of non-profit community development corporations (Robinson 1996), of community land trusts securing protection from speculative forces (Choi et al. 2017), and of exercising the right to the city to counter gentrifi-cation (Balzarini and Shlay 2016). In places like the Penghu Islands (Pescadores) in the Taiwan Strait, as with much of the South Pacific, traditional forms of land tenure, together with memory literally vested in the land (ancestors being buried on family plots), have constituted the main barriers to gentrification (e.g. Rodman 1987).

Each of these cases exemplifies in various ways a broader form of self-organised community stewardship with deep roots in human history: commons and commoning. As Nobel laureate

Elinor Ostrom and colleagues (1999, p. 278) put it, "for thousands of years, people have self-organised to manage common-pool resources, and users often do devise long-term, sustainable institutions for governing these resources".

A commons is a "self-organised system by which communities manage resources . . . with minimal or no reliance on the Market or State". The commons "*is not a resource*. It is a resource *plus* a defined community *and* the protocols, values and norms devised by the community to manage needed resources" (Bollier 2014, pp. 175–176, emphasis in original). Consequently:

> *There is no commons without commoning* – the social practices and norms that help a community manage a resource for collective benefit. . . . A commons must be animated by bottom-up participation, personal responsibility, transparency and self-policing accountability.
>
> *(Bollier 2014, p. 176)*

With deep roots in human history, commons and commoning have had many names. Today, many will associate social life beyond markets and states with 'civil society'. Predating and distinguished from markets and states, we rather find it helpful to relate commons to Polanyi's (2001) 'substantivist' approach to economic history, in which he distinguishes between three fundamental 'mechanisms of integration': community reciprocity, market exchange and state redistribution. Polanyi's substantivism involves a focus on materiality such that instituted processes consist of movements of people and material things, whereby relations between things and persons are manifested in time–space. The patterns of movement generated in the specific contexts of economic integration reflect societal metabolism in that the movements are about supplying people with flows of material goods, while the "social relations in which the process is embedded invest it with a measure of unity and stability" (Polanyi 1968, p. 307). Polanyi identifies three ideal patterns of movements corresponding to the configurations of economic coordination and institutional structure associated with each of the three basic mechanisms of economic integration. Dale (2010, p. 115) summarises: "The movements of redistribution are centric, its typical institutional locus is the state; those of reciprocity are symmetrical, its locus the community; those of exchange are polydirectional, its locus the market." These forms of economic integration are not mutually exclusive, but rather enmeshed and "cannot be understood in isolation from each other. Each is defined with respect to the role it plays vis-à-vis the others" (Harvey 1973, p. 283). As such, we can only imagine (and barely that) societies entirely integrated through one mechanism alone. In practice, state redistribution, market exchange and community reciprocity co-exist in all societies. Historically, however, state redistribution and market exchange and their associated patterns of movements of material flows have expanded with urbanisation and the increasing density and complexity of societal formations, at the cost of (though never totally displacing) community reciprocity. Importantly, while Polanyi emphasised their intrinsic interdependencies, the error of formalist economics – still in ascendency at the time of Polanyi's writing, and today a widely accepted myth – "was in equating the human economy in general with its market form" (Polanyi 1977, p. 6). Markets are, however, "not a kind of ether but are embedded to varying degrees in social relations" (Sayer and Walker 1992, p. 230). Similarly, although in the long stretch of human history commons *were* economic integration, commons today are seldom without connections to markets or (local) governments.

The recent renaissance of interest and activities in civil society and commons/commoning should be seen against the background of recent and ongoing experiences of totalitarian, oligarchic or plutocratic states, predatory capital in neo-liberalised 'free market' economies, and various combinations of the two. The difficulties faced by welfare states in providing security for large parts of the population have escalated with the advance of neo–liberal politics based on market fundamentalism (Peck 2010, Michaels 2011, Block and Somers 2014). With ecological and financial crises – whose frequency and intensity are both enhanced by market fundamental-ist politics – leaving many homeless and jobless, alternatives "beyond market and state" (Bollier and Helfrich 2012) attract growing attention.

Island commons, commoning islands

Islands have rich histories that can be usefully examined under the lens of Polanyi's three mecha-nisms of integration. It could be tempting to think that (particularly small) islands, commonly distinguished from continents or mainlands by their size, may be particularly well suited for com-munity reciprocity. While islands are not unlike mainlands in the sense that the integrating mecha-nisms of community reciprocity, market exchange and state redistribution mesh in complex ways, a case can be made that "small islands come with an 'amplification by compression' where eco-nomic monopoly, political totality and social intimacy come together in rather particular ways" (Baldacchino, personal communication; see also the editorial introduction to this volume). With-out waxing lyrical on the presumed solidarity and harmony of island communities, thick layers of bonding social capital on small islands can build clannish and resilient communities which deploy intensive networks; these habitually offer opportunities for satisfaction beyond state or market (Baldacchino 2005, Bray and Packer 1993, Lowenthal 1987, pp. 38–39). However that may be, we can look to islands for examples of how the creation of commons and commoning can provide strategies for development that manoeuvre through the false choice of gentrification or decline.

We mentioned gentrification research highlighting rehabilitation of built environments with-out displacement, non–profit community development corporations, community land trusts, millenary traditions of communal land tenure, and exercising the right to the city. In a critique of island development models, Clark (2013) argued for claiming and exercising "the right to the island" as an alternative to gentrification (associated with the PROFIT and SITE models; Baldacchino 2006b, McElroy 2006) or decay (associated with the MIRAB model; Bertram and Watters 1985). Here, we will briefly draw attention to some examples of island commons and commoning that, having avoided both gentrification/displacement and decay, do not fit easily into either of these models.

Community Land Trusts and Heritage Trusts have been central to the commoning of a grow-ing number of islands, most notably in the USA (Davis 2010, Reedy-Maschner and Maschner 2013), Canada (Bunce and Aslam 2016) and Scotland (Mackenzie et al. 2004). It should be noted however that related phenomena exist in other places under a variety of names and legal formats. Community Land Trusts often focus on, but are not necessarily limited to, affordable housing. Protection of livelihoods and culture from speculative displacement often go hand in hand with the protection of affordable housing. Community-held land protects the land from both market forces and the whim of current governments, as the land is, in a sense, de-commodified. Decisions on how to use the land are consequently very different from contexts where land is either treated above all as a financial asset (i.e. under constant pressure from financial interests to put it under its so-called 'highest and best use'), or under the control of an engineering, high-modernist state (Scott 1998).

Eigg

The Isle of Eigg (current population about 100) off the west coast of Scotland, is one of the more high-profile examples of community buyouts to establish a form of community trust ownership.

The process of this successful community struggle against the false choice of gentrification or decline has been thoroughly described and analysed in a doctoral thesis (Morgan 1998) and a highly praised book (McIntosh 2001). The process on Eigg supports the insight that commons require commoning. It cannot be the work of one passionate soul, but rather build on the bottom-up practices of community cooperation and solidarity, including people and organisations beyond the community. Twenty years on, the island community has clearly managed to navigate a development path based on community ownership and deep local democracy that avoids both externally driven gentrification and economic decline. Housing costs are about half the market level of affordable housing elsewhere in the region, providing residents with the freedom "to pursue less money-oriented goals" (Barkham 2017). Island residents enjoy security of tenure, encouraging use value oriented investments in the built environment. Tourism is a key livelihood, but remains under local control and is not under pressure to grow or to provide returns to absentee owners or dividends to shareholders: Eigg would not fit well in the SITE (Small Island Tourism Economies) model of island development conducive to island gentrification.

North Harris, Gigha, Ulva. . .

The example of Eigg has been followed by other islands in the Hebrides with considerably larger populations. North Harris, with a population of about 800, was well on a path of decline as population halved between 1951 and 2001, services were discontinued or closed, schools closed and jobs disappeared. The island was purchased by the community in 2003, forming the North Harris Trust, the community body which now manages the estate on behalf of the people of North Harris. The trust is engaged in projects dealing with provision of affordable housing, provision of office accommodation, employment opportunities, land and deer-management agreements, infrastructure development, improving access to more remote areas for visitors and locals, and installing renewable energy (Community Land Scotland 2017). The Isle of Gigha followed on the heels of North Harris, founding a Community Land Trust in 2004. Currently a community buyout aiming for land trust is being pursued for the Isle of Ulva (McIntosh 2017). The word seems to be out: communities thrive controlling their own development, avoiding continued decline and displacement by gentrification.

Orcas, Lopez, Waldron, San Juan

Similar uses of community trusts as a foundation for island community development are found in the San Juan Islands, Washington, USA, including the islands of Orcas, Lopez, Waldron and San Juan. These are "widely recognized as some of the most successful rural affordable housing CLTs in the U.S." (SIBAC 2017). Here, the focus is on affordable housing, reflected in the name 'Community Home Trusts' (e.g. OPAL CLT 2018). But the achievements regarding locally generated development are similar.

Pongso no Tao

Some islands have only recently been exposed to market exchange and state redistribution, having relied entirely on community reciprocity as the mechanism of integration for millennia. On the island of Pongso no Tao, southeast of Taiwan, there are elderly today who grew up with no experience or understanding of either money or state authority. Here, as in many other islands, commoning remains a dominant form of social reproduction amongst an indigenous community, in spite of colonialism having torn deeply into the fabric of Tau culture (Clark and

Tsai 2012, Tsai and Clark 2014). Learning from the experiences of other places, the Tau people are organising to reclaim control over development on the island in ways that secure cultural heritage and livelihoods.

Conclusion

Since market exchange and state redistribution are not capable of entirely displacing community reciprocity, the commons are not "something that existed once upon a time that has since been lost". Rather, they are "continuously being produced. The problem is that it is just as continuously being enclosed and appropriated by capital in its commodified and monetised form, even as it is being continuously produced by collective labour" (Harvey 2012, p. 77). This is what many island communities have experienced. Resistance to the false choice of gentrification (be it state-led or the ball-and-chain of the market) versus decay would do well to include strategies intended to secure viable traditions and form new ways of building community reciprocities through the practices of commoning.

There *are* alternatives to gentrification or decay, namely commons and commoning. Island communities do not have to choose between gentrification or decay, nor between the acronym models of PROFIT, SITE or MIRAB. There are development paths that (re)invigorate capacities that run deep in island histories, while benefiting from exchanges beyond the island. These paths will unavoidably need to include market relations and involve state redistribution, but do so in ways conducive to maintaining commons and the livelihoods these provide.

The examples briefly presented here do not represent mainstream island development. They do not fall easily into the most commonly advanced models for island development. But they do show that island communities are not bound to the false choice of gentrification or decline. They provide spaces of hope that community reciprocity is capable of integrating local economies in coevolution with – rather than being displaced by – market exchange and state redistribution.

References

Atkinson, R. (2000) The hidden costs of gentrification: displacement in Central London. *Journal of Housing and the Built Environment*, 15(4), 307–326.

Atkinson, R. (2002) Does gentrification help or harm urban neighbourhoods? An assessment of the evidence-base in the context of the new urban agenda. Centre for Neighbourhood Research, Paper 5. Retrieved from: www.bris.ac.uk/sps/cnrpapersword/cnr5pap.doc.

Baldacchino, G. (2004) The coming of age of island studies. *Tijdschrift voor Economische en Sociale Geografie*, 95(3), 272–284.

Baldacchino, G. (2005) The contribution of 'social capital' to economic growth: lessons from island jurisdictions. *The Round Table: Commonwealth Journal of International Affairs*, 94(378), 31–46.

Baldacchino, G. (2006a) Islands, island studies, Island Studies Journal. *Island Studies Journal*, 1(1), 3–18.

Baldacchino, G. (2006b) Managing the hinterland beyond: two ideal-type strategies of economic development for small island territories. *Asia Pacific Viewpoint*, 47(1), 45–60.

Balzarini, J., and Shlay, A. (2016) Gentrification and the right to the city: Community conflict and casinos. *Journal of Urban Affairs*, 38(4), 503–517.

Barkham, P. (2017) This island is not for sale: how Eigg fought back. *Guardian*, 26 September. Retrieved from: www.theguardian.com/uk-news/2017/sep/26/this-island-is-not-for-sale-how-eigg-fought-back.

Bauman, Z. (1998) *Globalisation: the human consequences*. Cambridge, Polity.

Bertram, G., and Watters, R.F. (1985) The MIRAB economy in South Pacific microstates. *Pacific Viewpoint*, 26(4), 497–519.

Block, F., and Somers, M.R. (2014) *The power of market fundamentalism: Karl Polanyi's Critique*. Cambridge, MA, Harvard University Press.

Blomley, N. (2002) Mud for the land. *Public Culture*, 14(3), 557–582.

Bollier, D. (2014) *Think like a commoner: a short introduction to the life of the commons.* Gabriola Island, Canada, New Society.

Bollier, D., and Helfrich, S. (2012) *The wealth of the commons: a world beyond market and state.* Amherst, MA, Levellers Press.

Bouchard, K. (2004, 26 April) A change in (home) values. *Portland Press Herald.*

Bray, M., and Packer, S. (1993) *Education in small states: concepts, challenges, and strategies.* London, Pergamon.

Bridge, G., and Dowling, R. (2001) Microgeographies of retailing and gentrification. *Australian Geographer,* 32(1), 93–107.

Bunce, S., and Aslam, F.C. (2016) Land trusts and the protection and stewardship of land in Canada: exploring non-governmental land trust practices and the role of urban community land trusts. *Canadian Journal of Urban Research,* 25(2), 23–34.

Cameron, S. (2003) Gentrification, housing redifferentiation and urban regeneration: 'going for growth' in Newcastle-upon-Tyne. *Urban Studies,* 20(12), 2367–2382.

Carmon, N., and Hill, M. (1988) Neighbourhood rehabilitation without relocation or gentrification. *Journal of the American Planning Association,* 54(4), 470–481.

Choi, M., Van Zandt, S., and Matarrita-Cascante, D. (2017) Can community land trusts slow gentrification? *Journal of Urban Affairs,* DOI:10.1080/07352166.2017.1362318.

Clark, E. (2005) The order and simplicity of gentrification: a political challenge. In R. Atkinson and G. Bridge (eds.), *Gentrification in a global context: the new urban colonialism,* London, Routledge, pp. 256–264.

Clark, E. (2009) Island development. In R. Kitchin and N. Thrift (eds.), *International Encyclopedia of Human Geography,* 5, Oxford, Elsevier, pp. 607–610.

Clark, E. (2013) Financialisation, sustainability and the right to the island: a critique of acronym models of island development. *Journal of Marine and Island Cultures,* 2(1), 128–136.

Clark, E., Johnson, K., Lundholm, E., and Malmberg, G. (2007) Island gentrification and space wars. In G. Baldacchino (ed.), *A world of islands: an island studies reader,* Canada and Malta, Institute of Island Studies and Agenda Academic, pp. 481–510.

Clark, E., and Tsai, H.-M. (2012) Islands: Ecologically unequal exchange and landesque capital. In A. Hornborg, B. Clark and K. Hermele (eds.), *Ecology and power: struggles over land and material resources in the past, present and future,* London, Routledge, pp. 52–67.

COHRE (2007) Fair play for housing rights. Mega-events, olympic games and housing rights: opportunities for the olympic movement and others. Centre on Housing Rights and Evictions. Retrieved from: http://www.ruig-gian.org/ressources/Report%20Fair%20Play%20FINAL%20FINAL%20070531.pdf

COHRE (2009) One world, whose dream? Housing rights violations and the Beijing olympic games. Centre on Housing Rights and Evictions Retrieved from: http://www.crin.org/en/docs/One_World_Whose_Dream_July08%5B1%5D.pdf.

Community Land Scotland (2017) North Harris Trust. Retrieved from: www.communitylandscotland.org.uk/members/north-harris-trust.

Dale, G. (2010) *Karl Polanyi: the limits of the market.* Cambridge, Polity.

Davidson, M., and Lees, L. (2005) New-build 'gentrification' and London's Riverside renaissance. *Environment & Planning A,* 37(7), 1165–1190.

Davis, J.E. (2010) *The community land trust reader.* Cambridge, MA, Lincoln Institute of Land Policy.

DeFilippis, J. (2004) *Unmaking Goliath: community control in the face of global capitalism.* London, Routledge.

Dreifus, C. (2004, March 21) Vieques: an island's ship comes in? *New York Times.*

The Eagles (1976) The Last Resort, *Hotel California,* Elektra/Wea.

Fighting the Gentrification of the North Shore of Staten Island (2017) Retrieved from: www.facebook.com/Fighting-the-Gentrification-of-the-North-Shore-of-Staten-Island-197593940414348/.

Garcia Herrera, L.M., and Smith, N. (2005) Gentrification and tourism in Santa Cruz de Tenerife. Paper presented at 4th International Conference of Critical Geography, Mexico City, January.

Glass, R. (1964) *London: aspects of change.* London, Centre for Urban Studies and MacGibbon and Kee.

Hägerstrand, T. (1986) Den geografiska traditionens kärnområde [The Core of the Geographical Tradition], *Sydsvenska Geografiska Årsbok,* 62(1), 38–43.

Hall-Arber, M., Dyer, C., Poggie, J., McNally, J., and Gagne, R. (2001) *New England's fishing communities.* Cambridge, MA, MIT Sea Grant College Program. Harvey, D. (1973). *Social justice and the city.* London, Edward Arnold.

Harvey, D. (1989) From managerialism to entrepreneurialism: the transformation in urban governance in late capitalism. *Geografiska Annaler Series B: Human Geography,* 71(1), 3–17.

Harvey, D. (2003a) *The new imperialism.* Oxford, Oxford University Press.

Harvey, D. (2003b) *Paris, capital of modernity*. London, Routledge.

Harvey, D. (2006) Neo-liberalism as creative destruction. *Geografiska Annaler: Series B, Human Geography*, 88(2), 145–158.

Harvey, D. (2010) *The enigma of capital and the crises of capitalism*. London, Profile.

Harvey, D. (2012) *Rebel cities: from the right to the city to the urban revolution*. London, Verso.

Horton, T. (2005, October 25) Smith Island harbour feels threat of gentrification. *Baltimore Sun*.

Jackson, R. (2006) Bruny on the brink: governance, gentrification and tourism on an Australian island. *Island Studies Journal*, 1(2), 201–222.

Jonsson, P. (2002, January 29) A fight to keep an island's Black heritage. *Christian Science Monitor*.

Kloosterman, R., and an der Leun, J. (1999) Just for starters: commercial gentrification by immigrant entrepreneurs in Amsterdam and Rotterdam neighbourhoods. *Housing Studies*, 14(5), 659–677.

Lees, L. (2003) Super-gentrification: The case of Brooklyn Heights, New York City. *Urban Studies*, 40(12), 2487–2509.

Lowenthal, D. (1987) Social features. In C. Clarke and A. Payne (eds.), *Politics, security and development in small states*, London, Allen and Unwin, pp. 26–49.

Lund Hansen, A. (2003) Urban space wars in 'wonderful Copenhagen': uneven development in the age of vagabond capitalism. In B. Petersson and E. Clark (eds.), *Identity dynamics and the construction of boundaries*, Lund, Nordic Academic Press, pp. 121–142.

McElroy, J.L. (2006) Small island tourist economies across the life cycle. *Asia Pacific Viewpoint*, 47(1), 61–77.

Mackenzie, A.F.D., MacAskill, J., Munro, G., and Seki, E. (2004) Contesting land, creating community, in the Highlands and Islands, Scotland. *Scottish Geographical Journal*, 120(3), 159–180.

McIntosh, A. (2001) *Soil and soul: people versus corporate power*. London, Aurum Press.

McIntosh, A. (2017, December 3) Why Ulva? Why land reform? *Bella Caledonia*, pp. 3–7.

Michaels, F.S. (2011) *Monoculture: how one story is changing everything*. Toronto, ON, Red Clover Press.

New Zealand Herald (2017, 5 January) White Man's island? Gentrification on Waiheke causes social tensions, board member says. Retrieved from: www.nzherald.co.nz/nz/news/article.cfm?c_id=1&objectid=11777178.

OPAL CLT (2018) OPAL (Of People And Land) Community Land Trust. Retrieved from: www.opalclt.org/about/history/.

Ostrom, E., Burger, J., Field, C.B., Norgaard, R.B., and Policansky, D. (1999) Revisiting the commons: local lessons, global challenges. *Science*, 284, 278–282.

Peck, J. (2010) *Constructions of neoliberal reason*. Oxford, Oxford University Press.

Phillips, M. (1993) Rural gentrification and the processes of class colonisation. *Journal of Rural Studies*, 9(2), 123–140.

Phillips, M. (2005) Differential productions of rural gentrification: illustrations from North and South Norfolk. *Geoforum*, 36(4), 477–494.

Polanyi, K. (1968) *Primitive, archaic and modern economies: essays of Karl Polanyi*. Boston, MA, Beacon Press.

Polanyi, K. (1977) *The livelihood of man*. New York, Academic Press.

Polanyi, K. (2001 [1944]) *The great transformation: The political and economic origins of our time*. Boston, MA, Beacon Press.

Porteous, J.D. (1988) Topocide: The annihilation of place. In J. Eyles and D.M. Smith (eds.), *Qualitative methods in human geography*, Cambridge, Polity Press, pp. 75–93.

Porteous, J.D., and Smith, S.E. (2001) *Domicide: the global destruction of home*. Montreal, QC, McGill-Queen's University Press.

Reedy-Maschner, K.L., and Maschner, H.D.G. (2013) Sustaining Sanak Island, Alaska: A cultural land trust. *Sustainability*, 5, 4406–4427.

Robinson, T. (1996) Inner-city innovator: the non-profit community development corporation. *Urban Studies*, 33(9), 1647–1670.

Rodman, M.C. (1987). *Masters of tradition: consequences of customary land tenure in Longana, Vanuatu*. Vancouver, BC, Canada, UBC Press.

Román, I. (2003, 4 June) New type of boom hits island. *Orlando Sentinel*.

Sayer, A., and Walker, R. (1992) *The new social economy: reworking the division of labour*. Oxford, Blackwell.

Scott, J.C. (1998) *Seeing like a state: how certain schemes to improve the human condition have failed*. New Haven, CN, Yale University Press.

Shao, Q. (2013) *Shanghai gone: domicide and defiance in a Chinese megacity*. Lanham, MA, Rowman & Littlefield.

Shilling, G. (2004) Yankee go home! *This Magazine*, July–August.

SIBAC (2017) Creating community: providing affordable housing on the San Juan Islands. Community Land Trusts & Rural Development Project. Retrieved from: www.bcruralcentre.org/wp-content/uploads/2017/01/San-Juan-Island-CLTs-Case-Studies.pdf.

Slater, T. (2004) North American gentrification? Revanchist and emancipatory perspectives explored. *Environment and Planning A*, 36(7), 1191–1213.

Slater, T. (2006) The eviction of critical perspectives from gentrification discourse. *International Journal of Urban and Regional Research*, 30(4), 737–757.

Slater, T. (2009) Missing Marcuse: on gentrification and displacement. *City*, 13(2), 292–311.

Smith, N. (2002) New globalism, new urbanism: Gentrification as global urban strategy. *Antipode*, 34(3), 427–450.

Snyder, R. (2006) Maine's working waterfront Coalition. Retrieved from: www.islandinstitute.org/programs.asp?section=workingwaterfront.

Strindberg, A. (2005 [1890]) *I Havsbandet* [By the open sea]. Stockholm, Natur & Kultur.

Tsai, H.-M., and Clark, E. (2014) Island commoning: social practices for surviving eco-colonialism on Pongso-no-Tau (Orchid Island, Taiwan). Paper presented at the American Association of Geographers Annual Meeting, 8–12 April, Tampa, Florida.

Yagley, J., Lance, G., Moore, C., and Pinder, J. (2005) *They paved Paradise: gentrification in rural communities*. Washington, DC, Housing Assistance Council.

20

ARTIFICIAL ISLANDS AND ISLOPHILIA

Klaus Dodds and Veronica della Dora

Introduction

In his accounts of pre-war life on Corfu and post-war Rhodes, the English writer Lawrence Durrell identifies a hitherto unrecognised condition called 'islomania', which has the effect of making islands irresistible to its sufferers. In *Reflections on a marine Venus*, the author offers his readers an 'anatomy of islomania', a journey of self-exploration and spiritual evaluation. The island was the perfect setting because it represented 'visionary intimations of solitude, of loneliness, of introspection because at heart everyone vaguely feels that the solitude they offer corresponds to his or her inner sense of aloneness' (Durrell 1943, p. 3). For the young and adventurous Durrell, the islands of Greece were places to experiment; not only in terms of self-development but also his craft as a poet, travel writer and novelist.

Fascinated by ancient Greece and Shakespeare's last play *The Tempest* (a play about a remote island, shipwreck, and the state of the human soul), Durrell's ruminations in *Prospero's Cell* capture well the author's tortured relationship with the island and the self, as he had to flee the islands of Greece in 1941 when news broke of a German invasion (Durrell 1945). Exiled for four years, like Prospero in *The Tempest*, and hungry for his former island life, Durrell eventually returned to Rhodes as a post-war information officer working for the British government who exercised 'artificial' sovereignty over the Greek islands between 1945 and 1947. Durrell's return to Greece formed the basis for *Reflections on a marine Venus*. Here he argues that the island *per se* holds out the promise for better things from life, and perhaps this became all the more poignant for the writer who lived with a daughter bedevilled with mental illness. In 1952, he moved to Cyprus to take up a position as a teacher. Local political events proved to be incommensurate with family life, as communal violence erupted between Greek and Turkish communities. British colonial authority became increasingly tenuous, as Durrell recorded in his 1957 memoir of the island, *Bitter lemons of Cyprus* (Durrell 1957). Unfortunately, Cyprus proved to offer only temporary sanctuary for Durrell and his young daughter. Their troubled and possibly incestuous relationship later on in life may well have contributed to her suicide at the age of 33 in London, on the island of Britain.

D. H. Lawrence had the Scottish writer Compton Mackenzie and his musings on the harsh Scottish Hebrides in mind when he penned his short story, 'The Man who loved islands'

(Lawrence 1928). In contrast, Durrell's novelist imagination offers us, his readers, a series of sensual and affecting evocations of social and cultural life with "flat alluvial coastlines" and the "long carved coastlines" separating other human activities transiting through the Mediterranean. In Lawrence's literary world, the island is a reassuringly fixed territory with largely clear-cut divisions. But, for Durrell, islands like Cyprus endured the bedlam of ethnic and communal strife, with deep social and geopolitical qualities of the place itself nested within a beguiling unitary geography (Sozen 2013). The battle over Cyprus was one that many states and communities recognised as about rival projects regarding self-determination rather than geo-formation. Communities may come and go, but Cyprus, as a particular geological outcome of a collision between the southern margin of the Anatolian Plate and the African Plate, will endure.

While tectonic forces, climate forcing, and episodes of advancing and retreating ice sheets would appear to have played little part in their creation, artificial islands have been integral to human settlement and development. Regardless of their material provenance, a medley of adventurers, dreamers, runaways, settlers, commercial opportunists and geopolitical plotters have been enchanted with their possibilities. Islands have often acted as natural and artificial laboratories: places to identify and audit life and its evolution (Dodds and Royle 2003). While many of us are well-versed with stories of islands that do not exist – including fantasy and 'fake' islands – the figure of the artificial island continues to be 'harvested' and peddled for its creative potential (Sheller 2007). Today, more than ever, it performs as a proverbial promised land.

From the artificial island projects that are to be found in such offshore gated communities as in Dubai, to such geopolitical posturing and land reclamation projects underway in the South China Sea, we find plenty of evidence of what Durrell terms as "indescribable intoxication" (Durrell 1943, Stephens 2016). As with their natural counterparts, or perhaps even more than them, human-made islands are spaces of desire and possession. Self-enclosure and separation from the continent feed dreams of escapism, control and autonomy. At the same time, however, artificial islands are also different from other islands. Hybrid creations of water and concrete, of sand and technology, they problematise traditional understandings of nature and wilderness as external to (or separated from) humans (Cronon 1995, Whatmore 2002). Artificial islands are expressions and icons of the Anthropocene, a term coined to denote the present epoch, in which many geologically significant conditions and processes are profoundly altered by human activities (Crutzen and Stoermer 2000).

This chapter initially attempts to define artificial islands as a category of geographical features, while at the same time acknowledging the fluidity of the boundaries between artificial and natural (islands). What makes the artificial island distinct and different from a natural island is a key theme of this section. Our second section turns to a more geopolitical-legal register and considers how artificial islands have been caught up in distinct projects designed to reinforce sovereign rights in the disputed waters of the South China Sea. At the heart of these earthly machinations are the entitlements that coastal states derive from the United Nations Law of the Sea Convention and the distinctions made between natural and artificial islands. It would be no exaggeration to claim that China, Philippines and Vietnam are gripped by the exclamatory potential of artificial islands: they may not love them, but they certainly see a great deal of infrastructural and legal-geopolitical value in their existence. Thereafter, we interrogate the role of artificial islands in primarily the Western cultural imagination and the roots of their promissory, utopian potential. Finally, we conclude the chapter by discussing why artificial islands continue to fascinate and inspire our geographical imaginations as we face the accumulating consequences of global warming and further earthly change.

What is an artificial island?

There seems to be nothing ambiguous about islands. Sixteenth-century Tuscan polygraph Thomaso Porcacchi called an island "that land that is surrounded by water all around; I mean, that which is separated and divided from the mainland and is encircled by the sea" (Porcacchi 1572, prohemio, n.p.). Challenging Isidore of Seville's 7th-century etymology, according to which the word island came from the phrase '*in salo*' (in the sea) (Is. XIV.6), Porcacchi nonetheless extended his definition to portions of land encompassed by other bodies of water, such as lakes. Today the Cambridge and Oxford English Dictionaries both define an island simply as "a piece of land completely surrounded by water". In other words, islands are unambiguously defined by their coastlines, by the interplay between the land and the water, between closure and openness (Ronström 2009).

When it comes to artificial islands, however, contours start to blur. An artificial island is commonly understood as "an island that has been constructed by people, rather than formed by natural means" (en.wikipedia.org). Yet, to what extent is human intervention needed to make an island artificial? Are artificial islands purely human-made objects, or are they pre-existing natural forms modified by humans? A brief search on Google, or even a simple glance at Wikipedia's list of artificial islands, reveals a variety of typologies, shapes and sizes: from the self-proclaimed principality of Sealand, a 0.004 km² concrete platform off the coast of Suffolk, UK (see Figures 20.1a and 20.1b), to the 970 km² Flevopolder in the Netherlands, the largest island formed by reclaimed land in the world (see Figure 20.2). Artificial islands can thus support a single building and entire cities alike. They can be inhabited or uninhabited. They can be permanent or temporary and they can involve both human and natural interventions, materials and processes. And while the shapes of their coastlines can be akin to those of their natural counterparts, in most cases artificial islands bear the imprint of human action: from the minimalist design of oil drilling platform-islands (e.g. Northstar, off Prudhoe Bay in Alaska), or the straight lines of airport islands such as Kansai in Japan and their modernist aesthetics of efficiency, to the playful fractal shapes of Dubai's Palms and their likes (see also Figure 20.11).

Figures 20.1a and 20.1b   The 'Principality of Sealand', and its location off the coast of Britain.

Sources: Photo by Richard Lazenby, Wikimedia Commons, https://commons.wikimedia.org/wiki/File:Sealand afterfire2.JPG. Locator map generated by DEMIS World Map Server™ and edited by Chris 73 / Wikimedia Commons, https://commons.wikimedia.org/wiki/File:Map_of_Sealand.png.

Figure 20.2   The Dutch island of Flevopolder, the world's largest artificial island.
Source: Corinne Schulze, www.cschulze.com.

More intriguingly, as a category, artificial islands encompass both islands created by humans 'from scratch' and pre-existing 'quasi-islands' – such as peninsulas, submerged seas or seamounts – transformed into islands through human interventions such as land reclamation, extension of pre-existing islands, or canal construction. The former category includes instances ranging from steel and concrete platforms like Sealand solidly anchored to the ocean floor, to the islands off the coast of Dubai made of sand pumped up from it, or Montreal's Île Notre-Dame constructed over ten months out of 15 million tonnes of rock excavated for the city's underground metro system to celebrate Expo '67 (Fischer 2012, p. 38). One of the earliest references to wholly human-made artificial islands, however, comes to us from the Roman writer Pliny the Elder who noted in his *Natural History* (AD 47) that Friesian communities (now in the northeast of the Netherlands) built artificial mounds in order to establish communities safe from the tidal vagaries of the North Sea (Fischer 2012).

The other category, that of pre-existing quasi-islands turned into islands, is just as varied. Once again, Pliny writes of the channels of ancient Leukas, a peninsula on the Greek Ionian coast that was made into an island by the Corinthians in the 7th century BC (Fischer 2012, pp. 36–37). Turning to modern times, while the Peloponnese is still considered a peninsula, it technically became an island in 1893, when it was cut off from mainland Greece through the excavation of the Canal of Corinth. Likewise, and yet at a totally different scale, the fluvial island of Donauinsel in central Vienna came into being in the Danube when a canal was dug as part of a flood prevention system in 1970. Artificial islands can also be manufactured through other techniques, such as flooding. For example, Barro Colorado Island was formed when the waters of the Chagres River in the Panama Canal were dammed to form an artificial lake in 1913. When the waters rose, they flooded much of the existing tropical forest, whilst leaving certain hilltops as islands in the middle of the lake: the reverse of reclaimed islands (see Berry and Gillespie, this volume).

Attempting a definition of an artificial island thus opens up a broader ontological question: where do we set the boundaries between natural and human-made? No matter what their shape or level of human intervention, as a category artificial islands transcend modern dualisms and invite us to think beyond 'nature' and 'culture' (or technology) as separate ontological zones. Bruno Latour (1993) calls the continuous effort to create such ontological zones 'purification' and ascribes it to a distinctively Western modern way of thinking, which in a way parallels the process of "conceptual islanding" described by John Gillis. Western culture, Gillis argues, "not only thinks about islands, but thinks *with* them. . . . Western thought has always preferred to assign meaning to nearly bounded, insulated things, regarding that which lies between as a void" (2004, p. 1).

This modernist tendency to 'think *with*' islands touches upon the intrinsic appeal of the 'insular imagination'. The geopolitical rubric of modernity is literally grounded in the idea of a stable geophysical order in which territorial states can make claims to distinct portions of the Earth's terrestrial environment. Later interventions in international law both championed the idea of the self-determination of peoples and their inalienable rights to live within recognised boundaries. While communities have experienced the violence of invasion and domestic upheavals, which might have altered those boundaries and corresponding 'lines on the map', it was nonetheless assumed that the modern nation-state could adjust and accommodate territorial change including ownership of islands (Steinberg 2005). Underpinning all this was the idea that the earth's surface would fall under the jurisdiction of one country or another and that those 'territories' were stable and identifiable (Elden 2013).

While international legal and political theory and practice can and does accommodate territorial change, it has faced challenges, which make relationships between territory, fixity and authority contingent. It is a delightful irony that artificial islands openly challenge this 'insular' way of geo-political thinking. They seem to float precisely in that 'void' or 'grey area' surrounding Gillis' reassuring 'islands of the mind'. Disrupting the conceptual boundaries between nature and technology, artificial islands call into question the very idea of 'nature' and the geographical distribution of 'territory'. In William Cronon's words, they remind us that "'nature' is not nearly so natural as it seems" (1995, p. 25). Artificial islands call for non-dualistic frameworks and new vocabularies to describe and make sense of our contemporary world and our relationship with the environment (Whatmore 2002). In a way, they give visual and material shape to post-environmentalist attempts to replace "stable taxonomies, fixed boundaries, and essential identities with more flexible, hybrid and permeable categories actively generated through performance and practice" (Cosgrove 2008, p. 1863). The very thing that we might think of as fixed – such as the distinction between land and sea – turns out to challenge our modern geographical/geopolitical imagination of boundaries, distinctions and fixities (Agnew 2004).

Performance and practice destabilise the idea of nature as a fixed entity and of geographical objects (like mountains, islands and rivers) as the most permanent and immutable of all objects (Spkyman 1942, Farinelli 1992). A splendid example of hybrid geographical object generated through 'performance and practice' is Tiber island in Rome (see Figure 20.3). Tradition holds that the nucleus of this small boat-shaped island was formed by the grain from the fields of the Tarquins, which was thrown into the Tiber in great quantities after the expulsion of the kings:

> There is even now a conspicuous monument of what happened on that occasion',' . . . an island of goodly size consecrated to Aesculapius and washed on all sides by the river; an island which was formed, they say, out of the heap of rotten straw and was further enlarged by the silt which the river kept adding.
>
> *(Dion. V. 13; also Liv. II. 5)*

Figure 20.3   Isola Tiberina (Tiber Island), the only island on the Tiber River, in Rome, Italy.
Source: www.zero.eu.

In the 1st century AD, Tiber island was modelled to resemble a ship. Walls were put around the island; travertine facing was added by the banks to resemble a prow and stern, and an obelisk was erected in the midst of it in the fashion of a mast.

Other examples of artificial islands built through slow collaborations between human and non-human action are found the world over. For example, the islet of Our Lady of the Rocks in Montenegro was formed by a pre-existing bulwark of rocks and by the carcasses of sunken ships, although local tradition ascribes its formation to the progressive accumulation of stones left by pilgrims over the centuries (sees Figure 20.4). Instances of reefs, atolls, and even icebergs transformed into proper or 'floating islands' by local inhabitants and military planners span the Arctic, Atlantic, Pacific and Indian Oceans. In the midst of the Second World War, plans were hatched under the moniker Project Habakkuk to construct an artificial aircraft carrier/landing strip somewhere in the North Atlantic Ocean in order to aid and abet the flying range of Allied airplanes (cited in Olovsson 2015, p. 34). Intriguingly, the plan involved a mixture of wood pulp and ice to construct the object concerned. The plan was eventually abandoned on cost grounds in 1943 but it did not prevent further interest in developing 'floating islands' for the purpose of accruing wartime advantage. Later, with the onset of the Cold War, floating icebergs were used as sites for scientific stations and landing strips, as the Soviet and American navies eagerly accumulated scientific knowledge about the Arctic Ocean and the polar atmosphere (Dodds and Nuttall 2016).

Artificiality takes on a more troubling presence when our geographical focus shifts away from projects and plans attempting to 'game' geographical forces and objects for strategic advantage. What happens when appeals to the artificial are rooted in long-term resilience, even survival? The people of Langa Langa Lagoon and Lau Lagoon in Malaita, Solomon Islands, built about 60 artificial islands on the reef for defensive purposes (Stanley 1999, p. 895). Today, as rising sea levels threaten the Maldives, the lowest country in the world at just two to three metres above the sea, plans are under discussion for a series of floating islands made of concrete

Figure 20.4   Gospa od Škrpjela (Our Lady of the Rocks), Bay of Kotor, Montenegro.

Source: Photo by Scott Liddell, Edinburgh, UK, Wikimedia Commons, https://commons.wikimedia.org/wiki/
File:Gospa_od_%C5%A0krpjela_(Our_Lady_of_the_Rocks)_-_Bay_of_Kotor,_Montenegro_-_5_Aug._2010.jpg.

and polystyrene foam (Rasheed 2015). Named the Five Lagoons Project, the floating islands
are intended to offer a partial solution to the prospect of further sea level rise. Working with
a Dutch company, Dutch Docklands, the 7.5 million m² project envisages a series of developments including executive villas, a golf course, a hotel complex, and homes for residents of
the country's capital city, Malé. The US$500 million venture is premised on the islands being
anchored to the seabed after being towed from India and the Middle East where they will be
built. And if that was not sufficient, we can also point to the remarkable development where
a reclaimed island called Hulhumale in the Maldives was created between 1997 and 2002,
involving a Belgian joint venture company, Singapore consultants, Japanese funding and Dutch
reclamation technology (see Figure 20.5).

In the case of low-lying island states, in the advent of fears that their worlds will simply
'drown', the geopolitical order of modernity faces elemental challenges. Becoming artificial not
only makes sense in terms of human endurance but also points to the legal and political rights
that accrue to recognised nation-states with accompanying sovereign rights over resources and
territories. Do states lose their sovereign rights when there is no longer, for example, a land
territory capable of generating offshore rights to resources such as fishing and minerals? What
happens when there is no longer a geographical baseline to establish a territorial sea, let alone
an exclusive economic zone? Can the artificial island take the place of a previous geological
entity identified as an island? In order to answer these questions, we need to turn to the way in
which international maritime law intervenes in and on the natural and artificial geographies of
the Earth (Elden 2013).

Figure 20.5   Hulhumale island, Maldives, on the top left of this photo, attached to the (natural) island
which contains Ibrahim Nasir Airport, the main international gateway to the Maldives.

Source: Photo courtesy of Maldives Airports Company www.macl.aero/plus/.

Legalising artificial islands

Islands, natural and artificial, have long attracted contention and dispute and acted as a major
source of tension between neighbouring coastal states, and extra-territorial parties who might
be simply navigating through the waters off disputed islands. Parties can and do dispute not only
the formal sovereignty of islands but also argue over the physical and human-assisted capacity of
islands to generate entitlements to maritime resources on the seabed and in the water column.
Japan, for example, has also experimented in fast-growing coral in order to enhance the human
colonisation of the remote rocks called Okinotorishima (literally: Remote Bird Island), located
some 1,900 km southwest of Tokyo (Sakhuia 2011) (see Figure 20.6). Artificial islands and structures add further complexity and frisson, as countries engage in artificial island-building projects
in highly disputed waters such as the South China Sea. What makes these disputes over artificial
island-building so vexatious is whether submarine topography can be augmented with other
materials, both natural and artificial, in order to terra-form. To paraphrase, D. H. Lawrence, we
have examples of 'countries that love (artificial) islands', and none more so than China, which
has invested heavily in dredging and reclamation projects designed to augment and build upon
subterranean elevations.

The United Nations Convention on the Law of the Sea (UNCLOS) (1982, entry into force
1994) identifies three features that are relevant to any discussion of artificial islands: low-tide
elevation (LTE), rock and island. A LTE is a landmass which is only above the waterline at low
tide and thus submerged during high tide. If located outside the territorial sea of a coastal state
(up 12 nautical miles from the baseline), it is not capable of generating a territorial sea and an
exclusive economic zone. A rock, however, is different from a LTE because it is permanently
exposed but is not capable of accommodating human or economic life without intervention.
While it is possible for a 'rock' to generate entitlement to a territorial sea, it generates neither an
exclusive economic zone (EEZ) nor acquires continental shelf rights. An example would be the
uninhabited islet of Rockall in the North Atlantic Ocean, which was incorporated by the UK
in 1955. Ireland continues to contest the UK's claim to the rock and the 12 nautical mile sea
surrounding it. An island, unlike a rock, can sustain human and economic life and is entitled to
a territorial sea, a contiguous zone and an EEZ and continental shelf rights.

In international legal terms, an artificial island is not defined explicitly by UNCLOS but
it is reasonable to conclude that it would fail the test posed by Article 121: "An island is a

Figure 20.6   Okinotorishima atoll with its three islets encased in concrete, plus constructed observation
platform on stilts, Japan.

Source: Japan Wikipedia, https://ja.wikipedia.org/wiki/%E3%83%95%E3%82%A1%E3%82%A4%E3%83%AB:
Okinotorishima20070602.jpg.

naturally formed area of land, surrounded by water, which is above water at high tide." The
expression 'naturally formed' is pivotal and, in Article 60, the Convention makes it clear that
"artificial islands, installations and structures do not possess the status of islands". Artificial
islands, even if composed of natural objects such as sand, silt, rock and organic matter rather
than concrete and steel, cannot transmogrify into 'natural islands'. What counts is the process
rather than the materiality of island formation, and whether as such earthly rather than human
forces have been involved in their genesis. The USA, in particular, was instrumental in ensuring that artificial islands (such as the self-styled Principality of Sealand in the North Sea),
were not granted any maritime rights, throughout the tortuous negotiations over UNCLOS
in the 1970s.

The most dramatic example of artificial island-building projects is to be found in the disputed waters of the South China Sea. A semi-enclosed sea, the reefs, shoals, cays, rocks and
low-tide elevations that litter its waters are claimed and counter-claimed by a medley of states
including Brunei, China (and Taiwan), Indonesia, Malaysia, the Philippines and Vietnam. In
terms of scale and extent, China's activities continue to attract the lion's share of legal and media
interest. Invoking historic rights to the South China Sea, and developing maps that depict what
is termed the 'Nine dash line', now 'Ten dash line', the Chinese authorities contend that the
country is asserting its legal entitlements (Kazianas 2014). Hundreds of millions of tons of sand,
silt, coral and rock have been dumped on coral reefs and low-tide elevations for the express
purpose of creating new 'islands' and in some cases so obliterating the natural topography, making it difficult to ascertain the pre-history of the reef and shoal. The reclamation is been carried

out by a fleet of advanced dredgers. Between 2006 and 2016, it has been estimated that Chinese construction teams might have been responsible for around 8 million cubic metres of land reclamation (Stephens 2016).

Such reclamation work is essentially an act of geopolitical assertion and suasion: the reclaimed islands are bristling with port and communication services, landing strips and search and rescue facilities. Once established, the islands enable further Chinese air and maritime force projection and harassment of other parties conducting fishing and other activities in the region. Establishing small mainly military communities on these artificial islands is also designed to add further Chinese 'presence' in a sensitive region through which passes one third of the world's commercial shipping by volume. Moreover, the ecosystems of the reefs and shoals concerned bear the indelible brunt of a calculated attempt to secure legal entitlements to the surrounding waters. Land reclamation as a process was designed to attract the legal entitlements noted in Articles 60 and 121.

In July 2016, the Tribunal established under the Law of the Sea Convention issued an arbitration award in the case of the South China Sea, involving China and the Philippines. The Tribunal concluded that China's historic rights and 'Nine dash line' did not comply with UNCLOS and exceeds any marine entitlements that the country is capable of possessing. As the Tribunal concluded:

> 305. With respect to low-tide elevations, several points necessarily follow from this pair of definitions. First, the inclusion of the term "naturally formed" in the definition of both a low-tide elevation and an island indicates that the status of a feature is to be evaluated on the basis of its natural condition. As a matter of law, human modification cannot change the seabed into a low-tide elevation or a low-tide elevation into an island. A low-tide elevation will remain a low-tide elevation under the Convention, regardless of the scale of the island or installation built atop it.
>
> 306. This point raises particular considerations in the present case. Many of the features in the South China Sea have been subjected to substantial human modification as large islands with installations and airstrips have been constructed on top of the coral reefs. In some cases, it would likely no longer be possible to directly observe the original status of the feature, as the contours of the reef platform have been entirely buried by millions of tons of landfill and concrete. In such circumstances, the Tribunal considers that the Convention requires that the status of a feature be ascertained on the basis of its earlier, natural condition, prior to the onset of significant human modification. The Tribunal will therefore reach its decision on the basis of the best available evidence of the previous status of what are now heavily modified coral reefs.
>
> *(United Nations 2016)*

Modification does not make an island; rocks are rocks regardless of how hard one tries to 'scale jump' and experiment with earth moving and human habitation. The high tide features in the highly disputed Spratly Islands (see Figure 20.7) were 'rocks' and thus, under the provisions of Article 121, are not capable of generating more than a 12 nautical mile (22.2 km) territorial sea.

The Tribunal also noted something rather significant for how we might think of artificial islands and their presence within the exclusive economic zones of other coastal states. In reflecting on China's construction of artificial islands, installations and structures on Hughes Reef,

Figure 20.7 Land reclamation by China underway at Hughes Reef, Spratly Islands, in the South China Sea.
Source: CSIS/Asia Maritime Transparency Initiative. Used with permission.

Mischief Reef, and Subi Reef, the Tribunal found that the scale of Chinese activity interfered with the sovereign rights of the Philippines. As low-tide elevations, the three reefs were incapable of being appropriated by another party and thus were integral to the continental shelf on which Philippines enjoyed sovereign rights. China did not receive formal authorisation from the Philippines for its construction and reclamation activities. China is, under UNCLOS, allowed to have a presence in the exclusive economic zones and continental shelf regions of other coastal states as the jurisdiction of coastal states over artificial islands and structures in these areas is not without its limits.

Third parties can build installations and structures on the continental shelf of coastal states provided that they do "not interfere with the exercise of the rights of the coastal state" over its resources (Stephens 2016). The interplay between Articles 56 and 60 is significant because it qualifies the overwhelming right of the coastal state to manage and regulate artificial islands and structures within its EEZ and continental shelf region. Foreign militaries are entitled to develop, for example, discreet installations and structures, but what they are not entitled to do is to create artificial islands with the explicit purpose of appropriating natural resources and acquiring sovereign rights. And it is important to bear in mind that, when UNCLOS was negotiated in the 1970s, the USA – which is not a party to UNCLOS – was nevertheless adamant in retaining extra-territorial entitlements to construct military assets without needing the formal consent of coastal states. The rights of coastal states over their EEZs and continental shelf are limited and extra-territorial parties including militaries are allowed to conduct operations in those spaces.

Thus, in their detailed 2016 ruling on the South China Arbitration, the Tribunal ruled that China's artificial island-building activities in the Philippines EEZ were not lawful, and indeed intended to interfere with the ability of the coastal state in question to exercise its sovereign rights in areas such as fishing and oil and gas exploration. China stood accused of worsening the situation by accelerating its dredging and reclamation programme and aggravating marine environmental disruption. Yet, it is a moot point as to whether this arbitration will be respected by China. Constructing artificial islands is not illegal, but what the Tribunal made clear, however, is that scale and intent matters. (Xinhua News 2016).

The international legalities attached to artificial islands are part of an intriguing story of how law, geophysics and geopolitics intersect with one another. 'Natural islands' are not artificial, they are visible, and they are as a consequence assumed to be above sea level. Earlier manifestations of the law of the sea did not distinguish terribly carefully between artificial and natural islands. As technological capacities including land reclamation and dredging shifted alongside the strategic/resource projection of coastal states and third parties, so international legal arguments followed. Since the 1970s, far greater attention has been given to an array of artificial islands and structures including oil rigs, ships and other mobile and immobile platforms and how their establishment and usage raises issues for how the entitlements and rights of coastal states and third parties are affected. International maritime law, overall, struggles with the interface between artificial and natural, as artificial intervention might be necessary to ensure that an 'island' does not lose its status and entitlements.

Dystopias and utopias: artificial islands and the geographical imagination

The contested geopolitics of the South China Sea provides a particularly jaundiced view of why the creation of islands as opposed to rocks is cloaked in promissory potential. International maritime law can be 'gamed' even if artificial islands call into question the matter of which they are made, and the way in which they are imagined. We might regard recent developments in the South China Sea as decidedly dystopian, as rival parties seek to create and stabilise islands and their communities in order to establish a 'baseline' for sovereign rights to the sea and its resources to thrive.

Circulating between dystopian and utopian registers lies an extraordinary range of possibilities and responses to the artificial island. In between the romantic origins of Montenegro's Our Lady of the Rocks, the contested present of the South China Sea and the apocalyptic futures impinging on the Maldivian atolls, float a number of other projects such as 'unwanted artificial islands'; or rather, islands made of unwanted matter. Thilafushi, an island created in the early 1990s as Malé's municipal landfill, shares little with the edenic imageries of the neighbouring islands and the fancy designs of prospective floating golf courses and luxury marinas. It also shares little with the artificial islands of the South China Sea for that matter. But in the making of more everyday geographies, the Maldives' 'rubbish island' receives more than 330 tonnes of waste daily, which are constantly transforming its size and morphology. Not only do Thilafushi's mountains of garbage keep reaching higher and higher altitudes, but it has been calculated that the size of the island itself is rapidly increasing in surface area (Ramesh 2009). The island's area now encompasses some 120 acres (0.5 km^2) and will continue to expand thereafter because of the supply of rubbish is not likely to abate in the short to medium term (see Figure 20.8).

While Thilafushi is deemed the largest 'rubbish island' in the world, a far larger 'rubbish archipelago' floats in the Pacific Ocean. The 'Great Pacific Garbage Patch', or 'Pacific Trash Vortex', is an enormous collection of flotsam accruing in the circular currents of the Pacific Gyre north of the Hawaiian Archipelago. 'Estimated at one hundred million tons in floating mass, the mostly non-biodegradable waste in the North Pacific Gyre coagulates in suspended

THILAFUSHI - 7 SEPT EMBER 2011   *photography by: AHMED SHAN*

Figure 20.8   The garbage island of Thilafushi, Maldives.

Source: Photo by Ahmed Shan/ EcoCare Maldives, www.ecocare.mv.

animation' (Jackson 2012, p. 209). Indeed, scientists have likened it to a living organism, moving around "like a big animal without a leash" and causing every year the death of more than a million seabirds and over 100,000 marine mammals (Marks 2008). Whilst difficult to quantify, estimates of size range from 700,000 km² (about the size of Borneo) to more than 15 million km², or, in some media reports, up to "twice the size of the continental United States" (ibid.). While not visible from outer space, and diffuse in nature, it is suggestive nonetheless that critics have sought refuge in an insular vocabulary in order to mobilise efforts to clean up and reduce the 'garbage patch'.

Sea plastic pollution, intriguingly, has been recently imagined in the form of a 'Recycled Island'. Over the past five years, Dutch architects and engineers have been exploring the possibility of turning plastic garbage polluting seas, coasts and estuaries into floating urban parks and residential islands. A prototype is being built in Rotterdam's bay:

> The prototype should be built from hollow building blocks made by recycling the plastic waste from coastlines. These hollow elements will be the components from which Recycled Island will be constructed. . . . The creation of the prototype will illustrate the potential of (sea) plastic recycling, flood-proof living, and sustainable and self-sufficient housing. The prototype can eventually be on display in any major (harbour) city.
>
> *(Recycled Island Foundation 2017)*

Described by critics as "a doomed modernist symbol of a myopic narrative that attempts to balance capitalist growth with environmental control" (Jackson 2012, p. 208), Recycled Island betrays a deeper anxiety, that is Western culture's incapacity to think 'beyond islands'. What better form to express control over pollution than the apparently self-bounded space of an island? Inherent in the island, Baldacchino observes, is an obsession to control and fix: a tendency we have already described as emblematic of the geopolitics of modernity:

> To embrace an island is to embrace something that is finite, that may be encapsulated by human strategy, design or desire. . . . Being geographically defined and circular, an island is easier to hold, to own, to manage or to manipulate, to embrace and caress.
>
> *(Baldacchino 2005, p. 247)*

And it also easier to reconcile with a geopolitical imagination that is premised on the fantasy of bounded territories, fixed spaces and lines on the map that neatly articulate where one jurisdiction begins and another ends. While the existence of trans-national pollution clearly reminds us that this territorial fixation is not adequate, compacted solid rubbish facilitate this fantastic projection.

Artificial islands exacerbate all these decidedly modernistic characteristics. Alongside natural islands transformed into prisons and *lazzaretti*, there is a history of artificial dystopian islands and islets manufactured to contain and control external agents and potential threats to the everyday lives of urban societies including their rubbish, but also suspicious humans. The Japanese island of Dejima in Nagasaki Bay (see Figure 20.9), for example, was constructed in 1634 to contain

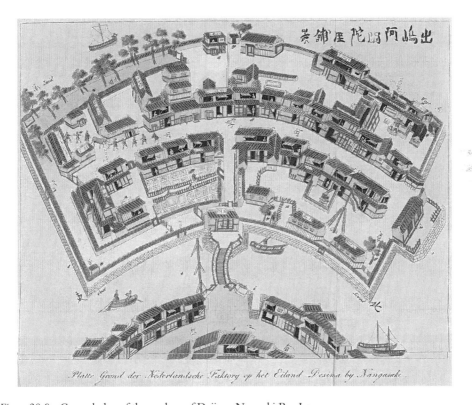

Figure 20.9   Ground plan of the enclave of Dejima, Nagasaki Bay, Japan.

Source: produced by Isaac Titsingh, 1824–25. Original in Koninklijke Bibliotheek, The Netherlands. Wikimedia Commons, https://commons.wikimedia.org/wiki/File:Plattegrond_van_Deshima.jpg.

Portuguese and then Dutch merchants who were forbidden to enter into direct contact with city-dwellers as part of the Edo Period's self-imposed isolationist policy; whereas Ellis Island in New York was created to serve as an immigration centre in the late 1800s and early 1900s, becoming in effect the portal to the USA (Fischer 2012, pp. 37–38). Bio-political projects were frequently geopolitical projects in the sense that the control of alien populations, refugees and displaced peoples was fundamentally spatial and territorial. Offshoring, however, has also become the preferred strategy of the wealthy as money travels away from the metropolitan territory in favour of a network of international financial outputs many of which being naturally formed islands with their decidedly artificial economies (Sheller 2007 and 2009, Urry 2014).

While some have worried about the bio-political and geopolitical purpose of human-made islands, others have sought hope in the manufacturing of communal virtue. Thomas More's Utopia itself was an artificial island (see Figure 20.10). Utopia, More explains, was originally

Figure 20.10 The island of Utopia, according to Thomas More.

Source: Libellus veer aureus ned minus salutaris quam festivus de optima reipublicae statu, deque nova insula Utopia, Louvain, Belgium, 1516. Original at the Wormsley Library, Oxford, UK. Available from Wikimedia Commons at: https://commons.wikimedia.org/wiki/File:Insel_Utopia.png.

attached to the mainland, but its king Utopus cut it off by digging a canal, so that its inhabitants could attain civic perfection. The story is likely to have been inspired by the ancient peninsula of Mount Athos, in the Aegean. According to Herodotus, after a massive shipwreck occurred off its stormy point in 492 BC, Xerxes, the king of the Persians, resolved to cut a canal on the neck of the peninsula, in order to remove the need for its circumnavigation. The canal collapsed soon after its excavation, leaving no visible trace. From being a temporary island, Athos returned to its original peninsular status. Nonetheless, the memory of its insular past survived through the centuries both through the secluded life of the Christian monks who have continuously inhabited Athos since the 9th century and through literary and graphic representations. Late Byzantine authors described the peninsula as a paradise on earth; Western Renaissance map-makers portrayed it in an oval insular shape, tenuously attached to the mainland through a thin isthmus, or even totally encircled by the sea, whereas the 'non-presence' of Xerxes' canal is still marked on modern maps (della Dora 2011, pp. 56–83). Strikingly, an island 'tenuously linked' 'to the mainland' would still appear to be a better prospect than the fate faced by low-lying islands in the Pacific and Indian Oceans.

As Baldacchino (2007, p. 166) notes, "a significant component of the contemporary intoxi-cating lure or fascination of islands has to do with the fact that they suggest themselves as *tabulae rasae*: potential laboratories for any conceivable human project, in thought or in action". Intriguingly, today similar experimental tropes pervade advertisements and promo-tional videos of the new artificial islands in the Arabian Gulf, which are in effect giant luxury gated communities. A 2013 advert of Palm Jumeirah properties in Dubai, for example, mar-kets the island as a social experiment that combines 'community living' with protection from perceived external threats:

> Community living is the trend of the moment, with people all over the world relocat-ing to gated or managed communities. . . . People move towards community living for various reasons: safety and security, leisure and recreation facilities, convenience and exclusivity.

Unlike the oft-cited tendency to seek solace in the vertical (Graham 2016), the artificiality of the horizontal is being appealed to either a proverbial 'ground zero' or a form of habitual regeneration.

Completed in 2006, Palm Jumeirah is the first and only currently inhabited of Dubai's three artificial palm-shaped islands originally planned (see Figure 20.11). The other projects were delayed or halted due to the collapse of property prices following the 2008 global financial cri-sis. A larger palm, Jebel Ali, is still under construction, whereas the third palm, Deira, originally designed to accommodate a population of 1.7 million by 2020, was scaled down to a remark-ably more modest 'Deira Island'. Measuring 5 by 5 km and connected to the mainland by a 300-metre bridge, Jumeirah is nonetheless already one of the world's largest artificial islands (Fischer 2012, p. 38). Over 500 families have already moved onto the island and become part of its exclusive experimental community.

Recent reporting suggests that at least one of Dubai's developments, The World, a complex of some 300 private artificial islands arranged in the shape of a world map and located 4 km off the coast, might be slipping back into the sea. The sand does not appear up to the task in question and the channels between the islands are collapsing. Remarkably, the company Dutch Dock-lands has sensed a business opportunity in the event that sand nourishment proves inadequate. They have, in 2016, offered up to the powerful Nakheel development company the prospect of building the 'Floating Proverb', a series of 89 floating islands that will in effect continue to

Figure 20.11   Palm Island, Dubai.

Source: Photo by Helmut Pfau, Wikimedia Commons, https://commons.wikimedia.org/wiki/File:PalmIslandDubai.JPG.

trade in the proposition that the territorial waters off Dubai can serve as a lively site of island experimentation. It remains a plan at this stage but might prosper as 'The World' project flounders in the water.

At a smaller scale, the Dutch Recycled Islands are also being imagined as social and environmental experiments. Intercepted by passive litter traps set on the estuaries of rivers, "the retrieved plastic is recycled to give new value to the river. With the plastics we make floating blocks to form new nature landscapes; floating parks." Intriguingly, variants on the project echo the 'social experiment' vision promoted by Dubai and include 'self-sufficient', 'sustainable' and 'flood-proof' habitats built around floating villas and family housing (Recycled Island Foundation 2017). A do-it-yourself version has already been built off the coast of Yucatan, Mexico. It took British artist Richart Sowa seven years and 150,000 recycled bottles to craft his own floating paradise, Joyxee island (see Figure 20.12).

Another project, Orsos, offers a choice of 'living together or alone' by changeable configurations of floating islands assembled along customised bridge systems:

> Island owners can use the park, for example, for an all year-round base station, with the summer months, however, spent alone in a remote bay. Hotel chains with a second bridge system can for example avoid rainy seasons and offer their guests an all year-round season.
>
> *(Orsos island 2017)*

Orsos islands synthesise the two island paradigms pioneered by Dubai: the Palms' 'experimental' communities and the seclusion of The World's islands. Each island in The World is available for

Figure 20.12   Joyxee, the floating island, made of and from trash.

Source: Richart Sowa. Still taken from video at: www.youtube.com/watch?v=GnLhWpy_nqI.

Figure 20.13   Orsos island hotel-residential parks. Available at: www.orsosisland.com/page/opportunities/
hotel-residential-parks.

purchase and transformation into one's private luxury Eden. Spectacular as seen from above, yet invisible from ground level, they are perhaps one of the most dramatic fulfilments of neo-liberal fantasies, and of their sinking (Jackson and della Dora 2009). Orsos combines these two visions – exclusive community living and individualism – by way of mobility (see Figure 20.13):

> The mobility of the Island, which is an excellent characteristic combined with all
> its unique and special features, allows the settling in seasonally restricted areas with

difficult weather conditions, such as the Caribbean known for its hurricanes, to relocate easily. The mobility assures the owner of safety and long-term functionality. Compared to a mainland real estate, the Islands retain their value long term and are cost-effective.

(Orsos island 2017)

What we are left with here is an incredibly diverse picture of how and where artificial islands have been put to work. For artists, however, this might also include opportunities to imagine artificial islands not as securitised archipelagos of privilege and wealth but as something rather than creative even disruptive to economic and political norms. Alex Hartley's Nowhereisland (Nowhereisland 2017) (see Figures 20.14a and 20.14b), for example, involved the discovery of a new island in the Svalbard archipelago in the Arctic. In 2012, a group of artists and writers organised to have the artificial island towed into international waters where it was declared as a new nation-state. After its journey back to UK waters, the island was dismantled in September 2013 and over 20,000 people took up the opportunity to sign up as 'citizens of Nowhereisland' – a contemporary Ou-topia. In a playful way, Hartley and his artistic team asked questions about how modern understandings of citizenship and territory are co-constituted by one another.

Predating Hartley's vision, in 1970 Robert Smithson conjured up a *Floating island to travel around Manhattan island* composed of a tug boat and barge planted with trees and rocks. Smithson's unrealised project was brought to life posthumously, 32 years after his death, by a non-profit artist organisation and the Whitney Museum. For some days in September 2005, this green artificial island orbited around Manhattan's artificial landscape of steel, glass and concrete. *Floating island* was no less subversive than Nowhere Island. If Hartley's island displaced territory, Smithson's displaced nature. The trees for his mobile miniature landscape, Smithson proposed, should be removed from Central Park and transferred to the barge. His project was 'a reversal of the island garden in the city, which moves to encircle the city' (Yusoff and Gabrys 2006, p. 445). In reversing the relationship between 'inside' and 'outside', Smithson's mobile artificial island once again called into question the boundaries between natural and artificial (see Figure 20.15).

The project has been likened to "voyages of discovery and loss, beginning with the Ark itself, which also put nature into a boat" (http://nymag.com/guides/fallpreview/2005/art/12862/). However, there is also a dystopian side to it. Barges are daily loaded with garbage from Manhattan and the surrounding areas. "To displace a miniature version of Central Park to a floating barge is to suggest the eventual slump of that park, and its possible disposability" (Yusoff and Gabrys 2006, p. 447): the reversal of the Dutch Recycled Island parks.

In their apparent playfulness, artificial islands can thus be subversive icons of the Anthropocene. At the same time, however, it is difficult not to conclude that for the rich and privileged the ultimate expression of earthly engineering is now living on one's own artificial island: a floating artificial island to be sure (Miéville 2007, Graham 2016). If we cannot get away from humanity by going either up and or underground terrestrial earth, then the only other choice baring space and underwater living appears to be to inhabit something human-made. As the anthropologist Michael Taussig (1999) might term it, artificial island constructions are a very 'public secret': communities like Orsos proffer a public vision of safety and privacy, rather than something that cannot be articulated in isolation and insulation from the rest of humanity.

Figures 20.14a and 20.14b Nowhereisland and 'Embassy' on Furzy cliff in Weymouth, UK, ready for the arrival of 'the Island' on 25 July 2012.

Sources: Alex Hartley/ https://gettingtoyoutidende.wordpress.com/2012/09/03/alex-hartley-nowhereisland-3/ and Chris Reay, Weymouth Beach B&B, www.pinterest.com/source/channelhotel.co.uk/.

Figure 20.15   Smithson's floating island garden idea brought to life by encircling Manhattan, USA, in 2005.

Source: Balmori Associates: www.balmori.com.

Conclusion

The artificial island as a feature is extraordinarily varied and foregrounds both the more-than-human qualities of earthly existence, as well as the sorts of phenomena that can materialise because of earthly forces themselves. Lakes, marshes and wetlands are capable of producing their own floating 'artificial islands' of mud, peat and aquatic communities. When sediment load and physical contours shift, rivers can manufacture their own islands of runoff and sand, some of which may endure or subsequently be overwhelmed by storms or rising water. Pliny the Elder referred to such changes in the physical environment as "dancing islands" (cited in Fischer 2012, p. 36), a term which delightfully pulls out the performativity, even playfulness, of Planet Earth. Perhaps, Smithson had Pliny in mind when he devised his plan for Floating Island.

Culturally, fancy artificial islands seem to be a typically 21st-century trend. Over the past decade, the uses, meanings and aesthetics of large artificial islands have shifted from pragmatic utilitarianism (e.g. the addition of a strip of land to accommodate an airport or further cultivable land, or to protect coastlines from erosion) towards a global market of cultural icons and building speculation. Between 2001 and 2008 over 20 countries put forth spectacular artificial island projects: from Qatar to Russia, from Thailand to Spain, from Korea to Slovenia, from the Netherlands to the Bahamas (Jackson and della Dora 2009). Only a tiny percentage of such projects has materialised. And yet, the trend continues. The point is not the realisation of these islands, but the idea, or rather their iconicity: coming in the most bizarre and eye-catching shapes, the goal of these speculative islands is to capture international attention and set aspiring global cities (or even countries) under the world's gaze, even if for a brief moment (Dematteis 2001). As Bella Dicks notes, "it is the symbol, not the information; [it is the medium and not the message,] which has become the actual resource of late modern economies" (2003, p. 34). As powerful cultural icons, artificial islands are key agents in contemporary global urban branding. One of the most glamorous recent examples of this speculative economy of 'phantom islands' is the UAE Real Madrid Resort Island, a US$1 billion soccer-themed resort on the artificial Marjan Island, including a 10,000-seat stadium by the sea, residential and retail properties, a 450-room five-star hotel, a Real Madrid Museum and, not least, a hologram stadium. The licensing agreement was scrapped soon after the launch of the project in 2013, making the resort island an immaterial, ephemeral fantasy; or rather, a hologram of itself (Fattah and Duff 2013).

On the other side of the spectrum is the solid concrete poured in tons over the contested rocks in the South China Sea. Geopolitically, any change in the physical landscape can provoke either expressions of anxiety or articulations of opportunism. We have for centuries accustomed ourselves, especially in the Western geopolitical imagination, to thinking about the Earth as a stable backdrop to the serious business of politics, law and society. Where political discourse has described a 'failing state', for example, it has done so when its domestic structures are alleged to have been overwhelmed by civil war, disease, and disorder. Nation-states are artificial creations but their artificiality was grounded in appeals to the 'natural borders' of rivers, mountains, seas, and other earthly features. Artificial islands have been used to garner legal and political advantage from a system premised on the notion that the environment could be used to sustain a national community. But now we face increasingly the prospect of some island states looking to artificiality as a way of holding on to their resources, sovereignty and territory. If a state fails environmentally rather than geopolitically, then what happens to its resource claims if that 'territory' is simply submerged? Going artificial becomes a perfectly rational response to a world in the geopolitical world of the anthropocene, where the natural and artificial increasingly depend on one another (Dalby 2017).

Artificial islands are thus one possible answer to the question: is there life after territory? Another might be to create artificial islands of humanity elsewhere. In the 1970s, futurists such as Carl Sagan imagined a world where human colonisation of Mars and even Venus might become a possibility. Intriguingly, 'terraforming', a word widely used by developers to describe the making of the artificial islands in the Gulf, was first coined in 1950 by American writer Jack Williamson and popularised by Sagan (Jackson and della Dora 2011, p. 99). As concerns mount about global warming, population increase, pollution and resource depletion, it is poignant how popular culture is returning again to the subject of planets such as Mars and the moon as extra-terrestrial *tabulae rasae* for humanity. The motion picture *The Martian* (2015), for example, explores the colonisation of Mars and how huge efforts are made to bring a remaining 'Martian' back home to earth (see review by Dargis 2015). While the film trades in the idea that Earth is worth returning to, the premise of the film remains that alternative spaces are needed for humanity. The allure of island territory does indeed continue, albeit this time imagined being elsewhere (Murphy 2013).

References

Agnew, J. (2004) *Geopolitics: re-visioning world politics*. London, Routledge.

Baldacchino, G. (2005) Islands: objects of representation. *Geografiska Annaler B*, 87(4), 247–251.

Baldacchino, G. (ed.) (2007) *A world of islands: an island studies reader*, Charlottetown, Canada and Luqa, Malta, Institute of Island Studies and Agenda Academic.

Cosgrove, D. (2008) Images and imagination in 20th-century environmentalism: from the Sierras to the Poles. *Environment and Planning A*, 40(8), 1862–1880.

Cronon, W. (1995) Introduction: in search of nature. In W. Cronon (ed.) *Uncommon ground: toward reinventing nature*, New York, W.W. Norton & Co., pp. 23–68.

Crutzen, P.J., and Stoermer, E.F. (2000) The 'anthropocene'. *IGBP Newsletter*, 41, 17–18.

Dalby, S. (2017) On 'not being persecuted': territory, security and climate. In A. Baldwin and G. Bettini (eds.), *Life adrift: climate change, migration, critique*, Lanham, MD, Rowman and Littlefield, pp. 41–58.

Dargis, M (2015) Review: In 'The Martian,' Marooned but Not Alone New York Times 1 October 2015 URL: https://www.nytimes.com/2015/10/02/movies/review-in-the-martian-marooned-but-not-alone.html (accessed 5th March 2018)

della Dora, V. (2011) *Imagining Mount Athos: visions of a holy place from Homer to World War II*. Charlottesville, VA, University of Virginia Press.

Dematteis, G. (2001) Shifting cities. In C. Minca (ed.), *Post-modern geography: theory and praxis*. Oxford, Blackwell, pp. 113–128.

Dicks, B. (2003) *Culture on display: the production of contemporary visitability*. London, Open University Press.

Dodds, K., and Nuttall, M. (2016) *The scramble for the Poles*. Cambridge, Polity.

Dodds, K., and Royle, S. (2003) Introduction: rethinking islands. *Journal of Historical Geography*, 29(4), 487–498.

Durrell, L. (1943) *Reflections on a marine Venus: a companion to the landscape of Rhodes*. London, Faber and Faber.

Durrell, L. (1945) *Prospero's cell: guide to the landscape and manners of the Island of Corfu*. London, Faber and Faber.

Durrell, L. (1957) *Bitter lemons of Cyprus*. London, Faber and Faber.

Elden, S. (2013) *The birth of territory*. Chicago, IL, University of Chicago Press.

Farinelli, F. (1992) *I segni del mondo: discorso geografico e immagine cartografica in età moderna*, Scandicci, Italy, Nuova Italia.

Fattah, Z., and Duff, A. (2013, 23 September) Real Madrid scraps $1 billion resort in U.A.E. sheikhdom. *Bloomberg*. Available from: www.bloomberg.com/news/articles/2013-09-22/real-madrid-scraps-1-billion-ras-al-khaimah-soccer-resort-plan.

Fischer, S. (2012) *Islands: from Atlantis to Zanzibar*. London, Reaktion Books.

Gillis, J. (2004) *Islands of the mind: how the human imagination created the Atlantic World*. New York, Palgrave Macmillan.

Graham, S. (2016) *Vertical*. London, Verso.

Isidore of Seville. (1911) *Isidori hispalensis episcopi etymologiarum sive originum*. English translation available from: http://penelope.uchicago.edu/Thayer/L/Roman/Texts/Isidore/14⋆.html.

Jackson, M. (2012) Plastic islands and processual grounds: ethics, ontology, and the matter of decay. *Cultural Geographies*, 20(2), 205–224.

Jackson, M., and della Dora, V. (2009) 'Dreams so big only the sea can hold them': man-made islands as anxious spaces, cultural icons, and travelling visions. *Environment and Planning A*, 41(9), 2086–2104.

Jackson, M., and della Dora, V. (2011) Spectacular enclosures of the hope: artificial islands in the Gulf and the present. In R. Shields, O. Park and T. Davidson (eds.), *Ecologies of affect: placing nostalgia, desire and hope*, Waterloo, ON, Wilfrid Laurier University Press, pp. 293–316.

Kazianas, H. (2014) China's 10 red lines in the South China Sea. *The Diplomat*, 1 July. Available from: http://thediplomat.com/2014/07/chinas-10-red-lines-in-the-south-china-sea/.

Latour, B. (1993) *We have never been modern*. Cambridge, MA, Harvard University Press.

Lawrence, D.H. (1928) *'The woman who rode away' and other stories*. London, Secker.

Marks, K. (2008, 5 February) The world's rubbish dump. *The Independent*. Available from: www.independent.co.uk/environment/green-living/the-worlds-rubbish-dump-a-tip-that-stretches-from-hawaii-to-japan-778016.html.

Miéville, C. (2007) Floating utopias: freedom and unfreedom of the seas. In M. Davis and D. Bertrand Monk (eds.), *Evil paradises: dreamworlds of neoliberalism*, New York, The New Press, pp. 251–263.

Murphy, A. (2013) Territory's continuing allure. *Annals of the Association of American Geographers*, 103(5), 1212–1226.

Nowhereisland (2017) An Arctic island travelled south and became a nation of global citizens. Available from: www.nowhereisland.org.

Olovsson, I. (2015) *Snow, ice and other wonders of water: a tribute to the hydrogen bond*. London, World Scientific.

Orsos island (2017) Orsos Island: your own little private island. Available from: www.youtube.com/watch?v=DmUr4Xqiopw.

Porcacchi, T. (1572) *L'isole più famose del mondo descritte da Thomaso Porcacchi da Castaglione Arretino e intagliate da Girolamo Porro Padovano*, Venice, Italy, Simon Gagliani.

Ramesh, R. (2009, 3 January) Paradise lost on Maldives' rubbish island. *The Guardian*. Available from: www.theguardian.com/environment/2009/jan/03/maldives-thilafushi-rubbish-landfill-pollution.

Rasheed, Z. (2015, 18 August). Plans underway to build floating luxury islands in the Maldives. *Maldives Independent*. Available from: http://maldivesindependent.com/business/plans-underway-to-build-floating-luxury-islands-in-the-maldives-116525.

Recycled Island Foundation (2017) Recycled island. Available from: http://recycledisland.com/.

Ronström, O. (2009) Island words, island worlds: the origin and meaning of words for islands in northwest Europe. *Island Studies Journal*, 4(2), 163–182.

Sakhuia, V. (2011) *Asian maritime power in the 21st century: strategic transactions in China, India and Southeast Asia*. Singapore, Institute of Southeast Asian Studies.

Sheller, M. (2007) Virtual islands: mobilities, connectivity and the New Caribbean spatialities. *Small Axe: A Caribbean Journal of Criticism*, 24(2), 16–33.

Sheller, M. (2009) Infrastructures of the imagined island: software, mobilities, and the architecture of Caribbean paradise. *Environment and Planning A*, 41(6), 1386–1403.

Sozen, A. (2013) Cyprus. In G. Baldacchino (ed.) *The political economy of divided islands: unified geographies, multiple polities*. Basingstoke, Palgrave Macmillan, pp. 102–118.

Spkyman, N. (1942) *America's strategy in world politics: the United States and the balance of power*. New York, Harcourt, Brace and Company.

Stanley, D. (1999) *The South Pacific handbook*. Chico, CA, Moon Publications.

Steinberg, P. (2005) Insularity, sovereignty and statehood: the representation of islands on portolan charts and the construction of the territorial state. *Geografiska Annaler B*, 87(4), 253–265.

Stephens, T. (2016) China's claims dashed in South China Sea arbitration. *LSJ: Law Society of NSW Journal*, 25, 73–75.

Taussig, M. (1999) *Defacement*. Stanford, CA, Stanford University Press.

UNCLOS (1982) Law of the Sea Convention. Available from: www.un.org/depts/los/convention_agreements/texts/unclos/unclos_e.pdf.

United Nations (2016, 12 July) In the matter of the South China Sea Arbitration: An arbitral tribunal constituted under Annex VII to the 1982 UNCLOS between the Republic of the Philippines and the People's Republic of China. Available from: https://pca-cpa.org/wp-content/uploads/sites/175/2016/07/PH-CN-20160712-Award.pdf

Urry, J. (2014) *Offshoring*. London, Polity.

Whatmore, S. (2002) *Hybrid geographies: natures, cultures, spaces*. London, Sage.

Xinhua News (2016, 13 July) Chinese foreign minister says South China Sea arbitration a political farce. Available from: http://news.xinhuanet.com/english/2016-07/13/c_135508275.htm.

Yusoff, K., and Gabrys, J. (2006) Time lapses: Robert Smithson's mobile landscapes. *Cultural Geographies*, 13(3), 444–450.

21

FUTURES

Green and blue

Graeme Robertson

Introduction

The 'green economy' is a form of development that addresses in a holistic way the multiple economic and environmental challenges confronting the planet. This concept has been rapidly picked up by many small island communities as evidenced by 50 Green Island case studies that the author compiled and presented on the **Global Islands Network (GIN)** website in 2012. (The websites for all references in **bold** are listed under Appendix 21.1.) These documented good practices related to renewable energy and eco housing; waste minimisation and recycling; water management and security; extensive agriculture and organic food production; transport; sustainable tourism and niche marketing; biodiversity and protected areas; integrated development planning; climate change mitigation and adaptation measures. The numerous online documents associated with these case studies are still averaging 1,000 downloads per month, indicating a high degree of interest. There is also growing anecdotal evidence from several European countries that, when an island starts taking some innovative approaches to developing sustainable economic activities and utilising the latest environmental technologies to combat climate change, then there is a corresponding upsurge and celebration of cultural awareness. These jointly create an even stronger identity and sense of place. This positive spin-off often leads to a complete turnaround in the fortunes of an island that was facing terminal decline due to out-migration and other adverse socio-economic factors.

Green investment can contribute to lowering demand for energy, water and other raw materials thus reducing the carbon footprint of production of goods and services. But some governments, academics and civil society organisations, especially in the developing world, would argue that the green economy is turning nature into merchandise, without addressing deeper problems like poverty and social inequality. Whilst many politicians laud the green economy, there are contradictions and U-turns in policy aplenty. The global energy industry is currently rushing headlong in two different directions, delivering record levels of low carbon investment at the same time as ploughing cash into continued oil exploration, Canadian tar sands extraction and fracking for shale gas. This tension between the competing visions of a low and high-carbon future (or, to put it another way, the fight between green growth and environmental catastrophe) has been ratcheted up since 2000 and is likely to escalate under newly elected US President Trump.

Several Small Island Developing States (SIDS), notably the Seychelles, have questioned the green economy concept and its applicability to their predicament (UN DESA 2014). Instead, they recognise that the oceans have a major role to play in their future and advocated that a blue economy offers an approach to sustainable development better suited to their circumstances, constraints and challenges. The blue economy conceptualises oceans as 'development spaces' where spatial planning can somehow integrate sustainable use of biodiversity and protected areas with fisheries, oil and mineral wealth extraction, bio-prospecting, marine energy production, aquaculture, tourism and shipping. But is this merely wishful thinking? The political and industrial stampede to claim and exploit natural resources in the Arctic and Southern oceans as their ice cover retreats due to global warming, as well as the start of deep sea mining in the Pacific, would tend to suggest so.

Baldacchino and Kelman (2014, p. 1) contend that:

> while inherently commendable, the thrust of many current initiatives related to the pursuit of island sustainability, especially those associated with climate change, promote an ethos which crowds out other pressing policy issues with more immediate relevance such as health, basic education, poverty reduction, and productive employment and livelihoods.

The authors acknowledge that long-term thinking and planning is needed; but, we may now have gone too far in the opposite direction in terms of aiming for sustainable development in, and for, a distant and unpredictable future that emphasises climate change, without better balancing that concern with the pressing needs of the moment.

Having lived on various Scottish islands for over 40 years working on numerous integrated development projects, I can state with confidence that whilst environmental concerns arising from global warming and a whole range of economic activities are fiercely debated (usually by a vociferous minority of the population), they are but nothing compared to the outcry from the vast majority of residents when a local hospital, school, transport link or major traditional employer is threatened with closure. The full gamut of public meetings, petitions, demonstrations, press campaigns and political lobbying are brought to bear in order to reverse what is perceived locally as a poor policy decision by a regional council and/or national government. Affordable housing, fuel poverty, freshwater shortages and access to high-speed broadband are other important issues that will be addressed long before seeking green, blue, eco or smart island status. The latter can just become trendy labels and are often further confused with adopting such buzz terms as bright spots, circular economy, closed loop processes, cradle to cradle and the degrowth movement that can rapidly turn into pure jargon.

Grydehøj and Kelman (2017, p. 106) argue more provocatively that:

> Small islands worldwide are increasingly turning to conspicuous sustainability as a development strategy. Island spatiality encourages renewable energy and sustainability initiatives that emphasise iconicity and are undertaken in order to gain competitive advantage, strengthen sustainable tourism or ecotourism, claim undue credit, distract from failures of governance or obviate the need for more comprehensive policy action.

A few islands – such as Chongming, China; El Hierro, Spain; Jeju, South Korea; Lanai, Hawai'i; and Saadiyat, Abu Dhabi – have probably jumped aboard the eco-island bandwagon and sometimes made spurious claims about their green credentials. However, these account for a tiny minority. More often than not, they are islands whose politicians, owners or simply passionate individuals adopt superficial green branding labels to try and attract favourable international publicity and external investment.

If we consider Scotland for the moment, there are, according to the 2011 official census, some 93 inhabited islands with 103,700 all-year residents: some 2 per cent of Scotland's entire population. There are now well over 100 community development groups operating on these islands, undertaking a myriad of sustainable development projects to improve their living conditions. I could choose any one of them to illustrate the amount of voluntary time and effort that goes into their formal establishment, undertaking feasibility studies, preparing business plans, submitting grant and loan applications in order to proceed and subsequently manage whatever their priority initiatives might be. These are lengthy 'bottom up' and not 'top down' processes that are being continually honed to reach a desired objective and are certainly not being undertaken merely to achieve conspicuous sustainability. Indeed, I would doubt whether any of these community development groups are even consciously aware that they are adopting the UNEP (2011) definition for a green economy as one that results in,

> improved human well-being and social equity, while significantly reducing environmental risks and ecological scarcities. In its simplest expression, a green economy can be thought of as one which is low carbon, resource efficient and socially inclusive. Furthermore, growth in income and employment are driven by public and private investments that reduce carbon emissions and pollution, enhance energy and resource efficiency, and prevent the loss of biodiversity and ecosystem services.

I will now embark on a whistle-stop tour to illustrate why I contend small islands have been in the vanguard of the growing green and blue economy movements and what has been achieved to date, starting first with international and regional initiatives followed by just a selection at a national level.

International and regional programmes

Led by the presidents of Palau and Seychelles, the prime minister of Grenada and the premier of the British Virgin Islands, the **Global Island Partnership (GLISPA)** assists islands in addressing one of the world's greatest challenges: to both conserve and utilise the invaluable natural resources that support people, cultures and livelihoods in their island homes around the world. GLISPA provides a global platform that enables all islands, regardless of size or political status, to work together to develop solutions to common problems, take high-level commitments and integrated actions to build resilient and sustainable island communities, and to drive implementation in line with the UN Sustainable Development Goals.

In 2016, GLISPA celebrated ten years of collective impact toward its mission through supporting over 30 countries to launch or strengthen major sustainable island commitments. These include the **Micronesia Challenge**, **Phoenix Islands Protected Area**, **Coral Triangle Initiative**, **Caribbean Challenge Initiative**, **Caribbean Biodiversity Fund**, **Western Indian Ocean Coastal Challenge**, **Aloha+ Challenge** and **Hawai'i Green Growth**, as well as engaging influential policy-makers to catalyse US$145 million for island action. GLISPA showcases these island commitments as bright spots on the international stage to inspire new leadership and encourage investment that can be scaled and replicated to address global challenges.

The Overseas Countries and Territories Association (OCTA) of the European Union (EU) runs the **Innovation for Sustainable Islands Growth** programme on 20 inhabited islands which have special links with Denmark, France, the Netherlands and the UK. This covers wide-ranging projects including providing more solar power to Anguilla and the Caymans; developing Aruba as a 'Green Gateway' between Latin America and the EU; establishing an algae production plant in Bonaire; advancing invasive alien species management on Montserrat; improving waste management and recycling on Saba; furthering organic agriculture, aquaponics and conch culture in the Turks and Caicos; diversification of the minerals industry in Greenland; social enterprise in gardening and horticultural sectors on Saint Pierre and Miquelon; developing ecotourism and harnessing niche markets for local produce in the Falkland Islands; building an anaerobic digestion plant for waste in St Helena; and installing port infrastructure on Pitcairn to welcome cruise ships. This OCTA programme is complemented with the voluntary scheme for **Biodiversity and Ecosystem Services in EU Outermost Regions and Overseas Countries and Territories (BEST)** that aims to support the conservation of biodiversity and sustainable use of ecosystem services including ecosystem-based approaches to climate change adaptation and mitigation

The **Small Islands Organisation (SMILO)** was created in 2016 with the objective of supporting islands with a land area of less than 150 km^2 that want to structure and federate measures to better manage resources and biodiversity. This approach is undertaken through a labelling process, providing international recognition for areas which are committed to resilient human development, compatible with the environment. The development phase of the programme (2017–2021), financed mainly by the French Facility for Global Environment and the European Union (via the Interreg Maritime Programme), is intended to target 18 small islands located in west Africa, the Indian Ocean, Europe and the Mediterranean.

In 2014, the Rocky Mountain Institute (RMI) and Carbon War Room (CWR) merged to form a strategic alliance and establish an **Islands Energy Programme** that delivers its impact through three mutually reinforcing pillars – energy transition planning; project development and implementation; and learning exchange and capacity building. Five islands have committed to ambitious renewable energy goals: Aruba (100 per cent by 2020), Montserrat (100 per cent by 2018), San Andrés, Colombia (35 per cent by 2020), St Lucia (35 per cent by 2020) and St Vincent and the Grenadines (80 per cent by 2018).

SIDS DOCK is another important initiative among member countries of the Alliance of Small Island States (AOSIS) providing SIDS with a collective institutional mechanism to assist them transform their national energy sectors into a catalyst for sustainable economic development and help generate financial resources to address adaptation to climate change.

The **Caribbean Natural Resources Institute (CANARI)** launched a Green Economy Action Learning Group in 2012 to strengthen economic opportunities for small and micro enterprises in rural communities through applying green economy principles. In addition, CANARI was contracted by the UK Joint Nature Conservation Committee to undertake green economy projects in the British Virgin Islands (2012) and Anguilla (2013) to establish what is needed to integrate environmental issues into the planning processes of these overseas territories and promote green economic growth. Another consultancy did a similar project for the Turks and Caicos Islands in 2014. Elsewhere in the Caribbean, UN DESA has produced a road map on building a green economy for sustainable development in Carriacou and Petite Martinique, Grenada in 2012 and UNEP published green economy scoping studies for Barbados (2014), Jamaica (2016) and St Lucia (2016).

RETI stands for *Reseau d'excellence des territoires insulaires*, the **excellence network of island universities**, set up following an initiative of the University of Corsica, France (hence the French acronym) in 2010. Initially set up with a membership largely consisting of francophone universities, RETI has expanded into an international network, and uses the English language as its medium of instruction and communication. Its main activity is an annual symposium and school for students and scholars, and during which rectors and presidents of the member universities meet as a 'board of governance'.

The **European Small Islands Federation (ESIN)** is the voice of 359,357 islanders on 1,640 small islands. Through their 11 national associations, it has been informing relevant EU institutions and influencing EU policies since 2001. In regional planning, small islands are mostly invisible or else being portrayed and judged by what they do *not* have, including people, natural resources and competitive markets. For too long, this rhetoric has been adopted even by islanders themselves to garner global attention and state aid. More recently, ESIN together with the **CPMR Islands Commission, B7 Baltic Islands Network, INSULEUR, Mediterranean Small Islands Initiative** and **Observatory on Tourism in the European Islands** have developed a countervailing narrative that sees island societies as resilient, nimble, flexible, connected and adaptable to external events. Rather than being poverty-stricken and destitute, many islands might be more accurately described as innovative and entrepreneurial, with a great potential to further advance their green and blue economies. In this regard, the European Union has already funded the **AGRISLES, BEAST, ESLAND, ISLENET, SMALLEST, Green Islands in the Baltic Sea** and **Cradle to Cradle Islands** projects that developed, managed and shared solutions in the fields of agriculture, cultural landscapes, renewable energy, waste, water and transport. ESIN itself successfully completed the most recent **SMILEGOV** project, which together with the **European Economic and Social Committee's (EESC) Smart Islands** project, helped to strengthen **ISLEPACT**. This political initiative with 117 EU island signatories became a key catalyst of islands' collaboration that reaches beyond the EU 2020 climate and energy targets by developing and implementing island sustainable energy action plans on their territories.

European national programmes

British Isles

England

The Isle of Wight Council made the bold claim in 2008 that it was to become an eco island through implementation of a sustainable community strategy and by 2020 have the smallest carbon footprint in England. This ambition was first dropped for lack of financial resources but was taken up again by David Green, described as a charismatic visionary, who had been selling the green dream for two years of making the island self-sufficient in terms of energy. He launched the Ecoisland Partnership Community Interest Company in June 2013 but this was to end in tragedy and ruin. Green's optimism masked a serious and familiar problem: there was no money coming in. Just four months after the launch of this company it had gone into voluntary liquidation and Green was arrested on suspicion of fraud after £115,000 of central government funding awarded via the council was unaccounted for. Unable to face the shame, he took his own life. At the subsequent inquest, Hampshire Constabulary confirmed there would not have been any criminal charges brought against Green and that the council had falsely accused him of stealing. The money had in fact gone into the running of the company and investigators found no evidence of fraud.

It would be easy and somewhat cynical to dismiss the Isle of Wight ever becoming an eco island. However, this would be wrong as over half the island is classed as an area of outstanding natural beauty and local groups such as the **Footprint Trust** and **Island 2000 Trust** have been diligently undertaking hands-on environmental projects for nearly 20 years. In 2016 the Isle of Wight Chamber of Commerce organised the **Green on Wight** conference that attracted delegates from across the community to re-energise the conversation around sustainability and the economy. Indeed, the future looks much brighter with local businesses, together with Portsmouth and Southampton universities on the mainland, keen to attract and develop green tech and innovation for the island. For example, Wight Community Energy finalised a renewable energy deal with the council that will bring a 3.95MW solar array into community ownership. In addition, the **Perpetuus Tidal Energy Centre** announced two major partnerships with global tidal turbine manufacturers and, having achieved all key project consents earlier that year and with a signed grid connection offer, is now a step closer to delivering 30MW of predictable electricity generated from tidal currents to the south of the Isle of Wight from 2020.

The Isles of Scilly, 45 km off the coast of Land's End, produced a strategic economic plan in 2014 that reflected a thriving, vibrant community rooted in nature, ready for change and excited about the future. The **Islands Partnership** that was established late in 2012 reaches out to all sectors of the business community – tourism, horticulture, the food and drink industry, fishing, transport, retail and so on – championing their needs and providing a shared platform to develop common goals. In 2016 an energy infrastructure plan, Smart Islands, was published and it was interesting to note that Hitachi, commissioned by the islands' council to develop the plan, cautioned how the whole project would be placed in jeopardy if the UK left the European Union. Like the Scottish islands, the Isles of Scilly have benefited from considerable grant funding (£33 million in the last seven years) through the European Structural Funds and the Common Agricultural Policy. These have enabled a wide range of social, economic and environmental improvements to be made – ranging from new workspace to farm equipment, from community centres to water harvesting. With Brexit now a reality and the prospect of no further European funding, the outlook for remote UK island communities to continue thriving is less secure.

Scotland

Scottish islands have some of the best wind, wave and tidal renewable energy resources in Europe. In 2016 the world's first grid-connected fully-operational tidal array comprising three 100Kw turbines was installed by **Nova Innovation** in Shetland at Bluemull Sound. Also in 2016, work started on the world's largest tidal stream project which will generate 86MW of energy. The **MeyGen project** is located in the Inner Sound of the Pentland Firth, the body of water that separates the north Scottish mainland from Stroma Island. With appropriate investment in grid infrastructure and generating assets, renewable energy deployment on the islands could grow rapidly by the early 2020s according to an independent report (Baringa 2016) and result in economic benefits up to £725 million for the island economies, including up to £225 million in community benefits.

The islands of Eigg, Gigha and Westray all featured in the original GIN case studies and have continued to innovate. The residents of **Green Eigg** celebrated 20 years of community control in 2017 and one of their early achievements is Eigg Electric. Completely off-grid, the island runs its own renewable energy company which since 2008 has sourced over 80 per cent of its electricity from a custom-designed hybrid system. A consultant from **Energy Matters** has reviewed the operating data and concluded that while the project has delivered good results it is inefficient and Eigg will probably never be able to do away entirely with diesel backup. The project also owes its existence to the fact that 94 per cent of the capital cost was financed by grants. Whilst it is economically unviable on a stand-alone basis, the overall impact of Eigg Electric has transformed island life. The biggest change has been the appearance of new self-built homes utilising local resources and a plethora of small businesses started by the children of the buyout pioneers, incomers, partners and friends. The islanders' shared equity model means quarter acre plots of land are free for individual house builders but they must pay the market value of the land to the Eigg Heritage Trust if the house is ever sold. There's also a loose cap of two new homes every year to ensure maximum involvement by locals in building projects (most Eigg men acquired construction skills when they became members of the Eigg Building Cooperative after the buyout to bring homes up to a habitable standard). This measured attitude towards development also ensures that the Eigg Electric system is not overloaded. These rules of island life have successfully stopped land speculation, avoided inundation by second homes and encouraged a sustainable mindset.

Whilst other islands like Canna, Muck and Rum have renewable energy systems to go off grid, the Western Isles are still reliant upon getting their electricity from the mainland. However, this has not stopped innovative thinking. **Stòras Uibhist** is the community company that manages 93,000 acres of land covering most of the islands of Benbecula, Eriskay and South Uist. It has a 6.9MW wind farm at Loch Carnan that derives income from selling power directly into the grid. Revenue from the three turbines until 2037 is projected to see over £20m being pumped into the local economy with funds already used to improve the outdated drainage infrastructure on the islands. Other major developments completed by Stòras Uibhist with separate funding include the Lochboisdale regeneration project with two causeways and road to Rhubha Bhuailt and Gasay Island, new fishery pier and 51 pontoon berths for fishermen and leisure sailors; restoring the Askernish Golf Club; and opening the Bonnie Prince Charlie Trail. The company is now looking to establish a significant food processing and retail facility, an abattoir and distillery.

Isle of Man

The Isle of Man, a self-governing crown dependency, together with the 14 UK overseas territories, is concerned about the implications of Brexit. In 2016, the **Isle of Man was awarded Biosphere Reserve** status by UNESCO, making it the first entire jurisdiction in the world to be awarded the accolade and therefore worthy of becoming another green island. This recognition complements other work being delivered through such high-level Government strategies as Vision 2020, the Destination Management Plan, Food Matters, Future Fisheries, Managing our Natural Wealth as well as forthcoming Amenity Strategy and plans to develop their clean tech and offshore renewable energy sectors. In addition, a group of local environmental organisations have partnered with key Government departments and the Manx Utilities Authority to spearhead the **EcoVannin** initiative aimed at securing a future for their island in which people, environment and economy can flourish.

Denmark

Danish companies and organisations are at the forefront of green technology with Aerø, Fur, Langeland and Lolland all termed as sustainable islands in various ways. **State of Green** is your gateway to their solutions and to Danish green policies – from renewable energy to clean water and resource efficiency. The **Samsø Energy Academy** and **Bright Green Island Bornholm** with its straw-fired and woodchip-fired district heating plants, biogas plant, waste-to-energy plant, water quality and treatment facilities, and electric vehicles, are probably the best known.

France

Association Les Iles du Ponant was formed in 1971 to maintain active communities on 15 islands off the Channel coast. Three of them – Molène, Ouessant and Sein – are especially energy-minded not being connected to the mainland grid. This led to their respective mayors signing a pledge to be using only renewable energy by 2030. The French minister in charge of energy and ecological transition visited Ouessant in May 2016 to support the islands with a programme worth €500,000. It allows for many actions starting in 2018 that include awareness-raising operations of residents and visitors to water and energy savings, distribution of efficient electrical appliances, insulation of public buildings, introduction of LED street lighting, charging stations for electric vehicles and reduction of food waste. In June 2015, a 1MW Sabella D10 tidal turbine was deployed in the Fromveur Passage and subsequently connected to the Ouessant grid via submarine cable. After a year of successful tests that delivered around 70MWh of clean power to the island, the turbine was retrieved but the company intends to reinstall the turbine at the same location, with additional plans to develop tidal energy projects in the Philippines and Indonesia.

Integrated within the Regional Natural Park of Armorica since 1969, Ouessant was declared a Biosphere Reserve of the Iroise Sea together with Molene by UNESCO in 1988. It also forms part of the Marine Natural Park of Iroise, created in 2007. The island counts over 500 plant species and is an important ornithological reserve. As such, the island has a vibrant eco-tourism industry and is actively trying to revive its traditional agriculture and native sheep breeding. There is an ecological museum, strong local customs together with an annual book fair that brings to the fore all forms of island writing and more recently the Ilophone music and song festival, both of which attract hundreds of visitors.

Germany

The **Island and Hallig Conference (IHKo)** founded in 2002 is an association in which the 26 municipalities of the North Frisian Islands and the Hallig islets, as well as Heligoland, work together for the interests of their region and population. The association deals with topics such as safety at sea, coastal protection, climate change, renewable energy, transport, sustainable regional development, waste management, product branding, tourism and international networking. All these German islands situated in the World Heritage National Park Wadden Sea have very strong green credentials but the third largest, Pellworm, has been a pioneer in many respects. In 1983 their first hybrid power plant was built and with further expansion now generates 22.5 million kWh/year. This is about three times the amount needed to supply the island itself so the excess is transmitted to the mainland. Although Pellworm both produces and controls energy from solar photovoltaics, wind turbines and biogas: E.ON's state-of-the-art management system links the power plants, storage batteries and smart meters in residents' homes together, so that E.ON can understand the island's power needs and predict the energy that needs to be generated. By using two types of batteries, **SmartRegion Pellworm** can store energy for both the long and short term.

Greece

Kythnos is a small Aegean island lying at the north of the Cycladic complex with a population of 2,000 inhabitants. It can lay claim to one of the oldest known Mesolithic settlements (10,000

BCE – 8,000 BCE) in the region but also for hosting the first wind farm in Europe back in 1982. This was followed by the installation and testing of a 100kW PV plant coupled with batteries, a hybrid station comprising of a 500kW wind turbine, battery storage and an automatic control system and finally the development of one of the first PV-powered autonomous microgrids with batteries and diesel generator back-up in 2001. For the municipality and citizens of Kythnos, the vision is to take stock of these past projects and incorporate smart and innovative technologies in renewable energy, water, waste and transport. The diversification of agriculture and development of sustainable tourism are also priorities. Elsewhere in the Aegean, the project **Green Island–Agios Efstratios** was selected in 2009 to be entirely powered by renewable energy sources.

Tilos, situated between Kos and Rhodes, is one of the smallest islands of the Dodecanese with a population of 500. On the renewable energy front, the innovative **TILOS** project will provide the island with a new hybrid power station but the ultimate goal is to demonstrate the potential of local/small-scale battery storage to serve a multipurpose role within an island microgrid that also interacts with a main electricity network. Tilos also leads the way in Greece for environmental protection and eco-tourism. The islanders realised years ago that the way forward was in sustainable tourism – becoming the first island in Greece to ban hunting in 1993 was a bold start – and have never looked back. The island's land area is 64 km², together with its surrounding archipelago of 16 uninhabited islets, are a Corine Biotopes Project region, Important Bird Area (IBA), EU registered Special Protection Area, Natura 2000 site, and 2005–2008 European Union LIFE Nature Project host. All this has resulted in new trees being planted, traditional cereal crops sown and nature interpretation trails created. Visitors can now explore the island and find a protected natural environment with a wonderful diversity of flora and fauna, especially birds like the Eleonora Falcon, to which Tilos is seasonal home to 10 per cent of the entire world population.

Ireland

The three Aran Islands – Inis Mór, Inis Meáin, and Inis Oírr – are located at the mouth of Galway Bay on the west coast of Ireland. They have supported farming communities for over 4,000 years and this has left behind a rich cultural legacy most dramatically seen in the spectacular great forts and the dense web of high field wall systems. Over 75 per cent of the total land area is designated as a Special Area of Conservation (SAC) under the EU Habitats Directive. **AranLIFE** is a demonstration project operating over a four-year period from 2014 to 2017, co-funded under the EU LIFE Nature programme. It involves 70 locals undertaking traditional extensive farming practices across more than 1,000 hectares of land on the three islands and is committed to maintaining and restoring habitats such as the limestone pavement, orchid rich grasslands and machair.

The **Aran Islands Energy Cooperative** won the Sustainable Energy Authority of Ireland awards in 2014 for its range of projects to achieve the goal of carbon neutrality by 2022. Some 353 homes and community buildings – two-thirds of the buildings on the islands – have already completed energy upgrades, improving insulation levels and installing efficient heating systems. This is resulting in more comfortable homes and lower energy bills with annual energy savings of €250,000 accruing to the islanders. Electric vehicles have been in use here since 2010, and have helped reduce reliance on imported energy for transport by 68 per cent. Analysis shows that replacing heating systems with heat pumps powered by wind or wave energy could reduce energy dependence further.

In 2010, the **West Cork Islands** Integrated Development Strategy was published following an intensive research and consultation process. It addressed the physical, economic, social and cultural development of the West Cork Islands of Chléire, Bere, Whiddy, Dursey, Long, Sherkin and Heir and sets out a framework of objectives and actions for the next 10+ years, with a view to making the islands a better place in which to live, work, visit and do business. The strategy report warned that the permanent, year-round habitation of some islands is "clearly under threat" and it recommended the development of projects such as inter-island sea kayaking and walks, innovative island-specific housing, branded food products, multi-island ferry tickets and cooperative approaches to farming, fishing and aquaculture.

Italy

The **Egadi Islands Marine Protected Area** encompasses a group of five small mountainous islands off the northwest coast of Sicily. Together, they constitute the largest marine reserve in the Mediterranean Sea stretching over 54,000 hectares. The intersection of three marine currents creates exceptional conditions of hydrodynamics, transparency and the presence of nutrients. This gives rise to a variety of habitats and rich biodiversity. Established in 1991, the MPA is divided into four zones with different levels of protection. Best practices include monitoring protected species like the Monk seal, placing anti-trawling bollards, introducing yacht mooring buoys and creating a label for the environmental certification of tourist services. For some time, the Italian National Agency for New Technologies, Energy and Sustainable Economic Development (ENEA) has been working on the archipelago and their **Egadi Project** developed sustainable tourism practices ranging from a pilot plant for composting to the treatment and reuse of wastewater. In 2016, a delegation of European Economic and Social Committee members visited the islands and identified a strongly diversified economy based on marine resources. They visited a turtle rescue centre, a cosmetic research company that uses an extract from seagrass, the conversion of an old tuna factory into a museum of archaeology and a project combining water supply and waste reduction. Also in 2016, the local power supplier on the main island of Favignana announced a drive towards a new, sustainable strategy through a mix of distributed solar PV, charge points for electric vehicles, conversion to a high-efficiency power generation unit, LED lighting and a smart grid.

Lampedusa is the most southern European island, located 60 miles east of the Tunisian coast, and is the biggest of three Pelagian Islands with some 5,000 inhabitants. Their **Smart Island** project aims to increase energy, economic and environmental efficiency of the island, in order to set an example for the many islands of the Mediterranean that are not connected to the mainland electrical network.

Netherlands

Whilst not adopting any fashionable environmental labels, the Dutch West Frisian island of Texel has been pursuing sustainable development policies and practices for many years. The latest manifestation of this was **Planet Texel Academy** founded in 2013 that brought together various island stakeholders to focus on how the different ambitions of Texel – growing a tourist-led economy, securing energy neutrality and a higher level of self-sufficiency, protecting the unique natural environment and maintaining the quality of social services – can be optimally integrated in a wide-ranging spatial plan. In many respects, Texel is already 'greener' than most islands. The renewable energy cooperative TexelEnergie with more than 3,000 members, the largest in the

Netherlands, focuses on the production of electricity and gas through solar installations and biomass. There is a highly efficient waste minimisation and recycling service, integrated water management, high uptake of agri-environment schemes, zoning for various tourism activities, restaurants serving locally produced food and drink, good public transport, extensive cycle path network and a third of the island is a national park.

Portugal

Located in the middle of the Atlantic, the Azores region of Portugal comprises nine volcanic islands, each with a unique endowment of renewable energy resources. For many years their **Green Islands Project** has been designing smart micro-grids and other energy network options that allow dynamic supplies and demands to co-exist. One such development on the island of Graciosa aiming to replace diesel generation with intermittent renewable energy has now become key for the credibility of island-based storage since the El Hierro plant in the Canary Islands failed to meet expectations. The Graciosa hybrid power system will feature a 4.5MW wind park and 1MW solar plant connected to a 2.6MW lithium-ion battery system equipped with Leclanché cells. The system should be able to cover an annual average of up to 65 per cent of the island's power demand with renewables.

The Azores lead all other European destinations in sustainable travel and frequently top the scoreboards in various green tourism awards. Besides two World Heritage Sites and three Biosphere Reserves, the **Azores Geopark** is based on a network of 121 geosites throughout the archipelago and is managed by a non-profit association established in 2010 with its headquarters in Horta on the island of Faial. The Association's mission is to ensure the geological conservation, environmental education and sustainable development of these sites, while promoting the well-being of the population and a respect for the environment. The **Azorean Biodiversity Group** has an international reputation and undertakes many projects, including organisation of the Island Biology 2016 congress that brought together 400 researchers in order to achieve a unified view of island biology.

Spain

The **Gorona del Viento (GdV)** plant on the Canary Island of El Hierro has often been portrayed in the media as a flagship project designed ultimately to provide the island with 100 per cent renewable electricity and to demonstrate that hybrid wind/pumped hydro systems can be used to generate 100 per cent renewable electricity in other parts of the world. GdV comprises a wind park with 11.5 MW capacity and a pumped hydro storage plant with 11.3MW gross (9.2MW net) capacity, installed at a total cost of €84 million. However, the current GdV plant can only supply an annual average of about 50 per cent renewable energy at best. As a result, El Hierro will remain dependent on diesel generation from the Llanos Blancos plant that GdV was designed to replace for a secure supply of electricity. So, instead of acting as a flagship project to demonstrate the feasibility of 100 per cent generation from intermittent renewable sources, GdV has in fact only succeeded in highlighting the severe, and in many cases insuperable difficulties involved in achieving this goal. However, El Hierro does have many other attractions that can be developed to better promote itself as a green island. The inhabitants have a strong culture passed down over the centuries from the original native settlers, the Bimbaches, in the form of sports, local festivals, handicrafts and an extensive network of traditional paths that are popular with those seeking an alternative to mass tourism found elsewhere in the Canary Islands.

North America and Canada

Canada

In Canada, we find Fogo Island, largest of the offshore islands of Newfoundland, that has undergone a process of community self-discovery since the late 1960s when inhabitants' resisted government plans to resettle them. In 2008, a strategic plan was published in response to the numerous challenges the community was facing, such as the decline of the fishing industry, an increase in out-migration, a decreasing birth rate, and an ageing population. One islander who fully embraced this plan was Zita Cobb, who left Fogo in 1979, and returned a multimillionaire in 2006 after helping California-based JDS Uniphase become a world leader in fibre-optics during the high-tech boom. Cobb and her brother established the **Shorefast Foundation** with the idea of revitalising the community by weaving together the fundamental components of Fogo's heritage – fishing, craftsmanship, nature and tourism. Shorefast put Can$6 million into the project, while the federal and provincial governments invested Can$5million each.

Shorefast began by putting Fogo on the map as an international art destination and hired Newfoundland-born, Norway-based architect Todd Saunders to design a series of studios where visiting artists could embark on projects and collaborate with local artisans. To date, six have been built and are completely off grid. Next, the Fogo Island Inn was opened in 2012. This five-star modernist building, also designed by Saunders, is perched on stilts and hugs the North Atlantic coastline, affording all 29 suites with floor-to-ceiling views of sea and sky. The rooms are filled with locally made furnishings – from hand-crafted bed quilts, hooked rugs and crocheted chair pads to sleek bar chandeliers that resemble nautical radar reflectors. Even the walls of the inn's library have special meaning, since they are painted in the same shade, 'Fishermen's Union green', that graces many of the 2,700 islanders' homes. At Can$1,575 plus a night to stay, the inn is targeting a wealthy, niche clientele attracted by local culinary delights featuring on the restaurant menu and a 'community host' programme that pairs guests with locals for island excursions. The Foundation also provides microfinance loans enabling new businesses to spring up that in turn support two other Shorefast initiatives, the Fogo Island Partridgeberry Harvest Festival and Great Fogo Island Punt Race.

Over on the west coast of Canada the **Islands Trust** was created in 1974 and is a unique federation of local island governments with a provincial mandate from the Islands Trust Act to make land use decisions that will "preserve and protect" British Columbia's southern Gulf Islands. The Trust Area is composed of 13 major islands and more than 450 smaller ones with outstanding scenery, wildlife and recreational resources. The islands also support strong communities characterised by a mix of lifestyles, livelihoods and individuals. Island residents join together to bring varied skills and experience to sustain a tradition of community involvement. For example, the **Gulf Islands Community Economic Sustainability Commission** is working toward achieving a resilient and sustainable local economy that improves and maintains the economic prosperity, social equity and environmental quality on the five island communities of Galiano, Mayne, Pender, Salt Spring and Saturna. Their various initiatives cover ways to improve broadband services, education, tourism, social finance, transport, food and agriculture.

USA

Kodiak Island, located on the southern coast of Alaska, is the second largest island in the USA. Home to approximately 14,000 people, the island has an isolated power grid that is almost entirely supplied by hydro and wind backed up by a 3 MW battery storage system. The local electricity cooperative, **Kodiak Electric Association**, is committed to maintaining a 95 per cent renewable electricity portfolio although in 2015–2016, diesel generators were used to generate only 0.3–0.5 per cent of the community's electricity, making the island powered by almost 100 per cent renewable sources. The main power source in the island's 28 MW electricity mix comes from three hydroelectric plants at Terror Lake. The community bought the facilities back from the federal government for US$38 million in 2007, which marked a major milestone in bringing the community's power supply under local control. Kodiak Island's renewable electricity generation is complemented by conservation efforts that include promotion of efficient lighting technologies like LEDs and a Home Energy Rebate Programme that provides up to US$10,000 to help homeowners improve the energy efficiency of their homes.

Kodiak's economy relies heavily on the relatively electricity-intense fishing industry. Stable electricity rates resulting from the new renewable system have led to an expansion in the fishing industry, creating more jobs and tax revenue for the local government. The recent addition of a flywheel to the system will enable a much larger crane to be operated at a major shipping pier without negatively impacting microgrid operation. Since the crane has a highly fluctuating load that can peak at 3 MW (18 per cent of the average KEA load), the flywheel provides spinning reserve that can handle these large, fast changes in load when the crane is in use.

The southwestern two-thirds of the island, like much of the Kodiak archipelago, is part of Kodiak National Wildlife Refuge that has a global conservation role being the largest intact, pristine island ecosystem in North America. Misty fjords, deep glacial valleys, and lofty mountains distinguish the 1.9 million-acre refuge. Diverse habitats encompass 117 salmon-bearing streams, 16 lakes, riparian wetlands, grasslands, shrub lands, spruce forest, tundra, and alpine meadows. Collectively these habitats sustain 3,000 bears, account for up to 30 million salmon caught by the Kodiak-based fishing fleet, support more than 400 breeding pairs of bald eagles, and provide essential migration and breeding habitat for another 250 species of fish, birds and mammals. Such natural abundance and spectacular scenery attracts thousands of visitors to the refuge annually and **Discover Kodiak**, a non-profit, promotes sustainable development of the tourism industry, thereby increasing economic opportunities, jobs and local tax revenues.

In 1983, the **Island Institute** was founded with a focus on supporting and maintaining the year-round communities living on any of Maine's 15 inhabited islands. That work has evolved over the years, and now encompasses six core programme areas: community development, economic development, education, energy, marine resources, and media. For example, their Archipelago store and gallery showcases a broad and unique collection of handmade art, craft and design. Since opening in 2000, it has returned nearly US$1.5 million to local artists and makers thus helping to sustain the economic viability of islands by diversifying local economies. The Institute also supports students in academic and professional development pursuits through scholarships; and island businesses are connected to local, regional, state, and federal resources to improve economic productivity and to expand and reach new markets. In particular, through their Aquaculture Business Development programme the Island Institute is working to help people from fishing communities diversify into shellfish or seaweed aquaculture.

High energy costs are a major challenge for Maine's island and coastal communities because they face some of the highest electric and heating fuel costs in the USA. To reduce this burden, the Institute helps island residents increase energy efficiency in homes, businesses, schools, and other public buildings, reducing costs and increasing comfort and safety. In addition, the Fox Islands Electric Cooperative has been assisted in its efforts to develop and operate their 4.5MW wind farm. Commissioned in 2009, this project has met strong community support and has lowered the cost of electricity on Vinalhaven and North Haven. The Monhegan Plantation Power District, Swan's Island Electric Cooperative and Matinicus Plantation Electric Company are also been given technical support to investigate community-based wind, solar, and microgrid projects. In 2014, the Island Institute and the Renewable Energy Alaska Project launched the **Islanded Grid Resource Center**, a collaborative initiative that partners with remote communities in Maine, Alaska and other US states and territories to promote fact-based decision making about wind power and other renewable energy sources as a way to lower energy costs.

Caribbean

Anguilla

In 2016, the Anguilla Electricity Company completed construction of its 1MW solar plant which was integrated into the main power grid. It produces one of the highest solar penetration percentages (over 10 per cent) in the region, and runs a spinning reserve to mitigate power loss and damage due to the intermittency of solar energy production. This project was a landmark development for Anguilla's future and supports the island's other green credentials like their Sustainable Tourism Master Plan. The **Anguilla National Trust** is a particularly strong NGO whose work is divided into five programme areas: terrestrial and wetlands conservation; marine and coastal conservation; protected areas; culture and heritage; and public awareness, education and stewardship.

Aruba

Green Aruba is an annual conference born in 2010 with the specific goal of making the island 100 per cent independent of fossil fuels for the generation of electricity and potable water by 2020. Over the past seven years, the conference has evolved into a practical and valuable well-known platform within the region for the exchange of information and applied knowledge on advanced green technologies such as floating offshore wind turbines, ocean thermal energy conversion and underwater compressed air energy storage. In 2015, Aruba was named the National Geographic World Legacy Award winner for 'Destination Leadership' at ITB Berlin, the world's leading travel trade show. The judges considered that Aruba's government was actively promoting the use of renewable energy through wind farms, airport solar park, waste-to-energy plant, electric vehicles and smart communities as well as continuing major investment largely focused on eco-tourism. More recently, Aruba is starting to develop an agriculture sector that makes the best use of water resources and their utility company is working to provide air conditioning using ice that is produced at night when electricity costs are lower. The Aruba Reef Care Project, the island's largest volunteer environmental initiative, has attracted more than 800 people annually since 1994 and results in cleaner reefs and public beaches.

Bonaire

In 2004, a fire destroyed Bonaire's existing diesel power plant. Although disruptive, the event afforded Bonaire and its residents an opportunity to design a new electricity generation system from scratch. Today, almost half of Bonaire's electricity comes from wind power and rates paid by local consumers have decreased substantially. The 12 wind turbines installed with a total of 11MW power capacity can provide up to 90 per cent of the island's electricity at times of peak wind. Battery storage of 6MWh is included in order to take advantage of available power in times of excess wind and provide that stored electricity in times of low wind. The island is also exploring local algae resources grown in the large salt flats on the island to create biofuel, which could then be used in existing generators. **Boneiru Duradero (Sustainable Bonaire)** is a small-scale initiative started in 2012 focusing on waste management, reduction of water use and local food production. The National Parks Foundation (STINAPA), besides managing Washington Slagbaai National Park and the Bonaire National Marine Park, is engaged in reforestation efforts to help combat the deleterious effects of non-indigenous goats and donkeys.

Grenada

Grenada, known by many as the 'Spice Isle' for its production of nutmeg, cloves and other exotic spices, is now setting its sights on being known as a world leader for innovation in the blue economy through events like **Grenada Blue Week**. Beyond its 345 km^2 of land territory, Grenada has 26,000 km^2 of blue ocean space and has already established four Marine Protected Areas and the world's first underwater sculpture park. However, it is the first country to initiate a national coastal master plan for blue growth that identifies opportunities for development in areas such as fisheries and aquaculture, blue biotechnology, renewable energy, research and innovation. The master plan proposes a Blue Innovation Institute as a key component of its strategy. The institute will aim to be a centre of excellence and a think tank on the blue economy, as well as seek to develop innovative blue financing instruments such as debt-for-nature swaps, blue bonds, blue insurance and blue impact investment schemes.

On the terrestrial front, the Grenada Electricity Services company is the sole provider of electricity in Grenada, Carriacou and Petit Martinique and is actively pursuing a strategic goal to provide for 20 per cent of demand through renewable energy generation by 2020 mostly through solar PV installations. The **Grenada Integrated Climate Change Adaptation Strategies** initiative has completed a range of community projects on soil protection, rainwater harvesting, organic farming practices, recycling waste and restoration of mangroves. The country also has a strong group of environmental NGOs including the Grenada Fund for Conservation, Grenada Green Group, Kido Foundation and Ocean Spirits whose work ranges from managing the tropical dry forests to protect the Grenada dove (their national bird), species monitoring, ecotourism and combating the litter problem on the island.

Nevis

Nevis has long been known for its hot springs and a geothermal project operated by West Indies Power began exploration drilling in 2008 and found three viable sites clustered around several faults. These wells confirmed the existence of a continuous 4 km geothermal reservoir on Nevis and the possibility of several hundred MW of production capacity. Following legal actions a new company, Nevis Renewable Energy International, won the tender in 2013 to further develop the geothermal power plant that is expected to supply 9MW by early 2018. It is hoped the facility can be upgraded soon after to expand its capacity by an estimated 40MW for export to neighbouring islands. With a need of only 10MW, Nevis is poised to become one of the most carbon-neutral islands in the world. In 2016 the Nevis Island Administration also signed off a deal that will see a waste-to-energy facility powered by a solar energy farm constructed on the island in partnership with an American company, Omni-Alpha. According to the administration, this plant will consume virtually all the island's household and commercial waste, tyres, plastics, paper, plant and vegetable material and provide much needed ease on the island's sole solid waste disposal site.

Indian Ocean

Seychelles

The Seychelles, a 115-island nation off the east coast of Africa whose mainland accounts for only 0.03 per cent of its full territory, is spearheading the shift towards a blue economy. The former president, James Alix Michel, has been instrumental in this move having published *Rethinking the Oceans – Towards the Blue Economy* in 2016 and establishing the **James Michel Foundation** in 2017 to promote the whole blue economy concept. He is also patron of **Blue Economy Research Institute** at the University of Seychelles. With tourism and fishing as the main economic sectors, Seychelles' large ocean territory provides most of the jobs and livelihoods; so maximising its full potential is key to the blue economy agenda. At its core, this involves changing the business model to focus on economic diversification, food security and the protection and sustainable use of marine resources.

During the Paris UNFCCC COP21 held in 2015, the Government of Seychelles completed a first of its kind debt-for-adaptation swap to enhance marine conservation and climate adaptation activities. The debt swap creates a sustainable source of funding to support Seychelles in the creation and management of 400,000 km² of new marine protected areas (the second largest in the Indian Ocean) to improve the resilience of coastal ecosystems. This landmark agreement reached between Seychelles and its Paris Club creditors, led by France and the Government of South Africa, resulted in a US$21.6 million debt swap. The Nature Conservancy designed the debt swap to enable Seychelles to redirect a portion of their current debt payments to fund nature-based solutions to climate change through the newly established **Seychelles Conservation and Climate Adaptation Trust (SeyCCAT)**. Over a 20-year period, the debt service payments will be used to finance marine and coastal management to increase resilience to the impacts of climate change; capitalise an endowment to finance work to support adaptation in the future; and repay impact investors.

Bonds are financial instruments to raise public and private capital for specific activities which can generate a return on investment. In 2017, the government of Seychelles issued a Blue Bond of US$15 million over 10 years with guarantees from the World Bank and the Global Environment Facility to support the transition to sustainable fisheries. The proceeds of the Blue Bond will be used as grants for fisheries management planning activities and as loans to encourage local public and private investment in activities consistent with sustainable fishing, such as post-harvest value-adding opportunities and jobs and the protection of ocean resources. The Blue Bond proceeds will be disbursed on a competitive basis through SeyCCAT and the Development Bank of Seychelles. Importantly, the Blue Bond proceeds complement other publicly funded sustainable fisheries projects and the implementation of Seychelles Marine Spatial Plan of its Exclusive Economic Zone (EEZ), a commitment of the Seychelles debt swap for conservation and climate adaptation, bringing efficiency and synergies between fisheries management, marine conservation and climate resilience.

The Seychelles are still highly dependent on imported fossil fuel. However, their Port Victoria 6MW wind farm is a major step toward meeting the targets of their energy policy, namely to produce 15 per cent of electricity generation from renewable energy sources by 2030. There are around 100 solar photovoltaic installations connected to the National Grid on Mahe, Praslin and La Digue and a good number of grid-independent systems on the smaller eco resort islands or special island reserves like Aride and Cousin. The Seychelles has a progressive waste management strategy and in 2016 approved a ban on the importation of styrofoam lunch boxes, plastic bags, plates, cups and cutlery. The NGO, Sustainability for Seychelles, seeks to promote green living through various waste, water, energy and ecosystem projects and the government Eco-Schools programme has over many years made several real changes in the way public schools operate.

Asia-Pacific Region

Australia

Many of the smaller islands surrounding the Australian continent are leading the way in terms of renewable energy technology. The **Flinders Island Hybrid Energy Hub**, **Garden Island Microgrid Project**, **Kangaroo Island Energy Project**, **King Island Renewable Energy Integration Project**, **Lord Howe Island Hybrid Renewable Energy System** and **Rottnest Island Water and Renewable Energy Nexus (WREN) project** are all solutions to a variety of local conditions. However, it is not just aiming to become carbon free that makes these islands greener. For example, if we take King Island we find that, besides a world-leading system that supplies over 65 per cent of the island's power needs using renewable energy generation, it is also renowned for its local beef, dairy and seafood products. A strong tourism brand has been developed that has led to a network of food and recreation trails as well as attracting golfers to three courses, surfers and those seeking cultural and historical interests. There is also the **King Island Natural Resource Management Group** whose mission is to promote co-ordinated and integrated management of natural resources, which will contribute to the economic and environmental sustainability of the island.

Cook Islands

While less than 1 per cent of the Cook Islands territory is land (240 sq km), the country's Exclusive Economic Zone (EEZ) spans a spectacular 1.8 million sq km of the Pacific Ocean, making it another nation keen to adopt a blue economy strategy. Rich with marine biodiversity, including rare seabirds, beaked whales, manta rays and several threatened shark species, the country's 15 small islands – home to about 15,000 people – host growing industries such as tourism, fishing and deep sea mining. The Cook Islands are also on track to become 100 per cent renewable by 2020. After careful planning, upgrades to the distribution grid and programmes to train and build local capacity, all these islands will soon be using solar PV and batteries, with diesel providing backup during longer periods of power deficiency.

At the 43rd Pacific Leaders Forum in 2012, Prime Minister Henry Puna announced the establishment of a multiple-use **Cook Islands Marine Park (Marae Moana)** that was initially to provide management to the southern half of the country's EEZ. After three years of public consultations it became clear that the northern group also wanted to be part of the Marae Moana so it will now span their entire EEZ. At the beginning of 2017 the size of marine protected areas or 'fishing exclusion zones' around islands was the final decision to be confirmed by Cabinet before Marae Moana is legally designated. These zones are areas where local artisanal fishermen will be allowed to fish, but large-scale commercial fishing vessels (longliners and purse seiners) will not be permitted. Seabed mining will also be banned within these protected areas. The size of the protected areas is measured in nautical miles by a line from the coast of each island and extending outwards towards the ocean. With public support, the government has since announced a 50 nautical mile protected area around all the Cook Islands to maintain, preserve and ensure fish stocks for the future. The government believe these protected areas, and implementing a catch-based harvest strategy, in the form of a quota management system (QMS) to better manage the longline fishery, can both be used as tools at the same time for conserving tuna. However, it is acknowledged a QMS requires a significant improvement in surveillance and reporting including electronic surveillance in order to be effective.

Federated States of Micronesia

FSM is an independent sovereign island nation made up of four states from west to east: Yap, Chuuk, Pohnpei, and Kosrae. Whilst Kosrae has at least five grid-connected PV systems it was a press report in 2012 that made headlines saying the island would soon have a new 1.5MW wave energy plant capable of providing 85 per cent of the island's power with solar and hydro contributing the remaining 15 per cent. Ocean Energy Kosrae described their WaveSurfer technology as a reliable, inexpensive and efficient offshore system with a 'point absorber' whose main power conversion and generation parts are completely submerged at the depth of between 8 and 25 metres thus protecting the device against damage from extreme storms. Whilst this joint venture with Kosrae Utilities Authority never materialised the island did have a different type of success on the terrestrial front. The Nature Conservancy helped establish the first income-generating conservation easement in the Asia-Pacific region when it joined with Kosraean landowners and other international partners to complete a unique land protection deal in 2014 that will safeguard 78 acres of forested wetland in the Yela Valley. The Yela conservation easement will preserve the largest remaining stand of *Terminalia carolinensis* ('ka') trees in the world, as well as the endangered Micronesian pigeon and several endemic plant species. Forests such as Yela are vital, because they act as natural water filters, trapping sediments that could harm nearby corals and mangroves – and the fish that depend on both.

The **Kosrae Conservation and Safety Organisation**, started in 1998 by a small group of island residents, has undertaken numerous environmental projects and the Women in Farming Group have created kitchen gardens to increase fresh produce as well as improve health to fight diabetes. All this helps to promote Kosrae as a true ecotourism destination if you take into account other attractions like trekking through forests and mountains, quietly canoeing in the mangrove channels, marvelling at Neolithic ruins, visiting a giant clam farm, or diving pristine reefs and sunken ships. One eco-resort, Pacific Treelodge, also runs the local recycling business, Micronesia Eco Corp (MEC). The Kosrae state government imposes a deposit fee on the importation of aluminium soft drink cans, plastic containers, bottled drinks and car batteries. The deposit fee collected on these items is put aside to fund the recycling operation. Every month, MEC goes around the communities on Kosrae to collect the trash and pay them for it. The trash is then taken to a facility where the cans and plastic containers are flattened and exported together with the discarded car batteries to Korea or China. The waste glass, however, stays on the island where it is crushed and used as an aggregate for cement and other construction materials. The glass crusher is run by solar power.

Indonesia

The island of Sumba is located in eastern Indonesia and is home to some 685,000 people. Megalithic stones, distinctive peaked bamboo and grass-topped homes and the intricate world-renowned Ikat cloth distinguish Sumba from much of Indonesia. Like many developing islands, not only are the effects of climate change felt more acutely, but also electricity is not widely available and any power source is most likely to be diesel and kerosene for cooking and lighting. In 2009, helped by recognition of ample, local renewable energy resources, the idea was born for **Sumba Iconic Island** whereby international donors working with the local government sought to bring electricity to all island residents using only renewable sources. Up to 37 MW of alternative renewable energy can be harvested from Sumba's sun, water, wind, biogas, and biomass potentials: enough to supply the island's electricity twice over until 2020. The project can generate 100 per cent green energy; increase the electrification ratio to 95 per cent by 2025; and bring about economic well-being for the island's population and women in particular. Between 2011 and 2014, a number of alternative renewable energy installations, with a total capacity of 4.87 MW, were commissioned; these contributed 9.8 per cent out of the total of 37.4 per cent of Sumba's current electrification ratio. This includes 4,158 households with access to electricity; 14,868 units of solar-powered plants; 100 units of wind-powered plants; 1,173 biogas home installations; 12 units micro-hydro plants; Rp.131 billion of total investments in renewable energy; and 16 training programmes and 27 R&Ds in alternative renewable energy.

Sumba villages depend heavily on natural resources to survive with most trying to eke out a living from farming. However, this main livelihood activity is affected by unreliable water supplies in the area and crops and livestock are often lost to locusts or cattle raiders. When crops fail, the Sumba people traditionally fall back on wild yams collected from the forest, which are protected by traditional laws – these forest resources have helped reduce people's vulnerability to natural climatic factors as well. However, when two national parks were declared in 1998, communities living around the forest found they were denied access to their ancestral land and forest resources. Burung Indonesia (BirdLife in Indonesia) has been working closely with local communities and government to conserve Sumba's forests and support local livelihoods. Their activities have focussed on improving the cultivation of perennial plants and making them more sustainable by terracing farmland, supporting women's groups to create kitchen gardens, and cultivating trees that reduce demand on the natural forest. BirdLife has also led the way in developing Village Conservation Agreements. These encompass the entire community, and are the outcome of negotiations between the community and the park's management to meet both groups' needs and aspirations. They cover access to resources (such as yams) inside the parks. All this has helped in reducing vulnerability to fluctuations in water supply that affect income-generating activities outside the parks; and agreements not to farm or graze in certain areas.

Japan

Bisected by the Kobe–Awaji–Naruto Expressway, which connects Honshu and Shikoku, Awaji Island (Hyogo Prefecture), the largest island in the Seto Inland Sea, is divided into three municipalities: Awaji, Sumoto and Minamiawaji. To get here, most people drive across the Akashi Kaikyo Bridge, a 3.9 km suspension bridge, the longest of its kind in the world. Once off that bridge, you are immediately immersed in centuries old traditions, historical landmarks and cultural gems. In fact, according to Japanese folklore, Awaji Island is the birthplace of Japan.

Japanese prefectures are now facing common issues, such as decreasing birthrates and an aging population, increasing income gaps between cities and farm villages, decreasing employment, and fewer business heirs. In an attempt to become a successful model in solving these issues, the **Awaji Green Future Project** was put together in 2010, aiming to make a 'Sustainable Island for Life'. There is no thermal power station on the island so 25 wind turbines were installed in 2014 that generate more than 55MW of renewable energy. Three photovoltaic power generation systems were also operational in 2014, providing a total output of roughly 4.6MW. Moreover, the Eurus Energy group Tsuna-Higashi solar park started commercial operation in 2015 with an output of 33.5MW and in 2017 their Kita-Awaji solar park added another 10.5MW, thus putting Awaji well on the way to achieving 100 per cent electric energy potential from renewable energy.

Japan's only small island prefecture, **Okinawa**, has developed various green technologies and innovations such as fruit fly eradication, the reuse and recycling of glass bottles, production of diesel from waste cooking oil, underground dams for rainfall storage, utilisation of deep ocean water and resource-based health foods. Miyako is the fourth-largest island in the prefecture and subsists on the key industries of agriculture and tourism. Some 400,000 tourists visit the island each year. As the 55,000 residents' lifestyles became increasingly affluent, their impact on the island's natural environment and growing dependence on outside resources began to appear in such forms as groundwater pollution and illegal dumping. In response, the 2008 **Eco-island Miyakojima** declaration expressed the goal of preserving a liveable island environment for future generations. In order to achieve lower carbon emissions the focus was upon becoming self-sufficient in green energy through a combination of wind turbines, solar panels and producing bio-ethanol from bagasse, a by-product of the process used to manufacture sugar from sugarcane, to replace diesel at existing power stations. To help eliminate CO_2 in the transport sector electric vehicles and recharging stations are being introduced. Other environmental actions like waste reduction, water conservation, mangrove forest protection and developing eco-tourism are all being actively pursued.

Palau

Whilst this nation does not particularly excel at using renewable energy it has made a name for itself through becoming a strong advocate of the blue economy approach to sustainable development. Palau is an archipelago of more than 250 islands and the nutrient-rich waters surrounding them are home to more than 1,300 species of fish and 700 species of coral. Island chiefs have acted for centuries to protect the local waters that serve as a critical food source and provide the means for trade and income. Following in this tradition and seeking to ensure a sustainable future for these rich waters, President Tommy Remengesau Jr. signed the **Palau National Marine Sanctuary** Act into law in 2015. The measure fully protects an area of about 500,000 sq km of ocean, covering 80 per cent of the country's Exclusive Economic Zone (EEZ). With this action, Palau dedicated a higher percentage of its waters for protection than any other country and created one of the largest fully protected ocean areas in the world. Under the law, local fishers can continue to work in the remaining 20 per cent of the EEZ for Palau's domestic market.

A number of Vietnamese 'Blue Boat' vessels carrying sea cucumbers, shark fins, black coral, shellfish and turtles have already been seized and burned by Palau's Division of Marine Law Enforcement. This aggressive response reinforced Palau's established zero-tolerance policy just weeks after the country released its five-year monitoring, control and surveillance plan in 2016. The report includes 25 detailed recommendations to help Palau fight illegal activities and manage emergency responses in the Pacific. To strengthen the monitoring and enforcement capacity provided by the Pacific Islands Forum Fisheries Agency, Australia, the USA, and other partners, Palau is working with Pew Charitable Trusts and the UK company Satellite Applications Catapult to begin using 'Project Eyes on the Seas': this is a state-of-the-art technology platform that integrates layers of satellite tracking data with fishing vessel databases and oceanographic information to monitor the country's vast waters.

South Korea

Jeju is the largest island off the coast of the Korean Peninsula and forms the Jeju Special Self-Governing Province of South Korea. Jeju was designated a World Biosphere Reserve in 2007 and registered as Korea's first World Natural Heritage Property with the title of Jeju Volcanic Island and Lava Tubes. The local government of Jeju and multi-national company LG Electronics signed a memorandum of understanding in 2015 to initiate the **Global Eco-Platform Jeju** project: a US$5.4 billion investment project that aims to ensure that the island is powered solely by renewable energy by 2030. The project's agenda includes expanding the island's wind turbines to reach a capacity of 2.35GW – more than 15 times the current energy output. Jeju and LG also plan to significantly increase the number of electric vehicles on the island – from 852 units now to 377,000 vehicles by 2030 – by providing subsidies for EV usage and running an EV battery lease programme. The two sides will build more rapid charging stations to operate roughly 15,000 stations by the deadline as well. At present, only 79 rapid-charging stations exist on the island. In addition, the project foresees the establishment of a new energy storage system which will conserve energy generated by solar and wind power plants for reliable future use. A smart grid system to manage efficient energy usage will also take root.

To achieve this ambitious target of becoming carbon-free by 2030, a pilot model was tested on the nearby island of Gapado. Here, the 177 residents have installed two wind power generators that create a total of 500kw of electricity and solar panels on top of their houses. They have replaced electricity poles with underground cables and the micro-grid operation centre with its energy storage system is designed to store the electricity when the demand is low and provide it at times of peak demand. There are only about nine cars on this island, and four of them are electric vehicles. Most tourists travel around on foot or on bicycles and are happy to do so in order to enjoy the uninterrupted views of the flat green barley fields, traditional stone walls and dolmen for which the island is famous.

Taiwan

The central government of Taiwan approved the **Penghu Low Carbon Island Project** in 2011 that had ambitious targets for energy saving, resources recycling, low carbon buildings, green transportation and forestation. There were already eight 600kW wind turbines and six 900kW turbines in two townships. By the end of 2013, the project had realised the replacement of 4,000 LED streetlamps, installed smart meters for 600 clients, purchased 3,346 electric scooters with subsidies, constructed 330 scooter charging stations, and provided 20 photovoltaic facilities; yet, this fell far short of the original vision. Penghu launched its first community wind farm in 2014 but, just nine months later, this government-backed operation went bust and shut down before residents had a chance to invest. Given its geographic isolation and challenging climate, Penghu has limited options to expand its economy of shaky seasonal tourism and fishing. The most feasible way forward would be to harness its abundant wind resources. Today, wind power supplies 30 per cent of Penghu's electricity. Speeding things up would generate revenue both for local residents and the government. However, whilst construction of a 66 km undersea cable to provide an electricity connection between the Penghu islands and Yunlin County on Taiwan proper has been completed, the above-ground work on the power line has been opposed by the residents of Yunlin. If this can be overcome then the cable system should spur further development of wind energy on Penghu and enable surplus power to be transmitted and sold to Taiwan in the winter when winds are strong.

Tokelau

Climate change is a priority for Tokelau, a non-self-governing territory of New Zealand which consists of three small coral atolls – Atafu, Nukunonu and Fakaofo that are 3–5 metres about sea level. Its total exclusive economic zone is approximately 300,000 sq km and it has a population of around 1,500. Tokelau is powered by at least 90 per cent solar energy and achieved this historical milestone in 2012, being the first in the world to do so. In 2014, some 250 homes in the villages of the three atolls received new water tanks and a community tank, or had their current water tanks repaired as part of the Pacific Adaptation to Climate Change Plus Project. In 2017, their *Living with change: Enhancing the resilience of Tokelau to climate change and related hazards, 2017–2030 (LivC)* strategy was launched after a consultation process to ensure inputs from all community members of all ages. The LivC identifies three inter-related strategic investment pathways that need to be supported and pursued in order for Tokelau to become a fully climate resilient and ready nation.

Tonga

In 2016, a national collective dialogue took place on an *Inclusive participatory approach towards a Blue–Green Economy*. Such an economy was seen as critical for all future developments and must include key elements for the mitigation of climate change. The **Tonga Environment and Climate Change Portal** and **Tonga Community Development Trust** detail numerous projects indicating a Blue–Green agenda is well on the way to being adopted. The government of Tonga set a renewable energy target in 2010 of reducing fossil fuel imports for power generation by 50 per cent by 2020. Tonga Power Company provides information on the full range of renewable energy projects including one for the construction of solar power plants on the eight outer islands. The 'on-grid' portion allocated to Ha'apai and 'Eua has recently been completed, while the 'off-grid' portion will incorporate the other six. The total installed capacity will be 1.3MWp. A biomass gasification plant has also been proposed for 'Eua using waste from log processing on the island. Geographically, 'Eua is the Kingdom's oldest and highest island and is a popular destination for travellers in search of adventure and eco experiences. There are spectacular cliffs, caves and underwater caverns to explore, well-marked hiking trails crisscrossing 'Eua National Park, as well as the opportunity to spot rare parrots and migrating Humpback whales.

Conclusion

Whilst GLISPA has undoubtedly been a catalyst in launching visionary commitments particularly amongst SIDS, there are still a great many other island communities wishing to become carbon neutral and seeking effective solutions to environmental problems but who do not know how or who to approach for practical help. The **Global Green Growth Institute**, **Green Economy Coalition**, **Green Growth Knowledge Platform** and **Partnership for Action on Green Economy** all provide data and networking opportunities on an international scale but there is no dedicated body just for supporting islands. There have been several attempts to address this significant gap in provision namely the first Eco-Islands Summit held on the Isle of Wight in 2012 followed by others on Bornholm in 2013 and Okinawa in 2015. An **International Green Island Forum** took place on Jeju in 2014 but another planned for the following year on the Indonesian island of Lombok was cancelled when the local Rinjani volcano erupted. The 2016 Forum took place again on Jeju in conjunction with the 3rd International Electric Vehicle Expo. Also in 2016 the governors of Hawai'i, Jeju and Okinawa signed a memorandum of understanding under the GLISPA umbrella to promote collaboration on sustainability issues, including co-hosting a Green Island Sustainability Summit in 2017. Whilst these different initiatives are welcome, actual events tend to be ephemeral in nature and do not necessarily address the problem of islanders' worldwide seeking answers to immediate technical questions that might arise in their respective communities on a daily basis. This could be partially addressed through using the most popular social networking sites to establish a new Green Islands Network that in turn might develop into a more permanent entity.

At a purely European level, the first **Greening the Islands** conference took place in 2014 on Pantelleria, an Italian island in the Strait of Sicily, followed by a second on Malta in 2015, and a third on Gran Canaria, Canary Islands, in 2016. Also in 2016, the **Promoting Renewable energy sources Integration for Smart Mediterranean Islands (PRISMI)** project started and the first Smart Islands Forum was hosted in Athens at the initiative of the **DAFNI Network of Sustainable Aegean and Ionian Islands** and Aegean Energy Agency. In January 2017 the **Islands of Innovation** project was started, with the Dutch Province of Fryslan in the lead, while in March 2017 representatives of island local and regional authorities from 13 countries met at the European Parliament to present the **Smart Islands Initiative**. Building on the first Smart Islands Forum, this event communicated the need to tap the significant, yet largely unexploited, potential of islands to function as living labs for technological, environmental, and social innovation. The main highlight of this event was the official signing ceremony of the Smart Islands Declaration, a manifesto to mobilise political and financial support so that European islands can transform into smart, inclusive and thriving societies.

Against the backdrop of the 'big four' global agreements in recent years – the SAMOA Pathway, the UN Sustainable Development Goals, the Paris Agreement on Climate Change, and the emerging New Urban Agenda of Habitat III – implementation of more resilient and equitable development at a local level supported by effective partnerships is critical if the 730 million plus people living on islands are to both survive and prosper. Already many SIDS are thinking differently, especially when it comes to the ocean. Their self-characterisation as *Large Ocean States* pursuing a blue economy is more than symbolic. It is a new mindset that represents a rethink on the opportunities and challenges facing them. As large ocean states, the focus shifts to a strengths-based approach rather than a deficit model based on their constraints. From being termed the canary in the coal mine facing the combined impacts of sea level rise and resource depletion, SIDS are now seen – and increasingly see themselves – as sentinels or guardians of the oceans.

References

Baldacchino, G., and Kelman, I. (2014) Critiquing the pursuit of island sustainability: blue and green, with hardly a colour in between. *Shima: The International Journal of Research into Island Cultures*, 8(1), 1–21.

Baringa (2016) *Economic opportunities of renewable energy for Scottish island communities*. London, Baringa. Available from: www.gov.scot/Resource/0049/00495193.pdf.

Grydehøj, A., and Kelman, I. (2017) The eco-island trap: climate change mitigation and conspicuous sustainability *Area: Journal of the Royal Geographical Society*, 49(1), 106–113.

UN DESA (2014) *Blue economy concept paper*. New York, United Nations Department of Economic and Social Affairs.

UNEP (2011) *Towards a green economy: pathways to sustainable development and poverty eradication*. Geneva, United Nations Environment Programme.

Appendix 21.1

AGRISLES Project www.agrisles.eu/en/accueil.html
Aloha+ Challenge https://hawaiigreengrowth.org/priorities
Anguilla National Trust www.axanationaltrust.com
Aran Islands Energy Cooperative www.facebook.com/aranislandsenergy
AranLIFE Project www.aranlife.ie
Association Les Iles du Ponant www.iles-du-ponant.com
Awaji Green Future Project www.awaji-kankyomiraijima.jp
Azorean Biodiversity Group www.gba.uac.pt
Azores Geopark www.azoresgeopark.com
B7 Baltic Islands Network www.b7.org
BEAST Project www.beastproject.eu
Biodiversity and Ecosystem Services in EU Outermost Regions and Overseas Countries and
    Territories (BEST) http://ec.europa.eu/environment/nature/biodiversity/best/
Blue Economy Research Institute www.unisey.ac.sc/research-consultancy/blue-economy-
    research-institute
Boneiru Duradero (Sustainable Bonaire) www.boneiruduradero.nl
Bright Green Island Bornholm www.brightgreenisland.dk
Caribbean Biodiversity Fund www.caribbeanbiodiversityfund.org/en/
Caribbean Challenge Initiative www.caribbeanchallengeinitiative.org
Caribbean Natural Resources Institute Green Economy Programme www.canari.org/
    programmes/issue-programmes/green-economy.php
Cook Islands Marine Park (Marae Moana) www.maraemoana.gov.ck/index.php
Coral Triangle Initiative www.coraltriangleinitiative.org
CPMR Islands Commission http://cpmr-islands.org/
Cradle to Cradle Islands www.c2cislands.org
DAFNI Network of Sustainable Aegean and Ionian Islands www.dafni.net.gr/en/home.htm
Discover Kodiak www.kodiak.org
Eco-island Miyakojima www.kantei.go.jp/jp/singi/tiiki/tkk2009/44miyakojima_PM_Eng.pdf
EcoVannin www.ecovannin.im
EESC Smart Islands Project www.eesc.europa.eu/?i=portal.en.smart-islands
Egadi Islands Marine Protected Area www.ampisoleegadi.it

ENEA Project Egadi http://progettoegadi.enea.it/it

Energy Matters – Green Eigg http://euanmearns.com/the-eigg-renewables-project-revisited/

ESLAND Project www.eslandproject.eu

European Small Islands Federation https://europeansmallislands.com/

Flinders Island Hybrid Energy Hub http://arena.gov.au/project/flinders-island-hybrid-energy-hub/

Footprint Trust www.footprint-trust.co.uk

Garden Island Microgrid Project http://arena.gov.au/project/garden-island-microgrid-project/

Global Eco-Platform Jeju Project https://smartgreencity2015.files.wordpress.com/2015/10/lintas_presentation_lg_pt_v1-1.pdf

Global Green Growth Institute http://gggi.org/

Global Islands Network (GIN) Green Island Case Studies www.globalislands.net/greenislands

Global Island Partnership (GLISPA) Bright Spots http://glispa.org/bright-spots

Gorona del Viento (GdV) www.goronadelviento.es

Green Aruba www.greenaruba.org/ga7/

Green Economy Coalition www.greeneconomycoalition.org

Green Eigg https://islandsgoinggreen.org/

Green Growth Knowledge Platform www.greengrowthknowledge.org

Green Island-Agios Efstratios www.ait.gr/export/sites/default/ait_web_site/conference/innoforum4/Assets/pdf_innoforum4/Session_3/1_Ayeridis___Green_Island_INNO-FORUM_2012.pdf

Green Islands Azores Project www.green-islands-azores.uac.pt/

Green Islands in the Baltic Sea http://greenislands.se/

Green on Wight www.iwchamber.co.uk/business-support/green-on-wight/

Greening the Islands www.greeningtheislands.com

Grenada Blue Week www.bluegrowth.org

Grenada Integrated Climate Change Adaptation Strategies Programme www.iccas.gd

Gulf Islands Community Economic Sustainability Commission www.sustainableislands.ca

Hawaii Green Growth https://hawaiigreengrowth.org/

INSULEUR (Network of the Insular Chambers of Commerce and Industry of the European Union) www.insuleur.com

International Green Islands Forum http://igiforumen.1940.co.kr/

Island 2000 Trust www.naturalenterprise.co.uk/pages/about

Islands Partnership www.islandspartnership.co.uk

Island and Hallig Conference (IHKo) www.ihko.de

Island Institute www.islandinstitute.org

Islanded Grid Resource Center http://islandedgrid.org/

Islands of Innovation www.interregeurope.eu/islandsofinnovation/

Islands Trust www.islandstrust.bc.ca/

ISLENET www.islenet.net

ISLEPACT Project www.islepact.eu

Isle of Man Biosphere Reserve www.biosphere.im

James Michel Foundation www.jamesmichelfoundation.org

Kangaroo Island Energy Project http://kangarooislandenergy.com/

King Island Natural Resource Management Group www.kingislandnaturalresources.org

King Island Renewable Energy Integration Project http://arena.gov.au/project/king-island-renewable-energy-integration-project/

Kodiak Electric Association www.kodiakelectric.com

Kosrae Conservation and Safety Organization http://kosraeconservation.org/

Lord Howe Island Hybrid Renewable Energy System http://arena.gov.au/project/
lord-howe-island-hybrid-renewable-energy-system/

Mediterranean Small Islands Initiative www.initiative-pim.org

MeyGen Project www.meygen.com

Micronesia Challenge www.micronesiachallenge.org

Nova Innovation www.novainnovation.com

Observatory on Tourism in the European Islands www.otie.org

OCTA Innovation for Sustainable Islands Growth www.octa-innovation.eu

Okinawa Green Technologies and Innovations http://announce.ndhu.edu.tw/message_1/
1430184666/Sustainalbe_Island_Technology.pdf

Palau National Marine Sanctuary http://palaumarine.org/

Partnership for Action on Green Economy www.un-page.org

Penghu Low Carbon Island Project http://penghu.re.org.tw/En

Perpetuus Tidal Energy Centre http://perpetuustidal.com/

Phoenix Islands Protected Area www.phoenixislands.org

Planet Texel Academy www.planettexelacademy.nl

Promoting Renewable energy sources Integration for Smart Mediterranean Islands (PRISMI)
https://prismi.interreg-med.eu/

Reseau d'excellence des territoires insulaires (RETI) https://reti.universita.corsica/

RMI-CWR Islands Energy Programme www.rmi.org/islands_programme_overview

Rottnest Island Water and Renewable Energy Nexus (WREN) Project http://arena.gov.au/
project/rottnest-island-water-and-renewable-energy-nexus-wren-project/

Samsø Energy Academy https://energiakademiet.dk/en/

Seychelles Conservation and Climate Adaptation Trust (SeyCCAT) https://seyccat.org/

Shorefast Foundation http://shorefast.org/

SIDS DOCK http://sidsdock.org/

SMALLEST Project www.smallestnpp.eu/index.htm

Small Islands Organisation (SMILO) www.smilo-program.org/en/

Smart Island www.smartisland.eu/en/

Smart Islands Initiative www.smartislandsinitiative.eu/en/index.php

Smart Region Pellworm www.smartregion-pellworm.de/home.html

SMILEGOV Project www.sustainableislands.eu

State of Green https://stateofgreen.com/en

Stòras Uibhist www.storasuibhist.com

Sumba Iconic Island http://en.sumbaiconicisland.org/

Tilos Project www.tiloshorizon.eu

Tonga Community Development Trust www.tcdt.to/index.html

Tonga Environment and Climate Change Portal http://ecc.gov.to/

West Cork Islands http://westcorkislands.com/

Western Indian Ocean Coastal Challenge www.wiocc.org

INDEX

Printed and bound by CPI Group (UK) Ltd, Croydon, CR0 4YY

30/10/2024

01781414-0002